WORLD RAILWAY ACCIDENTS 1881-1900

Francis Voisey

ISBN: 978-1-9162011-2-5

DEDICATION

This book is dedicated to signalman Brian Stone, who first awakened my lifelong interest in railways way back in 1966.

CONTENTS

ACKNOWLEDGMENTS

Thanks to the staff of the Exeter Reference Library and the University of Exeter Library for their assistance while I was researching this book, to my friend June Nightingale for encouraging me to go ahead and get it finished when my enthusiasm was waning a little, and to my publishing team, without whose guidance and assistance it is doubtful if I would have been able to navigate the mysteries of getting my books into print.

INTRODUCTION

This book is the second of what it is hoped will be a series detailing with railway accidents worldwide. A good many books have been written over the years about railway accidents, some concentrating on single disasters, others covering accidents in a given country or geographical area.

Accidents in countries such as Great Britain, Germany, the Netherlands, the United States, Australia, New Zealand and so on have been documented comprehensively in one or more books but no attempt has been made previously to detail every significant accident worldwide. This is clearly a very ambitious project, but an initial volume dealing with the earliest years having been completed, this book will attempt to detail the accidents which during latter part of the 19th century, between 1881 and 1900, with further volumes hopefully to follow dealing with the twentieth century. As before, an attempt will be made to include all accidents involving five or more fatalities to train passengers. A number of less serious accidents which have been considered of interest have also been included. The accidents have been grouped into chapters according to their primary cause and are dealt with as far as possible chronologically within each chapter.

The gathering of information about 19th-century accidents in distant parts of the world has not been easy, though it is much easier nowadays than when this project began in the days before the internet. Outside of Great Britain and Ireland few countries seem to have had a formal body charged with investigating all serious railway accidents before the twentieth century, though one or two major disasters such as the Bussey Bridge failure in the United States and the Baptiste Creek collision in Canada were the subject of full formal investigations. Otherwise reliance has largely had to be placed on contemporary newspaper reports and previously published works on the subject, though the author has tried as much as possible to keep to primary sources. If the book appears to be biased towards accidents which occurred in Great Britain and the United States this is probably partly because these are the countries about which information on accidents has been easiest to obtain, though these were also in any case the countries where many of the most serious accidents did occur.

Contemporary spellings of place names have been used where appropriate and for the many American accidents, terms such as railroad, engineer, caboose and so on have been used instead of railway, driver, brake van etc. The 24-hour clock was not used in the 19th

century except in Italy from 1893 and the 12-hour clock will therefore be used throughout.

By the beginning of 1881 railways had spread to much of the world and by 1900 they were to be found almost everywhere, though some countries only had very small systems or perhaps just one line.

The names and the boundaries of countries have undergone many changes over the years. German and Italian unification had been completed by 1881, but Eastern European countries such as the Czech Republic, Slovenia, Slovakia, Estonia and so on, did not exist before 1919 or later. Australia was a collection of separate colonies until 1901, likewise South Africa until 1910, whilst 19th-century India included the present countries of Pakistan and Bangladesh. The names of some other countries and of numerous places have changed. The names of locations which were in use at the time an accident occurred have been given, together with present names where appropriate, as for example Ürsküb (now Skopje).

CHAPTER 1

DERAILMENTS: TRACK DEFECTS

The tracks over which trains run are obviously a vital part of the railway and any problems with the track are likely to lead to the train being derailed. By 1881 the technology involved in laying railway track had much advanced since the early days, but train speeds and weights had also much increased, while accidents occurred from time to time due to failure to keep the track in a good state of maintenance, to insufficient allowance being made for hot weather expanding the rails, to damage caused to the track from floodwaters and so on. Also included in this chapter, though they were not strictly due to defects in the track, are accidents due to trains having become derailed on open points. Broken rails were another hazard which caused mishaps and worse from time to time and at this time were largely unpreventable.

In Germany, Sunday 3rd September 1882 was the day after Sedantag (the semi-official holiday held to commemorate the Prussian victory over France in the 1870 Battle of Sedan),[1] and a large number of people had travelled from Alsace to holiday in the town of Freiburg. By the evening it was time for the trippers to make their way homewards, but a severe rainstorm occurred in Freiburg, which delayed the departure of the 8 p.m. train to Colmar.

The Freiburg – Colmar line had originally been opened as a private railway, but it had been taken over by the state in 1878 and extended across the Rhine to Colmar in the same year. From Freiburg to the bank of the River Rhine the ruling gradient was 12‰ or about 1 in 83. The permanent way was unaltered from the opening of the line and it had not been thoroughly maintained. Very light rails were still in use, and the joints at the junctions of the lengths of rail were placed between pairs of sleepers. In 1881 alone, the line had had 31 cases of broken rails or fishplates. The sleepers were untreated. They lay in a bed of gravel rather than one of broken stone ballast. The track was in fact due for upgrading but the work had been postponed. An engine had become derailed due to the poor condition of the track six months earlier. The storm which had delayed the 8 p.m.

---

[1] Wikipedia – Sedantag

train probably also undermined the already poor track where it ran through the Mooswald Forest, west of Freiburg.[2]

The locomotive, named *Kniebis*, which was pulling the 8 p.m. train, was of the Stephenson long-boiler type, a design which was notoriously unstable at high speed, due to its extended boiler projecting beyond the outer axles, a feature which had been a factor in a derailment at Avenwedde in Westphalia, as long ago as 1851. Such engines were therefore little used in Germany by the 1880s and were certainly not employed hauling passenger trains. Why the Kniebis was nevertheless allocated to a passenger train on the day of the accident was not explained; an oversight perhaps, or possibly there was a shortage of motive power. The engine had a maximum authorised speed of 50 km/h.

The train consisted of 28 wooden carriages in which about 1,200 passengers were travelling. According to the regulations at least 25% of the carriages should have been equipped with brakes but only six brakesmen were provided. Worse, because the Colmar train was being used by ordinary passengers as well as by excursionists at the start of the journey, two of the brakesmen were charged with collecting tickets. Another brakesman was occupied instructing a new employee and the remaining two did not properly know the road, so one of these brakesmen did not apply his brakes at all when he should have done, at the commencement of a falling gradient, and the other did not apply them hard enough. Thus, the speed of the train began to get too high, the engine being pushed along by the almost unbraked carriages behind it despite the attempts of the engine crew to control the speed with the brakes on the engine and tender. The unstable engine began to rock and oscillate until the weak track could no longer resist the strain and with the train travelling at about 70 km/h the rails gave way beneath the locomotive, which was derailed at 8.20 p.m., ending up in the marshy ground adjoining the line. All but three or four of the carriages of the train were also derailed and they piled up into one another and were partially broken up. About 225 metres of track were damaged. The accident happened on a long straight stretch of track, only 5.4 kilometres from Freiburg and near the village of Hugstetten, from which the disaster takes its name. [3]

The rescue operation was begun by locals from Hugstetten, Hochdorf and Lehen using horses pulling carts and handcarts. Helped presumably by uninjured excursionists, the rescuers pulled 64 bodies from the wreckage and there were about 230 more seriously injured, five of whom died in the days after the disaster. The telegraph line had been destroyed by the accident and Freiburg station being unable to communicate with the next station down the line, a following train was despatched 15 minutes after the excursion. It was stopped short of the accident by a signalman (block posts were apparently provided at 1,000-metre intervals, which seems rather excessive). A message was then sent to Freiburg for a relief train, but they had no engine with which to send one until the locomotive of an arriving freight train could be commandeered. A relief train then left for the accident scene at about 10.20 but no one had thought to alert any doctors. Medical help finally arrived just before midnight, the doctors having got to the scene by horse and cart. It was the following morning before the last of the dead and injured left the site.

---

[2] Wikipedia – Eisenbahnunfall bei Hugstetten

[3] Ibid.

The disaster understandably caused much grief and horror throughout the German Reich. Flags were flown at half-mast around the country and all public celebrations were cancelled. Most of the victims' bodies were sent to their homes in towns and villages in Alsace, though five were buried in Freiburg. A cross was erected to commemorate the accident in 1885. Following the accident, the Baden State Railways abandoned the practice of giving locomotives names and gave them just numbers. The engine involved in the derailment was little damaged and was repaired and brought back into use.

In April 1883 the railway board, the Freiburg signalman, the train conductor, the train driver and the rolling stock inspector (Wagenwärter) were brought to trial at the Central Criminal Court in Freiburg charged with various errors and failures in connection with the accident. However, the accident was clearly the result of deficiencies in the track and the instability of the locomotive, for which the accused could not be held personally responsible, and they were all acquitted after a five-day trial. The higher management officials who might have been considered to be responsible were not prosecuted. [4] [5]

The London and South Western Railway branch from Salisbury to Wimborne was the perhaps rather unlikely location of a serious accident on 3rd June 1884 when the 4.33 p.m. passenger train from Salisbury was derailed between Downton and Breamore stations, England. The train was rather overpowered, having two tender engines pulling a brake van, two third-class, two first-class, a second-class and a third-class carriage and a brake van, just eight vehicles in all. It was a Tuesday, which was market day in Salisbury, and so the train was well-filled. The tender of the second engine and all the vehicles behind it left the rails and all but the tender and the leading brake van fell down a 6ft embankment. Four carriages overturned in a ditch and a field and were completely wrecked. Four passengers were killed outright, a fifth died later and another 41 were injured.

The five passengers who lost their lives were Mr George Waters, traction engine proprietor and farmer of Toyd Farm, near Fordingbridge; Mrs Lush, of Godshill; Lillian Chandler, daughter of the Fordingbridge stationmaster, who was on her way home from school; Mrs George Corbin, wife of a Ringwood tradesman; and Mr Matthew Dent, a Bournemouth hotel proprietor, who died the next day. [6]

The train had left Salisbury six minutes late at 4.39. It had apparently gained two minutes when it reached Downton, 8¾ miles from Salisbury, and Colonel F.H. Rich, the Board of Trade inspecting officer, considered that it must have been travelling at a speed of 40 to 44 miles an hour at the time of the accident.

The permanent way was in good order. The rails dated from the opening of the line, 18 years earlier, and had originally weighed 70lbs per yard. Many of them had been turned and Colonel Rich remarked that: 'I do not consider a permanent way of this class strong enough to carry trains safely, that are run at an average speed of 35 miles an hour, particularly as there are many curves and the gradients are steep. Drivers would frequently have to run at the rate of 45 miles an hour, if not more, down the inclines, to

---

[4] Ibid.
[5] *Eisenbahn-Katastrophen in Deutschland* by Hans Joachim Ritzau, pgs. 55-9
[6] *The Times*, London, England, 4th and 12th June 1884

keep time with trains.' [7]

William Adams, Chief Locomotive Engineer of the London and South Western Railway, attributed the derailment to the breaking of the coupling between the front brake van and the carriage next to it, but the inspecting officer disagreed, and stated that he had no hesitation in stating that the train had been run at too high a speed over a weak road. He did not believe the evidence of Driver Thomas Butler and Fireman Frank Saunders on the pilot engine or of Driver Robert Mills and Fireman Richard Scadding on the train engine. The two engine crews stated that they had not realised that the train was derailed until they had reached the cast-iron girder bridge, over 200 yards from the initial point of derailment. They admitted that they had been travelling at 35 miles an hour at the time and Colonel Rich believed that it had been the force and weight of the engines which had disturbed the permanent way and that the engine crews, feeling the oscillation, had applied the steam brakes, and jerked one or more carriages off of the line, then realising what had happened had re-applied steam to get away from the train.

In the closing paragraphs of his report Colonel Rich made some disparaging remarks about the London and South Western management: 'The numerous complaints which have been made, of the violent shaking which passengers experience when travelling on parts of the London and South Western Railway, leave no room to doubt, that a great deal of reform in the management, and improvement in the management of this railway, is required. I believe the complaints are caused in a great measure by bad driving, using inferior stock, and by the coaches in the trains not being properly coupled up.

'I would strongly urge the Company to make a thorough examination of their system and stock, to classify their drivers and parts of their system, and to classify their trains.

'It cannot be expected that the whole of a Company's stock and railway shall be of the best description; but the public has a right to expect, that old, weak, and inferior stock shall not be run over old and inferior parts of the railway at such a speed as to make it unpleasant and dangerous, to all that use it.'

The Salisbury and Wimborne line was single track and still worked by electric telegraph. The second engine had been attached to the train to avoid having to arrange extra crossings for a light engine.[8]

A mistake by a pointsman led to a bad accident on the recently opened Corinth – Kiata line in Greece. He inadvertently left open the points leading to a temporary gravel siding as a seven-carriage passenger train from Kiata to Corinth was approaching at full speed. The engine of the train left the rails as it was diverted into the siding, buried itself in the sand and almost at once its boiler exploded. Three of the carriages of the train were completely wrecked. There were about 300 passengers on the train, most of them peasants. Eight people were killed and 11 seriously injured; most of the injured were subsequently taken to Athens. The pointsman was arrested. [9]

On 8th April 1886 eastbound Fitchburg Railroad passenger train No. 35 left North

[7] Board of Trade report
[8] Ibid.
[9] *The Times*, London, England, 9th October 1885

Adams for Boston, Massachusetts, United States, at 4.45 p.m. It consisted of an engine, baggage car, mail car, smoker, two passenger cars and the Wagner Palace Car *Pavonia.* The train was in charge of Conductor John Foster with Engineer Herbert Littlejohn and Fireman Richardson. There were 45 passengers on it according to its conductor. Emerging from the long Hoosac Tunnel the train continued along the valley of the Deerfield River, past Zoar, Charlemont, Shelburne Falls and Bardwell stations. Beyond Bardwell the line ran on a ledge 100ft above the river and it was here that disaster struck. The train was only travelling at about 18 or 20 miles an hour when, without warning, it left the track and fell down the rocky bank to the river below, its cars tumbling over and over as they fell. Shortly after the initial accident fire broke out in the wreckage as well.[10]

A newspaper reporter described the scene after the accident: 'The mighty mass had scarcely ceased its dreadful quivering when that awful thing, fire, hurried to add its horrors to the desolation. In a few moments the wreckage was a blazing mass. Slowly but surely the flames crept from car to car, threatening the imprisoned people with a horrible death. It would be impossible to describe the awful scene with due impressiveness. To one standing on the road bed above and glancing down on the scene of destruction it was a picture that will remain forever. Hurrying to and fro were men and women frantic with despair and bewailing the loss of friends and relatives. One mother hurried from place to place in search of her little children torn from her care and protection in that moment of peril. When she found her darlings, they were nearly beyond need of care, for on their brows the chill of death was set. With cries of anguish such as come only from a mother's heart, the poor woman hastened to thank the man who had brought her children from the place of death. Tenderly removed to the heights above, the woman with many others was placed on board a train and taken away.

'Soon here and there hurried the attending surgeons, examining the many patients, while beyond were the sturdy train men working on the wreck. Much assistance was offered by those who came, and, considering the circumstances, everyone was as comfortable as medical skill would allow.

'Later on, it was a weird and fascinating scene. By the blaze of the burning coaches one was enabled to see for hundreds of feet around. In the rear was the abrupt rise of the embankment and above that eye could not penetrate the darkness of night. Here and there the train lights of wreckers on the up track, the ringing of bells and the frequent whistles that went out on the night air gave, seemingly, the only consolation to the victims still confined in the waters below. Above and beyond them rolled the mountain stream, swollen and turbulent, shallow and filled with whirlpools. The water rushed and eddied round the submerged train.

'Hour after hour passed, and when daylight came all the living had been attended to. Still beneath the stream were the remains of others, how many no one knew. Superintendent Adams was present directing the work the last portion of the night. He arrived accompanied by his son and a corps of Boston reporters about 5 o'clock. The returning light showed that the cars that had been hurled down to the bottom of the embankment, with one exception, were burned to the water's edge. On the track above was the engine, No. 12, turned on its side and badly wrecked. One car alone did not reach the bottom, the Troy and Greenfield car, which stuck halfway down. The mail car has

---

[10] *Boston Daily Globe News*, Boston, Massachusetts, United States, 9th April 1886

washed away, and the others are breaking up.

'The cause of the accident was not exactly a washout, but rather the loosening of the road-bed under the rail nearest the river, letting down that rail. The engine's impetus put it over the dangerous spot, but the weight of the first car drove down the rail and over went the train.'

Nine passengers and Engineer Littlejohn were killed in the accident and about 20 passengers injured. Conductor Foster and baggagemaster J.B. Sturtevant were also badly hurt and Aaron Lewis, coloured porter of the Wagner car, was hurt about the head, but Fireman Richardson escaped with very slight injuries.[11]

An inquiry into how the accident happened was opened already the following Monday (12th April)[12] but was twice postponed because some of the witnesses had not yet recovered from their injuries and it has not been possible to find a report of its final conclusions.

A derailment in Ireland on 30th June 1886 arose mainly from a cause which we have not come across before, namely the buckling of the track in hot weather. It is perhaps surprising that no bad accidents from this cause had occurred earlier, or at any rate none that were clearly attributed to it.

The train involved in the accident was the 1.30 p.m. Great Northern (Ireland) Railway up passenger train from Belfast to Dublin, which left Portadown at 2.31 p.m. in charge of Guards Robert Davis and George Finnegan, and which was derailed between there and the next station at Tandaragee at about 2.35 p.m. The train consisted of an engine and tender, five carriages and a brake van and the whole of it was derailed except the last carriage and the brake van. Most of the casualties occurred in the third-class carriage third from the front of the train. Three passengers were killed outright, three more died of their injuries and 29 others were more or less badly hurt. Five of those who lost their lives were Mrs Margaret Law, of Gilford; her daughter, Elizabeth; James M'Foy; William M' Neill and William Galt. The name of the sixth victim, one of those who died later, was not given in the *Times* newspaper, which only carried two brief reports of the accident.[13]

Major-General C.S. Hutchinson, who investigated the derailment on behalf of the Board of Trade, was pleased that the railway company had left the permanent way untouched after the accident until he had had time to examine it but said that despite this, the fact that the district engineer of the company was himself a passenger on the train and that the derailment was witnessed by two gangs of platelayers working nearby, its causes were still in some doubt.

Save perhaps as regards speed there was nothing, the inspecting officer stated, in the behaviour or condition of the engine or train to account for the accident. He considered that the guards were mistaken in the time that had elapsed between the accident and their looking at their watches, and that the train had taken three rather than two minutes to cover the preceding two miles and had thus been travelling at 40 rather than 60 miles an hour. The track itself was safe for a speed of 50 miles an hour, except that in the 100

[11] Ibid.
[12] Ibid, 12th April 1886
[13] *The Times*, London, England, 1st and 2nd July 1886

yards preceding the point of derailment there was a deficiency of ballast, and he considered that with the track in that condition the trains should certainly have been cautioned, especially as they were known to run at high speed at this point.[14]

There was, however, a third factor, without which the high speed of the train and the lack of ballast would probably not have been sufficient to cause a derailment. This was the fact that the rails on the curve had been in a high state of compression due to the heat of the day and the want of proper room for expansion between the rail joints. When he had examined the line five days after the accident on a cool morning over a stretch of about 86 yards in the rear of the point of derailment, all the joints but one had been quite close together. It had been much hotter on the day of the accident and the rails had clearly been in a state of great tension. Ganger Peter Robinson stated that he could see nothing wrong with the track immediately before the derailment, but it was easy to understand how the curve could have been suddenly forced outwards in front of the engine, derailing it, the track giving way more easily due to the deficiency of ballast. That it had been an exceptionally hot day was shown by the fact that another stretch of track had buckled near Newton Stewart, 40 miles away.

Major-General Hutchinson considered that Permanent Way Inspector Michael McCarthy, an old servant of the company with over 20 years' service, was deserving of censure for satisfying himself that the important work of lifting and re-ballasting over half a mile of track over boggy ground was being properly carried out merely by occasionally passing over it in a train. It was also his duty to have noticed and remedied the lack of expansion joints at the scene of the accident. It was also much to be regretted that no speed restriction had been imposed until the re-ballasting had been completed.[15]

The second brief report on the accident in the *Times* stated that Driver Thomas Tiernan and Gangers Peter Robinson and Thomas Galway had been charged by the Portadown magistrates with contributing to the accident by culpable negligence. The two gangers were remanded, Driver Tiernan was admitted to bail.[16]

Fifteen minutes after midnight on Thursday 28th October 1886 disaster struck at East Rio, Wisconsin, United States. East Rio was a small station on the Chicago Milwaukee and St Paul Railroad with a siding on each side of the single main track. Just before the accident the two sidings had both been occupied by freight trains, one by a wild freight train which had arrived some time earlier and the other by eastbound freight No. 14, in charge of Conductor H.C. Hankey, which arrived there about midnight. This train had to take the siding to cross the westbound St Paul Limited express passenger train. The latter train shortly arrived and as it was not due to call at East Rio was running at 50 miles an hour down a falling gradient as it reached the place. It ran through the yard but according to the initial news reports the switch (points) at the west end of the yard had not been closed, as they should have been, by the rear brakesman, either because he had thought that the conductor was going to close them or because the passenger train had arrived before he had had time to do so. As a result the express was derailed on this set of points and went down the embankment. The author had been under the impression that the train

[14] Board of Trade report
[15] Ibid.
[16] *The Times*, London, England, 2nd July 1886

must have become derailed at the facing points at the beginning of the sidings (crossing loops) but a newspaper account, which includes a diagram of the layout, gives the impression that the train was somehow derailed at the far end of the loops at what should have been trailing points, instead of just running through and splitting them.[17]

The express consisted of an engine and tender, a baggage car, a single day coach and three sleeping cars in that order. Engineer Little saw the open switch at the last moment but far too late to be able to prevent the accident. The whole of the train was thrown down an embankment apart from the rear sleeping car, and the day coach was crushed between the heavy well-loaded baggage car and the sleeping cars behind it. Fire, caused as usual by an overturned stove or stoves, then broke out in the wrecked day coach and only three passengers in it survived; a Chicago man who managed to climb out of the wreckage and two children of Mrs Margaret Scherever, of Winona, Minnesota, whom she was able to hand out of a window. Pinned in the wreckage herself, she was among those who perished, along with her mother, Mrs Rosena Johns. Engineer Little and Fireman Eagan were bruised and scalded but survived.

A passenger from one of the sleeping cars, described the accident: 'B Lowenbach, a job printer of this city {Milwaukee, Wisconsin} who was on the wrecked train, says that the passenger coach, which contained between fifteen and twenty people, was telescoped at both ends and the fire and smoke that enveloped the wreck prevented the imprisoned and injured passengers from escaping. The passengers from the sleepers gathered around the blazing cars but were unable to render assistance. Men and women could be seen tearing their hair in the agony of the moment and frightful screams issued from the death-trap. One heavy woman, in particular, he says, tore up one of the seats and endeavoured to break her way out, but her strength failed, and she fell to the floor and met a horrible death. Only three persons escaped from the passenger coach.

'Lowenbach saw a man and two children. The man forced himself through the ventilator on top of the car with all his clothing, from the waist down, and his flesh was roasted and bleeding from cuts inflicted by broken glass. Every one of the wrecked cars was consumed, with the exception of the last sleeper, which was cut away from the burning wreck. All the bodies of the victims were burned in the wreck.' [18]

Charles Smith, the Chicago medical student who escaped from the day coach, was able to give an interview to a newspaper reporter from his sickbed: 'I had a double seat to myself about the middle of the car and was sleeping at the time the crash came. My first thought was that someone had thrown pounded glass into my face and eyes. My mouth was full of it. I only knew that my face and head were bleeding badly and then became partially unconscious for a moment or two. On opening my eyes again, I saw that the lamps were out and that a lurid blaze lighted up the whole car and was rapidly approaching me. By a desperate effort I released myself from the rubbish and the broken seats which held me fast to the floor of the car and could see passengers all about the car so thoroughly wedged in among the seats which were smashed, that it was impossible for them to extricate themselves.

'The top of the car seemed to have fallen in at either side and almost or quite rested on the ruins beneath and pinned the passengers to the floor. As soon as I was released, I

---

[17] *Madison Wisconsin State Journal*, Madison, Wisconsin, United States, 28th October 1886
[18] Ibid.

attempted to extricate those near me but found my right arm helpless. A woman just in front of me who had a little child begged me to help her and I put my left arm around her shoulders and pulled with all my might but was unable to move her. She and all the rest seemed to be so wedged in that even if I had had both arms, I don't think I could have helped them in the moment I had to spare.

'The flames spread so rapidly that I could not remain longer, and with the greatest difficulty escaped through a crack between the top and the edge of the car and then dropped to the ground, and as I climbed the bank close to the car I could see the flames were already surrounding the screaming women and children just where I was but one moment before. The sights and sounds were terrible to see and hear, and the memory of that hour is even worse to me than the pain which racks my body. I think there must have been twenty or twenty-five passengers in all. Two small children were taken out of a broken window by some men from one of the sleeping cars, and I am the only other one who escaped. Under the excitement I attempted to aid in carrying the mail bags up on the bank and succeeded with four or five, and then sank down from utter exhaustion and have not been able to walk at all since then.' [19]

With the bodies of the victims having been consumed by the fire their identification proved very difficult and it was not even certain exactly how many had lost their lives, but the generally accepted figure appears to be 17.

Clarence. H. Wells, the brakesman who allegedly should have closed the switch, subsequently gave his version of the events leading up to the disaster, which the newspaper paraphrased as follows: - 'His train arrived at East Rio several minutes before the limited put in an appearance. The freight train was too long for one side-track, so it was cut in two, switching one part onto one side-track, and detaching the engine, switching the engine onto the other side-track. It was not his duty to close the switch. He had never closed it before. The conductor always did so. On the night of the accident, after they had reached the siding and side-tracked the train, he had to remain on top of one of the cars. Conductor Hankey had gone to the switch, and shortly afterwards had passed the car on which he (the brakeman) was standing. He told the brakesman that everything was all right and that he was going to the head of the train. He (the brakesman) had looked at his watch and discovered, or thought that he had, that the limited was late. Just then the limited had come in sight. He had been horrified to see it run off the track shortly after it had passed his car. He had been badly frightened and thinking that the freight train hands would be lynched had fled. He realised that he had made a mistake in running away, but it had, he considered, been a very natural thing to do under the circumstances.' [20]

By the conclusion of the inquest on the victims of the disaster the newspapers seem to have lost interest in the story, but the final verdict of the jury was that Brakesman Wells had been responsible and he was arrested and charged with criminal negligence.[21]

On 19th January 1887 it took the jury at Portage, Wisconsin, only 15 minutes to find Brakesman Wells not guilty of manslaughter.[22]

---

[19] Ibid, 30th October 1886
[20] Ibid.
[21] Ibid, 10th November 1886
[22] *Newark Daily Advocate*, Newark, Ohio, United States, 20th January 1887

In contrast to the East Rio disaster an accident at Steamburg, New York, United States, on 5th February 1888 was not really anyone's fault. New York Pennsylvania and Ohio Railroad eastbound freight train No. 34 received orders at Randolph to run the five miles to Steamburg and there take the siding to allow express passenger train No. 8 to overtake it. It reached Steamburg without any problem and there took the siding as instructed. While waiting there for the express to pass Conductor George Ellis and Brakesmen James Dean and Henry Shaffer took the opportunity to eat their lunch. The express duly appeared, running at 45 miles an hour, and its engine and the baggage and smoking cars passed over the points safely, but the frog of the points then broke and the ladies' coach following the smoking car was hurled with great force against the caboose of the freight train, and the ladies' coach, the caboose and two freight waggons were wrecked, half of the ladies' coach being telescoped.

The engineer of the express pulled up as quickly as possible and hurried back with other trainmen and passengers to assist the victims, while medical assistance was sent for and soon arrived. Brakesman Dean was killed outright, as was one passenger, 16-year-old Hattie Abbott, of Sheffield, Illinois. Conductor Ellis died the same night; George Kimball, who had been returning home from Garden City, Kansas, to Bradford, Pennsylvania, and was clearly seriously injured, died a day or so later; and Brakeman Shaffer, who had apparently not been badly hurt, died later as well. Half a dozen other passengers were hurt.[23]

The precise cause of a train derailment near Granada, Nicaragua, on the morning of 3rd June 1888, was not stated in the brief newspaper reports of the accident but the area had just been subject to a cyclone of unusual severity with heavy rain, sufficient to destroy 18 homes, so it was almost certainly due to the track having been washed away. Five people lost their lives and five more were injured.[24]

An accident in France on 5th September 1888 involved an initial derailment followed by a collision. Train No. 11 had left Paris Lyon at 9.25 p.m. on the evening of 4th September for Geneva via Culoz. It had passed through Blaisy-Bas Tunnel and was negotiating a left-hand curve 1½ miles short of Velars station at about 2.20 a.m. when its engine left the rails and overturning, slid 50 metres along the track, fouling the opposite running line. Its driver, named Stengel, tried vainly to sound the engine whistle as a warning but the whistle had been damaged in the derailment and would not work.[25] Shortly afterwards the derailed Geneva train was struck by international express No. 276 from Italy via Modane to Paris. The engine of the colliding train was derailed in turn and the carriages of both trains were piled up in a heap of wreckage, though the four rear vehicles of train No. 11 and the three rear vehicles of train No. 276 remained on the rails. A wooden second-class carriage of train No. 11 was crushed by a following 17 ton

[23] *Bradford Era*, Bradford, Pennsylvania, United States, 6th February 1888
[24] *Dubuque Daily Herald*, Dubuque, Iowa, United States, 9th June 1888
[25] *The Times*, London, England, 7th September 1888

sleeping car.[26]

The telegraph wires had been brought down by the accident but the night officer at Velars had heard the noise of the accident and found that train No. 11, which had been announced to him, did not arrive. He went and woke up the stationmaster and explained the matter. The latter set off down the track but had not gone far when he met the injured fireman of train No. 11. The stationmaster returned to his station and alerted Dijon, while the villagers of Velars were also alerted, though the *Times* reporter in Paris moaned that the villagers had not arrived until 4.30 a.m. and the first relief train from Dijon was not there until nearly 5. Earlier assistance was provided by a military doctor who chanced to be on train No. 276.

Driver Stengel was killed in the secondary collision and eight passengers lost their lives. Seven of the dead passengers had been in the crushed second-class carriage, three of them of the same Paris family, the Bachets. The other passenger who lost her life had been looking out of her first-class compartment after the initial derailment and was decapitated in the collision. There was also an English passenger who lost her life, a Miss Edith Marriott, who had been travelling to Geneva in the company of her brother, Captain Marriott, of the Indian Army, for the benefit of her health.[27] The accident thus received more coverage than it might otherwise have done in the London *Times*. Ten passengers and the fireman and chief conductor of train No. 11 were more or less badly hurt. The driver and fireman of train No. 276 escaped with bruises.[28]

As to the cause of the initial derailment, it was at first reported that the track had been under repair and that trains had been instructed to run at reduced speed at the accident location, but this was later shown not to be true and to be based on the claims of the editor of a Paris newspaper who was dissatisfied with the railway company. Investigations carried out by the Chemin de fer de Paris à Lyon et à la Méditerranée (PLM) Company revealed, however, that the track was slightly distorted at the point of derailment and it appeared that track defects had already been present before the arrival of train No. 11. The driver of train No. 7, which had passed at 11.30 p.m., had noticed a problem and had reduced speed, whilst passengers on train No. 121 had felt jolts which they had reported upon arrival at Dijon at 2.15 a.m., five minutes before the accident happened. Despite this the official investigation into the accident decided that it could not have been foreseen and that all the company servants had done their duty. The PLM Company had to pay compensation to the families of the victims though.

The accident led to suggestions such as the provision of red flares for use in emergencies and for the improvement of the construction of all passenger carriages on the lines of the heavy sleeping cars, which had withstood the shock of the accident well. The main proposals, however, concerned the permanent way, which had clearly become too weak for the increased weight, speed and frequency of the trains. Steel rails had been substituted for iron in 1868 and the length and the weight of rails had first been increased in 1883 to 12 metres and a weight of 43.5 kg/m. In 1889 the weight was increased to 47 kg/m with a broader head of rail, though other companies were still using lower values. There was also a proposal to substitute metal for wooden sleepers for increased lateral

[26] Wikipedia – Catastrophe ferroviaire de Velars du 5 Septembre 1888

[27] *The Times*, London, England, 7th September 1888

[28] Wikipedia – Catastrophe ferroviaire de Velars du 5 Septembre 1888

stability, but such sleepers were not much used in France.

The design of locomotives used on the two trains, which were of type 121, with two large coupled driving wheels and leading and trailing axles, was also shown to be damaging to the track and it was recommended that the company adopt American bogies, subsequent studies in 1889 and 1890 confirming the superiority of engines equipped with such bogies. The first locomotives of this type were thus introduced on the PLM in 1892.[29]

Tsar Alexander III of Russia and his family had a narrow escape when the Imperial train in which they were travelling was derailed near Borki, on the Kursk-Charkoff-Azoff Railway, on 29th October 1888. The *Times* correspondent in St Petersburg described the accident in detail: - 'The Imperial train, which left Taranovka at noon on the 29th inst., ran off the rails between that place and Borki, at a point where the line passes through a very deep gorge. At the time of the accident the Emperor and Empress, with all the members of the Imperial family and suite, were assembled at breakfast in the dining saloon.

'It seems incredible that with such havoc as was wrought any of the passengers in the train should have escaped unhurt. But God protected the Czar and his family, and they left the debris of the saloon without injury. All its other occupants were also saved, having only received some slight contusions, with the exception of Aide-de-Camp General Scheremtieff, who suffered more than the others, but was not seriously injured.

'It is to be regretted, however, that the wrecking of the other portions of the train was accompanied by loss of life. The killed included Captain Bresch, Staff Captain in the Courier Service, a physician, a secretary, and another official, two couriers, a chamber Cossack, a Chasseur, five railway servants, and six soldiers of the Railway Battalion. Eighteen other persons were injured. Baron Stjernval, Chief Inspector of Railways, received a severe shock.

'The Czarina herself superintended the arrangements for rendering help to the injured. Notwithstanding the extremely bad weather, rain falling continuously and the ground being in a slippery condition, the Czar several times descended the incline to look after the dead and injured, and only entered the reserve train which was held in readiness when the last of the injured had been placed in the ambulance train. The injured persons were conveyed to Charkoff. An officer was left at the scene of the accident to superintend the recovery of the bodies and the collection of property from the shattered carriages.

'The Emperor has ordered the conveyance of the killed to St Petersburg and has directed provision to be made for their relatives. On account of the block on the line caused by the accident, the train conveying the Imperial family went back to Losovoje. There the Czar directed a requiem mass to be said by the village clergy for the victims of the disaster. A thanksgiving service was also celebrated for the miraculous escape of the survivors. After the services were concluded the Czar invited all those who had escaped injury in the wrecked train to a dinner, which was served in the station-hall.' [30]

Given the conservative rule of the Tsar, who had reversed the liberalisations of his

[29] Ibid.

[30] *The Times*, London, England, 1st November 1888

father,[31] it was not surprising that there were rumours that the derailment had been the result of sabotage by Nihilists, but the *Times* correspondent added that: 'An examination will explain the cause of the catastrophe, but the idea of malicious design may be considered quite out of the question.' The train had been running downhill at a generally accepted speed of 64 versts an hour (about 42 miles an hour) at the moment of the derailment. It seemed likely that the track had just not been strong enough to support the weight of the heavy locomotives and large saloon carriages of the Imperial train and there were reports that old track had merely been patched up rather than worn rails being replaced by new ones,[32] and that the tsar himself had picked up a piece of rotten sleeper at the accident site.[33] An English railwayman commented that the accident was the result of 'Eight coupled running fast downhill on bloody awful permanent way!' [34] Meanwhile the *Neue Freie Presse* newspaper in Vienna continued to insist, despite official denials, that the disaster was the result of a Nihilist plot.[35]

On 2nd November the *Times* correspondent in St Petersburg quoted the Russian magazine *Grashdanin* on the cause of derailment: 'It was not one of the engines that ran off the rails, but the massive carriage in which the Minister of Ways and Communications was travelling. Although the Imperial train only proceeded at the rate of 40 versts an hour {about 26 miles an hour – as noted above most sources stated 64 versts an hour} the wrench caused by the swerving of the Minister's heavy carriage was so violent as to shatter the carriages preceding and following it. The carriage itself was wrecked.' [36]

The initial death toll in the accident was 21, including five railway servants, and 37 people were injured, one of whom died later.[37] The final death toll was therefore 22, with 36 serious injured. Those killed included the Tsar's principal servant and his favourite dog.[38]

Following the disaster Admiral Possiet, the Minister of Ways and Communications, was forced to resign and it was announced that the condition of the Russian railways was to be examined by a special commission.[39]

An extra passenger train was put on by the Savannah and Western Railroad early on the morning of Wednesday 7th November 1888 to carry passengers to the state fair at Macon, in charge of Conductor W.C. Singleton with Engineer Stokes. Due to a broken rail it got no further than Walker's Crossing, near Ellaville, Georgia, United States. A local negro man noticed the broken rail and being unaware of the special intended to warn the first regular train, but the special arrived earlier and he had no time to warn it. The engine, tender and two or three box cars passed over the break, but the baggage and smoking car and the passenger coach left the track and overturned after running a short distance on the sleepers. The body of the smoking car was dragged over a lightwood

[31] Wikipedia – Alexander III of Russia
[32] *The Times*, London, England, 2nd November 1888
[33] Ibid, 5th November 1888
[34] *The Royal Trains* by C. Hamilton-Ellis, pg. 86
[35] *The Times*, London, England, 2nd November 1888
[36] Ibid, 3rd November 1888
[37] Ibid, 5th November 1888
[38] *The Royal Trains* by C. Hamilton-Ellis, pg. 86
[39] Ibid, 8th and 10th November 1888

stump and its side was torn out. The derailment extinguished all the lamps in the carriages and the stove was unlit, so no fire broke out. The train was only travelling at about 18 miles an hour at the moment of the derailment. After the accident the engine ran to Americus, nine miles distant, for assistance.

Five passengers were killed outright in the accident. They were Charley Scoville and George Tison, both of Ellaville and both aged about 15; and three negros named George Davis, John Vanover and a man identified simply as 'Harry'. Passenger Willie Stewart was seriously, and Conductor Singleton was probably fatally injured. A number of others were less badly hurt.

The experiences of a passenger were detailed by a newspaper reporter: 'Visiting Miss Violet Wrigley in this city {Macon} is Miss Lucy Lumpkin, a plucky little lady who was on the train. When the accident occurred, and Miss Lumpkin found herself a prisoner in a car that had been turned on its side, and the lower side ripped out by a pine stump upon which it fell and was dragged, she broke a glass window with her fist and crawled out. Then she hired a horse and buggy and drove four miles to Ellaville, where she sent the following telegram to her father at Buena Vista. "Terrible wreck. I am not scratched. Lucy."' [40]

A very heavy rainstorm had affected parts of the state of Virginia, United States, during the days preceding Tuesday 2nd July 1889, but on the afternoon of Monday 1st the rain had first eased and then stopped. Norfolk and Western Railroad Section Master Tandy Jones and his men at Thaxton had had an exhausting time labouring to keep culverts clear and the track safe to run on during the previous 24 hours and that night Jones retired to bed early to get some well-earned rest, leaving one of his men, John Johnson, to watch the track overnight. After he had fallen asleep the area was struck by a tremendous cloudburst. Johnson left his home at 11 p.m. in pouring rain to inspect the line, collected his tools and went to nearby Thaxton station, from where he could hear the frightening sound of water rushing down a branch of Wolf Creek. He returned to his house, which adjoined the creek, and found it inundated by the rising water. He hastened to get his pregnant wife and three-year-old upstairs to a safer spot and tried to salvage some personal possessions only to then find that he was trapped in his house by the floodwater. And no one was left to keep a watch on the track.[41]

Eastbound train No. 2 consisted of an engine and tender, postal car, baggage and express car, second-class coach, first-class coach and three Pullman sleeping cars, *Beverly, Toboco* and *Calmar*. The train was in charge of Conductor Roland Johnson, with Engineer Pat Donovan and Fireman James Bruce. It left Roanoke on time at 11.56 p.m. for Lynchburg. At Blue Ridge its crew were cautioned by the operator, who had received an earlier telegram from Thaxton about flooding at Liberty (renamed Bedford 1890) the next station east of Thaxton. Midway between Blue Ridge and Buford (renamed Montvale about 1890) a watchman flagged down the train with a red lantern because the track was flooded. Superintendent James Cassell, who was riding on the train with fellow Superintendent Edmund DuBarry, decided that he had seen worse before, and the train

[40] *Macon Weekly Telegraph*, Macon, Georgia, United States, 13th November 1888
[41] *Lost at Thaxton* by Michael E. Jones, pgs. 3-7

proceeded cautiously until it was clear of the flooded area.[42]

There were a number of crossings of creeks in the vicinity of Thaxton and Superintendent Cassell decided to ride on the engine from Buford to ensure that the engineer and fireman drove cautiously but, in the end, let Road Foreman Alvin James do so instead, in what was to be a fateful decision for both of them.[43]

The train left Buford 32 minutes late and two miles further on crossed what was believed to be the most dangerous spot, the 100-foot-long bridge over Goose Creek. The bridge was crossed safely and a mile further on Engineer Donovan saw a watchman named Stephen Hurt and asked him how high the water was. Hurt replied that it had been very high at midnight, when a sudden flash flood might well have drowned him had he not been rescued by another man, but that the water had since gone down. Hurt was the last person to see the train before disaster struck.[44] The train then continued its journey towards Thaxton, with only a few minor creeks to cross, one of which was at Newman's Fill embankment, where there was a stone culvert over Wolf's Creek, only half a mile from Thaxton. It was here, at 1.25 a.m., that the culvert collapsed, and the embankment gave way as the engine was crossing, an enormous body of water having proved just too much for the structure to bear. Having almost cleared the gap the engine and tender slid back into it, the tender folding on top of the boiler. The baggage car telescoped the postal car and the second and first-class coaches followed them into the gap, the first-class coach breaking in two. The *Beverly* Pullman car half fell into the chasm and the weight of the cars in front pulled the *Toboco* car apart. The *Calmar* car remained on the rails.[45]

Some of the newspaper reports of the accident spoke of the boiler of the engine exploding following the accident but this does not seem to have been the case. Fire broke out in the mass of wreckage, however, and further injured a number of the passengers before they could be released. There was in addition a danger that a following freight train would run into the wreck, and to prevent this and to call for help Rear Flagman Samuel Boyd was sent back the down the line by an off-duty baggagemaster, followed by Flagman John Read, from a freight train stopped at Thaxton; the latter having been sent separately by one of the superintendents. The two managed to stop the freight and alerted the keeper at Bocock's Crossing, who was able to organise some of the locals to go and help at the wreck. Meanwhile Conductor Johnson had evacuated the passengers from the *Calmar* car for fear that it too would fall into the chasm.[46]

It was almost 4 a.m. before news of the accident reached Buford, brought by men on horseback despatched by Superintendent Cassell. Shortly afterwards off-duty baggage-handler J.H. Elam, who had made his way on foot from the wreck, arrived at Buford to add the news that it was on fire. The alarm was raised at Roanoke at 4.30 a.m., and a relief train was despatched from there at 5 a.m.[47] At about the same time horseback riders arrived at Liberty with news of the accident (the telegraph wires were down between Thaxton and Liberty, so no help could be called from Thaxton). Doctors from Roanoke finally arrived at the accident site at 6.25 a.m., five hours after the disaster had

[42] Ibid, pgs. 34-5
[43] Ibid, pg. 37
[44] Ibid, pg. 40
[45] Ibid, pgs. 41-2
[46] Ibid, pg. 49
[47] Ibid, pg. 63

happened.[48]

The total number killed in the accident was impossible to determine as a number of the bodies had been consumed in the fire. The minimum death toll was 18, consisting of 15 passengers, together with Engineer Donovan, Fireman Bruce and Road Foreman James. Twenty-one others were injured.[49]

A Bedford County Grand Jury found that the disaster had resulted from an 'Act of God' and the Railroad Commissioner reached the same conclusion.[50] Others disagreed. The Fredericksburg, Virginia, newspaper somehow managed to call the accident 'murder' and other papers demanded that every mile of railroad should be patrolled every night, and alleged that the train had been travelling at high speed when the accident occurred, Engineer Donovan having opened the throttle for the last stretch into Thaxton, but Michael E. Jones, the author of the definitive work on the disaster, considers this unlikely.[51]

On 28th October 1889 'A pointsman neglected to lock the points' at Hatras (now Hathras), Uttar Pradesh, India, with the result that an engine, a brake van and a carriage 'filled with women and children' were derailed. Eighteen people were killed, and 24 women and children injured.[52]

A derailment at Blowitz, near Pilsen (now Pizeň) in Bohemia, Austria-Hungary, on Tuesday 5th August 1890, was attributed to the sudden subsidence of a rain-sodden embankment. The third of three brief reports on the accident stated that four people had lost their lives in the accident, while 10 had been seriously and 32 slightly injured. An inquiry into the cause of the derailment concluded that it had been purely accidental and was not due to negligence on the part of any official.[53]

A major disaster in Macedonia (which until 1912 was still a part of the Ottoman Empire) was briefly noted in the London *Times*. It occurred on 14th November 1890. A train from Üsküb (now Skopje), carrying 'time-expired' soldiers, ran off the line following damage to the permanent way from heavy rain. The accident happened at Topsin, near Salonica (Thessaloniki from 1912). Eight carriages were smashed to pieces, and it was initially stated that 37 men had been killed and 15 seriously injured, a second report putting the number of dead as high as 40 with an equal number of injured. The Orient express was delayed by at least 16 hours while the line was cleared and repaired.[54]

On 22nd July 1891 a workman's train was derailed by track subsidence near Presburg (Bratislava from 1919), Hungary, and precipitated down an embankment. Ten men were

[48] Ibid, pg. 66
[49] Ibid, pg. 89
[50] Ibid, pgs. 107 and 111
[51] Ibid, pgs. 111-117
[52] *The Times*, London, England, c31st October 1889
[53] Ibid, 7th, 8th and 9th August 1890
[54] Ibid, 17th November 1890

killed, and a large number injured, some very seriously.[55]

Down western mail train No. 67, which left Sydney for Bourke at 8 p.m. on Tuesday 26th April 1892, was wrecked between Sodwalls and Tarana, New South Wales, at about 2 a.m. on Wednesday 28th April. The train consisted of a Baldwin engine (No. 452) and tender, a horse box, a louvred fruit van, a second-class lavatory carriage, a first-class lavatory carriage, a sleeping car, a postal van and a lavatory brake van. It was a charge of Guard Thomas Stone, with Assistant Guard Thomas Bates, Driver Henry Moses, Fireman George Evans and Assistant Fireman Robert Chadwick and was travelling at about 25 miles an hour on a falling gradient of 1 in 150, around a 12 chain curve and through a rock cutting when the rear brake van, the mail van and one bogie of the sleeping car left the track. One side of the brake van and the end and part of one side of the mail van were torn away by contact with the rock cutting and it was in the postal van that the casualties occurred. Instead of being purely a postal van this vehicle had compartments at each end for second-class passengers, and these were well filled. Nine passengers in the van were killed outright and others were injured.

The cause of the accident was quickly discovered when Driver Moses went back and found a rail which had broken in several places.

At the inquest on the victims of the accident, which concluded on Friday 29th, the jury returned a verdict that the accident had been caused by a break in a rail which contained a flaw, and that there was no evidence before them of neglect on the part of any of the officials. The accident was the subject of the usual departmental inquiry, and also of an independent inquiry, which concluded like the jury that the derailment had been purely accidental and unavoidable. This inquiry was also keen to stress the advantage of the train having been drawn by a single engine, so that the whole of the brake power was under the control of a single man, rather than requiring the co-ordinated action of two drivers.[56 57]

West Shore Railroad day express No. 1 was wrecked at Newburg, New York, United States, on 13th July 1893. It consisted of engine No. 200 and its tender, a baggage and express car, smoking car, two day coaches and a parlour car in charge of Conductor Richard Anderson with Engineer Daniel Shelby and Fireman Charles Willis. Four miles south of Newburg, at Cornwall, the train was running several minutes late and as it neared Newburg it was reportedly running very fast. When it reached the facing connection with the freight yard a mile south of the passenger station the locomotive reportedly left the rails at the points and continued running on the sleepers for the length of its train. Meanwhile standing in the yard waiting for the express to pass was a freight train in charge of Engineer Fred Holland. The express engine collided violently with the engine of the waiting freight train and was thrown to one side and its tender torn from it.[58]

---

[55] *Newcastle Morning Herald and Miner's Advocate*, Newcastle, New South Wales, 5th September 1891

[56] *Sydney Morning Herald*, Sydney, New South Wales, 28th April 1892

[57] *Bathurst Free Press*, Bathurst, New South Wales, 27th, 28th, 29th and 30th April and 20th May 1892

[58] *Syracuse Courier*, Syracuse, New York, United States, 14th July 1893

When his engine left the track Engineer Shelby had shut off steam and whistled for brakes, then he and his fireman had jumped from their engine; he escaped with slight injuries, Fireman Willis was rather more badly hurt. Engineer Holland of the freight train suffered a fractured skull, but it was thought that he would recover.

Behind the engine coach No. 71, the second passenger coach, was fully occupied and it as here that most of the casualties occurred, for the coach swayed across the track, struck the projecting end of a boxcar and its entire side was ripped off. Five passengers were killed more or less outright in this coach and a number injured, three of whom it was feared would die. Those killed were: Miss Eliza Klamm, of Highland Falls, New York; Rose Reilly, domestic servant, of Westpark, New York; Mrs Pauline Wright, of Brooklyn; three-year-old Addie Elberson, of Setauket, Long Island, whose parents were injured; and a woman who remained unidentified at the time of the report, probably of New York, well-dressed and wearing a lot of jewellery.

A report in a Boston newspaper gave further insight into how the accident had occurred. The points leading into the yard were under the control of a flagman named Michael Donoghue, who also had charge of a level crossing. A new switch (set of points) was being installed that morning and due to a mistake by Donoghue the connection had been left open to the yard. An automatic target 500 yards in the rear of the points should have indicated their position to Engineer Shelby and shown that they were wrongly set, but the rod connecting it to the new switch had been disconnected at 11 a.m. by the workmen and hence was wrongly indicating that the switch was open for the through road. Shelby thus received no warning that the switch was wrongly set. According to this report it was unclear whether the engine was derailed at the first connection before it struck the freight engine, but the second passenger coach certainly was.[59]

At the inquest at Newburg on 15th July Engineer Shelby testified as to the target showing "All right" for straight ahead but his engine then being diverted into the yard. He was not badly hurt in the accident and 20 minutes afterwards had gone back to crossing, thinking that his engine must have jumped the points, but when he had examined them there was no sign of the engine having damaged them and he realised that they must have been set for the yard, though when questioned Donoghue 'swore up and down that the switch was all right.'

The inquest jury did not believe Michael Donoghue and found that the accident was due to his negligence in leaving the switch open. They also censured the railway company for employing an incompetent man. Donoghue was arrested on a charge of second-degree manslaughter and not being able to afford the cost of bail was lodged in jail.[60]

Two of the injured passengers died a few days later, bringing the number of dead up to seven.[61]

Japan and China had been at war in Korea from August 1894 until April 1895 and a major railway disaster, possibly a major catastrophe, in Japan, was an indirect result of

---

59 *Boston Daily Globe News*, Boston, Massachusetts, United States, 14th July 1893
60 *Boston Sunday Globe News*, Boston, Massachusetts, 16th July 1893
61 *Warren Mail*, Warren, Pennsylvania, United States, 19th July 1893

this conflict. The accident happened very early on in the conflict, even before the official declaration of war on 1st August. Early on 27th July 1894 (or possibly late on 26th) a northbound train carrying 380 passengers, comprising wounded soldiers, doctors and nurses, left the port of Hiroshima during a severe gale. The line from Hiroshima to Kobe ran along the coast and unbeknown to anyone on the railway, heavy seas caused by the storm had washed away an embankment just beyond Mihara station. The safety of the line had not been checked despite the storm, perhaps due to insufficient staff, and at about 1 a.m. on 27th July, in conditions of poor visibility the train reached the spot where the embankment had collapsed and its front engine (there was a second locomotive at the rear of the train) and 13 of the 25 carriages left the track and plunged into the sea. The driver, fireman and an undetermined number of passengers lost their lives. A telegram from Yokohoma on 28th July put the death toll as high as 140 but a recent book has stated that only 16 people died.[62] [63]

The washing away of the track, this time by torrential rain, was also the cause of a catastrophe in Russia on 13th May 1897. The catastrophe was apparently the culmination of a series of railway accidents in the country for the *Times* reporter in St Petersburg commented that: 'Railway accidents have been so frequent in Russia this year that only something quite extraordinary could cause the sensation which has been produced by the terrible disaster of the 13 inst. on the line between Dorpat and Riga.'

The accident involved a troop train conveying two battalions of the Krasnoyarsk Regiment to their summer camp at Tsarskoe Solo, which ran off the line between Rockenhof and Elwa stations on the Walk-Yurieff line. A severe storm of rain and thunder was raging, it was very dark although still early in the day, and there was torrential rain. The driver applied his brakes as soon as he saw that the track was damaged, but he was too late, and the train was derailed, about 16 trucks of men and horses being crushed together and splintered to pieces. A passenger train had passed the site of the accident without encountering any problem only 20 minutes earlier.

The accident resulted in 56 fatalities among the soldiers, with another 43 seriously injured and three officers and 37 privates slightly injured. Two railway guards were also killed and one hurt. 'A number of high officials soon arrived to make investigations, and the professors and students of Dorpat University helped to attend to the sufferers.' [64]

Westbound Central Pacific Railroad passenger train No. 2 left Reno, Nevada, United States just after 9 a.m. on Thursday 29th July 1897. Less than an hour later it had been wrecked at a location known as Twenty-five-mile-hill, half a mile west of Laughton's Springs. The train was double-headed, and engine No. 1825, the pilot engine, with Engineer John Wright and Fireman Harvey, passed the danger point unscathed, but the train engine, No. 1813, was thrown down the embankment. After the derailment engine No. 1825 continued to Verdi to raise the alarm, and telegrams were sent to Reno and the company headquarters at San Francisco.

---

[62] *Early Japanese Railways 1853-1914* by Dan Free, pg. 177

[63] *The Times*, London, England, 29th July 1895

[64] Ibid, 16th and 17th May 1897

When news of the accident reached Reno a switch engine with a flatcar attached was despatched to the accident site with Doctors Thomas and Gibson, newspaper reporters and section hands to repair the track. When the relief train reached Twenty-five-mile-hill its crew found the mail, baggage and express cars overturned, the day coach lying across the track, followed by a Pullman sleeping car, dining car and two more sleeping cars, all damaged. Two sleeping cars at the rear of the train had remained on the track and were brought back to Reno. Despite the damage none of the passengers lost their lives and Engineer Irving (or Irwin) and Fireman Rollins of the train engine were not that badly hurt either. Less fortunate though were 15 Indians (Native Americans), men and women, riding 'blind baggage' en route to Sacramento to pick hops. Six of them were killed outright and several others badly hurt, two probably fatally so. Their wailings were said to be heartrending.[65] One of the Indians killed, the Reno reporter added, was known as "Sawmill Sam", and he had given officers much trouble in the past. 'Sam was known as a "bad Injun" in town, consequently his taking off is not considered such a loss as it would otherwise have been.'

Fireman Rollins from the train engine was helped a quarter mile to a Mrs W. Hackman's yard by her son and there his injuries were treated. In answer to questions he said, "I felt that the engine was off of the track and felt that she would stop, but before I had time for much thinking I was thrown from the engine, on my head, I suppose. I was thrown on one side of the track and the engine was thrown on the other. Just how this happened I do not understand."

"Were you ever in a wreck before?" he was asked.

"Yes," was the answer. "About two years and a half ago I was in a wreck at Mystic. We went into the water then, and my engineer had to have twenty-two stitches to fix him up straight again."

"Have you telegraphed your friends?"

"No, they would be too much startled by a dispatch. I am not hurt enough to telegraph. I live in Sacramento, and my wife would be scared to death by a dispatch. They will take me to Sacramento to the hospital as soon as they can get a train through and I'll soon be all right."

Also at Mrs Hackman's were a number of women passengers, 'Among them were Miss Lottie Gale of Truckee, who was en route from Wadsworth, where she had been to attend to her music class, to her home in Truckee. Miss Gale was in the day coach. She said, "It seemed an age from the time the car began to rock until it turned over. There were about fifteen in the car, but no one was hurt except a little girl who had her foot hurt. Myself and Miss Hunington were sitting together. Miss Cooper, on her way to Colfax, was also in the coach. We were badly shaken up, both mentally and physically, and climbed out of the coach the best way we could."

'Frank M. Edgar, wife and little daughter were on their way from Salt Lake to their home in Santa Paula. The little girl, aged 8 years, was in the toilet room when the wreck occurred, and the mother not seeing her was frightened into a fainting fit. The little one, before being released, kept crying: "I'm all right, mama."'

Dr Stubbs of Nevada University was a passenger on the train and described the

[65] *Reno Weekly Gazette and Stockman*, Reno, Nevada, 29th July 1897

accident to the reporter: 'I was sitting in on my way to San Francisco, having been wired to come down by my brother. I had invited Mr Farrington of Elko to occupy the seat with me. He came in from the smoking car, which was badly wrecked, and probably escaped serious injury. When the car began to rock one of the ladies in the coach, Mrs Jenkins, of San Francisco, fainted away, and we were all badly shaken up. When I found the head engine was all right, I went on with some of the train men to Verdi, where despatches were sent to Reno and Wadsworth for help.' [66]

An indirect casualty of the accident was a Mrs Finn, in her buggy on the way to Verdi, who chanced to be passing when the train was derailed. The horse, frightened by the wrecking of the train, began to run away, and she was thrown out of the buggy and badly shaken up, but not seriously hurt.[67]

It had been very hot in the two weeks preceding the accident and it was believed that the abnormal heat had expanded the rails, putting them under considerable stress. The passage of the leading engine of the train had then caused them to buckle and form a "sun link", derailing the train.[68]

At 7.30 p.m. on the evening of Saturday 23rd October 1897 the New York Central and Hudson River Railroad overnight express known as the Buffalo Special left Buffalo, New York, United States, for New York City, where it was due to arrive at 7 a.m. on Sunday morning. The train was in charge of Conductor E.O. Parish with Engineer John Foyle and Fireman John Tompkins and consisted of engine No. 872, an American express car, a composite baggage and smoking car, a day coach and the sleeping cars *Glenalpine, Hermes, Niobe, Diana, Anita* and *Racket River*. There were 20 passengers riding in the day coach and about 80 in the sleeping cars. The train had called as usual at Poughkeepsie and was at a rather isolated location known as King's Cut, three miles south of Garrison, New York, and five miles north of Peekskill, when disaster struck. [69]

Augustus Carr, a German living near the scene, described how he had been awoken early that Sunday morning: 'It was about five minutes before six when I was awakened by someone in my yard yelling for help. Looking out of my window I saw a sleeping car porter, who shouted, "For God's sake man, if you own a boat come quickly. Our train is in the river and people are drowning."

'I dressed myself and, accompanied by the porter, got into my rowboat and rowed around the curve to where the train was in the river.

'When we reached the cars, which were submerged nearly to their roofs, the crews of the tugboats were making efforts to save the passengers.

'The first man I saw them take out was, I think, the agent of the express car. The first persons whom we succeeded in rescuing were two Chinamen who were sitting on the roof of the smoker. One had his arm broken. We put them ashore and then took three more passengers off the top of another car.

'At the same time people in their underclothes were being taken out of the sleeping

[66] Ibid.
[67] Ibid.
[68] *Daily Nevada State Journal*, Reno, Nevada, United States, 29th July 1897
[69] *Wellsville Alleghany County Reporter*, Wellsville, New York, United States, 26th October 1897

cars by the crews of the several tugs.

'One man on shore with an arm cut off was dying, and we made his last moments as comfortable as possible. I want to say that the porters, although frightened, showed great bravery and saved many lives.'

The only witnesses to the accident had been the crew of a tugboat passing along the river at the time with a tow. They had seen the lights of the train coming around the curves in the line and then suddenly the majority of the train plunged down the embankment into the river. The whole of the train would probably have done so, but a coupling snapped and as result the last three sleeping cars remained on the track. Some of the cars with closed windows that had gone into the river floated and the tugboat crew had whistled for help, cast off the hawser and immediately begun the rescue work. Some of the passengers in the sleeping cars were thus rescued but the day coach and smoker had gone down into the deeper water and the rescue of the passengers from them was impossible. 'In the latter coach {the smoker} the conditions must have been terrible. The car turned completely over and the passenger end of it was in the deep water, while the baggage end stood up towards the surface.

'The men in the lower end must have fought like fiends for a brief period, for the bodies when taken out were a mass of wounds.'

An early report gave the number of confirmed dead as 19, including Engineer Foyle and Fireman Tompkins, with a likely final total of 28. The final number of dead was in fact 21.[70] [71]

There were allegations that the train had been deliberately wrecked by dynamite, but General Manager Toucey stated that: 'The accident was caused by the bed of the railroad being washed out in some inexplicable manner. In this undetermined condition the track sank as soon as the weight of the train was put on it, and the embankment giving way, the train was, of course, precipitated into the river.

'Such a condition as this we have never looked for. Trains have been running over this spot for years and years without accident or difficulty of any kind, and this piece of track was considered as good as any section of the road. Not only was the roadbed the hardest kind of an embankment, but it was strengthened by a retaining wall of masonry three feet thick.'

Mr Toucey lived by chance at Garrisons and had been one of the first officials on the scene. A trainman from the rear of the train had run to the nearest signal tower to raise the alarm and the engines of two other express trains had been despatched and one sent to Peekskill for doctors and the other for Mr Toucey.

Other railroad officials were of the opinion that a quicksand foundation of some kind below the water line was responsible for the sinking of the roadbed.[72]

That the track had been in good condition at the location of the accident was disputed by former section boss Michel Clare, of Cold Spring, who told a reporter: 'It is a wonder that such a disaster did not happen there before. The condition of the roadbed at the point where the disaster occurred has long been notoriously unsafe. While I was overseeing

---

[70] Ibid.

[71] *Syracuse Sunday Herald*, Syracuse, New York, United States, 28th November 1897

[72] *Wellsville Alleghany County Reporter*, Wellsville, New York, United States, 26th October 1897

that portion of the road, I found that every two days about the track at this very point was sagged, not only out of surface, but out of line as well.

'The condition of the roadbed of the west track at this and other points, and the condition of the ice last winter, compelled the Central officers in January or February of this year to build a sea wall on the river side, about an eight of a mile in length. When the soft weather of the spring set in and the frost left the ground the trouble began again, and in April last, I think it was, about sixty feet of this wall crumbled away at a point only a few feet north of where the Buffalo express plunged into the river.

'During my employment on the road, there was never an operator anywhere near this place, which is halfway between two 'blind towers,' Nos 42 and 43. After Sunday's disaster was the first time an operator or 'working' signal station was ever operated there.

'This wall was three feet wide at the top, and although seven feet wide at the bottom, was built on the shelving or sliding rock. They could never get a good foundation there, and, besides, the wall was what is known as 'dry wall' – uncemented.

'Time and again, as my reports to my superiors will prove, I reported not only this very spot as the most dangerous between New York and Buffalo, but also called attention to two other very weak spots, where there is likely to be a repetition of this horror. The wall was built by the company's own labourers, and when that portion of the wall where the train dashed into the river fell away last April General Roadmaster Sparrow came down and viewed it.

'It was never repaired up to the time that I left their employ. It was built under the supervision of General Roadmaster Caldwell of East Albany. The other dangerous places which I reported were above and below King's Dock and above Beverly Bridge. The man who followed me upon the section is a man named Fitzpatrick. As I have stated, my opinion that this was the most unsafe spot on the line between New York and Buffalo has been confirmed.'

As might be expected Clare's allegations were vehemently denied by officers of the New York Central, as was his statement that he had resigned voluntarily as section foreman because he knew that the line was unsafe and did not wish to take responsibility for it. They stated that he had in fact been discharged for drunkenness and neglect of duty.[73]

On 3rd December an inquest jury brought in the unsatisfactory verdict that the cause of the accident was unknown to them,[74] while the *Syracuse Evening Herald* commented sarcastically that the report of the investigating committee looking into the disaster was expected in four or five years.[75]

Westbound Central Pacific Railroad passenger train No. 2, which had been wrecked near Laughton Springs, Nevada, United States, in July 1897, was again in trouble on 19th May 1898. It had left Reno two hours late that day and was within a mile and a half of Boca, California, when its engine and mail, baggage and express cars went into the ditch and were badly wrecked. Engineer Trott was somewhat hurt but the main casualties were

[73] *Syracuse Evening Herald*, Syracuse, New York, United States, 27th October 1897
[74] *Fort Wayne Gazette*, Fort Wayne, Indiana, United States, 4th December 1897
[75] *Syracuse Evening Herald*, Syracuse, New York, United States, 4th December 1897

again Indians riding 'blind baggage'. Four of them who had been riding on the platform of the baggage car were killed and several others more or less badly injured. None of the passengers were hurt, other than being shaken up.

The derailment was blamed on the spreading of the rails. Wrecking (breakdown) trains were hurried to the scene from Wadsworth and Truckee. The train involved in the derailment was strictly the first section of passenger train No. 2, the following second section was carrying 329 officers and men of the First Regiment Nebraska United States Volunteers, en route to San Francisco from where they would sail to the Philippines to participate in the brief war between the United States and Spain. As the result of the blockage of the line caused by the accident the second section was held at Reno until early the next morning.[76]

At 4 a.m. on 14th August 1898 Chemins de fer de l'Ouest train No. 97, which had left Paris at 11.16 p.m. for Trouville became derailed between Saint-Mardes-de-Fresne and Lisieux, or more precisely between Beuvilliers and Glos, France, at a point where the line was under repair. The double-headed train was reportedly running fast down an incline at the time. The leading engine kept the track and was able to proceed to Lisieux for assistance, but 17 of the 19 carriages of the train were derailed, two third-class coaches in the middle of the train being badly damaged. Seven passengers were killed outright and 40 injured, 27 of them severely enough to be admitted to Lisieux Hospital. Five of them were not expected to survive. After the accident assistance was provided by railwaymen from Lisieux, Sisters of Charity (nuns) and soldiers from the local garrison.[77]

The cause of the accident was not entirely clear. The *Times* correspondent in Paris wrote the day after the accident that 'The evidence received thus far would go to show that the drivers had done their duty in slowing down considerably.' The Minister of Public Works, who hastened to the scene of the disaster, has stated to a journalist that, 'as far as he can judge, the double engines were the cause of the accident.' It would seem, however, that the train had approached the site where the permanent way was being repaired at excessive speed and that the train's brakes had then been applied in emergency, causing a traction shock which combined with the poor state of the track had been enough to cause the derailment. At any rate on 12th July 1899 the Lisieux Criminal Court found the Driver Mahov, who had been driving of the pilot engine, responsible for the accident and he was given a four-month suspended prison sentence.[78] [79] [80]

Excessive rain caused an accident to a Cedar Rapids Burlington and Northern Railroad train No. 5 at Sink Creek, four miles south of Waterloo, Iowa, United States, on 29th May 1899. A cloudburst had washed away the sand and gravel roadbed, leaving the track unsupported over a length of 20 feet. The rails and sleepers remained intact, so the engineer had little or no warning of any problem. An estimated five inches of rain had fallen in nearby Washburn in only five minutes just before the accident.

---

[76] *Daily Nevada State Journal*, Reno, Nevada, United States, 20th May 1898
[77] *The Times*, London, England, 15th August 1898
[78] Ibid, 16th August 1898
[79] *Le Petit Parisien*, Paris, France, 16th August 1898
[80] *Le Temps*, Paris, France, 13th July 1899

The train involved in the accident was a through one from Chicago to Minneapolis. It consisted of mail, baggage and smoking cars, two day coaches and a sleeping car, six vehicles in all. When the engine of the train hit the washout, it was thrown into a ditch and the coaches were piled up in a confused heap. The mail car tipped over and was telescoped by the baggage car, and the rear end of the baggage car was telescoped by the smoker. The roof of the sleeper ploughed its way through the adjoining day coach and the lower half of the day coach crashed through the sleeper. Engineer William Dinley and Fireman John Gibney jumped and escaped with slight injuries, but Conductor George Wainwright; Pullman Conductor F.S. Carpenter and passengers David Hallo, W.A. McLaughlin and E.L. Arnold, all of Minneapolis; E.J. Barker, of Hot Springs, Arkansas; and O.R. Hopkins, address unknown, were killed according to one report (the names of some of the passengers who lost their lives were variously given in different newspaper reports). About 15 passengers were injured, one of whom, R.H. Schneppe, of Alton, Illinois, had to undergo an amputation at the shoulder to release him from wreckage and died later. A good proportion of the injured passengers were immigrants from Norway.[81]

The first news of the accident was provided by uninjured mail clerks from the train who hurried along the line to Washburn station, where they alerted the operator. A special train was immediately made up at Waterloo upon news of the accident being telegraphed to there, and two special trains were sent from Cedar Rapids.[82] [83]

The same day as the Sink Creek accident, 29th May, a Chicago Great Western passenger train was wrecked by a washout near Oelwin, Iowa, United States, and a porter was killed.

The Ottawa Express which had left Montreal, Quebec, Canada at 9.30 a.m. on 9th August 1899, was wrecked while running at high speed at Ste-Polycarpe, near Coteau Junction, Quebec, on the Canadian Atlantic Railway. The engine, baggage car and a second-class coach left the track, the first-class coach and two Pullman cars remained on the rails. The second-class car mounted the engine and the victims of the accident were scalded by escaping steam. Fireman George McCuiag and passengers Wilson O'Connor and Edward Stairs of Ottawa and Joseph and Mrs Rochelon and their daughter, of Montreal, Quebec, were killed more or less outright. A number of passengers were injured, some seriously, and by the next day the Rochelons' eight-year-old son had died as well, wiping the whole family out, while little hope was entertained for Ellen McDougall and Miss Ryan, of Maniwaki, Quebec. It was not known what had caused the derailment, but it was believed that the track had spread.[84] Most of the victims were pilgrims returning from the shrine at St Anne de Beaupre.[85] [86]

Engineer Robert Orr suffered leg injuries and scalds and was confined to bed when

---

[81] *Burlington Evening Gazette*, Burlington, Iowa, United States, 29th May 1899
[82] *Des Moines Daily News*, Des Moines, Iowa, United States, 29th May 1899
[83] *Fairfield Tribune*, Fairfield, Iowa, United States, 31st May 1899
[84] *Bradford Era*, Bradford, Pennsylvania, United States, 10th August 1899
[85] *Golden Era*, Golden, British Columbia, Canada, 18th August 1899
[86] *Winnipeg Free Press*, Winnipeg, Manitoba, Canada, 11th August 1899

interviewed by a reporter the next day. Asked if he could explain the accident he replied, "I cannot account for it at all. We were making regular running time, not going as fast as we do at many other places on the line and not as fast as we do even at this point going the other way because we were going up grade."[87]

It was common at least in Great Britain and presumably abroad for speed restrictions to be imposed on lines passing through coal mining areas due to the possibility of subsidence caused by underground workings. No derailment of a passenger train seems to have occurred in the United Kingdom during the 19th century from such a cause, though on 22nd September 1892 the engine of a goods train which was shunting at Lindal, England, on the Furness Railway, had been swallowed up by a hole which suddenly appeared beneath the track. The driver and fireman were able to jump clear and later the tender of the engine was recovered but before the 35-ton locomotive could be extricated the hole deepened and its recovery was abandoned, so that it remains underground to this day.[88] At the end of the century in the United States, at about 1.50 a.m. on 17th April 1900, the Missouri Pacific Railroad passenger train from Kansas City due in Joplin, Missouri, at that hour was derailed about a mile east of the place. The whole train, but not the engine, left the rails, and a number of passengers were injured, some seriously. The track had subsided where it ran over an abandoned zinc mine, the galleries of which had caved in.[89]

Camp Creek, a tributary of the Ocmulgee River, ran parallel to the Southern Railroad main line for some distance north of McDonough, Georgia, United States, before passing beneath the railroad by means of a 64-foot stone culvert. Excessive rain had fallen in this part of the southern states in the two weeks prior to Saturday 23rd June 1900, all the streams in the area were much swollen and several washouts had been reported.

At 7.10 p.m. on 23rd a passenger train left Macon and it was due in Atlanta at 9.45. In charge of Conductor William A. Barclay, with Engineer James T. Sullivan, it reached McDonough station on time, and there it was normally combined with a train from Columbus for the remainder of the run to Atlanta, but on this night, for the first time for some months, the Columbus train was two hours late due to a washout on the branch, and the Macon portion of the train continued without it.

A cloudburst had occurred at about 7.30 p.m. in the area and probably just after dark the floodwaters of the grossly swollen Camp Creek washed away 100 feet of the track a mile and a half beyond McDonough station. The storm was still raging as the train set off and the windows of the train cars were all shut, its passengers doubtless feeling comfortable and secure from the raging storm outside. A short time after the train had got going again it had plunged into the washout.[90]

Behind the engine the train consisted of only four vehicles, a baggage car, second-class car, first-class car and a sleeper. The vehicles were 'knocked into kindling wood' by

---

[87] Ibid.

[88] Wikipedia – Lindal Railway Incident

[89] *The Daily Journal*, Telluride, Colorado, 17th April 1900

[90] *Boston Daily Globe*, Boston, Massachusetts, United States, 25th June 1900

the fall and a few minutes later fire broke out in the wreckage, consuming all but the sleeper. The heavy sleeping car, which had also been the last part of the train to take the plunge, crushed the cars beneath it, and no one in them escaped, the few passengers in the sleeping car being unable to help them.

'For a brief time, there was silence. Then the occupants of the sleeper recovered from their bewilderment, and after hard work managed to get out of their car and found themselves on the track in the pouring rain.

'Flames were seen coming from that part of the wreckage not covered by the water. As the wreck began to go to pieces under the destructive work of both flames and the flood, bodies floated out of the mass and were carried downstream by the swift current.

'The storm did not abate in fury. Flashes of lightning added to the steady glow of the burning train and lit up the scene with fearful distinctness.

'Flagman Quinlan, who was one of the first to get out, at once started for the nearest telegraph station. Making his way as rapidly as possible in the face of the blinding storm, he stumbled into the telegraph office at McDonough and after telling the operator of the wreck, fall fainting to the floor.

'Word was quickly sent to both Atlanta and Macon, but no assistance was to be had except in the latter city and the interrupted track prevented the arrival of any train from there.' [91]

The aforementioned Flagman J.J. Quinlan was one of the heroes of the disaster. It had taken an enormous effort on his part to climb up the very steep embankment from the wreck to the track and he had then made his way back to McDonough as fast as he could, knowing that a northbound freight train was shortly due to follow the passenger train. He got there just in time to stop it and prevent an even worse disaster.[92]

On learning of the disaster almost the whole of the adult population of McDonough went to the scene of the accident, but they could do little until daybreak. Some of the bodies that had floated from the wreck were then found, many along the banks of the stream and one which had floated a mile away.

About 35 people died in the accident, possibly as many as 41. The dead included Conductor Barclay, Engineer Sullivan, Firemen John Brantley, Ed Byrd and Will Green, Pullman Conductor H.R. Cressman and Porter Robert Spencer, baggagemaster W. Bennett, Conductors J.E. Wood and J.H. Hunnicut and Flagman J.H. Rhodes. Some of these railwaymen, for example Conductors Wood and Hunnicut and two of the firemen, were probably travelling 'on the cushions' rather than on duty. The train was also carrying a section hand and eight negro labourers, on their way to repair a washout on the Georgia, Midland and Gulf road. All of them lost their lives.[93] Of those who had been on the train only Flagman Quinlan, Porter T.C. Carter and seven passengers from the sleeping car survived.[94]

[91] Ibid.
[92] *In Atlanta or Hell*, by Jeffery C. Wells, pgs. 42-5
[93] *Bryan Morning Eagle*, Bryan, Texas, United States, 26th June 1900
[94] *In Atlanta or Hell*, by Jeffery C. Wells, pg. 81

Parts of India were affected by flooding during the 1900 monsoon season, and the floodwater caused the subsidence of an embankment near Baripada, in Orissa, and the derailment of a mail train, on 23rd August 1900. Nine carriages were wrecked, and 11 people lost their lives, including a European, a Mr Henry Wood, of Bangalore. Twenty-five more were injured, among them two more Europeans. After the accident the engine driver walked five miles in the dead of night to summon assistance.[95]

At 11.30 p.m. on Sunday 28th October 1900 Northern Pacific Railroad eastbound passenger train No. 4 was wrecked at Dehart Siding, about midway between the stations of Springdale and Big Timber, Montana, United States. The train was running three hours late, its engine having broken down in the morning at Belgrade. It was in charge of Conductor George Ott, with Engineer Brouse and Fireman Mikesell. The train was running at a good speed at Dehart and the engine, mail and baggage cars passed over the facing points of the siding all right, but the connection rod to the point blades then broke, allowing the points to open beneath the wheels of the train. Three coaches left the track and after running 200 feet along the track overturned and were dragged some distance along the ground. Some of the passengers who were on the left-hand side of the cars were then thrown through the windows and crushed under them.

Seven passengers lost their lives. Five of them were Ed Eastman, of Raymond, South Dakota; W.J. Riefenrath, Northern Pacific express agent at Billings, Montana; Lester Pendleton, of Fennimore, Wisconsin; Dr Charles Hathhorn, of Livingston, Montana, and Miss M.F. Tracey, of Bozeman, Montana. The other two fatalities were two women who were severely mangled and had not been identified. Seven other passengers were injured. A particularly sad aspect of the accident was the plight of Miss Hathhorn, sister of Dr Hathhorn, who had just become engaged to W.J. Riefenrath, and had thus lost her brother and her fiancé at one stroke.

An inquest was held at Big Timber the day after the accident and its jury returned a verdict that the cause of the accident was unknown and exonerating the railway company from all blame.[96]

---

95 *The Times*, London, England, 25th August 1900

96 *Anaconda Standard*, Anaconda, Montana, 30th October 1900

CHAPTER 2

DERAILMENTS - BROKEN TYRES, BROKEN AXLES AND OTHER VEHICLE DEFECTS

One of the frequent causes of derailments of trains other of problems with the track was defects in the trains themselves, such as broken tyres, wheels and axles. Such accidents were not as frequent by the 1880s as in the early days of railways but they still happened quite often and were to a large degree unavoidable.

On 22nd January 1881 an Erie Railroad train was derailed near Tioga Center, New York, in the United States. On that Saturday evening eastbound train No. 12 from Buffalo to New York, driven by Engineer Thomas (Tommy) DeWitt, left Hornellsville two hours and 20 minutes late having had to wait for connections from the west. It had stopped at Elmira and was nearing Tioga Center when De Witt felt a terrific pull as his train left the track. He applied the air brakes and reversed his engine, but it smashed into a couple of gondola cars loaded with coal standing on a siding, demolishing them and the front of his engine. Behind the engine tender the postal and express cars plunged down an embankment, rolled over and over and then quickly caught fire.

Mail agents Joe Reidinger, George Ingraham, and Dan Leibold and mail weigher H.B. Fox were probably killed outright in the postal car. Express messenger Henry Brewer was trapped in the wreckage of the express car and burnt to death before he could be released. The bodies of all five were burnt to a cinder. The occupants of the rest of the train escaped with cuts and bruises although the whole of the train was derailed, and two baggage cars and two day coaches were completely burnt. The mail and express matter was all lost. Oil from the kerosene lamps in the postal car was thought to have contributed to the intensity of the fire which could have begun either from the smashing of one or more of the lamps, the upsetting of the stove or both.

The derailment was caused by the fracture of the forward driving axle of the engine.

'The axle was made of the finest material, and its breakage is unexplainable.' [97] [98]

On Tuesday 30th August 1881 the 8.54 a.m. express from Brighton Beach to Melbourne Flinders Street, Victoria, carrying its usual compliment of business and professional men to work in the city, and also the pupils of Mr R.H. Budd's ladies' school, was close to its destination, when after it had passed Richmond station the passengers riding in the first-class smoking carriage heard a sudden noise, and their ride was punctuated by a severe bumping. As the train was passing the Melbourne Cricket Ground at Jolimont the bumping became so severe that the passengers realised that an accident was almost inevitable, and hung on to the rafters, doors and windows. Communication cords were still not provided on trains in Victoria at this date, not even on express trains, so the passengers had no means of contacting the driver or guard. A couple of seconds later, as the train passed Jolimont gates, there was a terrific crash and three carriages were derailed and pitched down an embankment, while two more left the line. The first-class smoking car was the first to go down the embankment, followed by the second-class smoking car and an ordinary second-class car. The second-class smoking car fell onto and crushed the first-class one, in which most of the casualties occurred. Three passengers were killed outright; a Mr F.F. Balliere, a well-known Melbourne publisher; the Rev. Mr Garrett, a Brighton Baptist minister; and a Mr J.J. Jamieson, an articled clerk. A number of other passengers were injured, a couple very seriously. As is often the case, there was also at least one lucky escape, a Mr H. Hodgkinson was standing at the door of the first-class sleeping carriage at the moment it plunged over the embankment and was thrown out of a window into a pool of shallow water, escaping with slight injuries. The schoolchildren, in a carriage further back the train, all escaped unhurt.[99]

Whether the train would have been stopped any sooner had it been equipped with a communication cord is unclear as the derailed coach was noticed by Porter Dennis Coakly, who was acting as assistant guard, as the train was passing Jolimont gates, and he informed Guard John Maud, who immediately applied the brake and shouted from his van, the leading vehicle of the train, to Driver Thomas Hartigan, who heard his shouts and pulled up as soon as possible. Whether stopping immediately was the correct thing to do anyway seems debatable in view of Colonel Yolland's conclusions after the 1874 Shipton disaster in England, when the inspecting officer concluded that the sudden pulling up of the rein had made the accident much worse by forcing the carriages off the track. After the accident Guard Maud sent a porter back down the line to protect the obstruction and told Driver Hartigan to take his engine on to Flinders Street to inform them of the accident, which Hartigan did.

It quickly became clear that the cause of the accident was a broken tyre on the near front wheel of the first-class smoking car, which was a four-wheel carriage which had been running on the suburban lines for 16 to 18 years, having been imported from England in pieces by the old Hobson's Bay Company and reassembled at the Sandridge

---

[97] *Wellsville Daily Reporter*, Wellsville, Pennsylvania, United States, 24th January 1881

[98] *Canal Dover Weekly Argus*, Canal Dover, Ohio, United States, 28th January 1881

[99] *The Argus*, Melbourne, Victoria, 31st August 1881

workshops.[100]

Mr Elmslie Stephen, the most seriously hurt of the injured passengers, with a broken back, died on 2nd September, making the fourth fatality in the accident. He was an inspector of distilleries in the government Customs department.[101]

At the inquest into the victims Carriage Repairer Henry Johnson testified that he had tapped the wheels of the express on the morning of the day of the accident and they had all been sound. Evidence was also given that there appeared to be no regulation as to how long tyres should be allowed to remain in service before being replaced.

On the following day of the inquest George Smiles said he was acting locomotive superintendent at Sandridge since the death of Mr Higgins. He stated that there was no carriage and waggon examiner as such. The tyre which had failed was made of Krupp's patent steel. He had examined the broken pieces of tyre after the accident and had found the breaks all to be fresh, with no flaws. Had the tyre come into the workshops he would have condemned it, as it was down to ⅝ in thickness.

On the same day of the inquest Soloman Mirls, locomotive superintendent of Victorian Railways other than the suburban lines, stated that the minimum thickness for steel tyres on his lines was 1⅛in. Tyres on the Victorian main lines were neither screwed nor bolted and he considered that screwed tyres, such as that which broke, must inevitably be weaker.

William Elsdon, general manager and engineer-in-chief, disagreed with Mr Mirls and considered that tyres which were screwed on were just as safe, while on the subject of communication cords he considered that they would be frequently misused and having them fitted would risk accidents on lines where trains followed closely behind one another. He had nothing to do with the rolling stock, it was all under the locomotive superintendent. He could not say if there was a periodical inspection of rolling stock. He assumed that there was. He did not interfere in such matters, that was a matter for the locomotive department, nor in the running of trains, that was for the traffic department.

The next day the inquest heard the testimony of John Currie Johnson, engineer of the Tyne Foundry, who said that he had examined the broken tyre. He agreed that it was of good steel and that the fractures were new, except for a broken bolt. He stated that he had some knowledge of tyres. The most recent practice was for them to be put on without bolts, and to contract them. Those in current use on the Victorian mainline railways were shrunk on, like the broken tyre, but had two shoulders instead of one. The bolts were the weakest part of the tyre.

Henry Jonson, re-called, stated that he had no instructions to condemn thin tyres.

George Smiles, locomotive superintendent at Sandridge, re-called, stated that he considered himself Johnson's superior officer but had given him no instructions as he had been given none himself and he considered that Johnson knew his job better than he (Smiles) did.

On the final day of the inquest Thomas Bent, the new Commissioner of Railways, testified that since taking charge he had ordered 100 new carriages to be built to replace

---

100 Ibid, 1st September 1881

101 Ibid, 3rd September 1881

the old ones on the Hobsons's Bay line, had had one and a half miles of new rails laid down, had begun experiments with the Wood's brake on the Brighton line, had invited tenders for the doubling of Windsor-Brighton and South Yarra – Caulfield lines and so on. Mr Bent had only been in office a month when the accident occurred.

The coroner, in his summing up, was very critical of Mr Elsdon, the general manager, for not knowing about the general bad condition of the rolling stock on the line but told the jury that they did not have sufficient evidence to find anyone guilty of manslaughter. At the conclusion of the inquest the jury passed a verdict, "That the deceased, at Jolimont crossing, on the 30th of August last, were accidentally killed by the breaking of the tires of the wheel of the first class smoking carriage, and that such tire was worn so thin that it was dangerous to use. We find that had the brake-van been glazed and placed in the proper position at the tail of the train, the accident might have been noticed by the guard, in all probability, in time to have saved the loss of life. We find that the management of the Hobson's Bay line has been so culpably lax that it is impossible to fix the responsibility of allowing this carriage to leave the yard on the day of the accident, and we recommend that the whole of this evidence be submitted to the Commissioner of Railways."[102]

Chicago and North Western Railroad eastbound freight train No. 50 left the track near Round Grove, Illinois, United States on 7th October 1881, four or five cars being derailed due to a broken journal. It appeared at first to be just an everyday freight train derailment, disruptive to traffic until the line was cleared and somewhat expensive in terms of repairs to rolling stock and compensation for lost cargo but no more. During clearing up operations, however, the workmen found the bodies of four boys who had been stealing a ride in a car of grain and had been buried and suffocated when the cars overturned. Their names were Peter Herman, John Garey, William Grady and James Riley, all of Sterling, Illinois, only about six miles further on. They had only boarded the train at Morrison, the previous station to Round Grove, on their way home from visiting Barnum's Circus at Clinton, a few miles to the west.[103]

Three bad accidents happened in different parts of North London, England, within less than two months in December 1881 and January 1882. On 10th December 1881 there was a triple collision in Canonbury Tunnel, on 25th January 1882 a rear collision in fog at Hornsey and only three days later, on 28th January 1882, there was a bad accident between Old Ford and Bow, on the North London Railway. The latter was an unlucky affair, the 9.50 p.m. passenger train from London Broad Street to Poplar coming into collision with some derailed waggons belonging to the 10.05 p.m. empty coal train from Poplar to Brent.

Giving evidence at the subsequent Board of Trade inquiry into the Old Ford accident, Driver William Crabb of the coal train said that it had consisted of 50 empty waggons and two brake vans. He had started five minutes late and had clear signals till he got to the Old Ford distant signal, which was at "Danger". He had whistled for the signal and it was cleared as he reached the Fairfield Road overbridge, whereupon he had proceeded

---

102 Ibid, 8th, 9th, 14th and 21st September 1881

103 *Freeport Republican*, Freeport, Illinois, United States, 15th October 1881

towards Victoria Park. As he did so he looked around and found that he could not see the lights of his brake van. The Victoria Park signals were all against him, but the signalman had cleared the distant and home signals to let him down to the box, where the signalman had shouted to him, "Have you got the whole of your train?" and he had replied that he did not think that he had. He told his fireman, Joseph Holmes, to go back and see if their train was complete, and the fireman had returned and told him no, they had lost part of it. He subsequently found that they had lost 15 waggons and the two vans. The passenger train had been stopped at Old Ford station when he passed there at 10.18.

Brakesman Frank Line said that he had been riding in the rearmost of the two vans. He had thought that their train was divided as they approached the Fairfield Road overbridge, as they had applied the brakes on seeing the Old Ford distant signal "On" and released them when it had cleared, and the driver had whistled, but he had not then experienced the usual jerk. His van had still been moving, and had not reached the overbridge, when the waggons in front of it were struck by the engine of the down passenger train.

Brakesman Frank Day was in the front brake van, the penultimate vehicle in the train. He recalled hearing his mate shout, "Good God, there's a smash." They were still moving then. When they stopped, they both jumped out and found that the engine of the passenger train was derailed and up against the abutments of the overbridge. He could see that it was a serious affair, and at once ran back to Bow station to get the signalman to block the line and to get help.

Driver John Witherby was too badly hurt to attend the inquiry but submitted a brief written statement. His fireman, Edward Hayhoe, said that they had left Old Ford station normally and had seen the coal train approaching them when the thought had flashed through his mind that the rear side light of the brake van of the coal train was not coming towards them as quickly as its engine was. Just at that moment his driver had exclaimed, "Bill, I think that train has broken loose," and before he had time to reply they struck something, and their engine was thrown off the rails and against the bridge abutment. They had only been travelling at about 10 miles an hour at the time. There had been no warning and no time to do anything. The leading brake van reared up against the chimney of their engine and the third-class carriage behind telescoped into it.

Signalman Charles Porter at Old Ford had accepted the coal train from Tilbury Junction at 10.05 p.m. and received the "Train on line" for it at 10.17, but Victoria Park did not clear back the previous train until 10.18, when the signalman there had accepted the coal train. The down passenger train also passed his box at 10.18. As the coal train went by at 10.19, he saw that it had no tail or side lights and gave the stop signal to Victoria Park and the positive block to Tilbury Junction. His mate, who was due to relieve him at 11, then came into the box at 10.20 and told him that there was something wrong with the passenger train, whereupon he had also given the positive block to Victoria Park.

Five passengers died in the accident; Charlotte Miller, Ellen Shary, Charlotte Shary, Mrs Shary's two-month-old daughter, for whom she had just bought a new pair of shoes; Phillis Warwick and James May. Ten other passengers were more or less seriously injured.[104]

---

[104] *The Times*, London, England, 30th and 31st January 1882

Colonel W. Yolland, the inspecting officer, stated that from the evidence the accident had clearly resulted from the failure of the draw-bar of the 35th waggon of the coal train, one end of which had ploughed into the ballast, struck a sleeper, and thrown the 36th truck and two or three behind it off the rails and foul of the down line. The down passenger train had come along immediately afterwards and collided in the darkness with the derailed waggons. None of the railwaymen involved could have done anything to prevent the accident.

The accident was due, the colonel stated, to the faulty mode of construction of the waggon. The draw-bar involved evidently had a flaw, which could not be seen by an examiner, as it was completely covered by a cast iron washer. The iron was also of inferior quality. The truck belonged to the Metropolitan Carriage and Waggon Company, had been hired by Bedworth Coal and Iron Company, and had taken coal from the Charity Colliery, Bedworth, to J & C Harrison, Mark Lane, London. The number of continuous draw-bars, such as the one on the waggon involved, had been increasing, as had been the number of accidents due to the failure of them, such as that between Longwitton and Scots Gap in July 1875. There was, the inspecting officer continued, no difficulty in how this class of accident might be prevented in the future, though it would likely mean an alteration to the greater proportion of rolling stock in use. It consisted of having no draw-bar long enough that, if it be broken or pulled out, it was able to drop to the track and act as a lever to lift the truck off the rails. The continuity of the strength through the truck provided by the long draw-bars could be obtained by having two tie rods through the centre, passing through and ending at the outer side of the headstocks, or the framing of the trucks could be strengthened. The question was a very serious one, as a similar accident could happen at any time in front of one of the fastest trains in the country, with no warning, and the train be wrecked, with disastrous consequences.[105]

A derailment at Sweet Briar Creek, Dakota Territory, United States, on 21st March 1882, was complicated by the train falling from a bridge and has been dealt with in the Train Accidents Involving Bridges chapter.

On 17th July 1882 the axle of a passenger train on the Prince Edward Island Railway broke near Charlottetown, New Brunswick, Canada, derailing the train and killing four passengers, including a woman and child who fell beneath the car. Others were injured, including Conductor Reuel, of Boston.[106]

Disaster struck the 12.30 p.m. Manchester Sheffield and Lincolnshire Railway up passenger train from Manchester to Grimsby and London on 16th July 1884 just west of Penistone, England. At 1.21 p.m. the train was approaching Bullhouse Colliery signal box, running at about 45 miles an hour down a falling gradient of 1 in 124. Signalman Baxter in Bullhouse box had received the "Train on Line" signal from Hazlehead box at 1.19 and watched it approach two minutes later. Nothing appeared to be wrong with it until just as the engine was passing his box when he had heard a very heavy thud as if

[105] Board of Trade Report

[106] *Indianapolis Journal*, Indiana, United States, 19th July 1882

something had broken. The engine began to oscillate very violently and as the middle of the train was about halfway between his box and an underbridge, he saw the carriages swaying to the left, and they then left the rails and went down an embankment. The whole train went down the embankment on the west side of the underbridge except for a brake van which went down to the east of the bridge and the engine, tender and a horse-box which remained on the line. As soon as he saw the accident, he threw back his down signals, which he had cleared for a goods train, and tried to contact Penistone Goods box via the speaking telegraph but could get no response. He then telegraphed the signalman at Huddersfield Junction, Penistone, told him of the accident and asked for assistance to be sent; he also telegraphed Hazlehead. Subsequently he went down to the wrecked train, met Guard Irving and asked him to stop the down goods; the guard found it stopped at the down home signal.

Driver Sam Cawood was in charge of the engine of the passenger train. He had his regular engine, No. 434, and was to take the express to Retford, and from there continue with the Grimsby portion of it to the latter place before working a train back to Retford. He had examined his engine before leaving Manchester and there was nothing wrong with it. The train was fitted with the continuous vacuum brake except for the horse-box, which was piped. His train had left Godley about four minutes late, after attaching three vehicles there, and he had run at the normal speed under clear signals all the way. All had gone smoothly until when about 40 yards short of Bullhouse box he heard something like a crack in the engine motion and it immediately began to lurch. He had applied the vacuum brake as soon as he felt the lurch and his engine came to a stand in about 300 yards, when he found that only the horse-box was attached to the tender, and had said to his fireman, John Horne, “Oh dear! Wherever is the train?” He had not felt the train part. His fireman was unable to apply the tender brake as the engine was lurching so much. After the accident he got underneath his engine and found that its crank axle was broken. It had also broken about 12 months previously at Wadsley Bridge, but on that occasion, he had been able to ease the train into Sheffield station.

Guard James Irving said that he had been busy with the Penistone parcels as the train approached Bullhouse box at the usual speed. He had felt the continuous brake suddenly go on, and almost at once his van left the track. It went down the embankment and was struck by the wheels of a carriage. He was cut about the face but got out, clambered up the embankment and saw the engine and tender at a stand, and the remainder of the train wrecked. All the signals were then at danger. Going towards Bullhouse box he met the signalman, who said something about a coal train, which he found was stopped at the down home signal. He had then assisted in the wreckage and later walked to Penistone.[107]

The damage sustained to the engine and to the train was extensive. As far as the rolling stock of the train was concerned five carriages were so badly damaged as to require rebuilding and four others, a horse-box and a brake van were badly damaged. The death toll was correspondingly high; 19 passengers were killed outright, five more died of their injuries and some 60 were injured, many seriously. Two guards were also injured. Many of the passengers on the train were bound for the Continent via Grimsby, from where vessels sailed on Wednesdays (the day of the accident) for Antwerp, Rotterdam and Hamburg, the train being timed to connect with the boats, though it was not strictly a boat train as such.

---

[107] Board of Trade Report

A paragraph from *The Times* gives a flavour of the disaster, 'Mrs H. Dickinson, wife of Mr Henry Dickinson, of Atherstone-terrace, South Kensington was travelling from West Temporley {sic}, Cheshire, with a nurse and a child aged two and a half years. Her left leg was fractured and her face badly injured. The child escaped with slight injury, but the nurse, Ann Lee, was so badly hurt that she is not expected to recover {she did not}. Mrs Dickinson states that they were the only three occupants of a second-class compartment. They were suddenly thrown with great force against the cushions, and she remembered nothing more until she found herself lying on the embankment in an overturned carriage. Her nurse, a little further off, was crying "Oh, help, take me out." Mrs Dickinson replied that she would do so in a moment, and rose for that purpose, but found her leg broken and fell back. Seeing her child in a perilous position she scrambled towards and managed to save it. A Liverpool gentleman reported at Penistone that he had escaped by getting through the smashed roof of his carriage. He had just started on a pleasure excursion, which was to extend to the Rhine, but he was glad to go home again. He had lost in the accident the sum of £25 and the seal of his watch. A companion, also proceeding to the Rhine, stated that they were in the last carriage, and received severe bruises on the face and legs. Two foreigners, believed to be Germans, were also injured, but went on to London. The latest news that has reached Sheffield is that one of the injured died on the way to Manchester where two others have since succumbed to their injuries. This increases the number of lives lost to 23.' [108]

Major F.A. Marindin investigated the accident. He pointed out that the fracture of axles was by no means uncommon, 247 having broken in 1883, but such breakages did not usually have serious consequences. The seriousness of the disaster was due to the breakage having occurred under peculiar circumstances; with the train running at high speed down a falling gradient, negotiating a 40 chains radius curve, and having left a cutting was passing over an embankment across a narrow valley, in the middle of which was an underbridge, making it almost inevitable that the carriages would run off at a tangent.

The fracture of the axle was, he continued, difficult to account for and could not be foreseen. The cranks were examined once a week and the big ends taken down once a month, but it was quite credible that no flaw would have been detectable at the time of the last examination.

The proper position of the horse-box was at the rear rather than the front of the train, but he did not think that in this case its position had contributed to the magnitude of the accident.

The continuous brake did not seem to have been very effective and as it was Smith's simple vacuum brake it became ineffective as soon as the train parted. The accident was the second emphatic warning in six months that the company had the need for an automatic brake, following an accident at Dinting in February where a coach had almost fallen from a 100-foot-high viaduct.

Locomotive engineers should consider the relative advantages of engines with inside cylinders and crank axles, as compared to those with outside cylinders and straight axles; and of crank-axles hooped with wrought iron bands compared to those with more metals in the web itself. He suggested the preparation of a return showing the relative number of

---

[108] *The Times*, London, England, 17th, 18th, 19th, 21st, 24th, 26th and 31st July 1884

failures of crank and straight axles, and distinguishing between iron, steel and hooped axles, while big ends should be examined weekly.

In conclusion, Major Marindin stated that there were no grounds whatever for the suggestion that had been circulated and even published that Signalman Baxter had failed to restore his signals to danger until after warned by a passenger to do so, nor was there any unreasonable delay in the despatch of relief trains, and in fact no fault to be found with any servant of the Company either before or after the accident. He also recorded his thanks to the men of Mr Hinchcliffe's (Bullhouse) colliery, whose quick intervention after the accident had saved at least one life.[109]

Charles Sacre, the Chief Locomotive Engineer of the Manchester Sheffield and Lincolnshire, who had designed the engine, resigned his post after the accident, and plagued by grief over the accident, later committed suicide. The railway company, however, continued to use the non-automatic Smith's brake for some time afterwards because it was cheaper than the safer but more expensive automatic brake systems.[110]

Penistone was unlucky enough to have another bad accident less than six months later, on New Year's Day, 1885. Again the accident was due to a broken axle, and as at Newark in 1870 and Old Ford in 1882 it was an axle of a goods waggon which broke, derailing it and causing the fouling of the line on which the passenger train was running.

The passenger train involved in this second accident was No. 1 excursion train from Rotherham and Sheffield to Liverpool and Southport. It had left Rotherham at 7.15 a.m. and Sheffield at 7.53, eight minutes late, and was approaching Penistone, where it was due to stop at 8.20, passing Barnsley Junction, at the east end of the station, at 8.26. A minute later its engine struck an empty coal waggon forming part of up mineral train No. 28, which had left Ardwick at 5 a.m. for Kiveton Park and Shireoaks, and which had just become derailed. All but two of the passenger carriages were damaged, the sixth, seventh and eighth being practically broken up, and about 10 goods waggons were damaged. Passengers Albert Halliby, a Sheffield butcher and drover, and Tom Wood, a Sheffield grinder, were killed outright. Thomas Elliott, a waiter at the Bull and Mouth Inn, Sheffield, died the same evening and a fourth passenger, William Harrison, a wood turner, of Sharrow, died two days later. Forty-seven other passengers were injured, some seriously.

Driver John Coldwell told the Board of Trade inquiry into the accident that he was the driver of engine No. 229 of the special excursion train. The signals had been clear all the way from Deepcar to Penistone. When he first saw the up mineral train it was 100 yards away and there was nothing wrong with it. When its engine was about 20 yards away, he saw a waggon, he thought it was the fifth, in the six foot. He had at once shut off steam and applied the vacuum brake and the fireman had applied the tender brake. His engine collided with a waggon and the smoke-box door was knocked in, but the engine was not badly damaged. They came to a stand in about 100 yards and he saw that six carriages were off the rails.

Driver John Schofield of the goods train said that he had been running through

[109] Board of Trade Report

[110] *Railwaymen Politics and Money* by Adrian Vaughan, pgs. 208-213

Penistone station when his fireman, Edmund Peacock, had told him that there was a waggon off the line, when he had looked back and seen what he thought was the fifth waggon in the six foot. He had immediately applied the vacuum brake and his fireman had popped the whistle and applied the tender brake, all this happening just as the engine of the excursion train was passing. His train had stopped just down the line; it had been running at 10 or 12 miles an hour.

Goods Guard Robert Higgs stated that he was looking out on the left side of his van when at a point about 400 short of the point of collision he saw fire flying from a waggon, he thought the fifth from the engine. The waggon was oscillating badly and then fell into the six foot, followed by the waggon behind. The engines of his train and the excursion train were at that moment passing one another. He had applied the brake in his van as soon as he saw there was a problem. He saw the engine of the excursion train strike the waggon, which reared up in the air and then fell down against the middle of the excursion train. It all happened very quickly. He had examined the waggons before starting from Ardwick and those taken on en route. He noticed a broken axle after the accident.

Major F.A. Marindin was again the inspecting officer investigating the accident and as in the previous Penistone accident stated that none of the Company's servants could be blamed for it. It was nearly certain that the initial derailment had resulted from the fracture of the leading axle of the fifth waggon of the train, a private-owner waggon belonging to the Shireoaks Coal Company. The waggon appeared to have been of the usual construction and in a good state of repair. It had left the colliery on train No. 63 for Hadfield on 24th December, when it was examined at Bridgehouses, Sheffield, and Dunford Bridge, this being the last time that it was examined before the accident.

The axle which failed had a latent flaw which probably would not be detected except by a critical examination in the shops. A large proportion of phosphorous in the metal used for the axle meant that it was of inferior quality and likely to be brittle. The regulations for the construction of trader's (private owner) waggons and the inspection of rolling stock were, nevertheless, Major Marindin stated, far from satisfactory. There was no 'absolute hold' over waggon builders as regards the quality of the materials that they used and the constant interchange of waggons between companies made it impossible for a company to guard against a defective waggon running on its line. In addition, once a waggon started running it was only examined by waggon examiners and travelling inspectors and not sent to the shops until it needed repair.

Given that the axles of no less than 141 waggons had broken in 1883 the inspecting officer recommended that either all waggons should be owned by the railways companies or that they should be carefully inspected before being allowed to run on any line, that care should be taken that they were built of good quality material, and that they should be subject to a systematic periodical inspection, including all axles, after they had run a specified time. All waggons should bear a label showing that a railway company had examined them and the date when they were last thoroughly overhauled.

The placing of two passenger running lines between two goods lines instead of to one side of them doubled the risk of a passenger train being affected by an accident to a goods train on an adjoining line.[111]

---

[111] Board of Trade Report

An accident which in some respects paralleled the second one at Penistone occurred in the United States on 27th May 1887, near Kittanning Point, Pennsylvania, on the Pennsylvania Railroad. As a westbound fast passenger train was approaching Kittanning Point a waggon of a passing eastbound freight train suffered a broken wheel, was derailed and the truck smashed into two passenger cars. Dal Graham, of Allegheny City, Pennsylvania; J.H. Stauffer, of Louisville, Ohio; Wymer Snyder, of Shamokin, Pennsylvania; and newsboy John Dorris, of East Liberty, Pennsylvania were killed outright; Frank McCue, of New York City, 'will die' and Charles Beidelman, of Brenfield, Indiana, was 'dying'. About half a dozen other passengers were injured but no one in the sleeping or parlour cars was hurt. The accident, for which no one could be blamed, was described as the worst on the Pennsylvania system for some years. [112]

Another American accident due to a broken wheel occurred on 19th January 1888, between Bradford and Haverhill, Massachusetts, on the Boston and Maine Railroad. The train involved was the 1 p.m. Portland express from Boston, in charge of veteran Conductor Edward Weymouth and Engineer Warren French. All went well on the journey from Boston until the train was between Bradford and Haverhill stations. In the short distance between the two stations the line crossed the Merrimac River, and it was here that the smoking car left the rails at a set of points and crashed into a water tank house. The force of the collision demolished the foundations of the building and the heavy water tank on top of it which fell onto and crushed the car following the smoking car, while the next car again telescoped into the crushed one, adding to the destruction. The dead and injured passengers were mostly in these two coaches. Nine people were killed more or less outright and about 50 more or less seriously injured, five of whom died later, including probably at least one more of the section hands. At the time of the accident a number of section hands were in the tank house eating their dinner and two of those who were killed outright were William Taylor, one of the section hands, and John O'Brien, a retired Bradford merchant who was in the house talking with the railwaymen.

The entire force of Haverhill doctors was summoned by telephone and quickly attended the victims of the accident. The fire department was also called, and its men were doubtless busy freeing victims trapped in the wreckage. The cars were equipped with Baker steam heating, so a disastrous fire caused by an upset stove was avoided. The fact that the train was equipped with automatic air brakes, which applied as soon as it became divided, was also credited with preventing the smoking car from falling from the bridge into the river, 100 ft below.[113]

Initially the accident was attributed to the points giving way after the front part of the train had passed over them, but it soon became clear that the damage to the points was the result, not the cause, of the accident, which was actually due to a broken flange on the leading right-hand bogie wheel of the smoking car. 'General Manager Furber revealed the fact that a section of the flange was missing, the break evidently resulting from a defective casting.' This car had been immediately in front of the two cars which crashed

112 *Dunkirk Observer Journal*, Dunkirk, New York, United States, 28th May 1887
113 *Tampico Tornado*, Tampico, Illinois, United States, 14th January 1888

into the water tank. The passengers in the car had a very frightening experience. Following the breaking of the wheel at the points, the end of the car sideswiped the water tank house, bumped against the parapet of the bridge and then ran over the bridge on the sleepers, careening first to one side and then the other before the automatic action of the brakes brought it to a stand resting against the ironwork of what was presumably a lattice girder structure. The bridge itself was not damaged. The 40 passengers in the car were badly shaken up but not seriously hurt and afterwards went back to help their less fortunate fellow passengers. A newspaper reporter recorded that, 'Four men were playing cards in the smoking car. When the car careened, they involuntarily rose to their feet, and that is the last they could remember until they found themselves out of the car, still holding the cards in their hands.' The rear cars of the train were also derailed, the passengers in them receiving a bad shaking up.[114]

As might be expected the disaster received extensive coverage in local newspapers, the *Boston Daily Globe* devoting column after column of small print to every aspect of the accident.[115]

The last Manchester Sheffield and Lincolnshire Railway passenger train from Manchester to Glossop and Hadfield left London Road station at 10.40 p.m. on Saturday 14th July 1888, fairly well filled with passengers. It consisted of an engine and tender, a third-class brake composite carriage, a third-class, two composites, two third-class, a composite and a third-class brake carriage, all six-wheeled vehicles except for the sixth, carriage No. 673, a four-wheeled vehicle which was the cause of the accident. The train was in charge of Guard Peter Robinson with Driver George Harwood and Fireman John Gilmore. It left Guide Bridge station at about 11.25 and not long afterwards several of its passengers noticed that their carriage had run off the line, though their testimonies as to exactly where this had happened varied. Signalman William Arnold in Dewsnap box, a mile from Guide Bridge, noticed nothing wrong with the train as it passed him, but shortly afterwards Guard Robinson in the rear brake composite felt a shock and applied his brake. The engine crew noticed the brake application made by their guard and applied their brakes in turn, bringing the train to a stand close to Hyde Junction box at 11.31 p.m.

As the train was coming to a stand on the up line the 10.45 p.m. goods train from Glossop to Ardwick was passing on the down line, in charge of Guard Peter Mellor, with Driver Edwin Barber and Fireman John Hopkinson. Driver Barber was running under clear signals and had no warning of anything being wrong, before his engine struck carriage No. 673 of the passenger train, which had become derailed and ended up foul of the down main line. It was doubtless this unavoidable secondary collision, rather than the initial derailment, caused the casualties in the accident. Four women; Mrs Jeanett Middleton, Mrs Hannah Beard and Miss Jane Wilson, all of Glossop; and Mrs Maria Bastian, of Hadfield, were killed outright and a fifth passenger, Mr Frank Cullen, a painter from Glossop, was seriously injured and died in Manchester Infirmary some days later. Several other passengers were badly enough hurt to be taken to hospital.[116] [117] [118]

---

[114] *Hamilton Daily Democrat*, Hamilton, Ohio, United States, 11th January 1888

[115] *Boston Daily Globe*, Boston, Massachusetts, United States, 11th January 1888

[116] Board of Trade report

[117] *The Times*, London, England, 16th July 1888

Investigation after the accident showed that carriage No. 673 had become derailed due to the fracture of its leading axle. The Board of Trade inspecting officer, Major-General C.S. Hutchinson, noted that the axle had fractured as the result of a growing flaw, a defect which could not have been detected by examination without the drastic step of removing the axle from the wheel. A second flaw in the axle was found after the accident. The inspecting officer remarked that only 18 carriage axles had been returned as having broken during the previous five years, though four of them, or 22%, had belonged to the Manchester Sheffield and Lincolnshire Company.

Turning to the brake power of the train Major-General Hutchinson noted that the carriages of the train had all been fitted with the automatic vacuum brake, but the engine had not. If Guard Robinson had control of an automatic brake, he could have stopped his train before carriage No. 673 overturned and fouled the opposite running line. The Company were, however, making rapid progress in fitting automatic brakes to all passenger trains.

Some of the passengers had stated in their evidence to the inquiry that they had looked for a communication cord when they felt their carriage leave the rails but as the train was not timetabled to run 20 miles non-stop the Company was not under any statutory obligation to supply one. Fresh legislation, the inspecting officer hoped, might make such a means of communication compulsory on all trains.

Finally, it was clear that if carriage No. 673 had had six or more wheels it probably would not have overturned.[119]

The jury at the inquest at Glossop on four of the victims returned a verdict of "Accidental death" with a recommendation that all trains should be fitted with communication cords.[120]

An accident occurred in the United States to a Chicago and North Western Railroad passenger train No. 6 at Elmwood, Michigan at 1.30 p.m. on Saturday 19th January 1889. Killed outright in it were James H. McDonald of Escanaba, the Michigan Lieutenant-Governor; Horace Tuttle, an important Cleveland iron manufacturer and president of the Commonwealth Iron Company; and William F. Cochrane, treasurer of the Cochrane Milling Company of Escanaba. E.P. Foster, of Iron Mountain, superintendent of the Marquette Ore Company, was very seriously injured. Several other passengers and Conductor Henry Armstrong were seriously injured.

Mr O.C. Davidson, of the Commonwealth Iron Company, who was playing cards with the three men who were killed, said that the first indication that anything was wrong had been a jolting sensation, after which the car, which was the rear vehicle of the train, had swayed two or three times and overturned onto its side. He remembered nothing more until he came to, lying in the snow, badly bruised, about 200 feet from where the train had come to a stand.

The train had been travelling at high speed when the car had left the rails and slid along the ground for about 500 feet, until it hit a tree stump which penetrated the side of

---

[118] *St James Gazette*, London, England, 16th July 1888
[119] Board of Trade report
[120] *The Times*, London, England, 17th and 25th July 1888

the car where the victims were and then a second stump which tore out the side of the coach, and it was these which had caused the casualties.[121]

As soon as possible news of the accident was wired to Escanaba and a special train with doctors and surgeons reached the scene about two hours after the accident had happened. The bodies of the three dead were temporarily left in cottages at Elmwood while Conductor Armstrong was taken to Ironwood on an eastbound train. News of the death of James McDonald reportedly caused great gloom at Lansing, the state capital, where flags were flown at half-mast.

The accident was attributed to a broken truck (bogie) on the last car.[122]

One of the seriously injured passengers apparently died later, bringing the total number of fatalities to four.

A brief report in *The Times* from their correspondent in Vienna mentioned an accident the previous evening, 18th March 1889, at Cercan station, in Bohemia, at that time a part of Austria-Hungary. A train from Prague was derailed due to the breakage of an axle of a third-class carriage. Four passengers were killed and nine seriously injured.[123]

Stealing a ride on a train in the United States was always a very dangerous occupation, with a good possibility of the rider being assaulted or worse by the railway police or fellow tramps, or being killed or injured in a train wreck, but in such a big country it was a very common means of travel for those looking for work or having to travel for some other reason and who could not afford the train fare. Thus it was that three such tramps were the victims of an accident near Wilmerding, Pennsylvania, on the Pennsylvania Railroad, on 8th July 1889. The double-headed eastbound freight train involved, consisting of 37 loaded cars and a caboose, left Pittsburgh at 8.30 p.m. and the accident occurred at about 10 p.m. Thirteen cars left the track, due to a broken wheel, were completely demolished and caught fire. William Connelley, 15, a Pittsburgh bootblack, was killed outright, and John Hyde, a 15-year-old Pittsburgh newsboy, and an unknown man died on the way to hospital. Andrew Kennedy, a 14-year-old Pittsburgh newsboy, was probably fatally injured; Alfred Young, coloured, of Lima, Ohio, was seriously injured and John Kennedy, of Milwaukee, Wisconsin, was badly cut and bruised. All except John Kennedy and Young were bound for Johnstown. None of the railwaymen were hurt. A tramp who escaped spoke of 25 tramps having been on the train, but no other remains were found. The line was blocked for several hours and thousands of dollars of damage was caused.[124]

Westbound vestibule train No. 3 of the Chesapeake and Ohio Railroad was derailed at 7.20 a.m. on Saturday 28th October 1889 at Jerrys Run, two miles west of White Sulphur Springs, West Virginia, United States. The train was reportedly travelling at 40 or 50 miles an hour when it left the track on a fill (embankment) 190 ft high, supposedly the

---

[121] *Manitowoc Lake Shore Times*, Manitowoc, Wisconsin, United States, 22nd January 1888
[122] *Bloomington Daily Leader*, Bloomington, Illinois, United States, 21st January 1888
[123] *The Times*, London, England, 20th March 1889
[124] *Philadelphia Inquirer*, Philadelphia, Pennsylvania, United States, 10th July 1889

highest in the United States. The mail car, combination car, passenger coach and one sleeper car were derailed, and 10 people were killed and about 25 injured. All the fatalities occurred in the combination car, which was telescoped by the passenger coach. The engine was not derailed and the second sleeping car, *Rockbridge*, also escaped damage. There was also no fire, due in part no doubt to the fact that the cars were steam-heated. baggagemaster Barksdale, Mail Clerk Morrison and three other unnamed employees were among the dead, while Conductor Schweilert and Section Master Myers were seriously injured. Initially the cause of the derailment was attributed to the track spreading or to the rails being won out and the ties being rotten,[125] but a later report by Company officials stated that a broken axle had caused the accident.[126] The only physician in White Sulphur Springs, Dr W.P. Caldwell, went to the scene and for a while was the only doctor there; naturally he could only treat one patient at a time. Later other doctors joined him, while 'the ladies of the vicinity turned out en masse and too much cannot be said in praise of their noble efforts.' [127]

A broken axle derailment at St Joseph de Levis, Quebec, Canada on 18th December 1890 involved part of the train falling from a bridge and is dealt with in the Train Accidents Involving Bridges chapter.

Between 10 and 11 a.m. on 31st August 1891 mixed train No. 51 on the Louisville Evansville and St Louis Railroad, carrying about 35 passengers, was derailed by a broken flange between Tell City and Troy, Indiana, United States. The train had been in charge of Conductor James Gordon with Engineer Jacob App and Fireman John Tolle. The flange of one of the forward trucks of the engine broke as it was crossing a trestle two miles from Troy. The engine cleared the trestle but plunged down the approach embankment and the following passenger coach rolled down the slope and ended up on top of the engine. The steam pipes of the locomotive broke and as happened in a number of other accidents, searing steam entered the passenger coach and severely scalded its occupants before they could escape. The slightly injured passengers were able to extricate all their more seriously injured fellow travellers, but none escaped injury. 'Many were so badly burned that their flesh was cooked in places until it came off,' a newspaper reported in the graphic language of the time. Five passengers had died by the time of the initial newspaper reports: Mrs Cavanagh and her 10-year-old son Troy; Miss Barbara Nemier, aged 18, of Harrison County, Indiana; Emma Schue, aged 10, of St Meinrad, Indiana; and a farmer from Hancock County, Kentucky, whose name was unknown. Another 30 passengers were injured, and it was feared that a dozen of them were so badly scalded that they could not recover. The engineer and fireman were able to jump in time to escape with a few bruises but the conductor, who had been standing on the platform of the car, was taken by surprise and crushed between the engine and the car before he had time to jump. He was still living at the time of the report.[128]

---

[125] *The Butte Daily Miner*, Butte City, Montana, 29th December 1889

[126] *Logansport Daily Reporter*, Logansport, Indiana, 30th December 1889

[127] *The Kokomo Daily Gazette Tribune*, Kokomo, Indiana, United States, 30th December 1889

[128] *Leadville Herald Democrat*, Leadville, Colorado, United States, 1st September 1891

A heavy morning mail train from Bombay (now Mumbai), India, was travelling at a good speed near Nagpur on 5th November 1891 when one of its two engines suffered a broken wheel and the train was derailed. Initially it was reported that five British soldiers travelling on the train, both drivers, both fireman and the guard had been killed on the spot and 31 British and four Indian soldiers had been injured. Sir George Greaves, the Commander-in-Chief of the Bombay Presidency, was riding on the train with his staff, but they all escaped with a shaking.[129] A further report on 6th November, the day after the accident, spoke of the deaths of Privates Halstead, Smith, Willis, Hogarth, Mabby, Russell, Horn, Dawson and Mulloy and on 9th November it was reported that Privates Gorsley, Thornley and Clark had also died. Eight of the soldiers were from the North Lancashire Regiment. The damage to rolling stock was estimated at 130,000 rupees.[130]

A derailment near Domnino, Russia, on 23rd November 1891, possibly due to a broken tyre, is covered in the Train Accidents Involving Bridges chapter.

The Taff Vale Railway, a relatively small if prosperous Welsh company, was unlucky enough to suffer three major disasters in its history, two of them in the 19th century. In 1878 there was a collision at Pontypridd, and on 12th August 1893 a derailment occurred at Llantrisant Junction. On the latter date the 3.50 p.m. down passenger train from Merthyr to Cardiff (which included through Cambrian Railway coaches from Aberystwyth) had passed Treforest station and was approaching Llantrisant Junction at 4.46 p.m. when the leading brake van left the rails and passed through the junction off the line. Beyond the junction the van and five following carriages plunged down the embankment, the four rear carriages remaining on the track, while the engine had broken loose from its train and run ahead a short distance. In falling down the high embankment the coaches were wrecked, and eight passengers were killed outright and four more died from their injuries, while seven passengers and five railwaymen were injured, some very seriously. One of the rescuers, Ganger William Jones, also lost his life when he was crushed by an overturned carriage, and it would seem that one of the passengers who was trapped under an overturned carriage was probably killed by the carriage above bring lifted by the rescuers and then thoughtlessly allowed to drop back down again.[131] Many people were soon on the scene to help in the rescue and a Dr Howard Davies, who chanced to be at a farmhouse nearby, was also quickly on the spot.

A passenger on the train, Mr W. Brace, South Wales representative of the Miners' Federation of Great Britain, gave a newspaper reporter a vivid description of the accident. 'I looked into the Cambrian carriages when the train was at Pontypridd and saw that they were full up, so I had to go on till I came to a vacant place in one of the middle carriages. Had the first carriages been only partly filled I should undoubtedly have got into one of them. After we left Treforest, and perhaps half a mile this side of the station, the train going at a tremendous speed – we were travelling just as fast as if we were in an express train – one of the party in the same compartment as myself said in an alarmed manner, "I

---

[129] *The Times*, London, England, 6th November 1891
[130] Ibid, 7th and 10th November 1891
[131] Ibid, 14th August 1893

believe this train is going off the line." In a moment we felt that the train was off the metals, and I then realized that we were being carried along towards the edge of the embankment. I made frantic efforts to get out of the carriage on the opposite side but found that the door on that side was locked. We came to a standstill, and when I got out, I saw that the first three or four carriages had gone down the embankment and were lying in a confused mass of wreckage at the bottom, the wheels of some being up in the air, and the poor unfortunate passengers buried beneath the ruins. There were clouds of dust, and the whole scene was one of the most terrible and indescribable confusion. The air was rent with the shrieks of women and children. I saw at a glance by the awful manner in which the woodwork of the carriages had been smashed up that the loss of life must be deplorable. It seemed to me that in some overturned carriages the people were so wedged that even those around them could render no assistance until the necessary appliances had been brought to the spot and gangs of the company's workmen put on. I saw four of the dead while I was on the spot, but I do not know how many more there may have been. Some terrible torturing scenes were witnessed. There were thousands of people at the spot in an incredibly short space of time, and they, of course, lent every assistance in their power; but the carriages had literally to be smashed in with hatchets in order that the poor people might be liberated. It was plain to my mind that the carriages must have turned over three or four times on the embankment. The cause of the accident appears to me to be the breaking of the coupling of the first carriage and the speed at which we were travelling. The uninjured passengers were brought on later by a special train from Walnut Tree Bridge Station.' [132]

Signalman Thomas Preece at Llantrisant Junction told the Board of Trade inquiry of accepting the train from Treforest station box and watching its approach. There had been nothing wrong with it until it had passed his down home signal, when he saw that the leading wheels of the brake van, which was behind the engine, were off the rails towards the six foot. He had waved a white flag violently up and down at the driver and saw the fireman looking back at the train. The trailing wheels of the brake van then also left the rails at the crossing opposite his box. The train was travelling at about 30 miles an hour. He saw a portion of the train run over the embankment and at once blocked all roads and advised his mates at Treforest and Willowford on the telephone what had happened. The fireman of the train subsequently came to the box and asked if it was all right for them to proceed on the engine for assistance and doctors. He told the fireman that it was, and the engine went away at 4.58.

Driver William Durston said that the signals at Llantrisant Junction had been clear for him and he had been running at the usual speed of about 30 miles an hour. Somewhere between the footbridge and the Llantrisant Junction down home signal he felt a knock under the footplate. He was about to shut off steam when he saw that the wheels of the leading brake van were off the rails. He then shut off steam but did not apply the continuous brake for fear of pulling the train up too suddenly and he then briefly reapplied steam briefly to keep the van straight. Twice he succeeded in doing so and he was about to apply steam a third time when his engine separated from the train and came to a stand about 70 yards down the line, the brakes having come on automatically. Almost immediately after his train had parted, he saw the van and following carriages go down the embankment. After the accident they had run to Cardiff without any problem

[132] Ibid.

and it was not until he had taken his engine (No. 173) to the sheds and examined it in the engine pit that he had found that the spring of the left trailing coupled wheel was missing, one of the spring links being fractured. He and his fireman had examined their engine over a pit at Merthyr before they started and had seen nothing wrong with it. He thought that he had done the right thing in trying to keep the van upright and said that placed in a similar situation again he would do the same thing.[133]

Guard Thomas Thomas said that he was in charge of the train and was riding in a Taff Vale brake van which was the last vehicle but one in the train. The train was fitted with continuous automatic brakes. They had left Merthyr on time and Treforest 12 minutes late, mostly due to station work at Quakers Yard, Aberdare and Pontypridd. He was not aware of anything being wrong at Llantrisant Junction until the brakes came on and looking out, he saw coaches going down the embankment.

Major F.A. Marindin stated that there could be no doubt from the evidence of witnesses and marks on the permanent way that the accident was due to the fracture of the T link of the left-hand driving wheel spring of the locomotive, as a result of which the spring had fallen off and been dragged along under the engine until it got under the wheels of the leading brake van, causing them to leave the rails, followed by the trailing wheels when the van had reached the crossing opposite the signal box. The brake van had swerved across the line 143 yards south of the signal box when the coupling between it and the engine had parted and the carriages of the train had then gone down the embankment.

No blame attached to the company's servants for the accident, the major stated. The permanent way was in good condition, the engine was properly constructed and overhauled and the flaw in the T link which had caused it to fracture was such that it could not be discovered while the spring was in position. Nevertheless, although he did not blame Driver Durston, who still considered that he had acted for the best in not stopping immediately, he (the inspecting officer) considered that the driver had made an error of judgement in not gradually applying the continuous brake after whistling for the guard to apply the brakes from the rear of the train. No doubt in the days when trains were only fitted with hand brakes it was dangerous for drivers to apply the brakes on the engine too soon for fear of the vehicles behind overrunning and telescoping one another but with good continuous brakes operable throughout the train the situation was quite different, and no danger should arise from a judicious application of the brakes. He suggested that the Company draw up further instructions for the guidance of drivers in such circumstances.

Criticism had been raised as to the off-side doors of the carriages having been locked but he did not think that had they not been, any of the passengers would have had time to jump out safely, and they would have been very likely killed had they tried to do so.

At the conclusion to his report Major Marindin noted that Guard Thomas had not advised his driver of the composition of the train as per the rules (probably because they were anxious to get away on time) and that the Llantrisant Junction signalmen worked 11-hour day shifts and 13-hour night shifts, 'a division of time to which objection has often been taken.'[134]

---

[133] Board of Trade Report

[134] Ibid.

The second section of the Pennsylvania Railroad "Limited" express from Chicago to New York was wrecked at Bear Swamp, near Lawrence station, New Jersey, United States at 5.45 p.m. on Wednesday 25th October1893, when it ran into the derailed waggons of a freight train. The latter train was travelling at about 20 miles an hour when an axle of one of its waggons broke and wreckage was piled up over the four tracks. An effort was made to stop the passenger train, but it was too close at hand. Engineer Daniel Mahoney and Fireman Henry Matthews stuck to their posts and were seriously injured when the engine turned over onto its side in a ditch. Five tramps were believed to have been killed (other reports said three were killed and four seriously injured) and their bodies buried in the wreckage. A tramp who escaped stated that there had been at least 15 of them on the train. Three of the passenger coaches followed the engine into the ditch but those in them somehow escaped, apart from one passenger and three porters who were slightly hurt. The line was blocked for several hours and trains were diverted via Bordentown and Monmouth Junction. The unauthorised passengers were stated not to be 'professional tramps' but men travelling together in search of work.[135] Six days earlier, on 19th October, three tramps had died in a collision between two freight trains near Yardley, Pennsylvania, on the Philadelphia and Reading Railroad, only a couple of miles away.

Three men killed in the wreck of a coal train on the Beech Creek Railroad between Peale and Viaduct, Pennsylvania, United States, late on Monday 31st October 1894, were at first thought to have been ordinary tramps stealing a ride but two of them were armed with revolvers and they had a full kit of burglar's tools, masks and so on. It was supposed that they had been on their way to rob the county bank at Clearfield. A fourth man, Lewis Trump, of Williamsport, was seriously injured with a fractured skull. The newspaper report did not state if he was another of the supposed burglars or an ordinary tramp who was also on the train. None of the railwaymen on the train was hurt. A brake beam had dropped down from one of the trucks and 11 of them ended up piled up in a confused mass.[136 137]

Train No. 103, the Pittsburgh and Cincinnati express of the Baltimore and Ohio Railroad, known as the "Cannon Ball", was derailed at 10 a.m. on 3rd November 1895 at Elm Grove, five miles east of Wheeling, West Virginia, United States, the entire train except the engine, tender and leading baggage car going down a 15ft embankment. Mrs Myranda Hare, of West Mahoning, Pennsylvania, and an eight-month-old child of Lawrence Bartley, of Pittsburgh, but formerly of Wheeling, were killed outright and about 30 passengers were injured, several of whom 'cannot recover.' [138] The train was in charge of Conductor Charles Ball and Engineer Lynch and consisted of a baggage car, mail car, smoking car, two day coaches and a Pullman chair car in charge of Conductor George Perkins of Cincinnati. A broken flange on the smoking car had caused that

---

135 *The Times*, Trenton, New Jersey, United States, 26th October 1893
136 *Chariton Democrat*, Chariton, Iowa, United States, 2nd November 1894
137 *Huntingburg News*, Huntingburg, Indiana, United States, 2nd November 1894
138 *Philadelphia Times*, Philadelphia, Pennsylvania, United States, 4th November 1895

vehicle to leave the track, pulling the mail car with it, the following vehicles then becoming derailed in turn. The coupling at the front of the smoking car gave way so the engine and baggage car were not affected.

The accident happened close to Elm Grove station and within the town limits, so help was soon at hand, including a local doctor, and further doctors were telephoned for. The injured were taken to houses nearby for attention and the badly hurt were later conveyed in a special train to hospital in Wheeling.[139 140]

On 31st January 1896 part of a train fell from a bridge near Lawrenceburg Junction, Indiana, United States, after a derailment due to a broken wheel. (See Train Accidents Involving Bridges chapter.)

Four men stealing a ride on a Chicago Rock Island and Pacific Railroad freight train were killed in a derailment near Leighton, Iowa, United States, at about 5 a.m. on the morning of 7th August 1896. The accident resulted from a broken journal on the rear truck of the 12th car of the train, a double-decker box car loaded with sheep in which the men were riding. Eight or nine cars and the caboose behind the derailed vehicle piled up on it and the strain split the car in two. The men were buried under the sheep being carried in it and were suffocated. The train involved in the accident had been diverted over the Keokuk and Des Moines branch because the main line was blocked by an earlier accident between Grinnell and Malcolm. The train was in charge of Conductor Andy Ryder and Engineer Dyer.

The men killed had been employees of the Liverpool and Des Moines Packing Company and were trying to beat their way to Chicago in the expectation of better opportunities in the Windy City. The names of the three killed outright were William Kuehner and Ed and Robert Garry, all of Des Moines. A fourth man, William Shea, also of Des Moines, died of his injuries shortly after the accident. The newspaper report did not make it clear whether the men had lost their jobs or were just fatally ambitious.[141] Other reports spoke of six or seven killed and gave additional names but that more than four had died was denied by the railroad company.

On the afternoon of 4th November 1897, the Chesapeake and Ohio Railroad F.F.V. (Fast Flying Virginian) vestibule express, from Cincinnati to Washington D.C. was wrecked at Old Shadwell, five miles east of Charlottesville, Virginia, United States. Three passengers initially known to have died were identified as an Italian woman named Merara and her child, from Cincinnati, and a Phelix Mareno. Coloured porter Henry Burnett of Keswick, Virginia, also lost his life and a second Italian child was believed buried in the wreckage. A number of passengers were injured, as were Engineer Robert Duke, Conductor C.S. Buster, Baggageman Peers, Pullman Conductor J.N. Miller of Washington and express messenger Richard Berkley. Engineer Duke and Fireman Nick Bailey were thrown from their engine and the engineer somehow ended up beneath it. He

---

139 Ibid.

140 *Baltimore Sun*, Baltimore, Maryland, United States, 4th November 1895

141 *Daily Iowa Capital*, Des Moines, Iowa, United States, 7th August 1896

was both injured and scalded and was pulled out, very badly hurt, by the fireman and conductor. Senators Platt, of Connecticut, and Russell, and Commissioner of Pensions H. Clay Evans, of Tennessee, with their stenographers and clerks, were in the parlour car, and were unhurt.

The engine of the train had left the track on a curve and all the vehicles of the train except the Richmond sleeping car were derailed. Two cars plunged down an embankment and one of them, the express car, 'loaded mostly with turkeys and other fowls,' ended up partly submerged in the Rivanna River. The day passenger car and express car were badly damaged and the *Salola* sleeping car ended up on its side.[142]

Mr Legh R. Page, a passenger in the rear car of the train, described the accident to a newspaper reporter, 'The train left Charlottesville 50 minutes late. I felt a jostle, and in a moment the car came to a standstill. Six or eight other passengers were in the rear sleeper, all of whom immediately ran to the window, not thinking that anything had happened more serious than a simple derailment. Upon reaching the window I saw the locomotive on the left side of the track, while the baggage and express cars were thrown over into the river, and the first passenger coach was smashed into smithereens. The second car, which apparently contained mostly Italians, seemed to have met with more serious results. One Italian woman was found. She had been instantly killed, while her infant was taken from the wreck later fatally injured. Another Italian had been thrown partly through the window by the shock, and was caught beneath the legs by the coach, so that it was a very long time before the coach could be so lifted as to release him. The agony caused by the pressure on his legs was so great that his cries were heard from some distance. The engineer was seriously injured internally, and his arms broken. He was also severely scalded about the face. Some time after the wreck a man was discovered in the second coach with both his legs cut off beneath the knee.'

It was theorised initially that the derailment was caused by a problem with the track, but General Manager Stevens issued a statement denying this. 'From appearances the accident was caused by a broken axle under the front truck of the engine. The report that the track spread on account of poor ties is untrue. Engine, express car, postal car, combination car-coach and two sleepers are off track, and on their sides. Dining car has front truck off, otherwise all right. Rear sleeper did not leave track.'

A later report on the accident stated that four Italian passengers were killed, all of the same family. They were Fessia Guarara, his wife and two children, on their way from Cincinnati to Philadelphia, where they had secured work, and were travelling with another Italian family, also named Guarara; the husband Flippo, his wife Marie and children Filona and Josef. Fessia's wife and children died at the scene and his legs were severely crushed. He lived to be taken to Harrisonburg depot, where overwhelmed by the loss of his wife and children, he tried to commit suicide by choking himself. He died shortly afterwards anyway from his injuries. The Italians were described as being from Sicily, 'and of the humbler classes.' Flippo Guarara and his wife were injured and with their children had been taken to the Farrish Hotel in Harrisonburg. The man had sold his possessions in Cincinnati and had had the proceeds, a few hundred dollars in gold and some jewels, sewn into his overcoat which had been lost in the accident, but the railway

[142] *Boston Daily Globe*, Boston, Massachusetts, United States, 5th November 1897

authorities were confident that it would be found, and he would lose nothing.[143]

Dr William B. Dodge, of Stuart's Drift Virginia, who had been a passenger on the train and who treated many of the victims, also gave a history to a newspaper reporter. 'I was in the sleeper *Salola*, and was thrown half through the window. As soon as I could get out, I went at once to look after the injured. I found an Italian woman in the debris of the combination car. She was badly cut about both eyes and the skin from her forehead was hanging down over her face. She was Marie Guarara. I dressed her wounds and she was removed to the rear car. The cries of the injured in the combination car, though coming from a few persons, sounded as if from a dozen. They were crying out in Italian. I found one dead, so I left her for the living that were not beyond help. I heard two babies crying. The head of one was caught fast. It was bruised and cut. This was Philona Guarara. Another was found near the bottom of the wreckage. It had a slight wound and its side was bruised. This was Josef Guarara. In this debris I saw a part of the scalp of a negro, which proved to be that of the porter, Henry Burnett. While I was in the rear car dressing wounds, some one came in for me. I ran from {sic} where the men were at work in the wreck and said that an Italian was caught fast. I went to him. He was pinned between the rods and irons of the truck. I examined him, but as he was beyond surgery, I told the men to continue the work of rescue while I went elsewhere. I then went down to the engineer. He was covered with mud. I could not tell whether he was Italian or what. His left arm was terribly shattered. I temporarily bound it up with bandages torn from the sheets brought from the sleeper. An appeal to the ladies for pins and Vaseline brought a generous supply. I washed the engineer and applied the Vaseline. He was perfectly rational and told me the worst pain was in the region of his bowels. I suspected he was scalded, and cut away his clothing, but found no external evidence of scald.'

As for the clearing up operations the same newspaper reported that General Baggage Agent Lorraine, of Richmond, came up on one of the first trains from the east to reach the wreck, and he at once took charge of the baggage. Trucks were strewn all over the hillside in great confusion, but all of them had their checks on them, and but two were burst open. One of these two belonged to one of the lady passengers, and dainty feminine underwear was scattered everywhere. A bicycle in the baggage car was badly broken up, and pieces of this were carried off as trophies. Mr Lorraine did all in his power to get the baggage straightened out, and at a late hour last night he had gotten it all in order, with the exception of the two broken trunks, and had shipped it to its destinations. The mail, which was partially submerged in the river, was also cared for in the same manner, and forwarded as soon as possible. It is thought that every package in this car was saved. The express car was not touched last night. All the goods in it were left to be removed later, a responsible agent being placed in charge. One end of this car went into the river, but, strange to say, the only articles which got into the water were three coops containing geese, and when the *Dispatch* correspondent reached the scene the aquatic fowls seemed, from the fluttering and hissing that they kept up almost continuously, to enjoy their experience. The rescue of Express Agent Berkeley, who was pinned down in the water beneath a box, was very timely, as had he not been pulled out when he was, he must surely have been drowned. The hero of this incident, Mr Alexander, the express company's route agent, returned to Charlottesville after the wreck, but he very modestly

[143] *Harrisonburg Rockingham Register*, Harrisonburg, Virginia, United States, 12th November 1897

declined to receive the praise of his many friends in this city, saying he had done no more than anyone else would have done under the circumstances. Several of the city's photographers went down to the wreck yesterday and made pictures of the scene, for which there is great demand.'

Wreaths of chrysanthemums were sent to the undertakers where the bodies of the four Italians were stored by the Women's Christian Temperance Union.[144]

[144] Ibid.

CHAPTER 3

OTHER DERAILMENTS

Other chapters in this book cover derailments caused by track defects, defects with the trains themselves and those caused deliberately by acts of vandalism or sabotage. This chapter will deal with derailments due to other causes such as excessive speed, collisions with animals and other accidental obstructions, traction shocks and so on, together with those for which no cause was apparent.

A train consisting of an engine, baggage and mail cars and two passenger coaches was travelling at 25 miles an hour when it left the rails on Golden Rock Curve, Amour Mountain, near Puerto Cortes, British Honduras, on 17th January 1881. The whole train fell 70 feet into a gorge and 'nearly all on board were either killed or fatally wounded, a great many bleeding to death without any medical attendance.'[145] One report blamed the derailment on a misplaced rail without clarifying whether this meant track defects, displacement of the rail by a rockfall or sabotage. [146]

The Denver and Rio Grande Railroad train which left Pueblo, Colorado, United States, on Wednesday 20th April 1881 for Chama, New Mexico was delayed for a time at La Veta, Colorado, by washouts of the track, then resumed its journey. It was running at normal speed along the Gunnison branch when coach No. 11 left the track two miles on the Pueblo side of Rock Tunnel (probably in New Mexico, though the line hereabouts crossed and re-crossed the state border between New Mexico and Colorado). The derailed coach fell 60 feet down the side of the mountain, rolled over and over another 140 feet and was completely wrecked. Of the fourteen passengers in it five men and a woman were killed outright and nine seriously injured. Two of the seriously injured passengers later died. The only railwayman hurt was a fireman who had been riding in the coach. The remainder of the train remained on the rails. The cause of the derailment could not be determined. The accident happened on Wednesday 20th or possibly

[145] *The Fairfield Tribune*, Fairfield, Iowa, United States, 3rd February 1881
[146] New York Times, New York, United States, 24th January 1881

Thursday 21st April.[147]

The passengers killed were Mrs Decloedodt; D.G. Brewer; Mr D.C. Scholes; Mr Jas Lynch, Mr S. Hall; Mr D. C. Wilson and two labourers whose names were unknown. 'The excitement in the town (of Chama) was intense, when it was learned that an accident had occurred, as many citizens were expecting friends and relatives. Both newspaper offices were besieged with anxious enquirers, desirous of seeing the list of killed and wounded. Many of those seeking information went away relieved, upon finding none of their friends on the black list of the dead and dying; but the tragedy nevertheless shrouded the town with gloom, and every heart swelled with sympathy for those bereft of friends by a blow which each felt had come so near their own door. Further information is anxiously awaited.' [148]

An accident on the outskirts of Barcelona, Spain, on 24th May 1882, was caused by the carelessness of a pointsman. A new line was being constructed, having a junction with the Granollers Railway near Fuerte Pio. Passenger train No. 33 from the French border was approaching Barcelona when due to an error by the pointsman it was diverted at a set of facing points onto the new line, which was still incomplete. The train driver saw that the points were wrongly set and applied but the brakes but there was insufficient distance to stop and running off the end of the track the wheels of his locomotive buried themselves in the earth and it came to a sudden stop. The leading brake-van was forced up into an almost vertical position by the weight of the following carriages and two third-class carriages were badly smashed up, the other carriages of the train also being damaged to a lesser degree. A poor elderly woman from Granollers was killed instantly and four seriously injured passengers succumbed in the following hours. Twenty other passengers were hurt to a greater or lesser extent and were taken to hospital in Barcelona. As the accident had happened close to a major station a relief train with doctors was soon on the scene. The pointsman whose negligence had caused the accident disappeared afterwards.[149]

A telegram to *The Times* in London from a correspondent in St Petersburg derailed the derailment of a passenger train on the Moscow – Kursk line between Tcherny and Bastijeur stations on 12th July 1882. Eight carriages of the train ended up buried in the earth. Of the 217 people on the train 39 were extricated from the wreckage, more or less injured. It was stated that the remaining 178 had lost their lives, making this one of the worst railway catastrophes of the 19th century. No other details of the accident, such as a possible cause, seem to have emerged.[150]

A brief report from Illinois stated that a train carrying Company A of the Ninth Regiment of Infantry of the State Militia, returning home to Grayville from the encampment of the Second Brigade met with an accident on 3rd September 1883, on the

---

[147] *Leadville Daily Herald*, Leadville, Colorado, United States, 23rd April 1881
[148] *Rico Dolores News*, Rico, Colorado, United States, 30th April 1881
[149] *La Illustración Española y Americana*, Madrid, Spain, 8th June 1882
[150] *The Times*, London, England, 14th July 1882

Louisville Evansville and St Louis Railroad. The train ran into a herd of cattle between Carmi and Grayville, Illinois, United States. The car in which the militia were riding was derailed and overturned and nine of them were killed outright and fifteen injured.[151]

Initial reports of an accident on the Cincinnati Southern Railroad at New River, Tennessee, United States, on 11th Jun 1885, were that a construction train loaded with labourers had been passing through a tunnel when the structure's roof and wall had collapsed onto it. Six men had, it was stated, been killed outright and 20 others badly hurt, half it was feared fatally. Eighteen of the victims had reportedly had to be dug out of the wreckage.[152] A further report from the officers of the company denied that there had been any failure of a tunnel and stated that Roadmaster Simmond's construction train had been derailed when it had struck a cow two miles from New River, killing five men and injuring Foreman O'Donnell and 10 or 11 others. The caboose and seven cars had been ditched.[153]

Seven passengers died in the derailment and overturning of a carriage of a train at Omori station, Tokyo, Japan, on 13th October 1885. The train was apparently being propelled rather than pulled at the time of the derailment.[154]

A major disaster in Peru on 12th January 1886 involved a train carrying troops from Paita to Sullana. The running of the troop train was probably connected with the overthrow of the previous Peruvian president, Miguel Iglesias, by General Andrés Cáceres a month earlier. The train was between La Huaca and Viviate, and passing the haciendas of La Chira and Valdivia, when it collided with a bull which had decided to take a rest on the railway's right of way and was not seen by the engine driver. The train was derailed and wrecked, and four officers and 18 other ranks lost their lives and another 46 were injured. A subsequent report on the accident written by one of the army officers did not mention casualties to any civilians but did speak of the great assistance which had been provided to the injured soldiers by the local residents after the accident. A chapel named Las Ánimas Descarriladas was later erected at the site of the disaster.[155] [156]

Yet another serious derailment involving an animal on the line, and in this case also a bull, occurred on the Nebraska and Colorado line of the Burlington and Missouri River in Nebraska Railroad on 19th October 1886. A construction train carrying a whole gang of track layers, about a hundred men, ran into the bull close to a bridge between Lawrence and Deweesee, Nebraska, United States. Six box cars were thrown from the track and five men killed and nine injured. Those who lost their lives were listed as Robert Marven, a Deweesee hotel keeper; George Bark, of St Louis, Missouri; Daniel O'Connor, of

---

[151] *Evening Observer*, Dunkirk, New York, United States, 5th September 1883
[152] *Defiance County Express*, Defiance, Ohio, United States, 18th June 1885
[153] *Terre Haute Weekly Gazette*, Terre Haute, United States, 18th June 1885
[154] Webpage – Japanese Railway Safety and the Technology of the Day
[155] Webpage – La Noche Negra del 12 de Enero de 1886 by Pablo Enrico Medina Sanginés
[156] Facebook page – El Descarrilamiento de 1886

Weston, Missouri; Robert Collins, an Englishman, and a man whose body was still buried in the wreckage at the time of the report, supposed to be a Dennis Hamilton, of Michigan. The Deweesee hotel keeper had presumably begged a lift from the train, unless he was doing a spot of labouring on the side. Some of the injured were seriously and possibly fatally hurt. The accident had occurred at noon.[157]

The regular westbound overnight passenger train on the Chicago Milwaukee and St Paul

Railroad left New Hampton, Iowa, United States at 3.45 a.m. on the morning of Thursday 5th April 1888 with Conductor Cyrus Morris and Engineer James W. Seagel in charge of engine No. 231. The train was initially accelerated to 40 miles an hour but Engineer Seagel reduced speed to 10 miles an hour for what was known as the dry bridge, 20 rods east of the main span over the Middle Wapsic River. Having crossed the first bridge the engineer put on steam again, but unbeknown to him heavy overnight rain had loosened a large sheet of ice, an acre in area, and the massive block slid downhill to the railway embankment. When it struck the earthwork it overturned across the track between the two bridges. The engine struck the ice block in the darkness and was thrown from the track. It ran about 30 feet and then plunged down the 10-foot-high embankment into the floodwater at the base of it and was mostly submerged, only its smokestack, dome and part of the cab showing above the water. The leading baggage and express car ran past the engine and also went down the embankment. The smoker, carrying 40 passengers, broke loose from the baggage car and struck the cab of the engine, tearing its side away. The next two passenger cars left the track but were not badly damaged and the sleeping car with its 15 passengers was not derailed.

After the accident the uninjured passengers helped to release those trapped in the smoker and the alarm having been raised a wrecking train was sent from Mason City, Iowa, at 7 a.m. with Superintendent Moll, Assistant Superintendent Cable and a large body of men. The passengers in the smoker were scalded by steam escaping from the engine and one of the passengers, Augustus Sharp, was also injured by an axe carelessly wielded by one of the rescuers. Engineer Seagel was drowned, though his unnamed fireman somehow escaped. Passengers Willard Andrews, of Lamont, Wisconsin; Christian Sørensen, a Dane, with a ticket from Chicago to Dell Rapids, Dakota Territory; John Duelos, travelling to Kimball (probably the place in Dakota Territory); and Mary Heidecker, aged one year, were killed, while a Frank Heimmerman, of Walkon, Iowa, was missing at the time of the report, believed buried in the wreckage. About 25 other passengers were more or less injured, one or two of them possibly fatally. The baggagemaster and express messenger in the baggage car escaped serious injury and most of the mail was recovered, though thoroughly soaked.[158] [159]

Disaster struck a train on the Naples-Metaponto line in Italy which was crowded with excursionists who were returning from a naval review and fetes held at Naples in connection with the visit of the German Kaiser. As the train was passing through a

---

157 *Nebraska State Journal*, Lincoln, Nebraska, United States, 22nd October 1886
158 *Defiance Democrat*, Defiance, Ohio, United States, 12th April 1888
159 *State Herald*, Holyoke, Colorado, United States, 13th April 1888

mountain gorge between Salandra and Grassano stations at 6 a.m. on Saturday 20th October 1888 a large amount of rock and soil fell down the mountainside and covered the track over a distance of 50 metres. Due to a curve in the line which limited his view the train driver had no chance to stop before his locomotive ploughed into the obstruction. The engine, luggage van and three leading carriages were buried in the mass, and four other carriages were 'smashed to atoms' and their wreckage rolled down the slope to the valley below. The last two carriages were derailed but remained upright and undamaged. By chance these contained thirty Carabineers who were returning to their garrison town after duties as Naples in connection with the Kaiser's visit. Shaken but otherwise unhurt, they were able to assist the railwaymen in the rescue work. The driver and fireman were precipitated into the River Busento but escaped with their lives.

An unfortunate result of the accident was the bringing down of the telegraph wires by the falling earth, which prevented speedy calls for assistance, and the accident occurred in an uninhabited area. A passenger set off on foot to get help, but the first rescuers were a band of labourers who had been working in fields on the opposite side of the river, and who swam across it to reach the accident site at considerable risk to themselves, as the river was much swollen and the current very rapid.

As was often the case the first reports of the number of casualties were much exaggerated, initial telegrams speaking of 90 dead and 70 more injured. Matters were not helped by many passengers who were unhurt leaving the scene afterwards and the figure of 90 was believed to stem from an order that 90 coffins be sent to the scene. A later telegram stated that 19 bodies had been recovered from the wreckage and that an unidentified woman and two Belgian monks from the Naples Monastery of Pietra Santra had also lost their lives, while two other victims had died of their injuries in hospital. Those killed included two railwaymen and five professional opera singers on their way to Corfu. The number of injured was corrected to 55. There had been between 350 and 380 passengers on the train.[160 161]

Belgium suffered a major disaster on Sunday 3rd February 1889. The 9.30 a.m. passenger train from Brussels to Namur was near Gronendael station when it became derailed and dashed with great violence against a pillar supporting a footbridge, which then collapsed onto the front vehicles of the train. The engine, tender, luggage van and leading four carriages, the latter crowded with passengers, were 'reduced to atoms'. After the accident help came immediately from local officials and residents and as soon as news of the accident reached Brussels a relief train with doctors and officials was despatched to the scene. It soon became apparent that the accident was a very serious one and the rescue operation was carried out as speedily as possible. The bodies of 14 victims were recovered and there were at least 50 injured. Among those killed was the engine driver. The bodies of those killed and some of the injured were taken to Brussels, other injured passengers were taken to the nearby village of Hoellaart, where there reportedly lay 15 persons with mortal injuries, including several young children. The injured were visited by M. Van de Peereboom, the Minister of Railways. Meanwhile eight amputations

---

[160] *The Times*, London, England, 23rd and 24th October 1888

[161] *Railway Press*, London, England, 26th October 1888

had been carried out at the Hospital of St Jean in Brussels.[162] An investigation was being made to determine the cause of the derailment. One report stated that the line had been under repair at the spot where the train left the rails.[163] The final number of dead was apparently 22.

At the end of April 1889 Americans celebrated the centennial of the election of their first president, George Washington. Thus, it was that most of the 150-odd passengers on a Canadian and Grand Trunk Railway express from Windsor to Hamilton and Niagara Falls were Americans returning to their home country from Canada for the celebrations. On Sunday 28th April 1889 the express consisted of an engine and tender, two baggage cars, a smoking car, a Chicago and Grand Trunk through coach, a Wabash Railroad coach, a first-class Wagner car, a Pullman car and two Wagner sleeping cars. It was in charge of Conductor W.H. Poole with Engineer Joseph Watson and Fireman Edward Chapman.

The line from Dundas ran through limestone cuttings as it descended an escarpment. About two miles from Hamilton, Ontario, there was a switch where the line to Toronto diverged from the line to Hamilton and Niagara Falls. An eastbound train such as the express had to negotiate a sharp right-hand curve on the approach to this switch with another sharp right-hand curve beyond on the Hamilton line. The express was due in Hamilton at 6.55 a.m. but was running 15 minutes late as it came to the switch. Company rules forbade the engineer from running fast to make up lost time. As his train was nearing the junction switch it was, according to a newspaper report, travelling at a speed of 40 miles an hour or possibly more. It negotiated the first right-hand curve and ran through the switch only to jump the line as it entered the second curve beyond the junction. The momentum of the engine carried it almost 200 feet across a stretch of open ground, then smashed into a wooden water tank and overturned. The tender and leading baggage car were thrown over the engine and the second baggage car caught fire from the engine and set fire to the first. The following coaches were piled up against the front of the train and were eventually also involved in the fire, apart from the two rear sleeping cars.

Most of the passengers had been asleep when the accident happened. Those who were unhurt and not imprisoned were able to climb out of the coaches and assist in the rescue work. A relief train also soon arrived from Hamilton. The rescuers worked as hard as they could to free those passengers trapped in the front of the train, the smoking car and day coach containing the most victims, but were eventually forced to withdraw by the advancing flames. The water tank having been smashed, the only water available was provided by a bucket brigade fetching it from the adjoining swamp, but this and the rain which was falling did little to quench the flames and it was many hours before the fire was finally put out. The two Wagner sleeping cars escaped the blaze because it was possible to uncouple them from the rear of the train and push them out of the way. baggagemaster James Welch was injured but was praised by the passengers for his help in the rescue work and a large gang of workmen under Foreman John Hall laboured unceasingly to extinguish the fire. By 5 o'clock in the afternoon, 18 charred bodies had been recovered from the wreckage. The engineer and fireman were thrown clear and but little hurt. Their escape was described as 'miraculous'. About 20 passengers were injured,

[162] *The Echo*, London, England, 4th February 1889
[163] *The Times*, London, England, 4th February 1889

some seriously.[164]

As was usual at this time there was no delay in opening an inquest, which began the same day, though with adjournments it dragged on until 10th June. Opinions expressed at the inquest as to the train's speed varied. The conductor stated that it had not been travelling at more than 20 miles an hour, some of the passengers put the speed at 50 miles an hour, most witness thought it had been 25 miles an hour. Railway employees thought that speed perfectly safe; the jury did not.

The question of problems with the track or switches was raised but nothing could be demonstrated. Then the inquest turned to the possibility of a problem with the locomotive. Evidence had been lost by the efforts to clear the line and get it re-opened again, plus the activities of souvenir hunters, but the finding of a broken flange led the jury to blame the derailment on 'the breaking of the flange of the left leading wheel of the engine truck, which break allowed the wheel to leave the track at the switch', ignoring the possibility that the broken flange could have been a result rather than the cause of the accident. Upset by the destruction of evidence the jury also recommended, 'that the government take such action as may be necessary to ensure a full and complete inspection by a competent person in all cases where there has been loss of life through accident, before anything whatsoever has been removed or touched than what may be necessary to rescue bodies from the wreck. Had this been followed the jury would have come to a verdict with much more ease and certainty.' The jury made no comment as to the fire which had followed the derailment, but a local newspaper pointed out the dangers of oil lamps and coal-fired stoves and urged their replacement with electric lighting and steam heating.[165]

In early 1890 work was in progress on the building of a railway from Penshurst to Koroit in the Australian colony of Victoria. At about 7.15 on the evening of Saturday 1st March 1890 a train left Penshurst for the navies' camp about 18 miles down the line, carrying 25 men who were riding in a ballast truck in which had been placed about two tons of chaff. The ballast waggon was being propelled by the engine. The train had only covered a couple of miles when it ran into a bullock which had strayed onto the railway, the truck fell over onto the right-hand side of the line and the engine ended up on top of it. Two men were killed by falling under the engine, two by falling under the truck, two by being smothered by the chaff and one by breaking his neck in falling, making seven deaths in all. Seven others were injured, three seriously enough to need to be taken to hospital. The others escaped with a few scratches. Two doctors from Penshurst were soon in attendance at the scene and two more doctors were called from Hamilton. The bull had got onto the track because the lineside fencing was as yet incomplete.[166]

The jury at the inquest on the victims of the accident criticised the contractors, McNeil & Co., for failing to completely fence the line and for running a train on an unfenced line with the engine propelling its vehicles. They recommended 'that a stricter supervision be placed over lines under construction with regard to the preservation of life and property.'

---

164 *Winnipeg Free Press*, Winnipeg, Manitoba, Canada, 29th April 1889

165 *Wreck!* by Hugh A. Halliday, pgs. 76-81

166 *Hamilton Spectator*, Hamilton, Victoria, 4th March 1890

167

Just after 1 p.m. on the afternoon of Tuesday 19th August 1890, Engineer C.L. Baker had just passed Quincy station, Massachusetts, United States, with a southbound Old Colony Railroad gravel train. There was a gang of section men working on the opposite, inward track and they appeared not to have noticed express train No. 182, the Boston-bound *Vineyard, Nantucket & Hyannis Express* (referred to by a newspaper as the Woods Holl express), bearing down upon them. Fearing that the express would run the men down, Engineer Baker and his fireman both waved to Engineer Charley Babcock to indicate to him to sound his whistle to warn the labourers, and he did so. The next moment Engineer Baker saw the vehicles of the express pile up across both tracks just north of an overbridge.[168]

Section Master Joseph Welch and his men had just resumed work after dinner and were doing some re-ballasting work, which necessitated Section Hand Michael Hartney placing a heavy jack under the western rail to lift it up. As he was doing so Section Master Welch had moved to a position some distance south of his men, but only so as to be able to obtain a sighting along the track and not to protect the work on the line, although he knew that the express was due. The noise of the express was drowned out by that of the ballast train and it was only when they heard Engineer Babcock whistle that the men became aware that the express was close at hand. Now Section Hand Harney panicked as he attempted to remove the jack. He only had a few seconds to complete the operation, he was inexperienced, and it was an awkward job to do, requiring the throwing of a trip hook, and then raising the handle after the load had been lifted a little higher. After his initial attempts to remove the jack had failed and with the express coming nearer and nearer Hartney had abandoned the exercise and jumped for his life.

As the gang of labourers finally moved clear of the track at the last moment Engineer Babcock was horrified to find that there was a large jack still fixed to the outside rail. He immediately applied the brakes and could remember nothing else until he regained consciousness after the accident. The collision with the jack threw the engine from the rails but it ran a further 280 feet on the ballast before toppling over on its side with its front end driven into the embankment and the rear across the southbound outward track. The tender briefly remained coupled to the engine and the front vehicles of the train were forced to the right more or less onto the southbound track. The tender, leading baggage car, the *Puritan* Pullman car and a smoking car continued 200 feet down the line and sustained little damage. The action of the tender in forcing the cars to the right was then lost and the following six cars continued along the northbound track until the fourth car, No. 263, struck the projecting end of the locomotive. By now the train was only moving slowly and some of the passengers in this car were able to save themselves by leaping to the right side of the aisle, but this was the most crowded coach of the train, with 70 occupants, and those who were not able to move quickly enough were crushed by impact with the locomotive and then scalded by searing steam from broken steam pipes, something which was always an inherent risk in accidents involving steam locomotives. Car No. 263 was also set on fire by burning coals from the engine, but the fire was not

[167] The Leader, Melbourne, Victoria, 22nd and 29th March 1890
[168] *Boston Daily Globe*, Boston, Massachusetts, United States, 20th August 1890

serious and was soon extinguished. Twenty-three people in all died in the disaster, all but one of them had been passengers in the ill-fated car. Engineer Babcock's fireman, James Ryan, was the other fatality; the engineer himself suffered a broken wrist and cuts to his head and legs. Over 30 other passengers were seriously enough hurt to need admission to Quincy City Hospital.[169]

At the conclusion of the inquest on the victims of the disaster on Thursday 4th September Judge James Humphrey found that it had been caused by the action of Foreman Welch in ordering the use of the jack without putting out warning signals and that he was guilty of gross carelessness and criminal negligence amounting to manslaughter.[170]

The disaster was also investigated by the Massachusetts Board of Railroad Commissioners. It transpired that Foreman Welch, an employee with 20 years' service and a hitherto clear record, had not actually broken any company rules, as there were none which specifically stated that flag protection must be provided for work on the track, though it might be argued that he had failed to adhere to the spirit of the rules, which directed that foreman should provide 'adequate protection'. Welch's supervision of his men was also criticised; he should have realised that the inexperienced Hartley might have problems with a tricky piece of equipment, and he should have ensured that the trip hook had been previously released, which would have allowed the jack to have been released more quickly. The commissioners also criticised the general practice of section crews remaining on the track until approaching trains were close at hand, thus depriving engine crews of an unobstructed view of the track and any possible obstructions thereon.[171] [172]

Thirteen people died and 31 were injured in the derailment of a passenger train at Sonsonate, about 50 miles from San Salvador, El Salvador, on Wednesday 23rd March 1892. Several carriages were hurled with great force down a steep embankment. The cause of the derailment was not known.[173] [174]

A report from Budapest, Hungary, itself based on a despatch from St Petersburg, spoke of a train derailment near Pensa, Russia, on 20th October 1892, in which eight carriages were smashed into splinters and 20 persons killed.[175]

The *Times* correspondent in Madrid described an accident which had occurred late on the afternoon of Friday 14th July 1893 between Zummarraga and Madrid, Spain. A mail van which was the last vehicle of a narrow gauge train apparently was the first to leave

---

169 Ibid.

170 *Dubuque Times*, Dubuque, Iowa, United States, 6th September 1890

171 *Boston Daily Globe*, Boston, Massachusetts, United States, 22nd and 28th August 1890

172 *A History of Railroad Accidents, Safety Precautions and Operating Practices* by Robert B. Shaw, pgs. 309-310

173 *Philadelphia Inquirer*, Philadelphia, Pennsylvania, United States, 24th March 1892

174 *The Traralgon Record*, Traralgon, Victoria, 29th March 1892

175 *Davenport Tribune*, Davenport, Iowa, United States, 22nd October 1892

the rails and it dragged two third-class carriages after it. Before the train could be pulled up all three vehicles had plunged over a 25-foot-high embankment. Five or possibly six passengers lost their lives and about 30 more were injured. The victims were all Spaniards from Bilbao, Vittoria and neighbouring provinces.

The correspondent added that, 'This line, which was opened about three years ago, very materially shortens the distance between the frontier and Bilbao, besides affording easy access to the many thermal baths which abound in the Viscayan provinces and is consequently very largely used during the summer and autumn months. I have many times travelled along this line, and always with a disagreeable impression of the terrible consequences of an accident which might at any moment occur. The line is cut in the side of the mountain nearly all the way and abounds in sharp curves. In places the train travels along the very edge of a high precipice.' [176]

An accident in Argentina on 6th November 1893 involved the collision of a troop train with an unspecified animal at a point near Ramallo, on the Buenos Ayres and Rosario Railway. The animal, which was very likely a bull or a cow, caused the locomotive and 15 carriages (described as waggons in the newspaper report) to leave the rails, and 10 of the carriages were destroyed. Ten passengers were killed and 42 injured, four seriously. The line was also damaged over 100 metres but was nevertheless back in operation the following day. [177]

On 7th November 1893 a passenger train jumped the track near Mikailov, in the government of Biarkrazen, Russia. The engine and 24 carriages rolled down an embankment and were wrecked. Fire broke out in the wreckage and six passengers were burnt to death and 50 more seriously injured. The driver and fireman were fatally scalded. [178]

The line being built from the port of Matadi to Leopoldville (now Kinshasa) in the Congo Free State to permit the transport of exploited resources (mainly ivory and rubber), and which had already witnessed a very serious accident at the beginning of 1893, suffered another one on or about 28th November 1893 when a train carrying several Belgian officials and a number of black labourers left the rails on a curve some distance from Matadi and its carriages were thrown 300 ft down a precipice, the whole train being wrecked. Four Europeans and an unstated number of labourers were killed and 'many were terribly injured'. [179] The narrow gauge (nominally 750 mm but actually 765 mm) line was finally completed in 1898, apparently without any further disastrous train accidents, though probably not without much loss of life among those forced to construct it by their Belgian masters.[180]

---

[176] *The Times*, London, England, 17th July 1893
[177] Ibid, 11th November 1893
[178] *Syracuse Evening Herald*, Syracuse, New York, United States, 9th November 1893
[179] *The Times*, London, England, 28th December 1893
[180] Wikipedia – Matadi – Kinshasa Railway, Wikipedia – Congo Free State

On about 15th January 1894 a passenger train 'running at a good speed' collided with a cow eight miles from Cumanayagua, on the Timina – Matanzos line in Cuba. The result was disastrous. The engine was derailed, several of the cars were piled up on one another and one car was thrown 200 yards from the track. The bodies of 16 victims were extricated from the wreckage and nine others were badly injured.[181] Another report gave the speed of the train as 25 miles an hour and added that the cow, frightened by the approaching train, had run onto the track and into the engine.[182]

Later in 1894, on 17th September, a large bull elephant refused to budge from the track in the face of an oncoming passenger train, reportedly travelling at 50 miles an hour, near Teluk Anson (now Teluk Intan) in the Malay States despite the repeated whistling of the engine driver. The collision killed the stubborn elephant and derailed the engine and carriages of the train. Many of the passengers were injured, two Indian workers fatally so. It was theorised that the elephant was either seeking to avenge the death of a calf killed by a previous train or, perhaps more likely, was seeking to defend its herd from what it saw as a rival.[183]

Just over a month after the Malay States incident, on 28th September 1894, a train on the Nagpur – Bengal Railway was passing through the jungle between Gailkara and Monarpur, India, when it collided on a dark night with a large male elephant which was walking the line where it ran on an embankment approaching the Carro River. Again the train, which was travelling at 30 miles an hour, was derailed and its driver and a number of passengers were injured. The elephant was pushed some distance along the track before it rolled down the 60 ft embankment, at the foot of which it was found dead the next morning. Driver Bell was praised for having immediately applied the brakes and reversed his engine, thus preventing a worse accident. He had formerly been employed on the Caledonian Railway in Scotland. Following the accident assistance was summoned from Chakardarpur, from where a relief train was sent to the scene. Driver Bell and Deputy Locomotive Superintendent Foulds, who was also riding on the engine, each claimed one of the dead elephant's tusks and the guard of the train claimed the animal's tail, though the forestry department protested that the remains were government property. This was said to have been the first occasion of a collision with an elephant in India. It would not be the last.[184] [185]

A special 10-coach train on the Ferrocarril Interoceanico del Pacifico (Inter-Oceanic Railway), carrying pilgrims, many of them women and children, left Amecameca, Mexico, for Mexico City at 1.25 p.m. on 28th February 1895. The pilgrims had set out from Mexico City a few days earlier for the Catholic shrine at Sacre Monta and 1,200 of them were returning home on the train. The special proceeded slowly at first, but after taking water the engineer increased his speed to 30 miles an hour, although he had

---

181 *Elkhart Weekly Truth*, Elkhart, Indiana, United States, 18th January 1894
182 *Decatur Daily Republican*, Decatur, Indiana, 16th January 1894
183 Webpage – Heroic Elephant in Taman Batavia (Melor) in 1894
184 *Colonies and India*, London, England, 3rd November 1894
185 *The Morning Bulletin*, Rockhampton, Queensland, 29th December 1894

reportedly been under orders to run carefully. Midway between Temamatla and Tenango and about 28 miles from the Mexican capital the line descended steeply as it ran along a hillside, with a high rock wall on one side and a steep drop on the other. At the point where the gradient steepened to 30 feet per mile (about 1 in 180) there was also a sharp curve. Trains normally braked to half speed to negotiate this curve but for some reason the engineer failed to reduce speed at all. The train passengers had been laughing and joking, or sitting admiring the scenery, when it came to them that their train was going much too fast, the laughing and joking ceased, and they began to look at one another with fear on their faces. As the pilot wheels of the engine reached the curve it swayed outwards and then back. As the driving wheels met the curve the engine swayed again and then either the track spread, or it jumped the rails, and engine and tender rolled down the embankment. Three coaches followed the engine and tender and broke up at the foot of the slope, reduced to 'little more than piles of splintered boards and beams and pieces of twisted iron'. The following two coaches also became derailed but did not go down the embankment. The coupling behind the fifth coach snapped and the remaining coaches safely negotiated the curve and were eventually stopped by the use of the hand brakes after they had run a considerable distance. The passengers in the derailed coaches which had not gone over the precipice alighted as quickly as they could and many of them fell on their knees in prayer at their deliverance.

The uninjured passengers set to work to rescue the others trapped in the wreckage. 'Four women were found with their arms crushed and immovable under beams and twisted axles. They lay near the engine and were screaming for fear of the flames. After all efforts to clear them had proved vain they were torn loose, one of them losing her arm from the shoulder, another her arm from the elbow, and each of the other two having both arms crushed and legs terribly mutilated. Three of the four will die.'

The train crew were Americans and some of them set about telegraphing news of the accident to Mexico City, from where a relief train was despatched with surgeons, medical equipment and suchlike. When the rescuers reached the scene, they found 35 pilgrims sitting near the wreck and five others who were alive but still trapped, there not being the means for the surviving passengers to release them. It took five hours' work to free the five and get the injured passengers into a fit state for the journey back to Mexico City. Officially it was stated that there were 65 dead and 40 injured. An Associated Press report stated, however, that there were 104 dead and another 85 injured. The engineer and conductor of the train supposedly had escaped serious injury and had fled to the woods, though another and perhaps more reliable account stated that the bodies of the engineer and fireman were believed to be buried under the engine. Feeling against the railway company and the train crew ran high after the catastrophe and it was stated that the latter would be held strictly responsible.[186] [187]

Further details were provided in a Reuters telegram from Mexico City on 3rd March which stated that there were rumours that the engineer, an American, had failed to reduce speed for the curve because he was drunk. Also, that upwards of 130 bodies had been recovered from the wreckage, while many more waited to be recovered. As for the uninjured survivors no measures had been taken to bring them back to Mexico City, fully 1,000 of them being left to fend for themselves at the accident site, 30 miles out. Their

---

[186] *World*, New York, New York, United States, 2nd March 1895
[187] *Boston Post*, Boston, Massachusetts, United States, 2nd March 1895

plight only ended the day after the accident when they took matters into their own hands, forced an inbound train to stop, and rode on it into the capital.[188]

Another derailment on the same curve between Temamatla and Tenango, Mexico, reportedly occurred on about 27th April 1895. This time 14 lives were lost and several more injured. A freight train was involved, so some or all of the dead were presumably trespassers stealing a ride. The report added that there had been several other accidents, less serious but resulting in loss of life, at the same location.

A train crowded with Roman Catholic pilgrims was involved in another disaster on 26th July 1895, this time not in Mexico but in France. Train No. 148 from Pontivy to Saint Brieuc on the Chemins de fer de l'Ouest, strengthened to 14 carriages hauled by two engines, was carrying devotees who had visited the famous shrine at Ste Anne d'Auray. It had left Quintin station and was within about three kilometres of Plaintel when the leading engine left the rails on a curve and dragged the remainder of the train with it, the derailment happening so suddenly that the drivers had no time to reverse their engines or apply the brakes. Most of the carriages piled up upon one another and six or eight 'were reduced to splinters'. After the accident a number of the passengers who had escaped injury reportedly jumped from the carriages and fled across the fields in terror but when the initial panic had passed many of the uninjured passengers assisted in the rescue work.

Twelve bodies were taken from the wreckage, eight passengers, both drivers and both firemen, the engine crews being crushed beneath their locomotives. A French newspaper praised the engine crews for sticking to their posts and noted that Driver Lenchon had shut off steam and applied the Westinghouse brake. M. Gaudu, an instrument maker from Saint Briuec who was a passenger on the train, ran to Plaintel station to raise the alarm. Fifty more people were injured, 20 of them seriously enough to have to be taken to hospital. Upon news of the accident reaching Paris the Minister of Public Works sent his *Chef de Cabinet* to the scene. The cause of the accident was initially unclear, but it later transpired that the train had been late, and the probable cause of it was the drivers running too fast in an attempt to regain time. The train was reportedly running at 60 kilometres per hour at the time of the derailment.[189] [190] [191] [192]

The mail train which left Johannesburg for Durban on the evening of Monday 30th December 1895, crowded with passengers, was derailed and completely overturned while negotiating a sharp curve in the line between Dannhausen and Glencoe, in Natal colony, southern Africa. Seven carriages were smashed to pieces and the number of dead, first given as 18 and then as 32, finally climbed to 37. All but five of the dead were women and children and on 4th January five of the bodies had still not been identified. Those who lost their lives included 17 members of the Tandy family, from Ladysmith, and Mr

---

[188] *The Week*, Brisbane, Queensland, 26th April 1895
[189] *The Times*, London, England, 27th July 1895
[190] *St James Gazette*, London, England, 27th July 1895
[191] *The Standard*, London, England, 27th July 1895
[192] *Le Petit Journal*, Paris, France, 28th July 1895

Michael Myers, of Birmingham, England, a partner in the firm of Myers Brothers, jewellers. About 50 other people were injured. Most of those on the train were refugees from the Boer-controlled Transvaal, where there was a worsening confrontation between the mostly British uitlanders and the Boer settlers and where the abortive Jameson raid had begun on the day before the accident. The derailment was presumably due to excessive speed.[193 194 195]

The derailment of a freight train on the Great Northern Railroad near Spokane, Washington, United States, resulted in 12 deaths, but because they were all tramps it attracted little attention in the press. The exact date of the disaster was not even given, only that it had occurred 'recently'. It happened on about 28th July 1896. The ill-fated train was struck by a landslide and 12 cars were thrown down the hillside into the river and then over some falls, whereby they were smashed to pieces. All but one of the waggons sank, the remaining one, minus its bogies, was carried over the rapids and into a whirlpool below, where it was completely broken up. 'In one of the cars carried into the river below were twelve tramps, and not one escaped. Those of the luckless tourists who were uninjured succeeded in gaining the top of the car. Persons on the shore attempted to cast ropes to them but failed.'[196]

An excursion train organised by the United Sunday Schools of Royton came to grief on its return journey from Barmouth, on the Welsh coast, to Royton at 10 p.m. on Friday 11th June 1897 as it was passing through Welshampton station, England, on the Cambrian Railways, where it became almost completely derailed. As a result of the derailment nine of its 330 passengers were killed on the spot, two died of their injuries and 15 were injured, some very badly.

The train had consisted of two goods engines and tenders, a four-wheeled Lancashire and Yorkshire Railway van, three Cambrian composite carriages, two Cambrian third-class carriages, two Cambrian composite carriages, a Cambrian saloon carriage, five Cambrian third-class carriages and a Cambrian van. It was in charge of Guard Albert Proctor with Driver Edwin Jones and Fireman George McKay on the leading (pilot) engine and Driver John Williams and Fireman William Brentwall on the train engine. Driver Jones told the subsequent Board of Trade inquiry into the accident that the train had left Oswestry at about right time, but had had to wait at Oswestry North Junction box for the arrival of the down ordinary train off the single line from Whitchurch and had left there 17 minutes late at 9.40 p.m. The vacuum brake had been in proper order. Tablets had been exchanged at Ellesmere South Loop, Junction and North boxes and he did not think that his train had travelled at a higher speed than 30 miles an hour at any point. The signals had been all off at Welshampton, which was not a tablet station, but when the train was about 100 yards from the station points the vacuum brake had suddenly come on. He had looked back and seen sparks flying from what appeared to be the first vehicle. His engine stopped in about 60 yards, by which time he had seen that the tender of the

---

193 *The Times*, London, England, 7th, 9th, 10th and 13th January 1896
194 *Mount Alexander Mail*, Mount Alexander, Victoria, 9th January 1896
195 Wikipedia – Jameson Raid
196 *Fort Wayne Sentinel*, Fort Wayne, Indiana, United States, 31st July 1896

second engine was off the rails and the following vehicles were scattered about both sides of the line. He had then uncoupled his engine and proceeded to Bettisfield for assistance. He had never had occasion to complain about the condition of the track. He had not been trying to regain time. Fireman McKay, Driver Williams and Fireman Brentwall did not really add a great deal to Driver Jones' evidence.

Guard Proctor said that he had noticed the rough riding of the four-wheeled Lancashire and Yorkshire van over the crossings on the trip down but had not had any fears of it becoming derailed. He estimated the speed of his train between Oswestry and Welshampton as 35 miles an hour. The first intimation that he had had of anything being wrong had been when he was thrown against the letter rack in his van and knocked to the floor. Getting up, he had gone to the vacuum brake, but the vacuum was all exhausted. He had found that his train was a complete wreck and the telegraph wires being down had been unable to summon assistance, so had despatched the engine to Bettisfield station.[197]

Welshampton stationmaster Harry Roberts put the speed of the train at 30 to 35 miles an hour as it passed through his station. It had not been running faster than trains usually did. After the accident he had asked John Humphreys, a London and North Western Railway shunter at Whitchurch who lived at Welshampton, to go back to Ellesmere for assistance, which Humphreys had done.

Carriage and waggon examiners John Astley at Barmouth Junction and James Howell at Oswestry stated that they had examined the vehicles of the excursion train and found nothing amiss.

Ganger John Rogers said that he had walked his length the morning and afternoon of the accident and it had then been in good condition. It had not been a very hot day and there was no sign of the road having been thrown out of line by heat expansion. The line had been partially re-sleepered two years previously. He had never known a broken rail on his length. He considered the line good enough for heavy engines to run over at 30 or 40 miles an hour. He had no opinion as to why the train had left the line.

Locomotive and Carriage Superintendent William Aston could see nothing wrong with the engines which he considered could have caused the accident and he was unable to form any definite opinion as to its cause.

Evidence given at the coroner's inquest by two Royton passengers, John Cooper and Thomas Burton, and Charles Dudley, a lineside resident, was also detailed in the Board of Trade report. All three agreed that the train had been travelling at high speed.[198]

The Board of Trade inquiry was conducted by Lt-Col H.A. Yorke and the report was extremely detailed and thorough, as such reports usually were. Having summarised the sequence of events leading up to and following the derailment the inspecting offer stated that he had examined the track and he was clearly not impressed with it, pointing out that the ballast was dirty, many of the sleepers which had not been renewed were much decayed, the chairs were very light, there were a number of loose spikes and so on. Old sleepers had also been used by the ganger to support joints making it impossible to properly pack ballast at those points. In summary he found that although the rails were good the road was so light and its maintenance so unsatisfactory that it was only suitable

[197] Board of Trade report
[198] Ibid.

for light trains running at low speed.

Turning to the speed of the train he stated that it was timed to leave Oswestry at 9.25p.m. and arrive at Whitchurch at 10p.m., this giving an average speed of 31.9 miles an hour over the 18¼ miles between the two points. However, speed had had to be slackened for tablet exchanges at Oswestry North, Ellesmere South Loop, Junction and North boxes; Bettisfield and Fenn's Bank stations, and Whitchurch Junction box so that a higher speed of at least 37 miles an hour would have to be maintained between the boxes and stations. Furthermore, the speed would not be uniform, with a higher speed on the falling gradients, and the point of derailment was approached by a descent of 16 chains at 1 in 80 followed by 10 chains at 1 in 268. The drivers and guards put the trains speed at 30 to 35 miles an hour, but passengers John Cooper and Thomas Burton and lineside resident Charles Dudley had put it considerably higher. Such estimates should not be too much relied upon of course, but coupled with the speed which the train needed to run at to maintain time, the fact that it was running down a falling gradient and the amount of damage caused by the derailment the inspecting officer had no doubt that the train had been travelling at 40 to 45 miles an hour at least. In the absence of any speed restriction the inspecting officer did not think the drivers too blameworthy, but he considered that such a speed was far too high for the poor track that he had described.

As regards the rolling stock there was little to say about the Cambrian coaches. The Lancashire and Yorkshire Railway van he described as of antiquated type and ugly appearance. The van had been used regularly on Lancashire and Yorkshire trains, including at speeds from 55 to 60 miles an hour in places between Manchester and Liverpool. Four-wheeled vehicles were necessarily more sensitive to imperfections in the track than those with six or more wheels, especially when empty.

Theorising as to how the derailment had occurred, Lt-Col Yorke concluded that the train, a heavy one, had been running down the incline at a speed of 40 to 45 miles an hour over weak permanent way. The oscillation of the engine and tender would have been considerable, especially as the tender was full of coal and water. The oscillation had knocked the permanent way out of shape and perhaps burst it, and probably caused the second tender to leave the rails, followed at once by the remainder of the train. The drawhook behind the second tender had snapped and the automatic vacuum brake had come into action. The wheels of the leading van had dug into the ballast, the following coach had overturned to the right and the next five to the left. The combined action of the brake, the ballast and the overturning had brought these carriages to a sudden stop and the following carriages had piled up against them, forcing the centre coaches into a confused heap. The detached engines and tenders had run ahead for some distance.

It had been suggested, the inspecting officer stated, that the Lancashire and Yorkshire van had been the first vehicle to leave the rails, but if the permanent way had been strong enough to resist the oscillation of the engines and tenders it was not likely to be disturbed by the light (10 ton) van.

It was evident, Lt-Col Yorke continued, that the condition of the permanent way demanded the immediate attention of the Cambrian Company. No expense should be spared to make the track strong enough to support the heavier engines and rolling stock now in use. Obviously, the whole line could not be re-laid at once but in the meantime heavier chairs and improved fishplates could be introduced, decayed sleepers replaced, and bad or deficient ballast remedied. A methodical system of permanent way maintenance and inspection should also be introduced. [199]

The two juries at the inquests on the victims of the disaster returned verdicts of 'Accidental death', those at the inquest upon the nine passengers who had initially died adding that the Lancashire and Yorkshire van should not have been at the front of the train or attached to vehicles of a more modern and heavier type.[200 201]

The Midland Railway suffered an unlucky accident at Wellingborough, England, on 2nd September 1898. The train concerned was the 6.45 p.m. express from London St Pancras to Manchester, in charge of Guard Jesse Atfield, with Driver Edwin Meadows and Fireman George Joyce. It consisted of an engine and tender, a Glasgow and South Western Railway saloon carriage, a third-class brake composite, a dining car, a first-class dining car and a brake composite, the last four all Midland Railway vehicles. The express was fitted throughout with the automatic vacuum brake. Not being timed to stop at Wellingborough the express was travelling at about 60 miles an hour when it collided with a luggage trolley which had rolled off the down platform. Part of the train left the rails as soon as it struck the trolley and the remainder of it left the rails at an angle crossing at the north end of the station. The engine overturned onto its left side, the tender onto its right side and the following vehicles were damaged in passing between them. Engine, tender and all the vehicles were damaged to a greater or lesser extent and the driver, fireman and five passengers lost their lives, while 60 passengers and five railwaymen were injured. Fire broke out in the wreckage, caused by the ignition of escaping gas, but was not a significant feature of the accident. The passengers who died were Percy Richardson, a Leicester shoe manufacturer; Rosa Butler, of Leicester, wife of a Midland Railway porter; Miss Mary Clarke, of Kings Norton; Arthur Dix, of Derby, a commercial traveller, and James Thornton, of Leicester, a boot and shoe manufacturer, who survived until 19th September.[202]

Like the Welshampton disaster the previous year, this one was investigated by Lt-Col H.A. Yorke. The first witness at the Board of Trade inquiry was George Smethurst, an auctioneer, who told the inquiry that he had been on the down platform on the evening of the accident and had seen a postman, Thomas Smith, pulling a four-wheeled trolley along the platform, with three or four boys jumping on and off it. His attention was diverted elsewhere for a moment when he heard someone shout, "Look out there, the down express is pegged, or signalled." He could not recall which. Turning round he saw the trolley tumble onto the line, and ran towards it, but it was too late. Postman Smith and a railway foreman had tried to up-end the trolley between the lines but there was a train standing at the up platform and they could not do so. Then he saw that the down express was very close at hand and shouting to the postman to get clear he had seized his (Smethurst's) young son and rushed through the booking hall out of the way. He had not seen the boys near the trolley when it ran away.

Postman Thomas Smith stated that he had arrived at the station with a cart loaded with the mail and parcels for the 8.22 p.m. down train, had gone to the platform for a barrow

---

[199] Ibid.

[200] *The Times*, London, England, 14th, 15th, 16th, 18th and 19th June 1897

[201] *St James Gazette*, London, England, 29th June 1897

[202] *The Times*, London, England, 5th and 20th September 1898

to put them on and had found a four-wheeled baggage trolley which four or five lads were sitting on. Shouting, "Look up here," to the lads he had begun to pull the trolley towards the gate as the lads were jumping up and down on it. He had reached the gate and turned the handle of the trolley towards the wall as he went to get the key to unlock the gate. Fetching the key from a nail on the wall nearby, he had unlocked the gate and then turned around to see the trolley rolling away and falling from the platform. He had not noticed it moving before. Seeing that the express was signalled he had at once jumped from the platform and with the assistance of acting foreman Henry Richardson had tried to get the trolley clear of the four-foot, but it was too heavy to lift back onto the platform and there was an up train on the opposite side. The express getting very close, they had had to abandon their efforts and he had just had time to roll back onto the platform, while the foreman had crawled under the carriages of the up train. He did not see the boys when the trolley was running away.

Three of the boys involved, George Houghton, Robert Anderson and Arthur Waite, were interrogated by the inspecting officer. Their stories more or less agreed. They had come to the station to collect newspapers from the newspaper train and said that they had been helping the postman to push the trolley, though one and then another of them had jumped onto the trolley alternately for a ride while the others pushed. They had left the trolley before the postman reached the gate and had had nothing to do with its running away.

Lt-Col Yorke said that some surprise had been expressed that so small an object as a trolley could derail a modern express hauled by a heavy engine, but the explanation was, he thought, simple enough. In the first place the trolley was a fairly substantial one, 6 ft long and 3 ft 8 ins wide and weighing about 4½ cwt. The trolley had been broken up by the collision and a fragment of iron had by chance probably got under and derailed the leading bogie wheels of the engine. Nevertheless, the remainder of the locomotive and the vehicles of the train had remained on the track for 140 yards and had there been no obstruction it was possible that the driver might have been able to pull the train up, but at that point the derailed wheels had struck a diamond crossing and the whole train was derailed.[203]

The only question to be decided, the inspecting officer continued, was why did the trolley run away? Postman Smith stated that he had locked the front wheels of the trolley before letting go of its handle and that it could not have moved unless somebody or something had touched it. He had implied, though he did not specifically say so, that the boys had meddled with it. The boys denied this and auctioneer Smethurst, Foreman Richardson, Driver Knighton of the up train and Ticket Collector Marchant all unhesitatingly stated that they had not seen the boys near the trolley when it ran away. Postman Smith's suspicion that the boys were involved rested on a very slight foundation and it was impossible to hold the boys in any way responsible. In any case, whether the boys had been present or not, Smith had taken charge of the trolley and was therefore responsible for its safe use.

It was the slope of the platform which had caused the trolley to run away. The point where the trolley had been standing had a gradient of 1 in 82. To the inspecting officer it had not appeared that the slope was sufficient for the trolley to run away due to the action

[203] Board of Trade report

of gravity alone but by experiment with a trolley similar to that involved in the accident it was found that unless it was placed at the point where the postman said he had left it with its handle turned right or left to almost a right angle it immediately began to run away. The experiment was repeated several times with the same result. It was clear that the postman, considering that the platform was level, had not given much thought to parking the trolley, and as soon as he had turned to get the key the trolley had quietly begun to run backwards. The accident was therefore due to the postman's want of care, but both he and Foreman Richardson deserved credit for their efforts to try and shift the trolley at the risk of their lives.

Since the accident, Lt-Col Yorke stated, he had received letters regarding similar accidents of luggage barrows or trolleys rolling off platforms at different places around the country, and from enquiries he had made it seemed that such incidents were by no means rare, though not previously attended with such serious results. Two solutions presented themselves; namely either to make platforms level or to provide three or four-wheeled trolleys with brakes. Providing level platforms on unroofed platforms might pose problems with drainage but a slight inclination away from the track would solve the problem. At stations with platforms sloping towards the track braked trolleys should certainly be provided. The slope of the down platform at Wellingborough was excessive and it was only 2 0½ ft high. The company had already raised the up platform and had informed him that they now intended to improve the down platform as well, which should serve to prevent a repetition of the accident.

Finally, the inspecting officer rejected suggestions made to him that a cowcatcher would have prevented the accident, pointing out that these were generally used in countries where the railway lines were unfenced, and that in the present case a cowcatcher, which could not be fixed less than four inches above the height of the rail, would not have caught the piece of iron responsible for the derailment.[204]

Three inquests were held on the different victims of the accident. Each returned a verdict of 'Accidental death', with recommendations that the sloping down platform should be improved and that the postmen should get more help from railway employees.[205]

A serious accident occurred in Rhodesia on Saturday night, 19th November 1898, on the Mashonaland Railway, '33 miles from Umtali'. A more precise location was not given. Whilst rounding a curve at full speed the engine and some trucks of the train were thrown off the rails and upset. Driver William McNeill, Fireman Patrick Ray, Timekeeper Gerald Conrad Hoffman and four 'natives' were killed and Mr A.R. Ross, Native Commissioner for the Makoni District, suffered a fractured skull. Mrs Ross was badly bruised. Mr and Mrs Richard O'Reilly and 18 'natives' were slightly hurt. The injured passengers (probably not including any of the 'natives') 'were at once removed to Umtali Hospital, where they are receiving every attention.' The accident was blamed on the disobedience of Driver McNeill, presumably because he had been driving at an excessive speed.[206]

---

204 Ibid.

205 *The Times*, London, England, 3rd, 5th, 6th, 8th, 14th, 15th, 20th and 22nd September 1898

206 Ibid, 25th November and 3rd December 1898

A very brief report from Berlin, Germany, spoke of the derailment of a workman's train near Brüggen on 12th November 1900. Six men died and several others were injured.[207]

The French Chemins de fer du Midi company's Sud express had left Irun in Spain at 10 a.m. on 15th November 1900 and was due at Bordeaux in France at 1.30 p.m. and Paris at 7.15. It never got any further than a point between Saint-Geours-de-Maremne and Saubusse stations, in France, near St-Vincent-de-Tyrosse, where it left the rails on a high embankment while travelling at an estimated 120 kilometres an hour, with disastrous consequences. Thirteen people died and 26 were injured. The driver and fireman were unhurt.

The accident occurred at 11.46 a.m. Most of the passengers had taken their seats in the restaurant car and had almost finished an early luncheon when a violent shock threw them against one another. Having left the rails, the locomotive reared up in the air and then fell back onto the kitchen part of the restaurant car, and the waiters, cook, cook's assistant and washer-up were crushed to death. A number of the passengers in the restaurant car were also killed and others seriously injured and only four of the 27 of them who had been in the car escaped unhurt. Those who lost their lives included the Duke of Canevaro, Peruvian Minister in Paris; Mlle Marie Valencoala, maid to the Duchess of Canevaro; M Lacloche, a jeweller of Biarritz; André Lamles, aged 14, of Biarritz; Mr Albert Stangoe, a British subject and Signor Carnera, the Italian Ambassador who had been to Madrid to notify the King of Spain of the accession of King Victor Emmanuel to the Italian throne. Among the injured was a M. Dombrowski, of Kieff, who was only slightly hurt but whose wife, sitting opposite to him, was killed. After the accident he chartered a special train to take her body to Paris.[208 209 210 211]

As to the cause of the derailment the driver stated that he had reported the bad condition of the track, due to damage caused by heavy rain, two days before the accident, but the official investigation found that the derailment was due solely to the excessive lightness of the van marshalled at the front of the train, but that the regulations had been complied with and the company was therefore absolved of responsibility.[212]

The accident was also referred to, at least in the British press, as the Dax derailment, the town of that name being a few kilometres distant.

---

207 *Brooklyn Daily Eagle*, Brooklyn, New York, United States, 12th November 1900
208 *The Standard*, London, England, 16th November 1900
209 *The Times*, London, 16th and 17th November 1900
210 *Le Petit Journal*, 16th November 1900
211 *Le Matin*, Paris, France, 17th November 1900
212 Ibid, 26th January 1901

CHAPTER 4

REAR COLLISIONS

The risks associated with one train running into the back of another had been low in the formative years of the railways when trains were few and train speeds low, but as the 19th century progressed train speeds and traffic intensity grew faster than advances in brake power, rolling stock construction and signalling and as a result by the 1880s serious rear collisions had become common. This chapter is therefore a long one, even though a great number of less serious accidents have not been included.

Blackburn, England, on the Lancashire and Yorkshire Railway, was the scene of what the Board of Trade report described, rather melodramatically for an official report, as a 'terrible collision' on the afternoon of 8th August 1881. An express train from Liverpool to Todmorden was standing at the station platform when it was run into by a Manchester to Hellifield express at a speed variously estimated at between 15 and 30 miles an hour. Two passengers died on the spot, another five were fatally injured and 64 were less seriously hurt. The fireman and guard of the Manchester express and the driver of a shunting engine which was at the rear of the Todmorden train were also seriously hurt. Two Lancashire and Yorkshire composite carriages ahead of a Midland carriage at the rear of the Todmorden train were those most badly damaged, along with the shunting engine, but various other vehicles of both trains and the colliding engine were also damaged to a lesser extent.

Blackburn station was approached from Manchester by a steep descent at 1 in 75 for over a mile from Over Darwen, followed by a fall at 1 in 100 for rather more than 2½ miles to Bolton Junction, just outside the station, thence level until a final short rise. The approach to the station was protected by signals worked from Locomotive, Bolton Junction, Old Junction and West Junction cabins with absolute block working as far as the Old Junction and permissive working thenceforth through the station itself.

The 2.10 p.m. express from Liverpool arrived at Blackburn at 3.9 p.m., three minutes late. The train engine then removed a carriage from the front of the train and a shunting engine passed via a mains crossover to the rear of the train, to remove the Midland carriage, but had not had time to do so before the collision. The 2.25 p.m. express from

Manchester was due in Blackburn at 3.14 and arrived there a minute or so later. It was equipped with the Westinghouse brake and was booked to stop at Bolton and Over Darwen before Blackburn. There had been no problems with the brakes during the journey either stopping at the stations or slowing down at each end of South Tunnel to pick up and drop off the pilotman, single line working being in operation through the tunnel, but Driver William Stansfield stated that the Westinghouse brake did not work when he applied it at Bolton Junction. He had applied the hand brakes, which reduced the speed from 40 to 25 or 30 miles an hour but the speed was not much less at the Old Junction, where he reversed the engine and applied steam, and he had not finished reversing as the engine passed the West Junction and the collision shortly occurred. Driver Stansfield's evidence was corroborated by his fireman, Mark Barker, and by Guard Charles Wade.[213]

The Westinghouse Brake Company were naturally very keen to disprove that there had been any failure of their brake system and co-operated as much as possible with the Board of Trade inspector, Colonel W. Yolland, in investigating the accident. During the course of the inquiry a number of tests were carried out, running trains into Blackburn station, to try to determine what had really happened.

In the conclusions to his report Colonel Yolland first remarked upon the system of permissive working employed at the Blackburn cabins, by which when a train was standing in the station the home signals at the Bolton Junction, Old Junction and West Junction cabins would all be cleared with just the Old Junction and West Junction distant signals kept at "Danger" and the driver of a second train shown a green hand signal from the West Junction, only 172 yards from the centre of the station. The inspector was especially critical of the evidence of Signalman Robert Thompson in the West Junction cabin, who had testified that, "I did what was right in lowering the home signal for the Manchester express to run past my box at 40 miles an hour when a train (the Liverpool) was still standing at the platform," and, "I have never had any instructions, written or verbal, to keep my home signal at 'danger' when a train is standing in the station against an approaching train running at a high rate of speed, until that train has either stopped or nearly come to a standstill." The inspecting officer said that there was a considerable difference of opinion among the signalmen employed at Blackburn station as to what the company's rules and regulations actually required them to do, but if they were generally applied as interpreted by Signalman Thompson, he was only surprised that many serious collisions had not previously occurred at Blackburn station. As the West Junction cabin was only 172 yards from the spot where the collision took place the safety of trains standing at the platform and those arriving from the south entirely depended on what was done with the West Junction home signal. Colonel Yolland added that he did not think the rules as definite and positive as they might be and would not say that traffic could not be safely worked through the station permissively with explicit obedience to explicit rules and regulations, but he considered that it would be far better to increase the accommodation at the station and if possible do away with the permissive working altogether. According to the evidence of the crew of the Manchester train the improper working of the Old Junction and West Junction home signals had not caused the collision, but he thought it highly probable that if the home signals at the Bolton, Old and West Junction cabins had all been at "Danger" the driver of the Manchester train would

[213] Board of Trade report

have begun to brake earlier and the collision would have been avoided or its results mitigated.[214]

Colonel Yolland stated that after very careful consideration he had come to the conclusion that the accident had been caused by the failure of the Westinghouse brakes to act behind the engine on the approach to Blackburn. He was critical of the continuous brakes being "self-acting" when this meant that they sometimes applied independent of the action of a driver, stopping a train unnecessarily in mid-section, thus incurring the risk of a rear collision should there be a mistake in block working or the line be worked permissively. He continued, 'The Westinghouse automatic air-pressure break is a very clever, ingenious piece of mechanism, made up of a very great number of separate parts, to give it the automatic action, but as is well known, the greater the number of parts in any piece of mechanism, the greater is the liability of failure, and I consider that the liability to apply itself when not required, and when there is no accident, and to fail to act or go on when absolutely required by an engine-driver, constitute two grave defects in its present construction. These defects are caused:

'1st by automatic action in the hose pipe connexions by which they frequently become uncoupled by a twisting of the pipe when air is admitted from the engine, when there is no accident of any kind, and no indication is conveyed to the engine-driver of the hose connexion having become uncoupled; and as a result, the action of the breaks when applied by the driver is limited to the break on the engine alone or to those on the engine and such other vehicles as remain connected by the hose pipes to the engine.

'2nd, the automatic action, by which sometimes the breaks go on, is probably caused by leakage from the pipes when there has been no accident, nor any separation of the train into two or more parts, which alone requires the proper automatic action of the brakes to be brought into play.

'Whether it is possible to get rid of these defects I offer no opinion.' He also recommended a plate to show the correct manner to connect up the brake pipes to prevent them being wrongly connected.

Colonel Yolland concluded that until the defects in the Westinghouse brakes which he had pointed out had been remedied it should not be used on trains running over lines not worked on the absolute block system.[215] [216]

The jury at the adjourned coroner's inquest on 16th September found that 'such collision was caused by the loose working of the signals and the excessive speed at which the Manchester train was being driven into the station; and the jury are further of opinion that there ought to be more protection to the station than the present system of signalling at the east and west cabins affords.' [217]

A bad accident at Charenton station, France, on Monday 5th September 1881, brought to light deficiencies in the working of trains on the Paris Lyon Méditerranée system. On the morning of the accident two trains bound for Paris were running out of order. Train

---

214 Ibid.
215 Ibid.
216 *The Times*, London, England, 10th, 12th, 15th and 16th August 1881
217 Ibid, 17th September 1881

No. 584 should have stopped at Villeneuve-Saint-Georges to set down passengers for several intermediate stations, including Charenton, and then run non-stop to Paris, the passengers for the intermediate stations catching the following omnibus train, No. 18. However, train No. 584 was running half an hour late on this morning and the omnibus train was thus allowed to precede it, train No. 584, making extra stops at the intermediate stations concerned. At 9.40 a.m. train No. 584 stopped at the platform at Charenton, to discharge its passengers. It was still there when a third train, No. 10, the Marseilles express, was heard approaching. The stationmaster ran to the driver of train No. 584 and shouted for him to get going quickly, while the signalman hurriedly tried to extract a tablet for the forward section, the line being single. The poorly maintained instrument would not let him have one and he then ran along the platform waving a flag at the approaching express. Meanwhile the driver of train No. 584 had heeded the stationmaster's warning and had hurriedly set off, sharply enough, unfortunately, to cause the couplings to snap so that the rear carriages of the train were left behind. Some of the more alert passengers were able to jump out of the front train when they realised that there was danger but many others dd not. The express was still travelling at about 25 kilometres an hour when it smashed into the wooden second and third-class wooden carriages, at least two of which reportedly exploded when the train hit them. Twelve bodies, badly disfigured, were removed from the wreckage and 25 or 30 others were badly injured; the final number of dead was at least 22, with 81 injured.[218 219]

The investigation into the accident revealed a combination of causes. The Charenton signalman, a youth of 17, had failed to replace his signal to protect the stationary train, the stationmaster at Maisons-Alfort had failed to advise the conductor and engineer of the express of the passing time of train No. 584, and the Maisons-Alfort signalman had let the express follow at less than the minimum five-minute interval (although there was tablet working to prevent single-line collisions following trains were apparently still regulated by the time interval system). The stationmaster apparently had problems at home and the signalman was feeling the after-effects of a paternal feast (a feast held to commemorate the patron saint of a parish – with considerable alcohol consumption no doubt).

There was a considerable public outcry following the accident with the PLM company being criticised for employing 17-year-old signalmen because they could be paid less, and for working its trains with obsolete wooden rolling stock.

The Maisons-Alfort stationmaster was subsequently sentenced to 12 months' imprisonment, the Maisons-Alfort signalman to six months and the Charenton signalman to three months.

The Ministry of Transport issued a series of recommendations: wooden coaches should be relegated to short local trains, the 'advanced disc. signals should be altered, the line between Paris and Villeneuve-Saint-Georges should be doubled and there should be an improved recruitment of signalmen'. A circular of 2nd November 1881 also called for the employment of brake vans on passenger trains, the interlocking of the Tyer's instruments and the lineside signals so that a signal could not be cleared unless the train had been accepted, modifications to the Tyer's instruments and the employment of continuous brakes on fast trains. For their part the PLM implemented the

---

218 Wikipedia – Collision ferroviaire de Charenton

219 *The Magnet*, London, England, 12th September 1881

recommendations as early as 1882.[220]

On 10th December 1881 a series of signalman's errors led to a probably unique series of three rear collisions in Canonbury Tunnel, England, which killed four passengers and a guard. The collisions are described in detail in the chapter on Tunnel Accidents.

The year 1882 was not quite a fortnight old when there was a rear collision at Spuyten Duyvil, New York, United States, on the New York Central and Hudson River Railroad, on the evening of 13th January. A session of the New York State legislature had concluded at Albany that afternoon, and a large crowd of legislators were anxious to catch the *Atlantic Express* to New York City. To cope with the extra traffic the express, which was running 23 minutes late on its arrival at Albany, was strengthened by the addition of 15 extra coaches, eight of them palace cars (luxury coaches). Two extra locomotives were also attached, and the train departed, 'filled with all the men of note and prominence of both the Republican and Democratic parties.' The Tammany Hall (a New York City Democratic Party organisation) delegation was about the last to arrive at the station, and the rest of the train being full they took seats in the rear car, the *Empire*. The train set off, but was shortly stopped at East Albany, where the palace car *Idlewild* from Troy was attached to its rear. There were about 12 passengers in the *Idlewild*, which had come from Bennington, Vermont.

The journey to New York then went smoothly as far as Spuyten Duyvil, where the train came to a halt just after 7 p.m., according to one newspaper report because of an overheated journal, though the cause of the stoppage was apparently the subject of much debate, it also being alleged that an inebriated party in the smoking car had applied the brakes or, by opponents of the Westinghouse brake with which the train was equipped, that the brakes had come on of their own (see Colonel Yolland's report on the 1881 Blackburn accident above). When the train came to a stand Conductor George Hanford assumed that Brakeman George Melius would go back and protect the train, a measure which was all the more necessary as the line was not protected by any sort of block system, either manual or automatic, and a following train, the *Tarrytown Special*, was shortly due, as it was booked to leave Tarrytown at 6.35 and Spuyten Duyvil at 7.7. Melius afterwards claimed that he had in fact gone back seven or eight car lengths, but too many witnesses stated that he had in fact merely stood at the rear of the *Idlewild* until he saw the Tarrytown Special coming around a curve and emerging from a cutting, when he had belatedly run back 50 feet waving his hand lamp. The Tarrytown Special ploughed into the *Idlewild* car and drove it into the *Empire*, the two cars being telescoped. Stoves and lamps in the parlour cars were upset and set fire to the woodwork and upholstery. The passengers were trapped between the seats and the sides of the cars and unable to escape the fire. A party consisting of assemblymen Sheeby, Robb and Chapin, Commissioner Ulrich and Messrs Spinney and Lyman was trapped in the telescoped cars, but axes were procured, and efforts made to free them, which finally succeeded after assemblyman Robb was able to break a window through which the whole party were dragged. Two women were also rescued in this manner, but others were not so lucky and nine died, including a Mr and Mrs Park Valentine of Bennington, married only

[220] Wikipedia – Collision ferroviaire de Charenton

the day before.[221] [222]

Another victim was Senator Webster Wagner, the founder of the Wagner Palace Car Company, who had been seated in the *Empire*, but after the train had been stopped about a quarter of an hour had gone back to the *Idlewild*, doubtless to ascertain the cause of the hold-up, and 'was burned to a crisp in one of his own luxurious cars.'

Brakeman Melius was clearly principally to blame for the accident but Conductor Hanford was blamed for not checking that he had in fact gone back and Engineer Archibald Buchanan, of the rear engine, was also found fault with for not having operated the valve to conserve air in the main reservoir as soon as the Westinghouse brakes were applied, which would have allowed the stalled train to restart more quickly. Finally Engineer Francis Burr of the Tarrytown train was criticised for not having noticed Melius' signal earlier and Superintendent John M. Troucey for allowing trains to pass through Spuyten Duyvil cut, 'the most dangerous place on the railroad' as fast as 20 mph. Rather more to the point was the lack of automatic block signals, installed on other sections of line but not here because the superintendent thought them 'at present impracticable'. [223]

George Melius subsequently stood trial for manslaughter and a witness against him was an Alonzo Valentine, whose son and daughter-in-law had died in the accident, and who had seen Melius leaning against the rear end of the *Idlewild* instead of going back down the line. However, a clever defence attorney stated that Melius had done his duty, the company rule merely stating that 'When a train stops on the road the conductor shall send a man to stop any train approaching.' Melius, the attorney claimed, was waiting for the conductor to tell him to go back. In addition, the attorney pointed out that there were only two instead of the customary three brakemen with the train and hinted that it should be Mr Vanderbilt rather than a poor brakesman who was on trial. The judge rejected all the assertions of the defence attorney, stating that if it was the conductor's duty to send a brakesman out it was equally the brakeman's duty to go without needing to be told. The jury, after an hour's deliberation, and despite the judge's strong hint that they should find Melius guilty, instead acquitted him, to the surprise of everyone. And doubtless the disgust of Alonzo Valentine.[224]

On 17th January 1882 there was an accident at Shikohabad, Uttar Pradesh, India, on the East Indian Railway. The up mail train, 'running under a caution message', as the annual report to the British government on the railways of India stated, ran into a goods train a mile from Shikohabad station, during a thick fog. Thirteen 'native' passengers were killed and 15 more injured. The European guard of the goods train saved his life by jumping from his van just before the collision but was seriously injured. The *Times* correspondent wrote that, 'The accident is officially ascribed to the foggy state of the morning, rendering the signals invisible, but it is difficult to avoid the suspicion that there

[221] *The Fitchburg Sentinel*, Fitchburg, Massachusetts, United States, 14th January 1882
[222] *The Bangor Daily Whig and Courier*, Bangor, Maine, United States, 16th January 1882
[223] *A History of Railroad Accidents, Safety Precautions and Operating Practices*, by Robert B. Shaw, pgs. 81-83
[224] Ibid, pg. 416

was culpable negligence in allowing one train to follow the other so closely.' [225] [226]

Fog caused accidents in England as well as in India in January 1882. On 25th January there was an accident on the Great Northern Railway main line at Hornsey in the early evening rush hour, when the 5.1 p.m. train from Moorgate to Barnet ran into the rear of the 4.23 p.m. passenger train from London Victoria to Enfield. The rear brake carriage of the Enfield train was completely broken up and a second-class carriage driven into the one in front. Two passengers were killed outright, a third died later and 15 other passengers were seriously and 53 slightly injured. There was a very poor visibility at the time of the accident, 5.30 p.m., an exceptionally thick fog (doubtless really a smog) having come on very quickly. The fog had commenced at Hornsey at 5.10, the fog bank moving southward to Finsbury Park down advanced starting signal at 5.25 or 5.26, after the Moorgate train had passed Finsbury Park No. 5 box but before it reached the advanced starting signal. The preceding London Victoria – Enfield train had arrived at Hornsey at 5.24 in fog so dense that the signals were invisible and tail lamps could not be seen more than three or four yards. There was also a hold-up to the down traffic at this moment with a temporary blockage at Wood Green, the next station, where a down goods train was delayed in shunting due to a broken coupling. The Hornsey stationmaster had asked his signalman, John Razzell, to send his booking boy to call a fogsignalman named Borringer, who fogged the down home signal, but the signalman did not realise the urgency of the situation, and delayed sending his booking boy for five minutes, so that Borringer arrived at the station just too late to prevent the collision.

Driver George Johnson of the Moorgate train started from Finsbury Park station at 5.25 p.m. two minutes late and found the starting and junction signals off for him. He then ran into the fog bank and could not see the advanced starting signal but took it for granted that it was off for him. He then became utterly lost in the dense fog and only knew where he was when he saw the glimmer of a gas lamp on the Hornsey down platform, when he applied the brake, but his engine being then only 45 yards from the rear of the Enfield rain it had little effect in reducing his speed. Driver Johnson stated that after leaving Finsbury Park he had run at a reduced speed all the way to Hornsey, but the Board of Trade inspector, Major F.A. Marindin said, evidently a far too liberal idea of what running at caution meant, as he had taken only four minutes to run over 2,400 yards, and must therefore at some point have been running at not less than 18 or 20 miles an hour, despite, according to his own statement, being able to see 'nothing off the engine'. He knew the line perfectly well, and that trains were often brought down to the advanced starting signal at "Danger" to wait there and must therefore be considered largely to blame for the accident despite the very sudden and exceptionally severe fog. His only possible excuse was that he knew that trains were not allowed, according to the rules, to be let down to the advanced starting signals in fog, but he should have realised, seeing the fog bank advancing towards him, that the fog was a sudden one, and got his fireman to check the position of the advanced starting signal either by asking the signalman or by feeling the weight upon the signal post.[227]

---

225 Report to the Secretary of State on the Railways in India 1882-3

226 *The Times*, London, England, 29th January 1882

227 Board of Trade report

It was impossible to avoid the conclusion, the inspecting officer continued, that signalman Razzell at Hornsey had contributed in some degree to causing the collision by his dilatoriness in sending the booking boy to call the fogsignalman out. He ought to have anticipated the danger and sent the boy at once, taking over the booking himself.

In conclusion Major Marindin stated that the lesson to be learned from the accident was the weakness of the whole system of fog signalling. In very thick fogs where the fogsignalman and his lamp were invisible to the driver the latter did not get any positive indication that the signal was clear, merely the negative indication of the non-explosion of the detonators. The rule requiring drivers to go cautiously during fog should therefore be rigidly enforced and it should be made clear to drivers that they were never to pass signals in fog without their being sure that the signal was off for them. Again trains stopped at stations in fogs before the fogsignalman had come on duty should be protected by detonators put down some distance in their rear as they would be if they were stopped in mid-section.

It was desirable, Major Marindin said, that some system should be devised whereby drivers could be told whether or no fogsignalmen were on duty. Better, though, would be a system where fog signals controlled by the signalman took the place of 'often wearied and half-frozen' fogsignalmen, and he hoped that a further trial might be given to some of the inventions already existing for the improvement of fog signalling, and the thoughts of many mechanical engineers turned to the subject.[228] [229]

The jury at the coroner's inquest on the victims of the accident found Driver Johnson guilty of great neglect in passing the Finsbury Park advanced starting signal without having seen it and continuing at the speed he had but stopped short of finding him guilty of manslaughter. They also considered that the Hornsey station staff should have protected the Enfield train while it was stood at the station.[230]

On 20th August 1882 W.C. Coup's circus train was en route to Cairo, Illinois, running in two sections.

As the trains were ascending an incline between New Burnside and Tunnel Hill, Illinois, United States, the heavily loaded first section was caught up by the lighter and faster running second section, the engine of which completely demolished the rear car of the first section, killing three waggon drivers and wounding about 25 or 30 others. The other vehicles of the two trains were undamaged. Relief trains and physicians were despatched to the scene and the two sections arrived in Cairo, 47 miles distant, at about 4 a.m.[231] About 80 men had been sleeping in the rear car.[232]

An accident occurred in Park Avenue Tunnel, New York, United States on 22nd September 1882. It is described in the chapter on Tunnel Accidents.

---

[228] Ibid.
[229] *The Times*, London, England, 26th and 28th January 1882
[230] Ibid, 30th January 1882
[231] *Chester Times*, Chester Pennsylvania, United States, 21st August 1882
[232] *Stevens Point Journal*, Stevens Point, Wisconsin, United States, 26th August 1882

'Killed by Carelessness' was the title of a brief newspaper report on an accident at Pittsford, Michigan, United States, on the Lake Shore and Michigan Southern Railroad at 6 a.m. on the morning of 6th October 1883. Local freight train No. 60 was standing in the station when freight train No. 50, 'which had been following at the proper distance', nevertheless came around a bend, down a steep incline west of the station and ran into the caboose of the first train. The local freight was several hours late and there were a number of passengers in its caboose. The collision killed a Mrs J. Rogers, of Hudson, Michigan, a Mr Milloick of West Unity, Ohio and Mr A. Rose, of Adrian, Michigan, while Mrs Mark Guy of Hudson, Michigan, had her legs crushed and 'died before noon, a raving maniac'. A Mr Harford of Pittsford suffered a broken leg – was he just setting out or just arriving home? Engineer Chris Nendhart of Toledo jumped from the engine of the colliding train and was seriously injured. The wreckage caught fire immediately after the collision and several cars were destroyed, while the engine of train No. 50 was badly damaged. The accident was blamed on the crew of the first train, which was in charge of Conductor Alcoupe of Elkhart, failing to send out protection far enough to stop the second – hence the carelessness.[233] [234]

An accident at Otter Creek, near Streator, Illinois, United States, on 16th November 1883 had its origins in the stalling of an extra freight train on a steep incline. The accident occurred on the Fox River branch of the Chicago Burlington and Quincy Railroad. A flagman went back and stopped the 12 noon passenger train, but the flagman of that train was only able to get back a short distance before a second freight train which had been closely following appeared around the corner. This third train crashed into the rear coach of the passenger train, crushing it. Six passengers were killed outright in the accident and about a dozen more were injured. Engineer Hans Young and Fireman Smith of the colliding freight train were also badly hurt, the fireman probably fatally. Immediately after the collision the boiler of the engine of the second freight train exploded, and it was stated that it was the boiler explosion which had killed three of the victims. On news of the accident being received an engine and caboose were despatched to the scene with a doctor and a relief corps and the victims were taken to Streator.[235]

A major disaster occurred at Hanau in Germany on 14th November 1884. The failure of a signalman – probably the one at Hanau though the brief reports do not make this very clear – to send a message resulted in passenger train P 26 being allowed to enter the Nieder Rodenbach – Hanau section before goods train Gz 304 had left it, resulting in a rear collision. Goods train Gz 511 travelling in the opposite direction then ran into the wreckage before it could be stopped. Fifteen passengers were initially reported as having been killed and 20 injured but the final casualty toll was apparently 22 dead and 26 injured.[236] [237]

---

[233] *Review*, Decatur, Illinois, United States, 8th October 1883
[234] *Davenport Gazette*, Davenport, Iowa, United States, 7th October 1883
[235] *The Dixon Sun*, Dixon, Illinois, United States, 21st November 1883
[236] *The Times*, London, England, 15th November 1884
[237] *Eisenbahn-Katastrophen in Deutschland*, by Hans Joachim Ritzau, pg. 68

An accident at Hackensack River, New Jersey, United States, on the Pennsylvania Railroad on 18th October 1885, had similarities to the Hanau disaster in that it was a double collision which resulted from a signalman's error. It will probably come as no surprise to the reader to learn that once again one of the trains involved was carrying immigrants from Europe travelling to new homes and a new life in the Mid-West. In this case it was a large party of Danish, German and Norwegian immigrants who had endured a slow trek by ship across the Atlantic Ocean to Jersey City, where they had boarded a Pennsylvania Railroad special train. The emigrant/immigrant train had left Jersey City at 7.50 p.m. or 8.2 p.m. (sources vary) and passing the Marion tower had crossed the Hackensack River and then stopped to take on fuel at a coaling tower on the opposite side. Having done so, it was just setting off again when a coupling gave way, leaving two cars behind. The train stopped and set back, and the crew were in the process of recoupling the detached vehicles when Conductor Brogan was shocked to find another train bearing down upon them at high speed. He grabbed a red lantern from the caboose and ran back down the track waving it at the approaching train, but his warning came too late to be of much use. The second train, the nine car *Pacific Express* which had left Jersey City at 8.15, slammed into the rear of the special train, throwing the caboose over to the eastbound running line and telescoping the next cars. To make matters worse, as at Hanau another train completed the work of destruction. Lehigh Valley Railroad train No. 3, which had left Newark at 8.10 p.m., ploughed into the wreckage and one of its cars caught fire. It was perhaps surprising that only nine or possibly 10 of the immigrants lost their lives. A number of others were more or less injured, as were Conductor Brogan and the engineer and fireman of the Lehigh Valley train. None of the passengers on the *Pacific Express* was very much hurt. The surviving immigrants continued their journey almost at once, and as they were strangers in a foreign country, and often also to one another, and many of them spoke little or no English, some of the victims were never identified.

A passenger travelling with his wife on the Lehigh Valley train, Mr Joseph Smith, gave a vivid description of the accident and its aftermath to a reporter. 'Our train was running, I should judge, thirty miles an hour. There were seven cars. The collision occurred a few moments after we had left East Newark. We felt a sudden shock that threw the passengers from their seats, and then there came the crashing of timbers, mingled with the screams and groans of women. A terrifying scene followed. No one knew exactly what had happened, and in the confusion a scramble was made for the doors. Our train immediately came to a standstill. Many were injured in getting out of the cars, but, so far as we could learn, none seriously. A terrible scene was presented by the wrecked cars. They lay in a heap across both tracks and down the embankment, which is fully twenty feet in height. The bodies of the dead and wounded lay about in the most horrible forms. Some were burned beneath heaps of broken timber. Others lay huddled about the track. One woman's head was cut completely off. The head rolled down the embankment, while the mangled body lay between the tracks. It was almost pitch dark and foggy, and the only light was that of the fire which enveloped the baggage car of the western Express. The wild shrieks of the sufferers could be heard for blocks away. In justice to our conductor I will say that he ran very carefully from East Newark.' [238]

The double track main line was worked on a manual block system with towers (signal

[238] *Chester Times*, Chester, Pennsylvania, United States, 19th October 1885

boxes) at Marion and Fish House. The accident was due to a fatal mistake by Thomas Pratt, the towerman (signalman) at Marion. The times of the sending of signals and the passing of trains were recorded in the equivalent of what in Great Britain would be called a train register book. When the *Pacific Express* was offered to Pratt the emigrant special had already been gone 13 minutes and he, unaware of the stoppage because of the broken coupling, assumed, as other signalmen have before and since, that because it should have cleared Fish Hook tower that it had actually done so. He said that he had looked at the entries in his book, which should have told him otherwise, but had misread them.[239] [240]

To understand the circumstances of the next accident, at Austell, Georgia, United States, on 14th December 1885, a little knowledge of local railroad geography is useful. The East Tennessee Virginia and Georgia and the Georgia Pacific Railroads used the same track from just outside Atlanta to Austell, Georgia, where they diverged, the one heading west and the other north. The two railroads used different depots in Atlanta and ran on separate tracks until they came together at the city limits. Situated about a mile before Austell was a water tank (Collier's or 17-mile tank) used by Georgia Pacific trains.

Georgia Pacific Conductor Anthony Stephens stated at the subsequent inquest that his train, eastbound night passenger train No. 52, had left the Simpson Street station in Atlanta at 10.15 p.m., on time, on the night of 14th December. His train had crossed a westbound East Tennessee Virginia and Georgia train, No. 11, at Mapleton, waiting there 10 minutes for the latter train to arrive, and his orders had indicated that the eastbound East Tennessee train, No. 12, which was following him, would cross its westbound counterpart at Chattahoochee, eight miles back towards Atlanta. As East Tennessee trains were limited to 30 miles per hour this would put him 32 minutes in front of train No. 12 at that point. He was unaware that train No. 12 had in fact crossed train No. 11 at Concord, only three miles from Mapleton, and was in fact only 15 rather than 32 minutes behind. His train had stopped for water as usual at the 17-mile tank, a mile and a half beyond Mapleton, when train No. 12, travelling he thought at 40 miles an hour, had run into them. He did not hear train No. 12 whistle. There had been two red tail lamps at the rear of his train. His train had been 20 minutes late but according to his orders train No. 12 should still have been 32 minutes behind them when they left Mapleton. He said that had he been aware that trains Nos. 11 and 12 were crossing at Concord he would have kept his train at Mapleton.

The cars of train No. 52 were standing on a 30-foot-high trestle. The East Tennessee train came quickly around a curve and its engineer, named Owens, and fireman saw the Georgia Pacific train in front of them too late to stop. The fireman jumped but the engineer stuck to his post, reversed his engine and applied the air brakes. The driver emerged unscathed and the fireman was only slightly hurt. The collision occurred at 11.52 p.m.

The rear sleeping car of the Georgia Pacific train, which was crowded with passengers, was crushed by the engine and many of the injured were pinned in the wreckage where they were scalded by escaping steam, which appears to have been what caused all the fatalities,

239 *The Herald and Torch Light*, Hagerstown, Maryland, United States, 22nd October 1885
240 *A History of Railroad Accidents, Safety Precautions and Operating Practices*, by Robert B. Shaw, pgs. 164-5

as a Dr Roy stated at the inquest that none of the bodies bore injuries sufficient to cause death, but all were scalded. There were the horrific scenes characteristic of such accidents with the cries of the injured cutting through the air in the darkness. As soon as possible the East Tennessee train was pulled back clear of the wreckage and the work of rescue began; assistance was telegraphed for from Atlanta and a relief train with doctors and railroad officials let there at 2.15 a.m. A train returned from Austell to Atlanta with the dead and injured at 7.30 a.m. and the line was cleared at 8.30 a.m.

As for the responsibility for the accident the East Tennessee Conductor, F.M. Duggar, stated that his train had left the Mitchell Street depot in Atlanta at 10.57, 12 minutes late. His initial orders had stated that his train was to meet train No. 11 at Chattahoochee, but they had then been superseded by fresh orders changing the meet to Concord. He had thought that the Georgia Pacific train was 30 minutes ahead of them. His train had been travelling at between 25 and 30 miles an hour at the time of the collision.

Conductor Stephens stated that he had no official knowledge of the running of the East Tennessee train. To an impartial observer much of the blame would appear to lie with him for failure to take steps to protect his train, regardless of whether or not he had any knowledge of train No. 12 being so close behind, the accident bearing similarities to that on the Housatonic Railroad at Bridgeport, Connecticut, 20 years earlier. Ten passengers were killed outright in the collision and the death of an 11th passenger and of Jack Scott, the coloured sleeping car porter, brought the number of fatalities to 12. It was thought that one or two others would be unlikely to recover. The final newspaper reports on the accident stated that the inquest was still in progress. Its findings have not come to light.[241 242]

On 11th May 1887 a serious accident happened near Windsor, Victoria, resulting in the death of four passengers and two railwaymen and injuries to a large number of passengers. The accident occurred during the Wednesday evening rush hour. What was known as an ordinary train left Melbourne for Balaclava, stopping at Richmond, South Yarra and Prahran but not at Windsor, the last station before Balaklava itself. This train was followed by an express at 5.40 p.m. which had Windsor as its first stop. The 5.30 train was brought to a stand some distance short of Windsor by a problem with its Westinghouse brakes and while standing there was run into by the express, which smashed its last four carriages, including a composite brake van and a smoking carriage. Fortunately, the first train was not very full, most of its passengers having alighted at one of the preceding three stations at which it had stopped. The passengers on the express, as was often the case in such accidents, escaped without any serious injuries, but its driver, Frederick Maskell and Fireman James M'Nab were among the fatalities.[243] Four passengers in the ordinary train also lost their lives and about 20 were injured. Those who died were Mrs Foster, wife of Dr Foster of Colac, who had been on a short visit to Melbourne, and Mr William Runtling, who were killed outright; Mr E.S. Parkes, manager of the Bank of Australia, who died at his home in Balaclava the same evening while his limbs were being amputated, and Mr Lempriere, who had been gravely hurt and

241 *Macon Weekly Telegraph*, Macon, Georgia, United States, 22nd December 1885
242 *Bismark Daily Tribune*, Bismark, Dakota Territory, United States, 16th December 1885
243 *The Kyneton Observer*, Kyneton, Victoria, 14th November 1887

succumbed to his injuries a couple of days later at the Alfred Hospital in Melbourne. [244]

The line was by this date worked upon the absolute block system and the collision was directly caused by a mistake by trainee J.C. Blackburne at Prahran station who wrongly believed that he had received "Line Clear" for the express train from Windsor signal box and had hence cleared the Prahran signals for the express. Blackburne denied having made such a mistake, but the weight of the evidence was that he had done so, rather than that the two experienced Windsor signalmen, Edward Tozer and Charles Berrie, had prematurely accepted the express. Blackburne was a probationer with only three weeks' service but under pressure of work the Prahran stationmaster had left him to do the signalling unsupervised, and this during the busy evening rush hour. Blackburne asserted afterwards that he had never received proper training in the operation of the block system and did not thoroughly understand it.[245] The 5.30 train had been initially brought to a stand by Driver James Keats a little in the rear of the Windsor outer home signal, which was at danger. Had he brought his train a little closer to the outer home signal his engine would have been more easily visible from Windsor box, while there was a suggestion that he had not whistled to alert the signalman of his presence. When the signal had been cleared Driver Keats had attempted to restart his train but could not, because a faulty Westinghouse brake pipe hose had burst, causing the brakes to apply.[246] Guard William Laing had relied upon the fact that the train was under the protection of the signals and had also assumed that releasing the brakes would only take a couple of minutes, so he did not go back and put down detonators to protect his train.

The inquest into the deaths of those killed in the accident was a very lengthy affair, occupying 16 days, but the coroner's jury finally reached a verdict on 7th June, in which they found Driver Keats and Guard Laing of the first train guilty of great neglect, the driver for failing to bring his train within the protection of the Windsor home signals and the guard for failing to go back and protect his train as per the regulations. They also placed great blame on the Prahran stationmaster, W.B. Gedge, for permitting a probationary railwayman, Blackburne, to work the signalling instruments unsupervised and added that, 'We also express our great surprise that Prahran station should have been in such a condition of disorganisation without the knowledge of the traffic department, and cannot but feel that the evidence given in this matter abundantly establishes the charge of laxity of supervision and systematic neglect of officers whose duty it was to see that the rules and regulations of the service were faithfully carried out. We also find that the public safety is seriously imperilled by placing at important stations inexperienced lads, and that before probationers are placed at stations, they should have careful preliminary training by a responsible officer, and gradually work their way from the lesser to the more important stations, so they may gain that experience which is necessary to guide them, should an emergency arise at any moment. We hope that immediate steps will be taken to ensure careful supervision of hose pipes in connect with the Westinghouse brake.' [247] Quite a mouthful!

Driver Keats, Guard Laing and Stationmaster Gedge were all dismissed from their jobs following the accident. No action was presumably taken against Blackburne, whose

---

[244] *The Age*, Melbourne, Victoria, 12th and 14th May 1887

[245] *The Argus*, Melbourne, Victoria, 18th May 1887

[246] *The Age*, Melbourne, Victoria, 1st June 1887

[247] *The Argus*, Melbourne, Victoria, 14th June 1887

actions had directly caused the accident, but who had been called upon to do a job well beyond his ability and experience.

Exceptionally heavy traffic, in this case, as at Egham some years earlier, on a race day, was the cause of a disaster at Hexthorpe, England, just outside of Doncaster on the Manchester Sheffield and Lincolnshire Railway, on 16th September 1887.

Various extra excursion trains were being run to Doncaster for the St Leger races, including on the day of the accident one provided by the Midland Railway, which reached Doncaster by way of running powers over the Manchester Sheffield and Lincolnshire Railway from Mexborough, and had left Sheffield at 11.20. The Manchester Sheffield and Lincolnshire line into Doncaster had been worked on the absolute block system since 1880, with signal boxes in the last few miles to the junction with the Great Northern Railway at South Yorkshire Junction (south of Doncaster station), Cherry Tree Junction, Hexthorpe Junction, Warmsworth and Conisborough East. In the 1 mile 23 chains between Hexthorpe and Cherry Tree Junctions there was also a wooden ticket platform, used instead of Conisborough station for the collection of tickets on race days. Although, as stated above, the line was normally worked on the absolute block system, to cope with the pressure of traffic on race days the block system was perversely suspended between Hexthorpe and South Yorkshire Junctions, traffic being controlled instead by flagmen. The drivers of trains were instructed to proceed cautiously between these points, keeping perfect control of their trains so as to be able to stop immediately upon seeing a train in front.

On the day of the accident the Midland excursion train was properly checked by signals by Signalman Thomas Welham at Hexthorpe Junction, ran under complete control past Flagmen George Coates and Joseph Frost and stopped at the ticket platform, close behind another train, at 12.13, drawing forward to the east end of the platform when the train ahead had moved on, and coming to a stand 506 yards ahead of Flagman Frost. It had been standing at the platform for about two minutes when it was run into by the 8.40 a.m. regular Manchester Sheffield and Lincolnshire passenger train from Liverpool to Hull, at 12.14, the colliding train probably moving at about 15 miles an hour. The force of the collision was sufficient to break up the four rear vehicles of the Midland excursion train, killing 24 passengers and injuring almost 100, one of whom died later.

The Liverpool train was also cautioned by the Hexthorpe Junction signalman but after passing the Hexthorpe starting signal its driver, Samuel Taylor, failed to check his train on passing the flagmen and continued, as he himself afterwards admitted, at about 35 miles an hour until he saw the tail of the Midland train about 350 yards ahead. He then did all in his power to stop, applying the continuous brakes, sanding the rails, sounding his whistle and reversing the engine, but all, of course, too late. He admitted that he had not looked out for any flag signals, believing that the tickets had been collected at Conisborough and that he would not need to stop at the ticket platform. His fireman, Robert Davis, stated at the Board of Trade inquiry that he had seen both flagmen, though he did not see the first one showing a red flag, but he was gesticulating in a manner which Davis could not understand. Davis said that he saw the red flag shown by Flagman Frost, but it was then too late to warn Driver Taylor before the collision took place. Major F.A. Marindin remarked that from the evidence there could be no doubt that the instructions had been faithfully carried out by Signalman Welham and Flagmen Coates and Frost and that both flagmen had been not only exhibiting but waving red flags, and that it was quite impossible to believe that if the driver, and even more the fireman, had been keeping a proper lookout they could not have seen these flag signals. Both driver and fireman acknowledged that they had received notice of the special race day traffic arrangements, though Fireman Davis said that he had not read the notice, itself a serious dereliction of duty. Had the instructions been obeyed the Liverpool train could easily have been brought to a stand behind the Midland one, and both men were to blame for not keeping a proper lookout. The driver's excuse that he thought that tickets were collected at Conisborough was not a very valid one given that the instructions stated otherwise, and that the train did not stop there long enough; also the guard had reminded the fireman that tickets would be collected at the ticket platform.

It was almost, Major Marindin said, as if Driver Taylor had entirely forgotten the special instructions and was driving his train as if it were an ordinary day. He had certainly been badly served by his fireman. The fireman probably thought that the flagman was repeating the advanced starting signal, and it would have been better had the flagman had stood a little away from the signal, but it was still the fireman's duty to tell his driver that the flagman was there.[248]

As to the brakes (Smith's non-automatic continuous vacuum on the Liverpool train as opposed to Layton-Sanders-Bolitho automatic continuous on the Midland train), the inspecting officer said that it was difficult to ascertain exactly where they had been applied but they did not seem to have been very effective in stopping the train and the case was one in which an automatic continuous brake would have been of inestimable value. It was satisfactory to know that the Manchester Sheffield and Lincolnshire Railway were now adopting one of that character.

Major Marindin then went on to consider the traffic arrangements and whether they would have ensured safety had they been strictly complied with by all the railway servants. He first considered whether it had been necessary to suspend absolute block working in the first place. He conceded that working the traffic through the 1¼ section between Hexthorpe and Cherry Tree boxes might be difficult but considered that given that the greatest number of up trains passing Hexthorpe Junction was 15, between 12 noon and 1 p.m. on St Leger day, and that there were two up lines in use (the other being the up goods line) he did not see that it would be impossible. He also pointed out the practice on the London and South Western Railway at Ascot, where to deal with the extra traffic extra temporary block posts were brought into use, and absolute block working maintained, and stated what one would have thought was obvious, 'If the block system is necessary as a safeguard under the ordinary condition of traffic, it is surely all the more necessary when the number of trains is largely increased, and in my opinion, block working should never be suspended upon a line where it is normally in work, if it can by any possibility be avoided.'

Given that the block working had been suspended, the inspecting officer did not think that the arrangements made had been all that they might have been. For example, by a blunder by someone, none of the Midland railwaymen working to Doncaster on the race days had even been issued with any instructions and could only rely on their memories of what had happened in previous years and by seeing the flagmen. Every train, the

[248] Board of Trade report

inspecting officer ruled, should have been not merely cautioned or checked at Hexthorpe Junction box, but stopped, as the regulations decreed, as they were to be should there be a failure of the block instruments, and the flagmen should have been supplementing their hand signals by placing two detonators on the line whenever they needed to exhibit a red flag, as they would do were they engaged in fog-signalling.[249 250]

The jury at the coroner's inquest on the victims of the accident indicted Driver Taylor and Fireman Davis on a charge of manslaughter.[251] They were committed for trial at the magistrate's court in October and tried at York Assizes in November, where after half an hour's deliberation by the jury both were found "Not Guilty". The recently formed Association of Locomotive Engineers and Firemen trade union had engaged an eminent barrister, Mr E. Tindal Atkinson, to defend them. At the end of the trial the presiding judge, the Lord Chief Justice, Lord Coleridge, remarked to the jury that, 'He was of opinion that the company were seriously to blame for not adopting a different kind of brake after the warnings they had received, and that the suspension of the block system when special care was needed required explanation. If the jury thought that the accident was due to a defective brake, they ought to acquit the prisoners.' [252] The acquittal 'was received with an outburst of applause, which was at once supressed'. Not everyone was pleased though, the Manchester Sheffield and Lincolnshire chairman, Sir Edward Watkin, remarked that, 'It was a misfortune that the Lord Chief Justice should have exonerated the driver and fireman.' [253]

By 1887 absolute block working was in force on most of the British railway system. In the United States such safeguards were still rare, with the safety of a stationary train depending much more on its crew protecting it themselves, and a failure to properly do so often resulted in an accident, as demonstrated by a collision at Kouts, Indiana, on 10th October 1887, on the Erie Railroad.

Express passenger train No. 12 had left the Chicago and Alton Railroad's Polk Street depot in Chicago at 7.45 p.m. that Monday evening in charge of Conductor Parks with Engineer Connor and Fireman Willets, and consisted of two Pullman sleeping cars, a day coach, a fruit car and a baggage car. It was a dark and misty night with slippery rails. The train had barely left Chicago when the engineer discovered that it had broken an eccentric strap, a common fault. He continued the journey, but the train could now only run at a very reduced speed, and by Boone Grove, Indiana, described as 'a dismal prairie station' it was already two hours late. Behind it, and 'thundering along at a great speed' was a heavy refrigerator train (No. 48) carrying tons of fresh meat, its engine manned by Engineer John Dorsey, who reportedly had orders to look out for the passenger train.

After passing Boone Grove Engineer Connor decided that he needed to stop and take water at a tank about midday between there and Kouts. He stopped his train and jumping down from the cab of his engine threw to "Danger" a semaphore signal, situated about half a mile in the rear, to protect his train, but it had been stopped only about two minutes

249 Ibid.
250 *The Times*, London, England, 16th, 17th, 19th and 20th September 1887
251 Ibid, 21st and 28th September 1887
252 Ibid, 13th, 14th and 15th October and 15th and 16th November 1887
253 Wikipedia – Hexthorpe Rail Accident

when train No. 48, running down a steep falling gradient, supposedly 'at a furious rate', crashed into the rear Pullman car, Engineer Dorsey having failed to see the semaphore signal through the fog and mist, and being unable to pull up on the slippery rails when he and his fireman saw the tail lamps of the passenger train ahead. The two men leapt from their engine just before the impact.

The engine of the refrigerator train penetrated as far as the middle of the rear Pullman car and the heavy Pullman car in front was driven into the lighter day coach and 'reduced it to splinters'. The upper part of the engine of the freight train was torn away, its tender was thrown across the track, and the freight cars thrown in every direction, with hundreds of pounds of meat strewn for 20 rods over the prairie.

The passenger train was fitted with Bakers improved heaters, supposed to lessen the risk of an overturned stove setting a train on fire, but the coals of the heater nevertheless set the upholstery of the rear Pullman and in five minutes a small flame had developed into a large fire which soon enveloped the coaches and their screaming trapped passengers. The light of the fire lit the prairie for miles around and attracted the attention of farmers who had not yet gone to bed. Rescuers reached the scene through rain and mud while the but slightly injured train crews were hard at work dragging passengers from the wreck. The newspaper reporter continued that, 'The fire burned with the most appalling rapidity of all railroad fires. Reinforced by the farmers and their sturdy boys the train men set about their work with a will. There was water close at hand, and pailful after pailful was dashed into the crackling mass. The railroad officials declare that there were but twenty passengers on the train, but the men who stood about the wrecks declared that the screams and appeals for mercy were from double that number of human beings. The Pullman, although the first to receive the shock, did not leave the track, but burned where they stood. So far as is known not a life was lost in these coaches. There were but five persons in the berths, and although all were hurled one upon the other, all escaped through the window before the flames burst through the flooring. It was in the day coach that the slaughter occurred. The frail box was smashed to atoms. The sides sprang out and the heavy roof flew into the air and then came down on the score or more of human beings who lay pinned beneath the seats.' [254]

The victims included a Bohemian family from Dundee, Illinois, en route back to Hungary, either because they were fed up with living in the United States or perhaps just going on a home visit. 'The father and husband lay dead with an iron rod through his throat, but the rest of the unfortunate family were still alive when the train men sought to rescue them. Everything was done to drag the children from the pile but without avail. The fire crept nearer and nearer, until finally a cushion of a seat blazed up just above the woman's head. The poor creature tried to speak, but no intelligible sound escaped her lips. Then she attempted to pass one of her girls, a wee mite of a thing, whose white apron was stained with blood, to the rescuers who were crowded about the terrible scene. When the woman saw death as inevitable, she closed her eyes as if in prayer, and wrapping her only free arm around the wrists of the children lying nearest her, awaited the coming of the flames. From this unfortunate family but one member escaped alive. He was one of the sons, Herman by name, a bright lad of fourteen years, who managed to break out of the furnace just as his parents perished. The boy now lies at a hotel in this city {Kouts}. He is crushed about the

---

[254] *The Marion Daily Star*, Marion, Ohio, United States, 11th and 12th October 1887

head, and it is thought he cannot live.' [255] Other stories of a similar nature followed in the newspaper account. The only one of the fatalities not to have died in the fire was thought to have been that of a tramp stealing a ride.

The coroner's jury returned a verdict at the inquest into the victims of the collision that, 'The accident was the result of negligence of the train despatcher in permitting the passenger train to attempt to run with a disabled engine, knowing that the freight train was but a few minutes behind it, and by the negligence of Engineer Dorsey, of the freight train, running his engine at a high rate of speed through a fog, knowing the passenger train was but a short distance in advance, with a disabled engine.' [256]

Three weeks after the accident it was reported that Charles Moore, a telegraph operator in the Chicago and Alton dispatcher's office at Huntington, Indiana, had gone insane. Moore had stated, before he showed signs of insanity, that on the day following the wreck a number of the railroad officials had met in the Huntington office, 'and they, with the telegraph operators and the trainmen, wept in each other's arms as the horrible news came in. The chief dispatcher sickened of reporting each hour additional deaths and horrors and called Moore to the key.' [257] In modern language it would probably be said that Moore had suffered a nervous breakdown.

Early on the morning of Sunday 13th November 1887 a regular southbound freight train on the St Paul Minneapolis and Manitoba Railroad received orders to take the siding at Averill, Minnesota, United States, to allow a following wild (special) cattle train to overtake it. In addition to its normal crew the freight had five section men riding in it, who were sleeping in the caboose. As it was heading into the siding its locomotive broke down and it came to a stand with the caboose still on the main line. The conductor of the freight train walked back down the track to warn the approaching cattle train (the report did not mention how quickly he went back) without disturbing the sleeping men, but the stock train was a heavy one, with 40 cars, and was approaching at over 30 miles an hour. The stock train's engineer had not reduced speed when he neared the siding and when he saw the conductor's signals, he was unable to pull up, and his engine smashed into the back of the freight train, splintering the caboose and wrecking four cars laden with wheat in front of it. The five section men were killed instantly, several of their bodies shockingly mangled. Their names were Charles Engel, Gottlieb Wolf, James Allen, Albert Evans and James Moore. All were single men and most of them were from St Cloud, Minnesota. The cattle train was little damaged.

A northbound express, which was delayed four hours by the accident, took up a jury at Glyndon, seven miles further south, and at Averill its passengers 'saw the crushed bodies of the victims laid out on planks beside the track and heard the account of the disaster.' Coroner Darrow of Moorhead and the jury laid the blame on Engineer O'Laughlin of the stock train for not reducing speed earlier on approaching Averill, and also criticised the despatcher for not allowing the heavy freight train of 51 cars more time to make its run. Nothing was apparently said about the freight train conductor's actions, so he was

---

[255] Ibid.
[256] *Reno Evening Gazette*, Reno, Nevada, United States, 14th October 1887
[257] *The Marion Daily Star*, Marion, Ohio, 31st October 1887

presumably not judged to have been at fault.[258]

The circumstances of the 1887 Kouts tragedy were repeated at Waynesville, Ohio, United States, on the Little Miami Railroad, on 9th September 1888, except that in this case there was no fire. The rear section of John Robinson's westbound circus train, consisting of 35 cars, had left Xenia, Ohio, doubtless after giving a performance there on Saturday evening, shortly after midnight, and at 2 a.m. on the Sunday morning had stopped for water ay the Waynesville tank, when it was run into by a following freight train. The newspaper report stated that there were 'no blocks' presumably meaning operators, between Xenia and Corwin (Waynesville), the safety of the trains depending on orders given at Xenia and the trainmen themselves. The stopped circus train was stated to have been 'protected by danger signals' but whether this just meant its rear lights or that a flagman went back the line was not stated. As the train was taking water its fireman saw the headlights of the freight train coming around the corner and shouted to his engineer, who at once gave his engine steam to try and pull away. Meanwhile Trainmaster Aiken had got out when the train stopped, and he also saw the freight coming, and ran through the first three sleeping cars, shouting, "Get out quick, a collision is coming!" Everyone in these three cars escaped. By the time that he had reached the fourth car the freight was very close, and he had to jump to save his own life.

The rear of the circus train was made up of the sleeping cars *Milford, Madisonville*, *Terrace Park*, *Gravelotte*, and *Walnut Hills*, with a caboose at the very end. The freight engine split the caboose in two and threw it down the embankment and the rear of the *Walnut Hills* car was crushed in and its front driven into *Gravelotte* which in turn was forced into *Terrace Park* which drove the *Madisonville* on top the *Milford*. The *Milford* was 'cut to kindling wood' with nothing remaining but its roof, and it was in this car that the victims were found.

After the collision, 'Trucks were procured and bonfires built, and the work of rescue commenced. A row of mattresses was laid upon the grass at the side of the track, and upon these the wounded were placed. As soon as the collision occurred the doctors from Waynesville were summoned and did all they could for the wounded'.

Ben Ceisceny, canvasman, Frank Smith and John Lacy, cooks, and John Churchill, 'whose right name is Fairbanks', were killed outright and Andy Smith, a contortionist, died at the Betts Street hospital, Cincinnati, from his injuries. (Another report gives the names of those who died at the scene as Frank Smith, Jack Lacy and John Clifford and does not mention Ben Ceisceny.) Seventeen other persons were hurt and treated by doctors but a number of others who were hurt refused surgical examination. There had been 150 people on the train. Five cars of the circus train and the engine of the freight train were badly wrecked. Property loss was estimated at $30,000, divided about equally between the railroad company and the circus.

Engineer Long of the freight train was blamed for the accident.[259] [260]

---

[258] *Winnipeg Free Press*, Winnipeg, Manitoba, Canada, 14th November 1887
[259] *Tyrone Daily Herald*, Tyrone, Pennsylvania, United States, 10th September 1888
[260] *The Elyria Democrat*, Elyria, Ohio, 13th September 1888

The very next day, 10th September 1888, the same thing happened again near Wadsworth, Ohio, on the New York Pennsylvania and Ohio Railroad, the accident this time involving an excursion train carrying 700 members of the G.A.R. (Grand Army of the Republic – an organisation of Union Army and Navy veterans of the American Civil War) and their families from Youngstown to an encampment at Columbus, Ohio. The excursion train was stopped 'when a rod on the engine broke' in a deep cutting and on a curve. A flagman was sent back to warn the following freight train (No. 37) but it was too close behind and coming down an incline at 25 miles an hour was unable to stop. The excursion passengers realised the danger they were in and almost all of them succeeded in getting out of the train but as they were scrambling up the side of the cutting the wrecked cars fell on them, nearly all of those who were killed and injured having been struck by flying debris. Two veterans were killed outright and 25 injured, three of whom died later. Engineer Barney Ballinger (or Conrad) of the freight train likewise jumped from his engine and lost his life when he fell back under its wheels, and Brakeman William Cochran of the freight train also jumped with the same result.[261 262]

The series of rear collisions in the United States in late 1888 was added to on 10th October by a major catastrophe at Mud Run, Pennsylvania, one of the worst accidents in American railroad history, certainly in the 19th century. The 10th October was the day of the annual parade of the Catholic temperance societies of the Scranton diocese at Hazleton and thousands of people travelled to Hazleton from the towns of Luzerne and Lackawanna counties, the Lehigh Valley Railroad putting on eight excursion trains, 'their cars filled to the doors'. All the excursions reached Hazleton safely and they began to depart on the return journey at about 7 p.m. The line was still being worked on the time interval system, with trains normally being despatched at five-minute intervals, but the interval was increased to 10 minutes in the hope of increased safety. All the trains were operated as sections of a single train in the usual American fashion. At about 8 p.m. the sixth section was stopped by signals at Mud Run station, at the end of the double track section. The train initially stopped at the station and then 170 feet beyond it. It was crowded and many of the passengers were somewhat the worse for drink, despite the fact that it was supposedly run in connection with a temperance event.

The following seventh section (the newspaper reports are very confused as to the number of sections, some reporting a collision between the third and fourth sections and others that the fifth and sixth sections were involved, but the inquest into the accident states that it was the sixth and seventh sections) was double-headed, with locomotives Nos. 452 and 466. The railroad here followed the bank of the Lehigh River, and both river and therefore railroad swept around to the left at the point of the collision, but there was a clear view across the bend of the river from the left side of the engine cab, while as an additional safeguard brakemen were riding on both engines specially to look out for signals and obstructions. The heavy pilot engine, No. 452, was fitted with steam brakes but the train engine, No. 466, controlled the train brakes, with communication between the two by whistle.

As the seventh section approached Mud Run station Brakeman Pohl on the pilot

[261] Ibid.

[262] *The Massillon Independent*, Massillon, Ohio, 14th September 1888

engine told Engineer Henry Cook that the road was clear, having seen the station signal showing white. About the same moment Fireman Murphy on the train engine saw that both the station and switch signals were clear and told his engineer, named Major. Engineer Major then blew two blasts on the whistle as a clear signal to Cook.

The rear brakeman of the sixth section, Hannigan, had gone back to protect his train, but only to the end of the station platform, 400 feet in its rear, and he stood there with his red lamp chatting to people on the platform. The lantern was visible from the approaching train, but less conspicuous than the station signal, and neither Mulhearn, Major's brakeman, nor Fireman Gallagher, who saw it about five telegraph posts away, realised that it was a danger signal until Hannigan, realising that the seventh section was not pulling up as he expected, began to wave it violently about. Fireman Gallagher then shouted to his engineer to reverse his engine, but the heavy train could not be stopped. The train engine ploughed 20 feet into the last car of the sixth section and that car was pushed violently forward and telescoped the car in front. There were about 200 passengers in the two cars, of whom 64 lost their lives.[263]

A newspaper reporter writing that evening from Wilkes-Barre stated, 'The rumour of a terrible accident near Penn Haven is confirmed. The fifth {sic} division of the Father Matthew excursion returning from Hazleton was wrecked and the cars piled in a shapeless mass, one being stood upon its end. All the cars were crowded, and the fatalities will run into awful figures. All the physicians from White Haven and nearby points are at the wreck, as also the Lehigh Valley officials from this city. The number of killed is variously reported from twenty-five to eighty persons. The only information is from passengers on the Central railroad trains, who report a horrible condition of affairs, the groans of the wounded being heard across the river where the tracks of the Central run. At the depot here hundreds of people are gathered on the platform, many weeping for the safety of their friends supposed to be on the wrecked train.' [264]

The coroner's jury at the inquest into the disaster passed their verdict at Mauch Chunk, Pennsylvania on 23rd October. They found that, "the engineers of locomotives 452 and 466, of the seventh section, were guilty of gross negligence, first in failing to discover the red signal in time, the evidence clearly showing that this signal was in full view as their train approached; and, secondly, for not approaching the station under full control as required by both the general and special orders. The evidence also shows that the air brakes of the entire seventh section were under the control of the engineer of locomotive 466, and that he could have stopped the train in spite of the locomotive ahead, No. 452.

"We find that the lookout men of Locomotives 452 and 466 were guilty of gross negligence in failing to report to their respective engineers the red light at the station as the train approached. These men were placed on their respective engines as an additional precaution, their special duty being to lookout for signals. The lookout man on No. 452 failed to see the signal. The other one on No. 466 testified that he saw it when about fifteen hundred feet from the station, and yet he reported all right to the engineer and claims that he did not think the red light meant anything, as nobody used it.

---

263 *A History of Railroad Accidents, Safety Precautions and Operating Practices* by Robert B. Shaw pgs. 87-8

264 *Tyrone Daily Herald*, Tyrone, Pennsylvania, United States, 11th October 1888

"We find that the rear brakeman of the sixth section was guilty of gross negligence, for when his train stopped, instead of promptly going back the proper distance to warn the approaching train, he stopped at the station. There was enough time for him to have gone back nearly half a mile, but he went back less than four hundred feet.

"We find that the conductors of the two sections failed in their duty. The first in not conforming to the rule requiring each conductor to see personally that his brakemen protect the rear of his train, and the second in not requiring his train to approach the station under control." [265]

Following the inquest Engineers Cook and Major and Brakemen Hannigan, Pohl and Mulhearn were all arrested on a charge of manslaughter. No action was taken, and indeed no criticism seems to have been made, of the Mud Run operator, J.W. Slocum, for leaving his train order signal showing "Clear" despite the sixth section of the train standing directly in front of it. In doing so he was technically correct as the signal was only used to indicate to train crews if the operator held any orders for their train, which he did not, but the result was the misleading display of a white clear signal at the same time as the brakeman was showing a red "Danger" signal. The operator was also required to regulate the intervals between trains at stations under the time interval system for which the use of the train order signal would seem the obvious method, but this apparently was not the practice on the Lehigh Valley. In Great Britain, for example, where the train order system was not used, such an anomaly would not have occurred, and if an official at a station had left his signal showing "Clear" with a train stood just in front of it he would most certainly have been guilty of gross negligence.[266]

What was relatively a minor accident compared to Mud Run occurred near Yellow Creek, Illinois, United States, on 18th October 1888, on the Chicago St Paul and Kansas City Railroad. Freight train No. 97 was running in two sections and the first section stopped, apparently because of an engine problem. Inadequate measures were taken to protect it in the rear and the second section ran into the caboose, killing three of its six occupants, James Orr, John Brown and Edward Hickey, and injuring two of the others. The three men killed were all stockmen in charge of livestock being carried on the train.[267] [268]

Residents of Forest Lawn, New York, United States, on the shore of Lake Ontario, who came into the city of Rochester to do business in 1889 usually caught a "stub" train of the Rome Watertown and Ogdensburg Railroad from Forest Lake west to Windsor Beach. On the morning of Saturday 10th August 1889 this "stub" train was standing at Forest Lawn station taking on passengers when it was run into by the Thousand Islands express, due out of Oswego at 4 a.m. The rear car of the "stub" train was completely telescoped in the collision, car and colliding engine were thrown up onto the station platform and 'the coaches of the express were thrown into a ditch and piled up in all

[265] *Newark Daily Advocate*, Newark, Ohio, United States, 23rd October 1888

[266] *A History of Railroad Accidents, Safety Precautions and Operating Practices* by Robert B. Shaw pg. 89

[267] *Mitchell Sunday Republican*, Mitchell, Dakota Territory, United States, 21st October 1888

[268] *The Ohio Democrat*, New Philadelphia, Ohio, United States, 25th October 1888

conceivable shapes.' The collision occurred at about 8 a.m.

Emma Perrin, aged 23, was killed outright and John Jay and Andrew Tiffany, engineer of the steamer Hazleton, of Oswego, were fatally injured. Seven other passengers were injured, some seriously. A special train was immediately despatched with the injured on cots to Rochester, where they were taken to the city hospital. 'An affecting scene occurred when Mrs Perrin, mother of the girl who was killed, asked for her daughter. She did not know her daughter was dead, and it was not thought wise to inform her, so she was told that her daughter had sustained injuries and could not get to her. Henry H. Perrin is 60 years old, and the shock, upon learning the sad death of his daughter, may prove fatal.'

As to who was responsible for the collision the newspaper wrote, 'One report has it that the Forest Lawn train had no flag out; another that the express had orders to run no further than Forest Lawn. The express was late. The railway officials here are very reticent and not inclined to give out any particulars.' [269]

St Louis and Chicago express No. 5 of the New York Central and Hudson River Railroad, which left Albany at 10 p.m. on 27th September 1889, met with an accident about two miles east of Palatine Bridge, New York, United States, at 11:40. The train was running in two sections, and the first section, which had left Fonda, about 10 miles east of Palatine Bridge, at 11.21 p.m., 10 minutes late, broke down and was stopped for repairs, Engineer Weeks 'having noticed a giving out of the steam chest'. The trains usually ran 10 minutes apart. A flagman was sent back to protect the failed train but Engineer Worth (or Horth in some newspaper reports), on the locomotive of the second section, said, according to an early newspaper report, that he did not see him and the first thing that he saw were the tail lights of the first train. The accident happened on a blind curve.

The first section consisted of baggage, mail, express and through passenger car, the Wagner sleeping car *New Mexico*, the *Kankakee* private car of President Ingalls of the Cleveland Cincinnati Chicago and St Louis Railroad and at the rear the private car of President H.B. Ledyard of the Michigan Central Railroad. The rear car was 'said to be the strongest and most perfect car ever constructed'. The crash was described as terrific. The *Kankakee* telescoped the packed Wagner sleeping car ahead of it. Porter Charles Franklin in the *Kankakee* was killed but everyone else in the car escaped serious injury. Those in the upper berths of the Wagner car escaped with slight injuries but in the lower berth recently married William Manning, setting off with his wife for a new home, the Mannings' maidservant Sadie Boyd and the Rev Deveuve were killed, Mrs Manning being one of those injured. All the lights in the cars were extinguished, leaving everything in darkness. The initial report on the accident stated that, 'Relief trains went to the scene. Up to this hour (2 a.m.) four bodies have been taken out of the sleeper and it is feared the total number of deaths will run up to twenty-five, and possibly more, as the car was full.' The initial fears of many fatalities were not realised, no more passengers seem to have died although a number were badly hurt. Engineer Worth was seriously injured, and it was feared that he would not survive the night. All four tracks were blocked.[270]

A further report in the same newspaper suggested that the trains were running rather

[269] *Semi-weekly Age*, Coshocton, Ohio, United States, 13th August 1889
[270] *Salem Daily News*, Salem, Ohio, United States, 28th September 1889

less than 10 minutes apart and quoted Engineer Worth, who had obviously survived the night, as saying that 'he made every effort to stop by applying the air brakes, but they did not work, and nothing was left for him to do but brace himself for the crash.' He was thrown from his engine and suffered two broken legs and other injuries. His fireman, John Slater, was thrown 20 feet into the air and landed on top of a Boston and Albany Railroad baggage car, which had half telescoped into the colliding engine. Baggageman Wilcox of Syracuse was thrown from one end to the other of the baggage car and suffered a head injury but was the first to rescue the fireman, who was badly shocked but not greatly hurt.[271]

Another account added that 'There was not a surgeon or physician on the train, and it was only after a long search that some brandy and a bottle of whisky were procured for the injured. Messengers were at once despatched to Palatine Bridge for assistance, and every physician was awakened and dispatched to the scene of the disaster as soon as conveyances could be secured.'

A report in a different newspaper, which is probably more reliable in suggesting the cause of the accident, states that the engine of the first section, No. 714, was disabled by the bursting of a cylinder head, but that Engineer Weeks, instead of stopping at once so that the flagman had ample time to run back and stop the second section, tried to keep going to the next station with the result that the second section gradually caught up with him, and by the time he did stop the second section, travelling at 30 miles an hour, was close behind. This paper adds that Conductor Abel stated most positively that he had ordered the rear brakeman to go back and flag but that the engineer of the second section denied having seen the flagman at all.[272] As is often the case the truth of the matter and who exactly was at fault are very hard to determine.

A bad accident happened near Cincinnati, Ohio, United States, on the evening of 17th January 1890. As the Glendale accommodation train on the Cincinnati Hamilton and Dayton Railroad, leaving there at 6.25 and due in Cincinnati at 7.5, was just leaving Winton Place station, near College Hill Junction at about 6.54, 12 minutes late. it was run into the Chicago vestibule train (No. 31). The accommodation train consisted of three coaches with about 75 passengers. The engine of the Chicago train smashed halfway through the rear car, wrecking it and setting it on fire, and the penultimate car was also badly damaged. Conductor F.W. Witherbee and two passengers, John Wilson of Cincinnati, superintendent of the Metropolitan Insurance Company, and an initially unknown man, later identified as Charles Keller, of Winton Place, were killed outright, and baggagemaster James Staley, who was terribly burned, and another passenger, William Klamits, died later in hospital. The engineer, William Coakley, of the Chicago train were injured. The Cincinnati fire department were called and were able to extinguish the blaze.

A newspaper stated that 'The block system is used there, and it is supposed that No. 77 was not out of her block when No. 31 entered.'[273] [274]

---

[271] Ibid, 30th September 1889

[272] *Tyrone Daily Herald*, Tyrone, Pennsylvania, United States, 30th September 1889

[273] *Newark Advocate*, Newark, Ohio, United States, 18th January 1890

[274] *Chillicothe Morning Constitution*, Chillicothe. Missouri, United States, 19th January 1890

A more detailed account of how the accident happened appeared in the press a couple of days later. 'General Superintendent Nelson gave the following account of the accident. 'The vestibule train was due here {Cincinnati} at 7.10 and the accommodation at 7.5. The operator at Elmwood signalled the express train that the block was clear, and the train proceeded on its way. It was drawn by a Strong locomotive in charge of engineer Coakley and an experienced engineer in the employment of the Strong company named George Magee. Just before the train reached College Hill Junction it rounded a curve and the engineer saw the accommodation a short distance ahead of him. Coakley slacked up the train and jumped and escaped with slight injuries and Magee remained on the engineer until it struck the rear car of the accommodation train but was not hurt. The express train was not damaged in the least and, with the exception of Coakley, no one was injured. Those killed and injured were on the accommodation train, which was almost completely wrecked. The rear car was a combination smoker and this, with two coaches, caught fire and {they} were reduced to ashes.' [275]

The fire was apparently caused by the overturning of a stove. The station agent at Winton Place had placed his signal against the express but it was then too close to the accommodation train to make much difference.

Torpedoes (detonators) laid on the track just before Shoemakersville station, Pennsylvania, United States on the Philadelphia and Reading Railroad were in a sense the initial factor in a disastrous double collision on 19th September 1890. Engineer A.B. Gehry, in charge of a southbound 150 waggon coal train, which was already running at a fairly low speed, shut off steam and let his train drift through the station after exploding them. When he opened the regulator again and began to take up the slack the train parted and the automatic brakes came into action, bringing both parts to a halt.[276] Flagman Henry Kemp unfortunately was not quick enough in going back to protect the rear of the disabled train and before it could be recoupled and restarted another coal train, with James Vail at the controls of its engine, had run into its rear, throwing coal trucks onto the opposite running line. The northbound Pottsville express, in charge of Conductor Harry Logan with Engineer John White and Fireman James Templin, which had left Reading at 6.5 p.m. was close at hand and ran into the wreckage before it could be warned. The engine and cars of the express were thrown down the embankment and five cars ended up in the Schuylkill River, some of the passengers losing their lives by drowning. Fire broke out in that part of the wreckage which did not go into the river but was extinguished with buckets of water and does not seem to have played a major role in the disaster.[277] Surviving passengers spread news of the accident through the neighbourhood and assistance was telegraphed for from Reading. 'Physicians and surgeons and a force of 300 workmen were taken to the spot by the company, and the work of clearing away the wreck was at once commenced.' The job was a slow and difficult one but by 2 a.m. the next (Saturday) morning 15 bodies and 30 injured had been recovered. Some of the injured were taken to Reading and others to the miners' hospital

---

275 *The Elyria Democrat*, Elyria, Ohio, United States, 22nd January 1890

276 *A History of Railroad Accidents, Safety Precautions and Operating Practices* by Robert B. Shaw pgs. 209-210

277 *Train Wrecks of Eastern Pennsylvania* by Charles J. Adams III and David J. Seibold, pgs. 85-94

at Ashland. Those killed included George B. Kaercher, a prominent Pottsville railroad lawyer who also had an office in Philadelphia, and William D. Shone, described as 'one of Reading's wealthiest citizens'. Engineer White, Fireman Templin and Conductor Logan were also among the dead. Twenty-one people seem to have lost their lives in the accident (Shaw states 23) and about 30 were injured.

A surviving passenger detailed his experiences to a newspaper reporter. "The train was going at a lively rate of speed. The passengers appeared a happy crowd, many of them ladies, chatting and laughing after a day's pleasure at the Berks county fair. I was viewing the country through which we were passing when there was a terrible crash. I was hurled from my seat, while the cars rolled down the twenty-feet embankment, and I was thrown from one end of the car to the other. One end of the car went into the water and I was thrown against the side of the car with a force that partially stunned me. I quickly recovered myself and managed to climb upon the seats on the side of the car which lay against the embankment. I was a prisoner in the car, unable to get out. Around me were human beings struggling in the water, screaming in their fright, and some almost dragged me back into the water again. A few saved themselves as I did, and the remainder struggled in the water and then quietly sunk out of sight." [278]

The Jury at the coroner's inquest into the victims of the Shoemakersville disaster returned a verdict that, 'In our opinion the cause of the accident was due to the negligence of the signal man of engine 847, who remained in his caboose while the train was slowing up instead of getting off and placing signals in such a manner as to enable the engineer of the train following, engine 921, sufficient time to bring his train to a stop. As a result, 921 crashed into the rear of the train drawn by 927, throwing the caboose and two eight-wheeled coal cars onto the northbound track, resulting in this great loss of life. It was shown by the testimony that the collision of the coal and that of the express train with the wrecked cars occurred about the same time." [279] Flagman Kemp did not have to indulge in the risky procedure of getting off his train while it was still moving but losing time as he could have thrown down flares to warn the following train. He was also slow in going back after his train had stopped as there were 10 minutes between the stoppage and the initial collision.

Only two days after the Shoemakersville disaster the United States had another bad rear collision, this time on the Chicago Burlington and Quincy Railroad line on the outskirts of Chicago, Illinois. At 8 p.m. on the evening of Sunday 21st September 1890 a Chicago Burlington and Quincy Railroad Downers Grove express ran into an excursion train near 18th Street. The excursion train was the last of four sections and consisted of 15 cars, all well filled. The Illinois Central Railroad had run the excursion trains to the orphan asylum at Addison. The train was carrying the excursionists back to homes in the city and was held up in traffic when it was run into. It had been switched to Chicago Burlington and Quincy Railroad tracks at Clyde, had stopped at Douglas Park station and was backing when the collision occurred. The impact telescoped the heavier rear coach of the excursion train into the lighter coach in front of it 'like an egg shell' for 10 feet.

Initially it was reported that there were four fatalities, but a revised list named Lily

[278] *Davenport Morning Tribune*, Davenport, Iowa, United States, 21st September 1890
[279] *The Fitchburg Sentinel*, Fitchburg, Massachusetts, United States, 23rd September 1890

Dierner, Martha Dierner, Otto Scholef and two unidentified men as having been killed and Lena Riesinger, Minnie Pilgrim, Eddie Turner, aged 16 and Tillie Burk as fatally wounded, together with two others less seriously injured.

One newspaper wrote that, 'No sooner had the accident happened than the conductor of the Illinois Central train uncoupled the telescoped cars and with the rest of his train drew into Chicago. The police at once sent word to the Central station, ordering the arrest of all the train hands, but only the conductor and brakeman could be found, the fireman and brakeman having fled.' Another newspaper report added that, 'One of the brakemen and the conductor {Carrington} of the passenger train were arrested. They refused to talk.' A third report stated that Engineer Dickson (or Dixon) of the Burlington train had also been arrested and had asserted that he had seen no lights on the excursion train until he was upon it, because they were burning so dimly. Contrary reports had it that the colliding train was behind time and was running at a reckless speed. The engineer and his fireman had jumped off when they saw that a collision was unavoidable and were unhurt. Fireman Jack Walsh and Conductor Ray of the Burlington train were also arrested. No one on the colliding train was hurt and the train, as opposed to its locomotive, was barely damaged [280] [281] [282]

A morning Atlantic and Pacific Railroad passenger train crashed into the caboose of a freight train near Needles, California, United States on 11th December 1890. Three men were killed, and five passengers fatally hurt.[283] This accident seems to have received very little media attention, probably because it happened out in the far west.

The United States suffered another major disaster due to a rear collision on 3rd July 1891.

Eastbound New York Lake Erie and Western Railroad train No. 8, a fast vestibule express for New York, due at Youngstown, Ohio, at 2.55 a.m., was running very late, and stopped at Ravenna station, Ohio, for orders, while its engineer and fireman took the opportunity to try and fix a faulty whistle. A flagman was sent back, but it was said that he only went a few feet, at the most two car lengths. Suddenly train No. 82, a freight train of meat cars, came around the corner, coming down an incline at 25 miles an hour. Its engineer whistled for brakes and reversed his engine, but the train's speed was very little reduced before the collision and there was no chance to warn the passengers in the front train. The engineer and fireman of the freight train jumped just before the collision. At the rear of the express was a day coach carrying workers from the Richardson Glass Works at Findlay, Ohio, who were returning home to Corning, New York, for the Independence Day celebrations. The day coach was telescoped in the collision and 19 of the workers were killed and another 12 injured. One of the dead men, Willie Kain, a glass blower, of Corning, had $300 on him which he was bring home as a surprise for his widowed mother. Following the collision, the day coach and the two sleeping cars next to it caught fire and were consumed, along with the bodies of the victims. Six other people

---

[280] *Sandusky Daily Register*, Sandusky, Ohio, United States, 22nd September 1890
[281] *Chillicothe Morning Constitution*, Chillicothe, Missouri, United States, 22nd September 1890
[282] *Hornellsville Weekly Tribune*, Hornellsville, New York, United States, 26th September 1890
[283] *The Fitchburg Sentinel*, Fitchburg, Massachusetts, United States, 13th December 1890

also died in the accident.

Superintendent Mattison, who investigated the accident, blamed it on the carelessness of Conductor Boynton of the passenger train and the rear flagman, who was the conductor's son. The younger Boynton, he alleged, had instead of going back just leaned against the rear car of the express until the meat train had come into view, by which time it was too late.[284] There was, it would seem, however, more to the accident than that. The line was still worked by the time interval system, with a permitted minimum interval between trains of five minutes, but the Erie rules did not require a flagman to go back to protect his train until it was two minutes late. In fact Engineer Halman of the freight train had left a siding east of Kent station, rather less than seven miles from Ravenna, only four minutes behind the passenger train and the speed recorder fitted to his engine showed that he had run at 20 miles an hour for the first three miles from Kent, then accelerated to 28 miles an hour for a mile and then to 30 miles an hour until his final attempt to pull up. He was of course aware that the express was in front of him but having been issued with an order to cross an opposing train at Freedom, beyond Ravenna, seemed to have got it into his head that he had a clear road to there. Approaching Ravenna he had first seen the train order signal showing "Clear" before he saw either the flagman's lamp or the brake lights of the train. Then, he pleaded, his brakes had seemed sluggish and the engine sand was caked. The Erie regulations apparently required the train order board at a station to remain at "Clear" until the rear of a train had passed 300 feet beyond it, so that the conductor and brakeman could satisfy themselves that there were no orders for the train, even though this meant that the signal would be standing at "Clear" with a train stopped in the station, potentially misleading the engineer of a closely following train, as here and at Mud Run in 1888 (qv).[285] The lack of a block system and the use of train order signals were of course distinctive features of the North American lines as opposed to say British ones.

The inquest on the victims of the Ravenna disaster heard testimony from S.C. Arnold, the operator at Kent, who stated that the express had only been 200 feet ahead when the freight started off, and from Barney Dyer, signalman at the Cleveland and Pittsburgh Railroad rail/rail crossing, about three miles short of Ravenna, who swore that the freight had only been a mile behind the express at that point.[286]

On 14th July M.J. McInarna, Ohio state inspector of tracks and bridges, reported to Railroad Commissioner Norton on the accident. He found the engineer, conductor and brakeman of the freight train guilty of gross negligence in not having their train under control and found that Fred Boynton, the flagman of the passenger train, had used poor judgement, and was guilty of neglect of duty in not dropping a fusee on the track as a sign of danger.'[287]

The report of Coroner Sherman, who conducted the inquest, was not published until 6th August. It was very lengthy, containing 4,000 words. Engineer Prendergast of the express was found guilty of carelessness, Conductor Young of the freight was severely criticised, Engineer Holman was censured, and Watchman Barney Dyer was found guilty

---

[284] *The News*, Frederick, Maryland, United States, 6th July 1891

[285] *A History of Railroad Accidents*, Safety Precautions and Operating Practices by Robert B. Shaw pgs. 90-1

[286] *The Salem Daily News*, Salem, Ohio, United States, 8th July 1891

[287] Ibid, 14th July 1891

of gross neglect! Conductor Boynton and his son Fred, who were surely much more to blame than Barney Dyer, were exonerated. One wonders if the coroner, presumably not a railwayman, was aware of the rather different conclusions of M.J. McInara. [288]

A railway accident in France equalled in magnitude up to this date only by the derailment at Meudon (Versailles) in the very early days of railways occurred at Saint Mandé, a Paris suburb, on the evening of 26th July 1891 on the Chemin de fer de L'Est. At 9 p.m. a passenger train of 23 carriages left Joinville-le-Pont, on the Vincennes line, for Paris. And after making several stops it arrived at Saint Mandé station, 'just outside of the Paris fortifications'. What was variously described as a fair, a fête, and a musical festival had been in progress at nearby Fontenay-sous-Bois, about a kilometre from Saint Mandé station, and its platforms were thronged with Parisians waiting for a train to take them back into town. When the 9 p.m. train arrived there was a rush for seats, 'as always happens on Sundays on the suburban line', the *Times* correspondent in Paris noted. The fortunate passengers, as it soon turned out, were those who had resigned themselves to waiting for the next train, but the scramble delayed the departure of the service, and further delay was occasioned by the appeals of two women in a ladies-only carriage to the conductor for some men who had boarded the carriage to be ordered out, and the refusal of the men to leave because there was no room elsewhere on the train. While this dispute was being settled, and just as the train was finally about to depart, it was run into by the following train, also from Joinville-le-Pont, which crushed the brake van and a first class and two second-class coaches, filled mostly with women and children, while the engine of the colliding train left the rails and overturned.

The *Times* correspondent in Paris gave a vivid description of the scene after the collision. 'There were piercing cries, and the occupants of the other carriages, feeling a terrible shock, got out in a panic. The engine fire of the second train set fire to the brake van and the hindmost carriage, and the occupants of them who were not crushed to death were burnt or suffocated by the flames. The uninjured passengers and the people waiting on the platform crowded round the crushed and burning carriage, members of families scattered in different parts of the train being in an agony of suspense as to the fate of husbands, wives or children. A party of police and soldiers, after some delay, arrived and commenced operations by removing the roof of the hindmost carriage. Bodies crushed or burnt almost beyond identification were taken out and conveyed to the Town-hall and to the school, followed by an agonised crowd. The work of taking out the killed and injured went on till 4 o'clock amid heartrending scenes. One of the first bodies was that of a woman with a bunch of wild flowers, which were covered in blood. Several of the bereaved are said to have lost their reason. In some cases children perished while their parents, though in the same carriage, escaped almost uninjured; in other cases the reverse was the case. Fourteen of the bodies have yet to be identified. The injured have been conveyed to the Paris hospitals. It is hoped that most of them will recover.' [289]

The next day the correspondent wrote, 'It was not till a late hour that the extent of the disaster became generally known in Paris, but before midnight the news had attracted an immense crowd to the scene. People flocked thither in thousands from Faubourg St

---

[288] *Tyrone Daily Herald*, Tyrone, Pennsylvania, 7th August 1891
[289] *The Times*, London, England, 28th July 1891

Antoine, Menilmontant and other eastern quarters of the city. Among them were many who had awaited, in vain the return of friends and relatives from the fête at Fontenay-sous-Bois, and who, on learning of the accident, had set out in search of them. By daybreak it is estimated no fewer than 20,000 persons had collected in and around the station. The local schoolroom and the Town-hall, to which the remains of the dead had been removed, were besieged with the distracted families of the missing. Inside these buildings a constant succession of the most heartrending scenes occurred. The bodies, charred and mangled in a manner impossible to describe, were laid out in rows on tables and on the floor. By the dim light afforded by three or four lamps the work of identification was carried on under the direction of the doctors, aided by half-a-dozen policemen. Nearly all the victims taken to the Town-hall were fearfully scorched and blackened by the flames, the bodies in some instances having lost all semblance of humanity. The members of the Municipal Council were in attendance and rendered great assistance in connexion with the disposal and identification of the dead. It was in the first-class carriage that the fire raged most fiercely, and from this coach at least ten bodies were recovered, so carbonized that their identification is impossible. They were taken to the schoolroom, whither 28 bodies had already been conveyed. Of the injured persons taken to the military hospital at St Mandé, the St Antoine Hospital, and the St Louis Hospital, a number died almost immediately after their admission, and many more will probably succumb.' [290] The final number of dead was probably 52, with over 150 injured.[291] Fifty-one of those killed were passengers, one seems to have a rescuer, for the *Times* correspondent reported on 28th July, 'During the work of rescue an Artillery Lieutenant climbed on to the top of one of the burning carriages in order to save a young woman who was travelling on the "imperiale", when the roof suddenly gave way and both the young woman and the gallant officer disappeared into the blazing interior of the carriage. Nothing was afterwards found of him except his sword and portions of his uniform.' [292]

A searching inquiry was begun to determine how the accident had happened. It would appear that the line was still worked on the time interval system, though it was also stated that the Vincennes assistant stationmaster (M. Deguerois) should not have let the second train leave there as St Mandé had not telegraphed back that the line was clear. St Mandé station was at any rate protected by fixed signals, which Driver Caron of the second train overran. After the collision the driver asserted that he had been the victim of an act of revenge, and that someone had tampered with the Westinghouse brakes, by means of which he would otherwise have been able to pull up his train on seeing the tail lights of the one in front.[293] There were also assertions that someone among the crowds on the platform at St Mandé had interfered with the signal wires 'for amusement' putting the signal to clear. The brake van of the first train, which according to the regulations should not have carried any passengers, was crammed full of them 'but regulations are habitually violated on Sunday nights.' [294]

On 26th September the Seine Criminal Court found Driver Carom guilty of negligence

[290] Ibid, 28th July 1891
[291] Ibid, 28th, 29th and 30th July 1891
[292] Ibid, 29th July 1891
[293] Ibid, 28th July 1891
[294] Ibid.

in having run past the St Mandé signals and Assistant Stationmaster Deguerois guilty of having permitted the second train to leave his station without having informed the driver that the line was blocked. They were sentenced to respectively two years' imprisonment and a fine of 500 francs and four months' imprisonment and a fine of 300 francs.[295]

On 31st July 1891 five persons reportedly died in a rear collision near Galion, Ohio, United States but the accident seems to have been barely mentioned in the press.[296]

Fire was a feature yet again of a disaster on the West Shore Railroad near Montezuma, New York, United States, at about 7 a.m. on 6th August 1891. A 40-car westbound freight train, in charge of Conductor Thomas Tobin and Engineer Thomas Whitcomb, was taking the siding at Montezuma, rather more than three miles west of Port Byron, out of the way of train No. 3, the *St Louis Express*, which was close behind. As it was moving off the main line a coupling broke, whereupon the conductor sent Brakeman Edward Connolly back with a lamp to stop the "Fast". Whether the flagman did not go far enough back or whether his signals were not seen in a fog was not immediately clear but as the freight was finally getting clear of the main line, its parts having been recoupled, the express crashed into it with disastrous results. The smoking car of the express, a less sturdily built vehicle perhaps than those on either side of it, was wrecked and all but one of the fatalities in the accident occurred among the Italian labourers travelling in it. Of 21 labourers in the car all but one was killed or badly hurt. Ten of them were killed outright and several others, probably four, died later from their injuries. Fireman Michael Burgen of the express was the other fatality while Engineer Patrick Ryan was seriously injured. The only casualty on the freight train was the conductor, who was standing on a car and was thrown to the ground.

Fire broke out after the collision and eventually consumed the two express cars, baggage car, smoking car, day coach and the *Marathon*, one of the four sleeping cars at the rear of the train, though fortunately the fire did not spread quickly enough to burn anyone before they could be released. The three sleeping cars at the rear of the train were pushed out of reach of the flames. 'John F. Bowdish, of Boston, was so pinned in by the debris that he could not escape. He was fully conscious of what was going on around him and knew that every effort was being made to rescue him, but at the same time he could see the flames breaking out in the wreckage, and there were many times before he was taken out that it was thought the flames would reach him before the rescue. The experience was thrilling in the extreme.' [297] 'Thrilling' was probably not the word that Mr Bowdish had in mind.

Twelve cars and the caboose of the freight train were thrown off the line and also caught fire. No one in the rear of the train was hurt and the attendants in the baggage and express cars had retreated there to snatch a little sleep and inadvertently probably saved their lives. The Italian labourers had come from Tomkins Cove New York and were on their way to Niagara Falls 'to work on the tunnel there' – doubtless in connection with the development of the hydroelectric plant at the falls which had just begun at this date.

---

295 Wikipedia – Liste des accidents ferroviaires en France au XIX siécle
296 *The Stevens Point Journal*, Stevens Point, Wisconsin, United States, 1st August 1891
297 *The News*, Frederick Maryland, United States, 7th August 1891

'They were in charge of an interpreter, Savarino Scozzavana.' [298]

The conductor and flagman were subsequently (11th August) charged with criminal negligence in connection with the accident. They pleaded "Not guilty". If the flagman did indeed go back far enough quickly enough in view of the fog, he should presumably have laid torpedoes on the line, not just relied on his lantern.

Having suffered the overwhelming catastrophe at Mönchenstein, by far the worst accident in Swiss railway history, only a few weeks earlier, the Jura-Simplon Company was unlucky enough to suffer another major disaster at Zollikofen, Switzerland, on 17th August 1891.The company was dealing with heavier than usual traffic at the time of the accident, with extra excursion trains being run in connection with fêtes at Berne, six miles from Zollikofen. One of these early morning excursion trains from the Biel line was brought to a stand short of the junction at Zollikofen to give priority to another train coming off the line from Burgdorf, which joined the Biel line at Zollikofen. While the excursion was still standing at this point a Calais-Berne express was, according to the initial account of the accident, mistakenly allowed to section from Münchenbuchsee, the preceding station to Zollikofen on the Biel line. The driver of the express saw the rear of the excursion train when he was some distance away and applied his brakes, while 30 or 40 passengers in the last carriage of the excursion train had time to jump out and save themselves, but even so the result was disastrous enough. Thirteen people in the second and third coaches from the rear, 11 of them women, were killed on the spot, 24 others were seriously injured and many other were slightly injured or shaken. The rear carriage and a van of the excursion train were smashed to pieces and the express engine was badly damaged. In the usual manner of such accidents the passengers on the international express escaped with a severe shaking.[299] One of the injured passengers died of her injuries bringing the final death toll to 14. 'The victims were principally tradespeople from Biel coming to Berne for the fêtes.'

The escape of the passengers from the rear vehicle of the first train was attributed by a surviving passenger to the fact that it was basically a covered goods van provided with seats, and with two large sliding doors, through which it was possible to quickly escape. Had it instead been one of the long third-class carriages commonly found on Swiss lines at the time, with very narrow corridors and small doors and windows, the accident would probably have been much worse. The passengers in the second and third coaches from the rear also saw the express coming but could not get out in time.

The *Times* correspondent in Lucerne noted the day after the accident that, 'The whole service seems, however, to have been deranged yesterday, and the company, which has had little or no experience of dealing with large masses of people such as have flocked into Berne during the last few days, was evidently incapable of dealing with them. Last night Berne Station was in a state of complete chaos, notwithstanding the fact that hundreds of visitors were afraid to return home on hearing the news of the accident, and to-day trains are still running much behind time.'

The initial report that the blame for the accident lay at Münchenbuchsee is confused by a statement the following day that, 'so far, it has not been possible to determine who is

[298] *Salem Daily News*, Salem, Ohio, United States, 7th August 1891
[299] *The Times*, London, England, 18th August 1891

to blame for the accident. The stationmaster at Zollikofen says that the signals were against the express, whilst the driver, who is much injured, maintains that they showed the line clear.'[300]

An accident occurred at Toledo, Ohio, United States, on the evening of 28th November 1891. The Flint and Pere Marquette Railroad reached Toledo by means of running powers of the Lake Shore and Michigan Southern route. On the outskirts of the city the railroad passed under a canal in a tunnel, and there was a signal at the east end of the tunnel. An eastbound Lake Shore Chicago and Boston express was stopped at this signal when it was run into from behind by a Flint and Pere Marquette express heading for the Union depot and travelling fast. The colliding train penetrated halfway through the day coach attached to the rear of the Lake Shore express, provided for way passengers, tearing up its seats and floor. In addition to their injuries the unfortunate passengers in this car were scalded by steam from the boiler of the locomotive of the Pere Marquette train.

The trains were due at Toledo only two minutes apart and the Lake Shore express was reportedly running behind time. The initial report on the accident said that the engineer of the Pere Marquette train saw the rear lights of the Lake Shore express while he was in the tunnel, shut off steam and seeing that a collision was unavoidable he and his fireman jumped. This report spoke of two dead and a number seriously injured.[301]

By 1st December eight passengers had died and an inquest had begun in which District Superintendent Whittlesey of the Lake Shore testified that trains were required to run at an average speed of 30 miles an hour through part of Toledo yard. Engineer Lyons and Fireman Schott of the Pere Marquette train were also examined and gave evidence that they had left Air Line Junction, about two miles short of the Union depot, with the proper signals but several minutes late. They had not known whether train No. 6, the Lake Shore express, was ahead of or behind them. Contradicting the initial report the engineer stated that the tunnel had been full of steam and he could not see through it. Emerging from the tunnel he had seen the back of the other train only three cars ahead and had been unable to pull up in time. [302]

A ninth victim of the accident died on 1st December and the inquest was continued. The evidence of Conductor Hunter and Brakeman Hall of the Pere Marquette train added little, while the evidence of Conductor Hall of the Lake Shore train was described as being very evasive and the only positive answer he would give was that the tail light of his train had been burning brightly and he had not seen the Pere Marquette train before the collision.[303]

A tenth victim died on 7th December.[304]

On 15th December State Railway Inspector McIneary blamed the accident on Engineer Lyons of the Pere Marquette train entering the tunnel at too high a speed, the failure of the brakeman of the Lake Shore train to go back and protect his train and the failure of Conductor Hall to see that he did so. He also remarked that the tunnel was not properly

[300] Ibid, 19th August 1891
[301] *Nevada State Journal*, Reno, Nevada, United States, 29th November 1891
[302] *Newark Daily Advocate*, Newark, Ohio, United States, 1st December 1891
[303] Ibid, 2nd December 1891
[304] Ibid, 8th December 1891

ventilated so that smoke obscured the view.[305] It has not been possible to discover the verdict of the coroner's inquest.

One accident allegedly led to another near Hastings, New York, United States, on the New York Central and Hudson River Railroad, on Christmas Eve, 1891. The initial accident was a collision in the tunnel at Sing Sing (Ossining) which blocked both lines. As a result, there was a queue of trains in both directions and train No. 45 for Buffalo and Niagara Falls was stopped for some time near Hastings station. Its rear brakeman, Edward Albert Herrick, should according to the regulations have gone back to protect it in the rear but he failed to do so, acting or rather failing to act on the assumption that the next train, No. 81, was a local one which would stop at Hastings station anyway. Unbeknown to him train No. 81 had been delayed at New York and the despatcher had decided to give priority instead to train No. 7, the *St Louis Express*, which forthwith collided with the rear of the Niagara Falls train.[306] This of course is an example of a feature of many railway accidents, namely the rash assumption, such as that of a signalman who dozes off and awakening clears out a train believing that it must have passed him because of the time that has elapsed.

The Wagner palace car *Gibraltar* at the end of the first train was telescoped in the collision and in an all too familiar scenario damage to the boiler of the engine of train No. 81 meant that the victims were exposed to scalding steam. A report on the accident stated that, 'The trainmen lost their heads and ran out in a frantic manner, peering into the wreck at those who lay groaning and dying, but doing very little towards reaching the victims. Finally a number of physicians and citizens of this place {Yonkers} reached the scene and the work of rescue began. The dead and injured were placed in a special train and brought to this place. Owing to the excitement it was impossible to get anything like a list of the killed and injured. It is stated that a number of the injured are likely to die.' Of the 18 passengers in the rear coach of the Niagara Falls train only three escaped unhurt. The accident happened about a mile north of Hastings at 8.45 p.m. The bodies of the victims and the injured were taken to Yonkers where they were expected to arrive at about 2 a.m. The final number of dead in the accident was reportedly 15, comprising 14 passengers and the coloured porter of the sleeping car.

During the examination of the bodies of the dead by the undertakers it was discovered that the pockets of a number of the victims had been either cut or torn out and valuables such as pocketbooks (wallets), jewellery and so on were missing. Subsequent inquiry at the hospital disclosed that a number of the injured had also been robbed. Several strangers had offered to help at the scene of the wreck, and it was believed that the opportunistic thief or thieves were amongst them.[307]

An official report on the circumstances of the accident from New York Central Third Vice President Webb gave a somewhat different picture of the events leading up to the accident. He made no mention of an earlier collision but stated that train No. 93, which had left New York at 6.45 p.m., had stopped at Dobbs' Ferry (the station immediately

---

305 *Lima Daily Times*, Lima, Ohio, United States, 18th December 1891

306 *A History of Railroad Accidents, Safety Precautions and Operating Practices* by Robert B. Shaw, pg. 179

307 *Decatur Daily Republican*, Decatur, Illinois, United States, 26th December 1891

beyond Hastings) while slight repairs were made to its engine. Signals had been displayed and the Buffalo and Niagara Falls special, which had left New York at 7.30 p.m., had been pulled up three-quarters of a mile south of Dobbs' Ferry. Brakeman Herrick had immediately been sent back by the conductor of train No. 45 and had gone as far as Hastings station, where he had gone inside and chatted to the stationmaster while waiting for the Cincinnati and St Louis express, which left New York at 8 p.m. While Herrick was standing at the station door the train had thundered past, travelling at 40 miles an hour. Herrick fled soon after the accident and had not been found a day or so later.[308]

Brakeman Herrick remained at large some days, telling his father that he did not wish to give himself up until public anger at the accident had abated. Through his father he issued a statement explaining his actions in which he wrote, 'when our train slowed up, but while yet moving, I dropped off and placed two torpedoes on the rail and then ran back to the train which was still moving, but as it slowed up more, I took my lights and started down the track to stop the Croton local, then nearly due. When half-way down I placed one torpedo on the rail and then walked on down to the station, thinking it better to notify the engineer there before he was fairly under way again. When I got to the station, I placed my lights on the platform and the red light showing south, and then stopped on the platform and inquired of the agent how the local was. He said he did not know, and I sat down and remained there three to five minutes when I heard the train coming and started for the door, believing it to be the local. As soon as I saw the train, I knew it was the express and I grabbed my light and tried to grab the engineer's attention. But it was too late. I followed on up the track, hoping that the train I had left twenty minutes before had pulled out of the way. I soon saw a man coming back with a red light. Soon as I got to him, I asked what had happened and he said they struck something, but he did not know what it was. I went on further and met another man and asked if anyone was hurt and he said all in the sleeper were killed. I did not deem it prudent for me to go to the wreck, so I turned about and came to New York, I fully realize the awful position I am placed in and can only say that I felt too sure the train to come first was the local.'

District Attorney Platt meanwhile announced that he would ask for Herrick to be indicted for manslaughter, also the dispatcher who had sent out the two express trains knowing the line to be blocked, and whom he considered to be as much to blame as Herrick.[309]

It was back to England for the next serious accident, at Bishopsgate, just outside of the London Liverpool Street terminus of the Great Eastern Railway, on 14th June 1892. The accident occurred in the early morning and involved two workman trains, the 6.4 a.m. from Walthamstow St James' Street being run into on the up suburban line by the 5.56 a.m. from Enfield Town, both destined for Liverpool Street, at 6.31 a.m. The third-class brake composite carriage at the rear of the Walthamstow train and the third-class carriage immediately in front of it were completely broken up in the collision and three workmen in that train were killed on the spot and two more fatally injured, with 52 being less badly hurt. Almost all the casualties were in the Walthamstow train.

[308] *Lincoln Evening News*, Lincoln, Nebraska, United States, 26th December 1891
[309] *Middletown Daily Times*, Middletown, New York, United States, 29th December 1891

The recently opened suburban lines were situated at Bishopsgate station beneath substantial girders which carried the goods lines over them, and as a result the suburban line platforms were constantly dark, with gas lights lit day and night. Under ideal circumstances the rear of a train standing at the platform could be seen for 150 yards but because of the dim surroundings this distance was very much reduced. Standing on the platform of the suburban lines was Bishopsgate North signal box, with 15 levers and two block instruments, which worked only those lines, while Bishopsgate Middle box, 310 yards east of the east end of the platforms, worked the suburban and local lines, with 13 levers, four block instruments and two train describers to Liverpool Street box. The line was worked on the 'open' block system, the block instruments normally indicating "Line Clear" and the outside signals being lowered when no train was in the section. A clearance bar in the rear of the Bishopsgate North up starting signal prevented the home signal being lowered when a train was standing on it, and a rail contact a train's length past the starting signal rang a bell in the box until a train had passed it.

At the Board of Trade inquiry into the collision Major-General C.S. Hutchinson decided that the immediate cause of the accident was the failure of Signalman George Meredith in the Middle box to signal forward the Walthamstow train to Signalman William Hughes at the North box, together with the failure of Hughes to notice the train standing for four or five minutes at the up suburban platform. Hughes had also failed to notice the Walthamstow train when he accepted the Enfield train and cleared his up distant signal, which he had inadvertently failed to lower after the clearing back of the previous train (also from Walthamstow), Meredith denied that he had failed to signal forward the 6.4 train and the inspecting officer said that there would have nothing to corroborate Hughes' denial of receiving the signal where it not for the fact that Meredith had also omitted to describe the very same train to Liverpool Street box.[310]

When the Walthamstow train ran into the platform Hughes had probably been busy at the back of the box boiling some water for his tea and the Walthamstow train not having been signalled to him had likely thought the noise of an up train was one on the up local line. The 6.22 a.m. down empty train and the 6.26 down train for Enfield had also passed at about the same time. Hughes had only realised that the Walthamstow train was there when he heard its rear guard shouting, by which time the Enfield train had passed his up home signal and was within 50 yards of the rear of the Walthamstow train. Major-General Hutchinson did not blame either of the train crews except that he remarked that Driver Sidney Woodcock of the Walthamstow train should have whistled when his train was held at the starting signal for more than two minutes.

The inspecting officer had several suggestions for preventing a repetition of the accident. Firstly, he thought that the Great Eastern Company should switch to a 'closed' system of block working, whereby the outdoor signals were normally kept at "Danger". Secondly, he thought that the lighting at Bishopsgate needed improving and stated that he was glad to learn that the station was shortly to be lit by electricity. Until then he suggested that trains up the up suburban lines should carry lit red tail lights during daytime as well as darkness. In addition, he suggested a rail contact in the rear of the Bishopsgate North up home signal which rang a bell in the box when a train passed it and that the contact in advance of the up starting signal should lock the up home signal at

[310] Board of Trade report

"Danger" until a train had passed it.[311]

The coroner's jury at the inquest at the London Hospital into the deaths of three of the workmen returned a verdict of "Accidental death", adding that Signalman Meredith was deserving of the highest censure and Signalman Hughes must have had great want of attention. They considered that two men should be provided in Bishopsgate Middle box and commended the actions of Driver Harwood and Fireman Santer of the Enfield train and Guard Howard of the Walthamstow train in trying to prevent or minimise the effects of the collision. The coroner then addressed the two signalmen and, 'told them that they might consider themselves fortunate in not being sent to take their trial for manslaughter. He did not further wish to harass their feelings, which must be painful enough, owing to the knowledge that they had caused the accident.' At a separate inquest at Shoreditch into the deaths of the other two victims a similar verdict and censure of the signalmen were returned.[312]

A disaster in the United States only 11 days after the Bishopsgate collision was also due, to a large extent, to a signalman's error. Just after midnight on Saturday 25th June 1892 the first section of the westbound Pennsylvania Railroad train No. 9 the Western Express had been stopped at Dock Street tower until there was a clear road for it into Harrisburg Union depot. While it was stood there, it was run into by the second section. It was raining very hard at the time. The engineer of the first section had whistled for Flagman Richard Brown to go back but had almost immediately recalled him as the signal was cleared for him to proceed into the station. With that, the collision occurred. The private Pullman car at the rear of the first train, occupied by none other than George Westinghouse, the inventor or the air brake, was badly damaged but none of his party were hurt. The two cars ahead of the private car, however, were telescoped into one another and as they were packed with passengers, the accident resulted in 12 fatalities. Engineer Hugh Kelly and Fireman Harry Neill, both of Philadelphia, remained on their engine and escaped unhurt. One of the cars caught fire after the collision, causing a panic, but the fire department soon managed to extinguish the blaze, helped no doubt by the cloudburst.

H.S. Hayes, the Steelton operator, was described as a 22-year-old farm boy and was an extra operator working wherever he was sent. On the evening of Friday 24th he had been assigned to Steelton tower in place of the regular operator, William Godd. Arrested after the accident, 'he spent the entire night in dreadful suspense, lying on his cot and sobbing bitterly. Several physicians say he is liable to become insane if confined much longer' {and} 'Representative Tinksberry, of Columbia county, offered bond for his release, which was refused.' Hayes had had two friends in the tower with him, contrary of course to the rules, and admitted that when the second section of the train had approached, he had contacted his colleague J.E. Painter at Dock Street and receiving no reply had rashly assumed that the line was clear. 'When informed of the wreck and loss of life {Hayes} became almost frantic and the day operator had to be ordered on duty.' At the inquest, begun on the day of the accident, Hayes 'withheld nothing from the jury, and cried as though his heart would break as he finished his testimony'.

---

[311] Ibid.

[312] *The Times*, London, England, 18th and 21st June 1892

Although he was clearly at fault in wrongly giving the second section a clear signal it turned out that Hayes was not alone to blame. The first section had in fact cleared the Steelton – Dock Street section, but only by about 180 feet. No distant signal was provided at Dock Street because of its proximity to Harrisburg station and the fact that all trains were expected to approach the station under control. Engineer Kelly of the second section admitted to the coroner's jury that he had been travelling at 40 miles an hour at the time of the collision and that had he been only going at the statutory 15 miles an hour the collision would have been avoided. With the heavy rain obscuring his view he could not see the danger signal at Dock Street until he was 200 feet from it. The flagman of the first section also stated that he had put down torpedoes before being recalled, Kelly denied having heard any go off. Kelly had 21 years' service with the company and had never been in an accident before.[313] [314]

The coroner's jury passed a verdict on 29th June finding not only Operator Hayes and Engineer Kelly guilty of gross neglect for their part in causing the accident but likewise Flagman Robert Brown for not going back far enough and placing torpedoes on the rails (although he stated that he had) and not staying back to flag the second section (although he had been whistled in by his engineer). In Great Britain at any rate the guard would not be expected to go back and protect a train at all if it were standing within the protection of fixed signals. The jury also censured the Pennsylvania company for allowing a local freight train to be on the track in the time of an approaching passenger train.[315]

On 10th September 1892 an accommodation train left Boston at 10.15 p.m. on the Fitchburg Railroad for Waltham. At West Cambridge Junction station, Massachusetts, United States, it had to wait several minutes for an incoming accommodation train to clear the single-track Watertown branch. Following the accommodation, freight train No. 907, consisting of an engine and 30 heavily laden cars, was approaching at 30 miles an hour. There was a thick fog, making it reportedly impossible to see 50 feet. Brakeman Noon went back to flag the freight train but Engineer Goodwin, whose train was not booked to stop at West Cambridge, failed to see the flagman's signals and believed that he had a clear road. The incoming train had just cleared the single line, and the flagman had been whistled in when the freight train suddenly loomed out of the fog, running at 30 miles an hour. When Engineer Goodwin saw the rear of the passenger train it was but 10 yards away and as he was in the act of reversing his engine and shouting to his fireman the collision occurred.

The freight engine penetrated two-thirds of the way through the rear coach of the accommodation train as the trucks piled up behind it. Scalding steam entered the car after the collision and the burning coals of the colliding engine set the woodwork ablaze to add fire to the horror. By Monday 12th it was reported that there were nine passengers dead and 26 injured, two of whom were dying. Most of the victims were from Watertown and Waltham.

The inquiry into the accident by the railroad commissioners was told of an allegedly defective signal between Cambridge and West Cambridge and that Engineer Goodwin

[313] *The Trenton Times*, Trenton, New Jersey, United States, 27th June 1892
[314] *The Marion Daily Star*, Marion, Ohio, United States, 27th June 1892
[315] *The Trenton Times*, Trenton, New Jersey, United States, 29th June 1892

had ignored a rule obliging him to slow down on failing to see a signal and a further rule restricting his speed to eight miles an hour through the junction. Brakeman Noon spoke of waving his lantern at the freight train and his signal being answered by two whistles from the freight train, but Goodwin denied giving such whistles.[316] The three-mile section of line east of Cambridge Junction was already protected by automatic signals, one of the earliest such installations.[317] Fitchburg Railroad General Manager Adams blamed Goodwin for running too fast trying to make up time.[318]

The next serious rear collision to occur was the rather well-known one at Manor House, England, on the North Eastern Railway main line between Otterington and Thirsk on 2nd November 1892. The 2 a.m. up goods train from Middlesbrough to Starbeck was just starting away from the up main home signal at Manor House Siding cabin at 4.2 or 4.3 a.m. when it was run into from behind by the second portion of the up Scotch express, which had left Edinburgh Waverly at 11.2 p.m. the night before. The brake van and one waggon of the goods train were broken to pieces, the engine and tender of the express were overturned, and the leading vehicles of the express were wrecked, while all but the last three of the 13 coaches of the train were damaged to some degree in the collision. Fire subsequently broke out in the wreckage and wholly or partly consumed eight of the carriages of the express and the guard's van and eight waggons of the goods train. Nine passengers died in the accident, all of whom were believed to have been in the leading third-class carriage, and the guard of the goods train, George Petch, also lost his life.

At 3.38 a.m. on the morning of the accident the signalman at Northallerton North Junction, on receiving "Line clear" from Hutton Wood cabin for the first portion of the up Scotch express, had decided to let the Starbeck goods train, waiting at Northallerton Low Junction, proceed towards Thirsk, as the second portion of the Scotch express, which had been delayed awaiting connections from the north, was not due to pass within the requisite 20-minute margin. This goods train duly left Northallerton at 3.39 a.m., passed Hutton Wood cabin at 3.43, Otterington station at 3.47 and came to a stand at the Manor House up main home signal at probably about 3.52. At 3.58 a.m. the Otterington signalman received the "Be ready" and at 3.59 the "Train on Line" signals from Hutton Wood for the second portion of the Scotch express and at 4 a.m. he gave the "Be ready" signal for the Scotch express to Manor House, although he had not received the "Train out of section" signal for the goods train (as he was justified in doing under the North Eastern regulations of the time). Manor House immediately gave the out of section signal for the Starbeck goods and accepted the express, but just as the latter train was passing Otterington the Manor House signalman called Otterington on the speaking telegraph instrument and asked, 'Is this the E.P?' The Otterington signalman replied that it was and shortly afterwards heard the sound of the collision.

James Holmes, the Manor House signalman, remembered accepting the Starbeck goods train from Otterington but did not ask forward for it to Avenue Junction or record any times in his train register book, for at that moment he was overcome by sleep and knew nothing of the arrival of the goods train at his home signal. When he received the

---

316 *The Fitchburg Sentinel*, Fitchburg, Massachusetts, United States, 18th September 1892

317 *A History of Railroad Accidents, Safety Precautions and Operating Practices* by Robert B. Shaw pgs. 237-8

318 *Indiana Progress*, Indiana, Pennsylvania, United States, 21st September 1892

"Call attention" signal for the second part of the Scotch express and saw that his block instrument to Otterington was still standing at "Train on Line" he assumed that he had omitted to clear back the first portion of the express, his subsequent acceptance of the goods train having completely gone out of his mind, and he gave the "Out of section" signal and accepted the second portion. Having pulled off his signals he saw the goods train instead of the express approaching his cabin and the thought must have occurred to him that he had made a mistake, hence his telegraph message to Otterington.[319]

Major F.A. Marindin, who conducted the Board of Trade inquiry into the accident, stated that the accident was clearly principally the result of breaches of the block regulations by Signalman Holmes, who had manifestly been unfit for duty, but there were also others responsible. In particular Joseph Barnes, the driver of the Starbeck goods, had stood at the up home signal for what he stated was a period of four to five minutes, but which was probably nearer seven, without having whistled, as he should have done according to Rule 275a, or, given that the weather was somewhat foggy, sent his fireman to the signal cabin.

The inspecting officer stated that the other railwaymen whose actions might be considered to have contributed to the accident were Driver Roland Ewart of the express and the Otterington and Northallerton North Junction signalmen. It might be supposed, Major Marindin declared, that with the continuous brake the driver might have been able to very much reduce the speed of his train if not entirely avoid the collision upon seeing the tail lights of the goods train in front of him, but the evidence as to the thickness of the fog and the distance at which signal lights could be seen was so conflicting that it would not be fair to blame him.

Turning to the signalmen, Major Marindin commented that Signalman Harry Eden at Otterington (whose train register book, he noted, was very badly kept) ought to have questioned why it took his colleague 13 minutes to clear back the goods train, given that the section was only 1 mile 30 chains long and that the goods train should have had a clear run to Avenue Junction, especially given the condition in which Holmes had been just before he took over. Signalman Eden had candidly admitted that he had sat down to read after sending on the goods train and that had it occurred to him, he would have inquired as to why it was taking so long to reach Manor House. Eden, the inspecting officer admitted, had not broken any rules and could hardly be blamed for what had occurred. As for the Northallerton North Junction signalman his action in letting the goods train run between the two portions of the express had been much criticised (presumably in the press) but the margin of over 20 minutes between the two portions of the express had been quite sufficient, the goods train had only just over six miles to run on the main line to Avenue Junction and the line was worked on the absolute block system. Were goods trains not permitted to take advantage of such an interval, it was hard to see how the traffic could be carried on.

Signalman Holmes had honestly admitted that the accident was due to his having fallen asleep. He had come on duty at 6 p.m. on Monday 31st October, leaving duty at 6 a.m. on Tuesday 1st, when he had walked 2½ miles home, to be told by Mrs Holmes that one of his children was ill. He had had some sleep between 8 a.m. and 12 but then his wife had roused him and told him that his youngest child, Rosie, had had a fit, whereupon

[319] Board of Trade reports

he had dressed and gone in search of a doctor. Walking a mile to Otterington station he had taken a train to Northallerton, but the doctor had been out on his rounds. He had then walked 4½ miles home, getting back at about 2 p.m. to find that Rosie was dead. He had then walked to Otterington again to telegraph for his mother to come and be with his wife, and on the way had met Mr Thomas Kirby, the Otterington stationmaster, and had asked if he might be relieved that night as he did not feel fit for duty. He had then walked home again, returning to Otterington station at 6 p.m. to be told by the stationmaster that he could not be relieved, whereupon he had arranged for his mate, Signalman Creed, to stay on until 8 p.m. while he walked home again to arrange for a neighbour to be with his wife until his mother arrived. He had then returned to Otterington to ask Signalman Eden to let him know if his mother arrived by a later train and as he was setting off to Manor House had remarked to Eden that he, Holmes, was "about done up". After being told by Holmes of his child's death and that he wished to be relieved that night Stationmaster Kirby had telegraphed to York, "Can you send relief to Manor House tonight, Holmes' child dead," but the reply had come back, "Cannot send relief." [320]

The inspecting officer noted that Holmes had had only four hours' sleep between two night shifts, had walked 12 miles and had been in considerable distress over the death of his daughter. He remarked that if those facts could not be accepted as sufficient excuse for Holmes' subsequent failure, they went a considerable way towards explaining it. He had no hesitation in saying that Holmes should have been relieved and was of the opinion that Stationmaster Kirby had committed a grave error of judgement in not stating positively that Holmes was unfit for duty. Having had his original request to York for relief turned down the stationmaster should have made a fresh appeal or failing that he could have had Manor House cabin, which was basically little more than a block post, switched out that night, though it was not advisable for cabins to be closed without notice if it could be avoided. When Holmes had returned at 6 p.m. he had commented to the stationmaster that he was mainly concerned about his wife, from which the stationmaster had wrongly concluded that Holmes was in a fit state to do his night shift. The decision by York not to send a relief was made by a clerk in the absence of the inspector but had been endorsed by the latter and had been necessary due to an excessive number of relief men being needed elsewhere, including five in connection with the washing away of a bridge.

Commenting on events after the collision Major Marindin criticised Driver Barnes of the goods train for not having proceeded on his engine to Avenue Junction after the accident as per Rule 281 and thus saving much valuable time, and with Barnes' fireman and guard and the conductor of the express all walking to Avenue Junction as there had as a result been no one left at the scene among the train crews for over an hour able to assist the passengers, except Pullman car conductor Andrew Baillie, who seemed to have done everything in his power. There also seemed to have been a lack of organisation at the accident scene, otherwise the fire, which had not broken out until almost an hour after the collision and was not apparently due to gas, would not have spread as it did, and caused burns to one child victim, although it probably had not contributed to her death.

With a view to preventing similar accidents in the future the inspecting officer recommended that the North Eastern and other railway companies install some sort of lock and block apparatus on all their important lines, such as had already been done by

[320] Ibid.

some companies, to prevent signalmen making the mistake that Holmes had made. Meanwhile he was critical of the fact that Signalman Eden had been quite in order to give the "Be ready" for the section portion of the express before Holmes had given the "Out of section" for the goods, also of that fact that the North Eastern considered having only 17 relief signalmen out of 300 to be sufficient. Finally he recommended that the hours of main line signalmen should be limited to 10 and that as far as possible they should live as close as possible to their place of work.[321]

Holmes' case attracted much sympathy.[322] The jury at the inquest into the deaths of the victims of the accident nonetheless found him guilty of manslaughter, while criticising the fact that he was not relieved, condemning the North Eastern directors for expecting main line signalmen to work 12-hour turns, criticising Driver Barnes for not whistling when his train was stopped at Manor House and the lack of a means on trains of extinguishing fires.[323]

Holmes subsequently stood trial for the manslaughter of George Petch, the goods train guard. At York Assizes on 5th December he was found "Guilty" with a very strong recommendation to mercy, and Mr Justice Charles ruled that he be bound over in the sum of £50. "The announcement of the sentence was received with a tremendous outburst of applause, which was with difficulty supressed. The prisoner then entered into the required recognizances and was discharged." [324]

Holmes subsequently went back to working on the railway, though not as a signalman.

A self-contradictory report appeared in more than one newspaper of an accident at Tadmore, Ohio, United States, on 6th May 1893, the report stating that five tramps had died in the accident and then in the next sentence that no lives had been lost! The lives of the tramps presumably did not count. Dayton and Michigan Railroad train No. 47 was unloading ties (sleepers) at Tadmore when a through freight train, travelling at 25 miles an hour, ran into it, smashing 20 cars.[325]

Tramps were again the main victims of a rear collision that occurred on a curve a mile from Welch, West Virginia, United States, on the Norfolk and Western Railroad, late on 26th or early on 27th July 1893. A freight train ran into the back of a passenger train. Five people again lost their lives, four of them being unknown men who were 'beating their way' in the first boxcar behind the tender of the freight train engine. A severe depression had begun in the United States in February 1893, the so-called Panic of 1893, so there were probably more tramps than usual riding trains and thus more potential accident victims.[326] The fifth fatality of the accident was Brakeman John Frankman of the passenger train, who was underneath it at the time of the accident, presumably trying to

---

321 Ibid.
322 *The Times*, London, England, 3rd, 4th, 5th, 7th, 8th, 9th, 11th and 12th November 1892
323 Ibid, 18th and 19th November 1892
324 Ibid, 6th December 1892
325 *Newark Daily Advocate*, Newark, Ohio, United States, 12th May 1893.
326 Wikipedia – Panic of 1893

fix the problem that had caused the train to stop out of course.[327]

The United States suffered another disastrous rear collision close to midnight on Saturday 26th or very early on Sunday 27th July 1893 on the Long Island Railroad near Bushwick Junction, New York. The train which had left Manhattan Beach at 11.15 p.m. on Saturday 26th was run into from behind by the train which left Rockaway Beach at 11.30. The two rear cars of the five that made up the Manhattan train were demolished in the collision and the middle one overturned. There were 14 dead by the next day and the final number was 16, while many others were injured. The Manhattan train had just restarted after an unintended stop when the Rockaway train ran into it.

The Manhattan train had been a special bringing away the pleasure seekers who had stayed late, the members of Sousa's band and many employees 'who were on their way to New York to stay over Sunday'. In charge of the train was Conductor James A. Mott with Engineer Otis Donaldson and Fireman William Shurr. There were also two brakemen, Thomas Finn and James McCormack. Finn was killed in the collision. The signal had just cleared when the collision occurred, and Engineer Donaldson reportedly immediately opened the regulator and made with all speed to Long Island City for help, leaving the rear three wrecked cars behind.

Among the fatalities was Colonel E.A. Buck, the editor of *The Spirit of The Times*, a weekly newspaper with an upper-class readership, mostly of sportsmen.[328] His son was seriously injured.

A newspaper report noted that, 'Dr Burke, who was with the hospital train, said that the three Manhattan Beach cars which were so terribly smashed contained a large party of excursionists who were of the better class. They were not of the rough element that usually comes up from the beach on the last trains.'

Conductor Frank E. Finley was in charge of the Rockaway train. No one on this train was seriously hurt, though all were severely shaken and some badly bruised.

The cause of the accident was not immediately clear as it boiled down to a question of veracity between Tower Operator Knott and Engineer Conkrite of the Rockaway train. Conkrite declared that there had been a heavy fog at the time, but he was able to pick out the signals and they were clear for him, so he had kept going without reducing speed. Then suddenly his fireman, Robert Price, had shouted, "My God, there's a train ahead!" Looking ahead he had then seen the two red tail lamps of the Manhattan train, had reversed his engine and applied the steam brakes and then he and his fireman had jumped. After the accident the signals had been at "Danger" and operator Nutt insisted that they had been so when the Rockaway train had passed them.

Engineer Donaldson said the signals had been clear for him and his engine and three cars of his train had passed the semaphore at Berlin when the train accidentally became divided. He had stopped, set back, recoupled the train and was just moving off again when the collision took place. Brakeman McCormack had been sent back as an extra precaution to protect his train while it was stopped.[329]

[327] *The Marion Daily Star*, Marion, Ohio, United States, 27th July 1893
[328] Wikipedia – Spirit of the Times
[329] *The Olean Democrat*, Olean, New York, United States, 29th August 1893

The railway commissioners announced on 15th September that the accident was primarily caused by Towerman Knott but that Rear Brakeman James McCormack was also partly responsible for failing to have gone back as far with his red lantern as the time would have allowed him to do.[330]

On the evening of Monday 18th September 1893 traffic for train No. 45 of the Cleveland Cincinnati Chicago and St Louis Railroad (the Big Four) was heavy because of people returning home from the Chicago World's Fair, and it was run in two sections, the first leaving Chicago at 8.10 p.m. and the second a little later. The first section, in charge of Conductor Tyner and Engineer Sam Phillips, encountered no problems until it reached Manteno, Illinois, United States, on the Illinois Central Railroad, where it had to stop for water, the train register at Chicago bearing a note that no water was available at Kankakee, five miles to the south. What happened next was the subject of controversy. The line at this point was dead straight for several miles, and it was maintained that no flagman could have gone back to warn the second section, the engineer of which could not have failed to see him. On the other hand it was maintained that the engineer had been asleep, and the fireman had been preoccupied with firing. The second section, in charge of Conductor Tom Eden, with Engineer Charley Ames, was stated to have been going at a terrific speed, and its engineer and fireman saw the first section ahead only in time to jump and save themselves, without making any attempt to stop their engine. The rear vehicle of the first section was an empty sleeping car, and the locomotive of the second section pushed this car into the day coach in front, which was telescoped from end to end. The 40 or 50 passengers in the day coach, who were awake and conversing, 'were swept from their feet in a twinkling, and crushed and mangled. Limbs were torn off, heads and bodies crushed, and the broken timbers dripped with the blood of the dead and dying. Of those killed seven are so badly mutilated as to be unrecognizable. Several of the unknown dead are believed to be citizens of Cincinnati and adjoining towns in Indiana, Kentucky and Ohio. Several of the injured cannot live.'[331]

A slightly injured passenger, J.C. Means, of Chester, South Carolina, described his experiences to a newspaper reporter. "I was just ready to go to bed, and had told the porter to fix my section, when our train stopped. We had been standing still perhaps four when I heard a rumble behind us and a moment after the whistle of a locomotive a half dozen times repeated. I turned towards the rear and there came a crash which put out every light in the car. I was hurled headlong to the aisle and under one of the lower berths. It seemed an age to me before I could get out. I heard cries all about me for help and answering cries from outside from the people who were beginning to come back from the day coaches and aid in the work of rescue. As I crawled out from the wrecked sleeper everything was in total darkness, but I could see outlined against the sky the cars that were wedged together, broken and splintered. I helped to pull out a number of the injured before I discovered that I too was hurt. Then my arm and back began to pain and I felt the blood running from my head."[332]

Another newspaper account stated that Conductor Tyner, knowing that the second

[330] *Middletown Daily Times*, Middletown, New York, United States, 15th September 1893.

[331] *Newark Daily Advocate*, Newark, Ohio, United States, 20th September 1893

[332] *The Ohio Democrat*, New Philadelphia, Ohio, United States, 21st September 1893

section was following close behind, had sent the flagman back with a red light, but the flagman had only got back 100 yards when he heard the second section rapidly approaching, though it was hidden from his view around a curve. According to this probably more reliable version, Engineer Ames saw his frantic signals, reversed his engine and applied the air brakes.

'The passengers in the rear cars escaped with nothing more than a shaking up. Many of them dressed and hurried forward to assist in rescuing the unfortunates who were still pinioned in the wreck. Several physicians were on the cars and they passed among the wounded alleviating with the few resources at their command the suffering of the wounded. Sheets and tablecloths were torn into bandages and wounds skilfully dressed while brake and kindly women hastened to their wants with coffee hastily prepared and with cups of water. The crash of the collision was plainly heard in Manteno and in a few minutes the residents of that village were upon the scene in force. Houses near the scene were at once thrown open and each became an improvised hospital. Local surgeons from the city were hastily summoned and several of them, together with other residents of Kankakee, hastened to Manteno. The arrivals were timely and their efforts much appreciated by the terror-stricken passengers, many of whom, otherwise uninjured, were suffering from the suddenness of the shock and going about wringing their hands and crying hysterically.' [333] Nine passengers were killed outright and about 25 injured, three of whom later died.

The coroner's inquest at Manteno held that Engineer Ames of the colliding train and Orville Duncan, flagman of the train, which was run into, were both guilty of criminal carelessness in connection with the accident.[334]

A disastrous collision at Jackson, Michigan, United States, on 13th October 1893 occurred within 100 yards of the site of a previous disaster almost 14 years earlier to the day. An excursion from Oswego, New York, had switched from the Delaware Lackawanna and Western Railroad at Buffalo to the Michigan Central Railroad and was standing at Jackson station when a second excursion train, also from the east, crashed into its rear, telescoping the rear three cars of the front train and killing outright a dozen passengers, the final death toll rising to 18. The accident occurred at 8.45 a.m.

Many of the Oswego passengers had got out of the train to have breakfast. Some of them had returned to their coaches, others were standing on the platform. The train was almost ready to leave when the second section appeared around a curve, apparently under control. When within a few hundred feet of the stationary train Engineer William Whelan of the second train applied the brakes and found that they would not work. His train failed to reduce speed and he reversed his engine and he and his fireman, Joseph Kimbling, jumped off. The collision happened in full view of several hundred people on the station.

The first train consisted of eight cars, seven of which were New York Central Railroad ones, much lighter than the Michigan Central ones. The only Michigan Central vehicle was the one fourth from the engine, which had been added at a station on the Canadian Southern division to replace a disabled one. The sixth and seventh cars were

---

[333] *The Marion Daily Star*, Marion, Ohio, United States, 19th September 1893

[334] *Decatur Daily Republican*, Decatur, Illinois, United States, 25th September 1893

completely telescoped, the one fitted snugly into the other. The fifth car was overturned across a siding and the Michigan Central car was smashed in at each end, but there were few passengers in it, and none of them was seriously hurt. The deaths and injuries were all in the sixth, seventh and eighth cars, from which hardly a passenger escaped unhurt. Because the accident happened at a major station physicians were at hand almost immediately and a breakdown train was also soon on the spot. The bodies of those killed were conveyed to the undertakers and the injured to local hotels.

Describing the scene after the accident a newspaper reporter wrote, 'While the dead were being taken out from under the debris there were heartrending scenes. George Stark of Elmira and Darwin Z. Gibbs of Wheeler, N.Y. and George Woodbury of Bath, N.Y. had left their wives sitting together in the coach while they stepped out to smoke during the stop. The three women were killed. In the undertaking rooms lay the body of a 2-year-old baby for four hours unidentified. It was finally identified as the child of Mrs Anson Harrington, of Elmira, who lay two miles away with injuries from which she may die. There were scores of other cases where travelling companions and relatives were separated, and some killed or injured seriously. Some of the wounded are in pitiable circumstances, cut into pieces, with death waiting almost to seize them in a few hours at most. The City Hospital is the scene of terrible suffering. Here Dr Gibson and a corps of assistants are in charge. A sad case is that of Johanna Hartley. She lies on a stretcher in one of the lower rooms shaking in an agonizing manner. Both legs are broken near the ankle, and her right arm is broken above the elbow and she is terribly injured about the body. Her recovery is doubtful. Nearby lies Mrs Anson Harrington of Elmira, N.Y. with her right leg broken below the knee and her body terribly bruised. Nearby lies her dead baby, aged 6 months. Though in dreadful agony she constantly askes for the child, and no one has the courage to tell her that she will never respond to her voice. A gateman in the Michigan Central yards here tells of the miraculous escape of a woman and her little child. They were in the car that was so completely telescoped and were among the first he extricated from the wreck. Every stitch of clothing had been torn from the child's body, but it had not received a scratch. The woman was also uninjured save for a cut on one of her arms.'[335]

On 17th October it was stated that an old railwayman, who had witnessed the accident, corroborated the story of Engineer Whelan in detail. A few seconds after the collision he had noticed a poorly dressed man, possibly a tramp, standing on the platform of the car adjoining the engine. The cock of the brake pipe immediately behind the engine had been found closed after the accident, preventing the air brakes from functioning. It was surmised that the tramp wanted to stop the train before it had reached the station so that he could get off unnoticed and had operated the cock wrongly thinking that it would make the train slacken up.[336]

The coroner's inquest into the deaths of those killed in the collision rejected the story of the tramp interfering with the brake connection and instead on Friday 20th October found Engineer Whelan, Conductor Webb and two brakemen guilty of gross negligence in having failed to brake the colliding train.[337]

---

[335] *The Marion Daily Star*, Marion, Ohio, United States, 14th October 1893

[336] Ibid, 17th October 1893

[337] *Elyria Republican*, Elyria. Ohio, United States, 26th October 1893

A thick fog shrouded the city of Chicago, Illinois, United States on the evening of 8th November 1893 as Chicago Rock Island and Pacific Railroad suburban train No. 127 pulled out of the Van Buren Street depot with Conductor N.W. Freeman and Engineer A.E. Gary. Engineer Gary ran cautiously through the fog and by the time his train had reached Englewood it had lost 10 minutes, and had lost another minute by Normal Park.

Train No. 11, the *Kansas City Limited* vestibule express started from Van Buren Street 10 minutes after the suburban train, in charge of Conductor A.P. Danforth and Engineer Nelson Dickerman. Engineer Dickerman was also concerned about the fog and ran slowly through the yards, the conductor telling him that he was three minutes behind time at Englewood. By this time the suburban train should have passed onto the Blue Island branch, leaving the main line clear for the express. Engineer Dickerman therefore opened his throttle and the train ran through Normal Park at 30 miles an hour. Suddenly Dickerman saw a red light right in front of him. He applied the brakes in emergency and shortly afterwards the express was in the back of the suburban train.

A local passenger, Charles McCreeken, who had just alighted from the suburban train at Normal Park, told a newspaper reporter that he had been walking along the platform when he saw the headlights of the approaching express, "I saw that a collision could not be averted, but with others I called to the engineer of the through train. The heavy train swept by and just then the engineer saw the lights on the rear coach. He reversed the engine and then came the shock. It was terrific. The engine seemed to lift the train and then drove right on. I think the fault was with the engineer in running so fast on such a night and so close to the dummy train." [338]

The suburban train was stopped at Eggleston station when the express ran into it. The engine of the express ploughed into the rear coach of the accommodation train, lifted it up and forced it into the end of the car in front. The explosion of an oil lamp set light to the shattered woodwork of the car and the fire soon began to spread rapidly. A dozen people saw the fire and raised the alarm and although the fire, which generated very little smoke, made rapid headway the quick arrival of the fire department meant that it was soon extinguished. Scalding steam escaping from the damaged boiler of the express locomotive also added to the difficulty of rescuing the injured passengers. As the accident occurred in the suburbs of a large city there were also problems with spectators hampering the rescue work. 'The crowd attracted by the news of the wreck surged in and about the little station. Curious women forced their way through the crowd to peer through the windows at the dead and dying in the stuffy room. So strong was the rush that the police were compelled to charge the crowd again and again. It was even impossible to remove the wounded from the front way and the rear was used. Backed up against the doors the ambulances and patrol waggons waited and received the victims.'

The passengers in the rear coach of the suburban train had had little warning of danger. A handful had alighted at Eggleston depot, situated about 200 feet south of 71st Street, and the train was just starting away again but only moving at about two miles an hour when it was struck. The passengers sitting at the very rear of the front train saw a flash a moment before the impact as the headlights of the express appeared through the fog.

---

[338] *The Daily Review*, Decatur, Illinois, United States, 10th November 1893

Newspaper reports also stated that, 'Eyewitnesses claim that soon after the accident occurred a portly man climbed into the cab {of the express locomotive} and addressing Dickerman said, "The railroad company won't want you to be seen here." "What's that?" Dickerman is said to have asked. "Why the best thing for you to do is to get away as fast as you can," the stranger is alleged to have replied. There are those by whom it is claimed that Dickerson hastily threw aside his overalls and jumped from the engine. While the work of rescuing the injured and searching for the dead was being carried on, he was nowhere to be seen, nor did he put in an appearance later in the evening.'

Neither Dickerman nor his fireman, William Senzky, had jumped from their engine but both escaped with slight injuries. Five passengers were killed outright and 11 injured, several being severely scalded.[339]

On 10th November it was announced that Flagman Ortman of the suburban train had been arrested for causing the collision by failing to display the proper lights at the rear of the train.[340]

On 15th November the coroner's jury at the inquest in the victims of the accident recommended that not only Flagman Ortman but also Conductor Freeman and a ticket collector (?) be held for manslaughter and censured the Rock Island company for running trains so close together. By this date the number of fatalities had climbed to 12, doubtless because of the number of severely scalded passengers who had died, and the number of stated injured had risen to 30.[341]

On 15th January 1894 the circumstances of the Chicago disaster the previous November were faithfully reproduced at Jersey Meadows, New Jersey, United States, except that in this case it was an express which was run into by an accommodation train. Again, it was a rear collision in fog, a fog which led to a collision because the engineer of the first train was driving more cautiously than the engineer of the second train, which thus overtook him.

The express train involved was the *Newark Express* of the Delaware Lackawanna and Western Railroad, which had pulled out of Newark, New Jersey, at 8.15 a.m., and consisting of eight cars, with a combination smoking and baggage car at the rear, filled with well-off commuters from Summit, Basking Ridge and Short Hills on their way to their daily jobs in New York. Engineer Stitzer was at the controls of the locomotive and Conductor Jeremiah George was in charge of the train. Like half a dozen other railroads in the area the Lackawanna crossed the Hackensack River on a drawbridge.[342] Drawbridges were of course a special hazard, especially when thick fogs made it impossible for engineers to check from any distance whether they were safely closed. Engineer Stitzer therefore approached the Lower Hackensack drawbridge very cautiously, at no more than walking pace. Rear Brakeman D.H. Wheaton probably should have got down to flag the following train, but the rules were unclear and left the matter to his judgement, while he could have run the risk of being left behind if the engineer had suddenly seen that the bridge was closed and put on steam. He could also assume that the

339 *Decatur Daily Republican*, Decatur Illinois, United States, 9th November 1893
340 *The Salem Daily News*, Salem, Ohio, United States,10th November 1893
341 *Decatur Daily Republican*, Decatur, Illinois, United States, 16th November 1893
342 SPV Comprehensive Atlas of North America – North East

engineer of the following train was also advancing cautiously.[343]

Unfortunately Engineer David Hoffman of the following South Orange accommodation train was not driving as cautiously as his colleague. Brakeman Wheaton did eventually jump off his train and was just in the act of putting down a torpedo when the accommodation train loomed up out of the fog in front of him and he was forced to jump down the embankment to save himself. Engineer Hoffman therefore received no warning of the express being in front of him until he saw it less than 200 feet away, when it was far too late to prevent the collision. The combination baggage and smoking car and the car next ahead of it were 'smashed into kindling wood'. Twelve passengers were killed more or less outright, the number of dead later rising to 15, and many others were injured. Engineer Hoffman jumped from the engine and was seriously injured while his fireman, Irving Metz, remained on it and was scalded. Although he survived Engineer Hoffman was thrown against an embankment and knocked unconscious so was not able to give any explanation of the accident, at least immediately afterwards. Fireman Metz said that the first warning that they had had of the express was seeing its lights in front of them at, according to him, only about 25 feet, and his engineer had had no time to apply the brakes. He complained that no signals had been given by the brakeman of the Dover express although it was three minutes behind time. The crew of the Dover express refused to make any comments after the accident. No one on the South Orange accommodation train was much hurt. The collision occurred at 8.30 a.m.[344]

The baggagemaster of the Dover express was in the rear car and chanced to glance through the window and see the accommodation train coming just in time to open a door and jump out. He told a newspaper reporter, "I think the accident was due in part to the recklessness of the engineer of the South Orange. Our train was moving at the rate of about five miles an hour and was pushed 300 feet before it came to a stop. The engine of the accommodation crashed into the rear car about 20 feet. This car, a combination baggage and sleeper {sic}, telescoped into the second car from the rear and split the roof of that car completely from the sides and end. The side of the car heaved in at an angle of about 45 degrees and the men who were forward on the left caught the worst of it. The side of the car ploughed right through them."[345]

Lackawanna President Samuel Sloan blamed Engineer Hoffman for running too fast in foggy weather[346] and stated that, "We have a system of safeguards which are sufficient if humanity can be depended upon to obey its rules." But even at this late date a busy main line on the outskirts of one of the country's major cities was still worked by telegraph and train orders, without any signal protection, either manual or automatic.[347] The inspecting officers of the contemporary British Board of Trade would have had a fit at such primitive working methods!

---

[343] *A History of Railroad Accidents, Safety Precautions and Operating Practices* by Robert B. Shaw, pgs. 174-6

[344] *Daily Advocate*, Newark, Ohio, United States, 16th January 1894

[345] *The Olean Democrat*, Olean, New York, United States, 16th January 1894

[346] *Newark Daily Advocate*, Newark, Ohio, United States, 18th January 1894

[347] *A History of Railroad Accidents, Safety Precautions and Operating Practices* by Robert B. Shaw, pg. 176

The next significant rear collision did not occur for a year and a half, and this time it was an accident in Canada. On the 8th July 1895 the Grand Trunk Railway organised two special trains to carry pilgrims from the town of Sherbrooke, Windsor Mills and Richmond, in southern Quebec, to Lévis, from where they could cross the St Lawrence River by ferry to Quebec City and thence make their way to the important Catholic shrine at Ste Anne de Beaupré, about 30 miles east of Quebec City. The first train, in charge of Conductor Abraham Dionne and Engineer Alexander Ferguson, carried 314 passengers. In charge of the engine of the second train was Engineer Hector McLeod. His train carried about 300 pilgrims.

The first train pulled into St Agapit station at 2.28 a.m. on 9th July and Engineer Ferguson was given orders to advance to Craig's Road, 18 miles before Lévis, and there to hold the main track and cross a westbound freight train. At 2.32 a.m. his train left St Agapit and proceeded towards Craig's Road.

The second train left St Agapit at about 2.50 a.m., so at that point the trains were 18 minutes apart.

The leading train reached Craig's Road, Quebec, Canada, at about 2.55, three minutes after the second train had left St Agapit. The station was unstaffed at night, but was equipped with signals, which were supposed to be put to "Danger" by the train crew themselves. The freight train arrived at about 3.5 a.m. and stopped, then was brought slowly into the siding, but the west end of the siding was blocked by some rolling stock and the freight train had to pull up with its last four boxcars still on the main line while this obstruction was removed. Meanwhile the first excursion train could of course not go on. Conductor Dionne looked back towards St Agapit and saw the second train coming towards the station. It did not seem to be slowing up and had not whistled. He shouted for his brakeman to run back down the track towards it to flag it and signalled for his engineer to draw forward. Engineer Ferguson could not see the danger from his engine but obeyed, though he could not move very far with the boxcars blocking his path. But his and the conductor's actions probably went some way to lessening the severity of the accident.[348]

Brakeman Wheeler managed to get as far as the western end of the siding where he stood swinging his lantern and shouting as the second train passed him at unreduced speed, its whistle and bell not making a sound. A few moments later it crashed into the Pullman sleeping car *Balmoral* at the rear of the first train, pushing it into the penultimate coach which in turn partly telescoped the coach ahead. Eleven passengers were killed instantly or died of their injuries. The engine of the colliding train overturned after the collision and Engineer McLeod and Fireman Richard Perkins also died, along with the coloured porter of the *Balmoral*, whom the newspaper did not consider worth giving a name to. About 40 other passengers and Pullman conductor Morewood were injured.[349]

The accident was clearly the fault of Engineer McLeod as there was ample evidence that the station signal at Craig's Road had been standing at "Danger", that red lights had been properly displayed at the rear of the first train and that the second train had not slowed up appreciably or whistled as it approached the station, as it was obliged to do according to the rules even if did not need to stop. A thorough inquest was held into the

---

348 *Wreck!* by Hugh A. Halliday, pgs. 81-5

349 *Bangor Daily Whig and Courier*, Bangor, Maine, United States, 10th July 1895

accident but necessarily lacked the two people who could have explained why the collision happened. Engineer McLeod was 50 years old, had many years' service with the railway and was regarded as a steady sober engineer. Evidence was however tendered at the inquest by Ludger Perrault, the owner of a hotel at Athabaska, that in the evening of the 8th the engineer had bought a quart of beer there and poured it into his tea jar, which given that he was on duty was a serious violation of the company rules. It was hardly likely that two pints of beer would have made him drunk, but it was also a very warm night and the heat in the engine cab must have been considerable. There was the possibility that the alcohol plus the heat and the fact that it was the early hours of the morning had combined to make the engineer doze off. There was evidence that he had applied the brakes and prepared to jump at the last moment. Fireman Perkins was perhaps preoccupied with attending to his fire or had also dozed off. An alternative possibility, given that Engineer McLeod suffered from asthma, was that he had had an acute asthmatic attack and that the fireman had been trying to assist him.

The inquest brought out the fact that the Grand Trunk did not at this date require regular eye examinations of its employees. Engineer McLeod had only had one in 23 years of service, while Conductor Dionne was colour-blind. The coroner's jury blamed the accident on the negligence and carelessness of Engineer McLeod and Fireman Perkins and recommended an improved system of signalling on the line and yearly eye tests for employees.[350]

A mistake by a signalman led to a collision at Oederan, Saxony, Germany, on 10th September 1895. A goods train was standing at the Einfahrtsignal (home signal) at Oederan station awaiting the arrival of fast train No. 235 to Dresden from the opposite direction. For reasons that were not explained the signalman at Memmendorf cleared his signals for a troop train to proceed towards Oederan despite the fact that the section was still occupied by the goods train. The troop special was carrying the 133rd Infantry Regiment from manoeuvres back to its garrison town of Zwickau.[351] The collision which ensued wrecked five carriages of the military train and a couple of waggons of the goods train. A moment later train No. 235 passed on the other track and was struck by bits of broken glass and woodwork but escaped much damage. Had it been a few seconds later it would have run into the mound of wreckage which by then obstructed the other line and the accident would have probably been a lot worse than it was. As it was, 10 soldiers were killed, 13 seriously and 34 slightly injured.

In December 1895 the Memmendorf signalman was sentenced to five years and a month's imprisonment for manslaughter through negligence and endangering the safety of people travelling on the railway. A memorial was erected to the dead soldiers.[352]

The next accident was the first big railway disaster to happen in Denmark. It was the warm Sunday evening of 11th July 1897. Gentofte station, just outside of København, was filled with happy if tired city dwellers who had been visiting its green spaces and were now on their way home. A train from Holte arrived just after 11 p.m. and the many

---

[350] *Wreck!* by Hugh A. Halliday, pg. 85

[351] *The Times*, London, England, 21st September 1895

[352] *Eisenbahnunfälle in Sachsen* by Claus Schlegel, pgs. 75-6

passengers got on, but because of their number there was some delay and it was 11.15 before the train was ready to depart. Its conductor was just checking that all the carriage doors were closed when he saw another train approaching the station. The home signal protecting the standing train was at "Danger" but the train coming down the incline to the station did not seem to be pulling up. By the time that he had realised the danger it was too late to do anything and the next moment the incoming train ploughed into the open carriages (skovvogner) which were at the rear of the Holte train. The impact crushed the flimsy rear carriages like matchboxes, the vehicles in front of them were lifted on top of one another and the front carriages of the colliding rain were apparently also telescoped. Thirty-two passengers died on the spot, eight more died later, 55 were seriously injured and many more slightly hurt.[353 354]

The train which had run into the Holte one was a special from Helsingør to København (Copenhagen), booked to run non-stop from Holte to København. Its driver, Carl Hansen, survived the crash uninjured and after the accident walked around the station telling anyone who would listen that the accident had not been his fault as his vacuum brakes had failed. He said later that when he had heard the angry voices of some of the passengers, he had wondered why he had not died along with the others.

After the accident the Gentofte stationmaster immediately tried to contact København for assistance, but it was a quarter hour before he could get through. Meanwhile the station staff and uninjured passengers did what they could to help the injured. After three hours an ambulance train arrived from the capital with 15 doctors, porters and police. Help was also provided by soldiers from nearby Jægersborg but they had difficulty in keeping order among the many spectators who flocked to the scene of the accident and there were reports of rescuers coming across looters. In København the local magistrate was advised of the disaster at 2 a.m. and ambulances were arranged to convey the injured from the station to various hospitals, but the operation was very slow because of lack of resources. It was only the next day that it was learnt that many more ambulances could have been provided by the military authorities had they been informed of the accident.

Those who lost their lives in the accident included the mother of singer Dagmar Hansen, while the writer Sophus Micha'lis and his wife were injured.

In view of Driver Hansen's allegation of a brake failure the locomotive was carefully examined after the accident, but nothing could be found wrong with it. On 27th October the court at Oldenburg sentenced him to four months' imprisonment. The state railway treated him harshly. He was dismissed, lost his pension, and was faced with a demand for 44,875 kroner to cover the cost of the damage – a sum probably far beyond what he was ever likely to be realistically able to pay.[355]

The accident created a lasting sensation in Denmark and a broadsheet published about it sold 3,000 copies. It was said that passenger numbers on all the Danish railways dropped considerably for a time after the accident, doubtless because people were afraid to travel.[356]

---

353 Webpage – Blodet flød på Gentofte Station
354 Webpage – Gentofteulykken 1897
355 Webpage – Blodet flød på Gentofte Station
356 Webpage – Gentofte 1897

Having escaped serious rear collisions for a while the United States had another on 21st August 1898 on the New York New Haven and Hartford Railroad. On that Sunday evening traffic was heavy and the 6.08 p.m. train from Providence, Rhode Island, to Boston, Massachusetts, which was usually combined with one from New Bedford, Massachusetts at Mansfield, was run as two sections. The first section, running as a local accommodation train, was 13 minutes late when it reached Sharon, Massachusetts at 7.15 p.m., having lost the time since leaving Mansfield. It was due at Canton Junction, its next stop, two minutes before the express, which was to pass it there. Sharon station was situated on a curve and protected by 'electric block signals'. While it was standing at the depot, however, it was run into by an express from New Bedford, which demolished all of the rear car of the local train but its roof. As had happened so often previously, escaping steam from the boiler of the express engine contributed to the injuries by scalding the occupants of the wrecked car. Six passengers were reportedly killed and 26 injured.[357]

The conductor of the first section had not sent back a flagman and Engineer Daniel Getchell of the express had apparently failed to realise that there was a train in front of him until he was 400 feet from it, when he had belatedly applied the brakes and whistled.

Engineer Getchell stood trial for manslaughter at Dedham on 16th December. Like Driver Hansen at Gentofte he admitted that the signals had been against him but claimed that his brakes had failed. The jury did not believe him, and he was found guilty.[358]

Four months after the accident, on 20th December, Getchell was sentenced to two years in the house of correction.[359]

It was the turn of Belgium to suffer a major disaster on Saturday, 18th February 1899. A train which had left Tournai at 5.22 a.m. arrived at Forest station, Brussels at 8.14, 10 minutes late. While it was standing at the platform it was run into from behind by the 6.47 a.m. departure from Mons, which ran fast from Hal, and which having overrun the signals in fog was reportedly still travelling at full speed when the collision took place. The locomotive of the colliding train reared up on the roofs of the last three carriages of the Tournai train, its driver not even having had time to shut off steam. As might be expected at that time of morning the carriages were filled with passengers, and it was in them that most of the casualties were found. The *Times* correspondent in Brussels described the scene. 'The greater number were third-class passengers. The heavy engine broke through the roofs of the crowded carriages and crumpled them up into an appalling and inextricable tangle of splintered wood, fractured metal and human bodies. Six bodies were found intertwined in the wheels of the locomotive.' [360]

A relief train was despatched as quickly as possible from Brussels-Midi while nine doctors, several members of the Red Cross Society and Sisters of Charity nuns and 10 medical students assisted with the rescue. Some of the injured were conveyed to hospital or Brussels-Midi station and others were reportedly taken to various estaminets (small cafés selling alcoholic drinks). Many of the victims refused to divulge their names for

---

[357] *North Adams Transcript*, North Adams, Massachusetts, United States, 22nd August 1898
[358] Ibid, 16th December 1898
[359] *The Fitchburg Sentinel*, Fitchburg, Massachusetts, Massachusetts, 21st December 1898
[360] *The Times*, London, England, 20th February 1899

fear of upsetting their families.

The *Times* correspondent continued that, 'The spectacle of the dead and wounded, many of them partially buried in the wrecked train, was appalling. Here one saw an old man with his head completely crushed, there a man with his body bent double. In another place lay a schoolboy with his class books strewn round him. A Government clerk and an official of the Cour des Comptes {an administrative court charged with conducting public audits} were terribly mutilated. The greatest difficulty was experienced in extricating the bodies crushed underneath the locomotive. Some of these were fearfully mangled. A governess was rescued alive after being buried for three hours after the engine, surrounded by three dead bodies. The engine driver and stoker of the Tournai train are said to have escaped injury, but they rushed away across country in a state bordering on madness and have not been seen since. All the officials of the telescoped train escaped, but the son of the stationmaster of Ruysbroek, who was among the passengers, was killed. The prefect of St Michaels College, finding that 15 of the students were missing this morning, went at once to the scene of the disaster. The train service from Brussels for Paris is being run by way of Baulers, and the Lille traffic is being sent via Enghien.' [361]

What was only the second railway accident of any magnitude on the small railway network of New Zealand occurred on 11th March 1899. Yet again, it involved an excursion train, or rather two trains. The employees of the Islington Freezing Works meat processing factory and their families and friends had gone on their annual picnic from Canterbury to Ashburton, and the number of passengers dictated that two trains were required to carry them all. The two trains reached Ashburton without any problems, the excursionists spent the day there, and the return journey was commenced, the two trains leaving Ashburton some minutes apart. When the first double-headed train reached Rakaia station it had to wait there to cross a southbound train from Canterbury to Ashburton, which was 20 minutes late, and fortunately a number of the passengers alighted during the wait for the Canterbury train to stroll along the platform. Some of these passengers then became aware that the second excursion train was approaching the station at a pace which made it obvious that it was not likely to be able to pull up in time. They shouted to their fellow passengers who were still on the train to jump for their lives and some of them got out. At the same time Guard J. Curson of the first train signalled to his drivers, Gardiner and Hyland, to set it in motion, though he could not have gone very far in any case before he reached the end of the crossing loop (cf Craig's Road). The guard's action was credited with considerably mitigating the effects of the collision that followed.

The second train had left Ashburton without Head Guard William Climpson. Having given his driver the signal to depart the guard had found that he was unable to get to his van because of the crowd of people on the platform and he was left behind. The Ashburton stationmaster tried to telegraph the station at Chertsey to tell them, but could get no answer, probably because the stationmaster was out on the platform, so the second train went on to Rakaia with its driver unaware that he lacked a guard. However, Assistant Guard Joseph Burrow was relieved of his ticket-collecting duties by the secretary of the picnic committee, arranging for two of his men to do the job, so the

[361] Ibid, 20th February 1899

assistant guard could attend to his guard's duties, while another passenger, V.J. Fitzpatrick, was assigned to help Burrow work the brakes. The secretary, Fisher, was also worried about the excessive speed of the train and mentioned this to Burrow, who showed a green light to the engine crew, which was apparently not observed through the rain and smoke.[362]

The wayside stations on the lines were not provided with fixed signals but station clerk James O'Neill was sent to the entrance points with a red lamp to flag the second train. He signalled to the excursion and Driver Carter gave three short sharp whistles to "down brakes" for his guards to assist him in stopping the train, which was not fitted with continuous brakes, only the engine and tender having Westinghouse brakes. Guard Burrow, assisted by passenger Fitzpatrick did so, Driver Carter applied his air brake and reversed the engine, sand was put down and his fireman, F.D. Mather, applied the screw brake on the tender, but the rails were wet and slippery from rain which had fallen and there was a strong wind behind the train. It could not be stopped, and the collision occurred, telescoping the guard's van and three carriages at the rear of the first train.

Panic reportedly ensued after the collision, with men and women literally fighting one another to get out of the wrecked carriages, though by then the danger had passed, unless they were afraid of fire breaking out, which was, no doubt, a very real possibility. It did not but a Miss Franks, a Miss Harwood, a boy named Bowden and a baby [363] whom one of the women had apparently been looking after on the journey home were killed and about 30 other passengers were injured, some of them seriously. In the same manner as today when accidents abroad are reported the reporter of a Melbourne newspaper reassured his readers that 'there were no Victorian people amongst the killed or wounded.' It was all right, it was only foreigners who were the victims.

Happening in a country which like Denmark did not suffer very many bad railway accidents, the collision caused much public concern, and in addition to the coroner's inquest the government ordered a public inquiry, which was to examine not just the circumstances of the accident but the whole management of the railway. The newspaper commented, 'A strong feeling prevails that too much consideration is paid to showing off a big credit balance in the railway accounts, and too little to the efficient equipment of the lines, both in regard to the permanent way and rolling stock. Experts are of opinion that had the train which caused the disaster been furnished with efficient brakes the accident would not have happened. What the colony will have to pay in damages would have gone a considerable way in providing brakes of the best description.' It was surmised that the driver of the second train, attempting to make up time, had got up too much speed and had then been unable to stop with wet rails and inadequate brake power.[364]

On 20th March the coroner's jury passed a verdict of manslaughter against Driver Carter, and they also made a series of recommendations. They called for improved signalling at Rakaia, that passenger trains should be supplied with continuous brakes, called for sufficient and capable guards and found fault with the Ashburton stationmaster for not informing Driver Carter that his train's head guard had not boarded the train. [365]

[362] *Tragedy on the Track*, by Geoff Conley and Graham Stewart, pg. 74
[363] *Launceston Examiner*, Launceston, Tasmania, 23rd March 1899
[364] *The Age*, Melbourne, Victoria, 24th March 1899
[365] Ibid, 22nd March 1899

In June it was reported that Driver Carter had been acquitted. 'He had not his train very well in hand, but he was proved to be sober, and in general was careful. The night was dark, the signalling system at Rakaia was elementary and bad; Carter was not responsible for the fact that the chief guard had been left behind; he has lost his billet, there is sympathy for him, and so he was acquitted. The Opposition will, no doubt, make something of the accident during the session, for there is a suspicion that the Railway department is the real culprit.' [366]

The public enquiry also blamed Driver Carter for the accident, and he lost his job, though it has been suggested that he was a scapegoat for a railway suffering growing pains and whose operational methods, which were certainly primitive by the standards of the colony's mother country, had not kept pace with increasing traffic.[367]

A major disaster on the Philadelphia and Reading Railroad, at Exeter, Pennsylvania, United States, on Friday 12th May 1899, once again involved an excursion train. The first section of train No. 12, an express train booked to leave Reading for Philadelphia, started late and was stopped at Exeter, only about four miles from Reading, because a northbound coal train had stopped due to problems with its brake rigging at Birdsboro, further down the line. While it was standing there, it was run into by an excursion train, running as the second section of train No. 12 and carrying people returning from Harrisburg, the Pennsylvania state capital, where a memorial to Union major general and later Republican politician John F. Hartranft had just been unveiled. [368] The excursion train was travelling at 45 miles per hour when it ploughed through the two rear day coaches and the Pullman parlour car *Orpheus* of the express, while the leading three coaches of the colliding train were telescoped against the locomotive. The death toll was high, and it was initially reported that 25 had been killed, a figure which later rose to 29. The passengers on the excursion were mostly from Norristown, and 16 of them lost their lives. Engineer Harry Orrell and Fireman Everett of the second section jumped at the last moment and escaped. The collision occurred at 8.51 p.m.[369 370 371]

A newspaper reported that, 'At least 10,000 persons visited the scene of the wreck yesterday {Sunday 14th}. They came from places 20 and 30 miles distant. Railroad trains brought crowds, others came by carriages and bicycles, and hundreds came on foot. Relic hunters kept the police busy. Almost everybody was bent on carrying away some kind of memento. Many of the people did not hesitate to take away with them pieces of wood thoroughly saturated with human blood.' [372]

The inquest into the deaths of those killed in the disaster opened at Reading on 22nd May and on its first day heard evidence that the trains were running but three minutes apart, while Rear Brakeman Charles Miller of the express testified that he had not had time enough to run back very far to protect his train. That he had gone back promptly was

---

366 *The Australasian*, Melbourne, Victoria, 24th June 1899
367 *Tragedy on the Track* by Geoff Conley and Graham Stewart, pg. 77
368 Wikipedia – John. F. Hartranft
369 *Tyrone Daily Herald*, Tyrone, Pennsylvania, United States, 13th May 1899
370 *New Oxford Item*, New Oxford, Pennsylvania, United States, 19th May 1899
371 *Train Wrecks of Eastern Pennsylvania* by Charles J. Adams III and David J. Seibold, pgs. 123-137
372 *New Oxford Item*, New Oxford, Pennsylvania, United States, 19th May 1899

confirmed by Conductor M.W. Brommer and Engineer Dan Wildermuth of the express and he was also seen running back by Robert A. Jackson, a passenger on the colliding train and Conductor Edward Sentman, in charge of a freight train stopped at Exeter. Evidence was also given of the robbing of dead and injured passengers by parties whom it was believed could be identified.[373]

The second day of the inquest included testimony from Engineer Orrell of the second section, who was asked if he thought that the collision could have been avoided? He had replied, "Yes, if the rear brakeman of the express had come back and signalled me, and then if the signal board indicated that the express train was still there why I should have stopped my train." Engineer Orrell denied that he had not been given red signals at any of the four towers between Reading and Exeter, but one of the towerman, Thomas H.B. Geiger, declared positively that he had given Orrell's train a red signal, and towards the end of the day's proceedings train dispatcher James Rourke directly accused Orrell of "reckless running". It also emerged that the first section had been setting back at the moment of the collision; after having overrun the station signal at Exeter.[374]

On 24th May, the third day of the inquest, the jury reached its verdict. It censured Philadelphia Trainmaster George Bowers for using poor judgement in equipping the excursion train with a crew unfamiliar with the main line division and condemned Engineer Orrell and Conductor Magee of the excursion for careless and reckless running, plus Engineer Wildermuth for violating the rules in running past the signal board at Exeter and then setting back northwards 236 feet. Upon the evidence submitted by Trainmaster Blaine and Assistant Trainmaster Bingaman they also found Dispatcher Rourke at fault for not notifying the crew of the second section of the fact that the first section was delayed. They emphatically condemned the independent tower system {?} operating below Reading at the time of the accident and recommended the cutting down of trees obscuring the view of the signalman in Tower 22. They recommended that the Reading company inaugurate a 10-minute interval between trains as five minutes was too dangerous, especially at night, and rigidly enforce the strict obedience of all rules by its employees.[375] Not surprisingly the jury took eight hours to reach its verdict.

Engineers Wildermuth and Orrell, Conductors Brommer and Magee and Train Dispatcher Rourke were subsequently all been arrested on a charge of misdemeanour to appear at the Berks county quarter sessions on 12th June.[376]

A couple of days later a different coroner's jury at Norristown, at an inquest on the deaths of one or more passengers who died later of their injuries, passed a verdict exonerating all the railwaymen involved in the accident and placing the entire blame on the Philadelphia and Reading company.[377]

In July Dispatcher Rourke was cleared of blame by the company and reinstated.[378] The eventual fate of the two engineers and two conductors is unclear.

Exeter was renamed Lorane after the accident.

---

373 *Tyrone Daily Herald*, Tyrone, Pennsylvania, United States, 23rd May 1899
374 Ibid, 24th May 1899
375 *Gettysburg Compiler*, Gettysburg, Pennsylvania, United States, 30th May 1899
376 *Tyrone Daily Herald*, Tyrone, Pennsylvania, United States, 26th May 1899
377 Ibid, 30th May 1899
378 *Gettysburg Compiler*, Gettysburg, Pennsylvania, 18th July 1899

France suffered a major disaster on Saturday 5th August 1899. It was the beginning of the school holidays, passenger traffic was heavy, and further increased by the sale of cheap bains-de-mer (sea bathing) tickets, so as usual at busy periods Chemin de fer de Paris à Orléans train No. 29, the 9.40 p.m. departure from Paris-Orléans station to Le Croisic on the Atlantic coast, was being run in two parts. In accordance with normal practice the relief train No. 29 bis, with Driver Batch and Fireman Meslê, departed at the normal departure time of 9.35 and train No. 29 followed it at 9.40, with Driver Paquet and Fireman Buron. There were four tracks as far as Juvisy-sur-Orge station, but only two beyond that point. and at 10 p.m. train 29 bis was brought to a stand at a carré (stop signal) 100 metres beyond the station to await the clearance of omnibus train No. 243, which had left Paris at 8.38 p.m., and which the two sections of train No. 29 were timetabled to overtake at Brêtigny. Delayed by half an hour because of traffic disruption caused by a severe thunderstorm which was raging at the time, this train had only just left Juvisy for the next station at Savigny-sur-Orge. Although it was running behind time Assistant Stationmaster Joseph Delaire had decided to let it continue as booked to Brêtigny in front of the two sections of the express.

Train No. 29 arrived at Juvisy-sur-Orge seven minutes later and it was only at the last moment that Driver Paquet, who had been running under clear signals, glimpsed the red tail lights of the first train through the heavy rain and flashes of lightning. An emergency brake application did not prevent his engine ramming the brake carriage at the rear of the relief train, crushing it and the carriage in front and throwing the vehicles ahead off the track. Seventeen passengers were killed and 85 injured, one of whom died later. Fireman Buron of the colliding train was the only railwayman injured. The sound of the thunderstorm reportedly drowned out the noise of the collision so that the station staff remained unaware of it until they were alerted by the conductor of the first train, who had had time to jump from his van before the collision and had gone back to protect the obstruction.[379]

The only fortunate feature of the disaster was that it happened on the outskirts of Paris, and close to a police station and a chemist where the staff were still working. Help therefore came quickly and a large number of rescuers, including three doctors, were soon on the scene. The chemist, which had closed at the end of the day's business, was re-opened, and the casualties were treated there and in a room at the station, before a special train or trains took 85 of the injured back to Paris-Orléans station and thence to various hospitals or to their homes. All the deaths and most of the serious injuries occurred to the passengers in the rear third-class carriage, which was also the most crowded of those on the train. However, somehow a young woman and her 15-month-old baby in this coach escaped unharmed.

The question of how the accident happened was the subject of both a judicial inquiry and an investigation by the Ministry of Public Works. The first train should have been doubly protected. Firstly, the Lartigue electro-semaphore block system had been installed between Paris and Brêtigny in 1881 and the Latigue signal at "Danger" should have stopped the second train. In addition, there was a carré empéchant (home signal) worked by the signalman at the entrance to the station.

[379] Wikipedia – Accident ferroviaire de Juvisy-sur-Orge le 5 août 1899

Investigation determined that it was not that Driver Paquet had overrun the signals, either because he had failed to see them due to the bad weather or because he had assumed that, as usual, they were clear, but rather that the accident was due to a wrong-side signal failure combined with a signalling error. A squall of wind had damaged the mechanism by which red or green spectacles were moved in front of the signal light such that only a white "All clear" light was shown to the driver. As for the carré empéchant this had, through an oversight by the assistant stationmaster and the signalman, not been put to "Danger" behind train No. 29 bis. Driver Paquet was therefore cleared of all responsibility for the accident, but Assistant Stationmaster Delaire and Signalmen Alexandre Derouet and Alfred Monjeaux (they had just been changing over shifts) were not, and they were charged with manslaughter and causing injury through negligence.

The trial of the three accused railwaymen took place at the Corbell Criminal Court at the end of December. The 27 witnesses included astronomer Camille Flammarion, who made the dubious assertion that the thunderstorm could not only have affected the signals but made the railwaymen more likely to make mistakes. The prosecution did not in any case press their case too hard. Signalman Monjeaux was acquitted on the grounds that at the time of the accident he had only just come on duty, and Assistant Stationmaster Delaire and Signalman Derouet were each sentenced to a suspended sentence of a year's imprisonment and a fine of 100 francs.

Following the accident Minister of Public Works Pierre Baudin issued orders which stipulated that stationmasters' hours were to be limited to a maximum of 12 in every 24 and that the training of employees should be improved. Measures were also taken to improve the working of the Latigue signals and to quadruple the line between Juvisy-sur-Orge and Brêtigny (completed in June 1904).[380] [381]

The (Second) Boer War had begun on 11th October 1899 with an offensive by the Boers into the British-held colonies of Natal and Cape Colony.[382] On the 13th the *Times* correspondent reported an accident on about 12th October involving two trains carrying refugees, mostly from the Orange Free State, at Seven Sisters Road in Cape Colony. The identity of the refugees was not stated but they were presumably non-Boer settlers (uitlanders). The second section of a refugee train had crashed into the rear of the first section in a siding. The last of three brief reports on the accident put the casualties at nine dead, 10 seriously injured and 'many others' slightly hurt.[383]

The Dutch Maatschappij tot Exploitatie van Staatsspoorwegen company had already had a bad year in 1899 with several relatively minor mishaps when on 15th November the worst Dutch railway accident of the 19th century occurred near Capelle.

On the morning of the day of the accident trains Nos. 80, 60 and 80 bis were following one another along the Capelle Aansluiting – Gouda line, trains Nos. 80 and 80 bis from Rotterdam Delftsche Poort and train No. 60 from Rotterdam Maas. The day was

[380] Ibid.
[381] *The Times*, London, England, 7th August 1899
[382] Wikipedia – Second Boer War.
[383] *The Times*, London, England, 16th October 1899

extremely foggy. East of the junction at Capelle Aansluiting, where the lines from Rotterdam DP and Rotterdam Maas met, was the simple block post of Capelle, working just a pair of stop signals at a level crossing. Train No. 80 passed Capelle without stopping and the signalman placed his signal against train No. 60, which was closely following. Due to the thick fog the driver of this train failed to see the Capelle signal until the last moment and ran past it, but at the same moment the signalman was advised by his colleague at Nieuwerkerk that train No. 80 had passed there and cleared the signal anyway. The signalman assumed that the train crew had seen the signal clear and continued, but instead the train had stopped, though with slippery wet rails its rear was 180 metres beyond the signal before it pulled up. The Capelle signalman replaced his signal and cleared back to his colleague at block post 14.

Meanwhile train No. 80 bis was drawing up to the signal at block post 14 as the signalman there received the "All clear" for train No. 60 from Capelle. He cleared his signal but had not time to pick up the detonators that he had put down, so gave the drive a wave to indicate that the line was clear and to ignore the detonators going off.[384]

The Capelle signalman had put down fresh detonators after train No. 60 had passed and his stop signal stood at "Danger". The driver of train No. 80 bis, heard the detonators explode, saw the signal at danger and heard someone whistling an alarm signal. He immediately applied the brakes, and the head conductor in his van screwed down his brake, but the slippery rails prevented the train being pulled up very quickly and it collided with the rear of train No. 60.

When train No. 60 came to a stand its head conductor had set off back to the signal box to find the cause of the stoppage. Just as he had got there he had heard train No. 80 bis approaching. Thoroughly alarmed, he had whistled and gave alarm signals, as had the porter at Capelle Halt, on the far side of the level crossing, where the latter was waiting with passengers for train No. 201 to Rotterdam.

The rear carriage of train No. 60 was completely destroyed in the collision and the end of the next coach was damaged. Of the passengers in the rear carriage five were killed outright and 15 injured, one of whom died while being taken to Rotterdam, another later the same day and an eighth the next day. None of the railwaymen were hurt.

A recent Dutch author has described the accident as having been basically the result of an unfortunate chain of circumstances, but it led to alterations in the signalling regulations for working during fog. The two main Dutch companies went in different directions, the Holladsche Ijzeren Spoorweg Maatschappij ordering signalmen, if visibility was less than 300 metres, not to accept a following train until the previous one was at least 400 metres in advance of the block post, while the Maatschappij tot Exploitatie van Staatsspoorwegen went in for what in Britain would be termed double block working, that is during foggy conditions no train was to be accepted from the box in the rear until the previous train had been cleared by the box in advance.[385]

Deficiencies in the fog-signalling arrangements also came to light as a result of a collision in England on 23rd December 1899. The accident occurred at Wivelsfield

[384] *Spoorwegongevallen in Nederland* by R.T. Jongerius, pgs. 79-81
[385] Ibid.

station, near Keymer Junction, on the main line from Brighton to London of the London Brighton and South Coast Railway. The 5.35 p.m. up boat train from Newhaven was standing at the Wivelsfield starting signal awaiting the road when it was run into at about 6.14 p.m. by the 5.35 p.m. up Pullman express from Brighton, which had started seven minutes late. There was a thick fog at the time. The brake van, second-class carriage and composite carriage which formed the last three vehicles of the boat express were almost completely broken up and part of the composite carriage fourth from the rear was also broken up. The engine of the Pullman express was overturned, the leading brake van was broken up and telescoped by the composite carriage behind it and further back the train a first-class saloon carriage was telescoped by the Pullman car behind it. A small fire broke out in the wreckage.[386] One passenger in each train and the Rear Guard Davis of the boat train were killed outright and three more of the boat train passengers died later. Seventeen passengers from one or other train, two guards and a fireman were injured.

The line was controlled from south to north, as far as the main line from Brighton was concerned, by signal boxes at Burgess Hill, Keymer Junction (where the Newhaven branch joined the main line) Wivelsfield and Folly Hill. The boat train had left Newhaven on time, came onto the Brighton line at Keymer Junction and was brought to a stand at Wivelsfield at about 6.9 p.m. as the previous up train, the 5.20 p.m. passenger train from Lewes, had not cleared the next block section to Folly Hill. The latter cabin cleared back the Lewes train and accepted the boat train at about 6.13 (6.14 in the Wivelsfield train register book), the Wivelsfield starting signal was then lowered and the Newhaven train was just setting off again when it was run into by the Brighton Pullman train.[387]

The Pullman train almost invariably preceded the boat train but because it had started late the Keymer Junction signalman had given the boat train from the Newhaven branch precedence. The delayed departure of the Pullman express was due to the very late running of the 3.50 down train from London, the vehicles of which were normally formed the Pullman train. That train had in turn been delayed by the fog and heavy Christmas traffic. Its lateness meant that the Pullman train had instead to be marshalled from other rolling stock at Brighton, resulting in delay. Driver Alfred Westlake of the Brighton train had received clear signals as far as Burgess Hill, but he failed to see, because of the fog, the Burgess Hill up starting signal, which was at "Clear", or the Keymer Junction up distant or home signals, which were without doubt at "Danger", and having failed to see them did not reduce speed until his fireman, William Franklin, shouted out upon seeing a red light waved by Charles Viney, the Wivelsfield signalman. Then he had braked hard and reversed his engine but could not, of course, prevent the collision. Driver Westlake estimated his speed between Burgess Hill and Wivelsfield as having been 35 miles an hour; the various signalman thought it was closer to 50 miles an hour. Colonel F.A. Marindin, the Board of Trade inspecting officer, calculated from the times recorded in the various signal box train register books that the driver's estimate was probably more accurate.[388]

The fog had come on at 3.30 p.m. and there should have been a fogsignalman at the Keymer Junction up distant signal, but there was none. As Keymer Junction box was under the supervision of Wivelsfield station it was technically the responsibility of the

---

[386] *The Times*, London, England, 25th December 1899

[387] Board of Trade report

[388] Ibid.

stationmaster there to see that a fogsignalman was provided, but as the man was a member of the Burgess Hill platelayers' gang it had become the practice for the Keymer Junction signalman to telegraph Burgess Hill station when he required one. Signalman Richard Edwards at Keymer Junction had duly wired Burgess Hill at 3.45 p.m. and three more times in the next two hours, but it was not until he telephoned the Wivelsfield stationmaster at 6 p.m. that a fogsignalman was finally provided, too late. The regular fogsignalman, Alfred Lewry, had been sick for some weeks and his substitute, a man named Moore, had gone home to Brighton at 1 p.m., it being a Saturday. In Moore's absence the job should have been done by Ganger Tullett, but he too was sick. Finding that none of these men were available the Burgess Hill stationmaster, Alfred Tranah, had then made attempts to find other staff, but there was a misunderstanding between him and a porter and as a result no fogsignalmen had been stationed at either the Keymer Junction or Burgess Hill up distant signals by the time of the collision.

Colonel Marindin found that Driver Westlake was primarily responsible for the accident as he had failed to treat the Keymer Junction up distant signal, which he could not see through the fog, as a danger signal, in contravention of Rules 47, 51 and 61A. He had relied upon his train having a clear road, as it normally had, and on the absence of any warning from the fogsignalman, whom he had assumed must be stationed at the Keymer Junction up distant signal. He had a reputation as one of the best drivers on the line.

The inspecting officer did not find fault with the Wivelsfield stationmaster for not seeing that the Keymer Junction up distant signal fogsignalman was in place as he was a relief stationmaster who had only been there five days, but he criticised Signalman Edwards at Keymer Junction for letting matters drift from 3.30 until 6 before contacting his own stationmaster. He could also have had someone positioned at his box to lay fog signals there prior to the arrival of the fogsignalman, a task which given his age and the height of the box he, Edwards, would have found difficulty in undertaking. The Burgess Hill stationmaster, had he been sufficiently alive to the dangers of not having a fogsignalman at the Keymer Junction up distant, should also have taken more pains to provide one, if need be using one of his porters, though he had pleaded that he was short staffed, and the station was busy with Christmas traffic. Finally, Ganger Tullett, knowing that he was the only man to take Lewry's place, ought when he had felt too ill to go out fog-signalling himself, have let the stationmaster know so that the latter could find someone else.

Turning to the lessons to be learnt from the accident Colonel Marindin stated that firstly the accident showed the need for the strictest possible discipline to ensure that all servants obeyed the rules and regulations, and secondly demonstrated the need for a better system of fog-signalling. He suggested also that the Brighton Company should introduce double block working during fogs, whereby a signalman did not give permission for a second train to approach until he had had the "Train out of section" signal for the previous train from the box in advance. Thus, in the case of the trains involved in the accident the Keymer Junction signalman would not have accepted the Pullman train from Burgess Hill while the boat train was standing at Wivelsfield. He also recommended improved arrangements for the calling out of fogsignalmen and their substitutes and an alteration to the signalling at Wivelsfield, where at the time of the accident the cabin had no separate up home or distant signals but merely controlled (slotted) the Keymer Junction up signals. Finally, he recommended the provision of a second lower arm on the tall Keymer Junction up home signal, so that it was more visible to a driver in a fog.[389]

Two inquests were held on the deaths of those killed in the accident, at Wivelsfield and (on those who had subsequently died of their injuries) at Guy's Hospital, London. Both returned verdicts of "Accidental death". The Guy's Hospital jury found Driver Westlake, Signalman Edwards and Stationmaster Tranah guilty of negligence not amounting to manslaughter. The Wivelsfield jury found only Driver Westlake guilty of negligence while criticising Ganger Tullett for not contacting the stationmaster. This jury considered that the stations at Wivelsfield and Burgess Hill were understaffed, that all home signals should have a low lamp, that the Keymer Junction up distant should be repositioned, that all important signal boxes should be provided with a second man or intelligent youth and that fogsignalmen should live close to where they were needed.[390]

Colour light signals, with their much greater visibility in fogs, would almost certainly have prevented the accident, and would have made fogsignalmen unnecessary, but in 1899 the technology had not advanced sufficiently for it to be possible to provide a sufficiently powerful light source for such a signal.[391]

On 28th March 1900 seven passengers died in a rear collision in a tunnel at Charing Cross, Glasgow, Scotland, Scotland, on the North British Railway. This accident is described in the Tunnel Accidents chapter, as is the accident at Philadelphia, Pennsylvania, United States on 11th May 1900.

A collision on the Illinois Central Railroad at Vaughan, Mississippi, United States, on 30th April 1900, received little attention at the time and resulted in only one fatality but the engineer who was killed, John Luther (Casey) Jones, was later to become almost the patron saint of American railroad engineers. Jones was driving engine No. 382 with passenger train No. 1, which he had taken over at Memphis Tennessee, along with his negro fireman, Simon Webb. The train left Memphis at 12.50 a.m. 95 minutes late, but Jones was determined to arrive at Canton, Mississippi, 189 miles to the south, on time at 4.45 a.m., even though this would entail some fast running over a single line with opposing trains to cross at various points. Jones did in fact manage to regain an hour over the uncongested line between Memphis and Grenada, Mississippi, but beyond there he had to cross passenger trains No. 2 and first and second No. 26 and freight train No. 72 while following passenger train No. 25 and freight train No. 83. Passenger train No. 2 was crossed at Goodman with the loss of five minutes and the train continued at speed towards Vaughan. Here, as Jones' train approached from the north there were four trains waiting; passengers trains Nos. first and second 26 northbound, freight train No. 72 northbound and freight train No. 83 southbound. In the course of the rather complex set of manoeuvres necessary for the northbound and southbound trains to cross one another, given that the combined length of the two freight trains exceeded that of the siding, train No. 83 had had to set back onto the single line at the north end of the station. It should have been able to move southwards again before the arrival of train No. 1 but the bursting

389 Ibid.

390 *The Times*, London, England, 26th, 28th, and 30th December 1899 and 2nd, 3rd and 6th January 1900

391 *Spoorwegongevallen in Nederland* by R.T. Jongerius, pg. 139

of an air brake hose delayed matters. Even so there should have been no accident, as Flagman J.M. Newberry had gone northwards to protect the obstruction. Fireman Webb testified at the subsequent investigation that train No. 1 had been travelling at 75 miles an hour when it exploded the torpedo (detonator) put down by the flagman and that he had also seen the flagman's lamp. On hearing the noise of the explosion Engineer Jones had belatedly applied the brakes but Fireman Webb, seeing that their train was not going to be able to pull up in time, had then jumped and he escaped with a severe bruising. Jones stayed on his engine, sounding its whistle to warn the conductor of freight train No. 83. Three other railwaymen were slightly hurt and engine No. 382 and the mail car of the passenger train plus the caboose and two box cars of the freight train were damaged.

Blame for the accident was placed entirely on Jones, who had clearly approached Vaughan much too fast in the expectation of having a clear road. No blame was attached to Flagman Newberry, who just previously had flagged train No. 25, which had easily been able to stop.[392 393]

A short time after the accident a negro engine wiper named Wallace Saunders, who was a close friend of Jones, wrote some verses about Jones' bravery in sticking to his post and sang them to his own melody. Then in 1902 the ballad came to the attention of a songwriter who changed and improved the words but kept the melody, and a legend was born.[394]

Six months after the Wivelsfield accident another bad collision occurred in England, this time on the Great Western Railway. On 16th June 1900 there was a collision at Slough station. The 1.05 p.m. train from London Paddington to Windsor was standing at the down main platform there when it was run into from behind at about 1.41 p.m. by the 1.15 p.m. down express from London Paddington to Falmouth. The two rear vehicles of the Windsor train were completely destroyed in the collision and the next two had their ends knocked in. The express engine was much damaged and five of its coaches were also damaged. A small fire broke out in the wreckage, one of the carriage roofs having ended up on top of the chimney of the express engine but was extinguished before it could do much harm. Five passengers were killed or died from their injuries, 35 were seriously injured and another 90 slightly hurt or suffered from shock.

Slough was the junction for the Windsor branch and the station was served by two running lines in each direction. Windsor branch trains normally used the down main line platform to avoid them having to cross both up and down main lines to reach the branch. Slough Middle signal box controlled the junction with the branch, there was another box at the London end of the station (Slough East) and the next block post in the rear of Slough East was Dolphin Junction, 1 mile and 68 yards closer to London. Absolute block working was in force between the three boxes but when the line through the station was occupied the Slough East signalman was authorised to accept trains from Dolphin Junction under the "Warning arrangement" whereby the train would be stopped at Dolphin box and its driver informed by the signalman that he had been so accepted, and

392 Webpage – Casey Jones Accident Report

393 *A History of Railroad Accidents, Safety Precautions and Operating Practices* by Robert B. Shaw, pgs. 290-3

394 Webpage – The True Story of Casey Jones

that he could only expect to find the line clear to the Slough East home signal.

It was Windsor Race day, and a large number of passengers arriving at Paddington for the Windsor train at the last moment it had to be strengthened by the addition of an extra carriage, so that it did not depart until 1.12 p.m., seven minutes late. Another minute was lost in running from London to Slough, and four minutes instead of two were occupied in the collection of tickets at Slough due to the large number of passengers. The starting signal had been lowered and the train was just about to depart when the collision occurred.[395]

Signalman William Wardell at Slough East said that Dolphin Junction had asked for the Falmouth express as soon as he had given the "Train out of Section" signal for the Windsor train at 1.37. He had not accepted it at once, but at 1.38 Dolphin box had given him the "Fast train approaching" signal. He had then accepted the Falmouth express under the "Warning arrangement". Almost at once Dolphin box had given him the "Train entering section" signal and a minute or so later he had been horrified to see the express approaching, apparently at speed, despite all his down main signals being at "Danger". He had seized a red flag and waved it at the express, at the same time shouting to the platform staff to get the passengers out of the Windsor train. As a result of his efforts a number of passengers were got out in time.

Signalman William Colbourn at Dolphin Junction confirmed Signalman Wardell's evidence and said that he had offered the express again to Slough East when Langley box had sent it on at 1.38. He had replaced all of his signals at "Danger" after the passing of the Windsor train, but the express had run past his box at 50 or 60 miles an hour, giving him no chance to warn its driver. He had tried to contact Slough East on the telephone but could get no answer before the collision.

In charge of the engine of the express was Driver Henry Woodman. He testified that he had left Paddington a minute late at 1.16. He could remember passing the distant, home and starting signals at Langley but had no recollection of passing Dolphin box, or of seeing any of the Dolphin signals or the Slough East distant signal. His first realisation of danger was his fireman, Henry Cann, calling out that the Slough East home signal was at "Danger". He had reached to close the regulator, but the fireman had just done so. He then went to apply the brake but again the fireman had beaten him to it. He had reversed the engine and given it steam. He could not understand how he had run past the signals. He had not felt sleepy, had nothing on his mind, was in good health and there was nothing wrong with his eyesight.[396]

Fireman Cann told the Board of Trade inspecting officer, Lt.-Col H.A. Yorke, that he had been firing between Langley and Dolphin Junction and so had not seen the Dolphin signals, but he had seen the Slough East distant at "Danger" and likewise the Slough East home, when he had shouted to his driver. Lt-Col Yorke was critical of the amount of praise which had been heaped upon the fireman by the public, remarking that he had only been doing his duty, and doing that indifferently, as even if he had been firing, he should have paused to check the signals as per Rules 139 and 143 – surely an unfair criticism?

In any case Driver Woodman took the whole blame for the accident upon himself and did not blame his fireman or allege that the Dolphin signals had been off. Why he had

[395] Board of Trade report
[396] Ibid.

failed to see the signals was a matter for speculation. His eyesight was tested after the accident and found to be good. He had not been intoxicated or obviously taken ill and said himself that he had not been feeling sleepy. The inspecting officer could only attribute the driver's failure to a temporary absence of mind. Woodman, Lt-Col Yorke remarked, was 60 years of age but looked considerably older. Possibly Driver Woodman suffered a transient ischaemic attack or transient global amnesia, though the symptoms of neither exactly fit the case.

Continuing his report, Lt-Col Yorke wrote that, 'The news of this catastrophe was received by the public in the first instance with bewilderment and consternation, which were not altogether unnaturally followed by feelings of indignation. It seemed hardly credible and quite inexplicable, that in broad daylight on one of the principal lines in the Kingdom, which is equipped with the most modern appliances for the purpose of ensuring safety, it should be possible for an express train to overtake and run into another train at an important station under cover of the signals. And notwithstanding the driver's admission of his fault, criticisms continued to be made in the public press upon the management of a railway, upon which such a thing could happen.' [397]

The inspecting officer went on to demolish criticisms that there had only been an interval of four minutes between the trains – irrelevant given that the line was worked on the absolute block system; that the train had been stopped on the main line at Slough while tickets were being collected – it should have been fully protected by the signals; and that the accommodation for Windsor trains at Slough was inconvenient and insufficient – true, perhaps, but it would be an expensive job to improve it.

The inspecting officer suggested that, in view of their demanding jobs, all drivers should have medical examinations at age 55, and yearly thereafter. He went on to criticise Guard Richard Kelly of the express, who said that he had not been looking out for signals as he had been attending to the luggage and mail bags at the time. To the annoyance of Lt-Col Yorke, who pointed out Rule 170 stated that guards must first attend to the safe working of trains and then attend to their luggage and parcels, the guard was supported by his superiors. The question presumably was whether the rule regarding the guard seeing to the safe working of the train meant his looking out for every signal, something which would have kept him fully occupied for much of the journey, to the detriment of any sorting of the luggage and parcels. Again the inspecting officer's comments seem unrealistic.

Although he conceded that it had not contributed to the Slough accident the inspecting officer was not very happy with the use of the "Warning arrangement" signal, and he pointed out the agreement of the railway companies to reduce its use after two serious collisions several years earlier (at Derby Junction, Birmingham and Esholt Junction in 1892). Finally he mentioned experiments being made with regard to the introduction of some sort of automatic train control, about which he did not seem very enthusiastic, commenting that it would require a lot of maintenance, have to perform well in all conditions and be very expensive to install at every signal.[398]

The juries of coroners' inquests at Windsor and London subsequently found Driver

---

[397] Ibid.
[398] Ibid.

Woodman guilty of manslaughter,[399] while in August the sum of £52 16s was distributed at Slough between six railwaymen whose actions had helped to mitigate the effects of the accident, namely Fireman Cann (to Lt-Col Yorke's displeasure no doubt), Signalman Wardell, Ticket Collector Wardell (a brother of the signalman?) and Porters Penn, Hatton and Davis. Wardell and Penn had already previously been awarded five guineas each by the Great Western directors.[400]

Driver Woodman duly stood trial for manslaughter at Reading Assizes at the beginning of November. He was acquitted.[401]

On 12th August 1900 a passenger train returning from the funeral of King Umberto I and the coronation of King Vittorio Emanuele III, and carrying Grand Duke and Duchess Peter Nikholaievitch of Russia, members of the Belgian Mission and many Senators and Deputies came to a stand in the section near Castel Guibileo, Italy, on the Rome-Florence line, 12 kilometres from Rome Central station. An overcrowded relief train, which had been despatched 10 minutes later, crashed into its rear, demolished its four rear carriages, killing at least 15 passengers and injuring at least 47. The Grand Duke and Duchess (née Princess Militza of Montenegro, sister of Queen Elena of Italy) escaped injury thanks to the strength of the saloon in which they were riding, but General Buffin, special envoy of the Belgian Government, suffered a broken leg and the Comte de Beaufort, attaché to the Belgian Mission, sustained a dislocated shoulder and other members of the mission were severely shaken.[402] The Grand Duke shut down the engine boiler to prevent any possibility of a boiler explosion.[403]

The new Italian King and Queen, on receiving news of the accident, attempted to charter a train to take them to the scene, but being unable to obtain one quickly enough set off there in two cabs, accompanied by officers on bicycles. Halfway there they were overtaken by a squadron of Cuirassiers with the Court carriages, in which they completed the journey. Arriving at the scene at 2 a.m. the King and Queen, having checked that their illustrious relatives were safe, assisted in the rescue work and the care of the victims, not returning to Rome with the Grand Duke and Duchess until 6 a.m. and visiting the victims in hospital in the afternoon.

The *Times* correspondent in Rome concluded his report by adding that, 'A searching inquiry has been ordered into the responsibility for the disaster. The conduct of the King and Queen is everywhere eulogized as a magnificent exemplification of the traditions of the House of Savoy.' [404]

A brief follow up report in *The Times* reported that the police had ordered the arrest of two men named Adani and Cellini because 'they did not make the customary signals to announce the approach of a train.' [405] Quite what this statement meant is unclear;

---

[399] *The Times*, London, England, 18th, 19th, 20th, 22nd, 23rd, 24th, 25th, 26th, 27th, 29th and 30th June and 3rd, 7th and 28th July 1900

[400] Ibid, 16th August 1900

[401] *The Mid-Surrey Times and General Advertiser*, London, 10th November 1900

[402] Ibid, 14th August 1900

[403] Wikipedia – Incidente ferroviari in Italia

[404] *The Times*, London, England, 14th August 1900

[405] Ibid, 15th August 1900

presumably it was either a mistake by a signalman, assuming the line was at this date worked on some sort of block system, or a failure to go back and protect the stalled train, or possibly both.

An excursion train and a milk train were involved in a collision at Hatfield, Pennsylvania, United States, on the Philadelphia and Reading Railroad on 2nd September 1900. The excursion train, No. 249, was made up of 10 day coaches and was the first section of a large excursion of people from Bethlehem, Allentown and the surrounding places to Atlantic City. This section carried only passengers from Bethlehem and Allentown. It departed from the Union depot in Bethlehem exactly 45 minutes after the milk train. The latter, which consisted of two milk cars and two passenger coaches, was booked to stop at every station between Bethlehem and Philadelphia. At 6.54 a.m. it drew up at the milk platform at Hatfield, and two minutes later the excursion train, travelling at 35 miles an hour, ran into its rear, crushing the two passenger coaches and wrecking the milk car in front of them. The engine and six of the cars of the excursion train were also wrecked. Four passengers in the milk train, which was fortunately nearly empty, and nine passengers in the leading car of the excursion train were killed and Engineer John Davis and Fireman Albert Wagner of that train were seriously injured.[406]

The second section of the excursion train had been despatched not long after the first, but it was flagged before it reached Hatfield and returned to Bethlehem. Fifteen hundred tickets had been sold for the excursion, which was to have been the last of the season.

Detailing the circumstances of the accident Superintendent Swelgert stated that the excursion train had left Bethlehem 45 minutes after the milk train. The engineers of both trains had duplex orders that the excursion train was to pass the milk train at Lansdale, the next station to Hatfield, three miles further on. At Perkassie the trains were 21 minutes apart, at Sellersville they should have been 17 minutes apart, but the gap there had fallen to 11 minutes and Dispatcher Groves tried to send a message to hold the excursion, but the wires were working badly due to a heavy fog. At Telford, rather more than four miles from Hatfield, the trains were six minutes apart and at Souderton they were only three minutes apart. Here the operator should have displayed a red signal to the excursion train because the interval between the trains was less than five minutes. The milk train was four minutes late at Hatfield and the excursion train a minute ahead of time, having gained 41 minutes on the milk train in a run of 30 miles.[407]

Winfield Wildoneter, the operator at Hatfield, stated that there had been a thick fog when the milk train had arrived at 6.55. He had then been stamping tickets when the excursion train had appeared two minutes later, traveling at about 40 miles an hour, at which point it had been too late for him to do anything to avert the accident.

On 7th September the coroner's jury at Lansdale passed a verdict blaming Engineer John Davis for driving recklessly and ahead of time and for passing the red flag at Souderton, Conductor Thomas Shelby of the excursion for permitting his engineer to do so, Train Dispatcher W.S. Groves for failing to keep in continuous touch with the two trains and Operator D.B. Beidler for failing to display the red flag which Engineer Davis had ignored! Clearly the engineer denied that any signal was displayed, and the operator

406 *The News*, Frederick, Maryland, United States, 3rd September 1900
407 *The Evening Democrat*, Warren, Pennsylvania, United States, 4th December 1900

swore that it had been. The jury, not knowing whom to believe, had condemned both.

The jury also recommended that the Reading Company strictly enforce the five-minute interval between trains on lines not worked by the block system and extend the system to Bethlehem.

The charge against Train Dispatcher Groves was dropped in early December.[408]

Due to a mistake by a signalman at Karlstor, Germany, a passenger train from Jaxtfeld overtook a Heidelberg local train near Karlstor on 7th October 1900. The rear carriage of the local train, full of Sunday excursionists, was smashed and the next two were badly damaged. Three passengers and a railwayman were killed outright and about 80 injured, three more of whom died later.[409]

Germany suffered another bad collision on 8th November 1900. The accident happened on the Offenbach side of Hanau, in circumstances very similar to the collision at Capelle in the Netherlands a year earlier. There was thick fog and a Berlin to Frankfurt express, train D42, overran a stop signal. At that moment, however, the signalman received permission for the express to proceed anyway and thinking, like the signalman at Capelle, that it had done so, had accepted a following passenger train from the signalman in the rear. The driver of the express had, however, glimpsed the signal at "Danger" at the last moment and had not only stopped but began to set back. The fog prevented the signalman from seeing that the express was reversing until the last moment, when he tried to warn its driver, but he was too late to prevent the second passenger train, which was possibly also running past the signals in the fog, from crashing into the rear of it. A gas cylinder in the rear corridor coach exploded and set that coach and the one next to it alight. A panic occurred amongst the passengers and many were burnt before they could escape. The two carriages were burnt out. Twelve passengers died, and a number were seriously injured.[410] The *Times* correspondent in Berlin noted the next day that, 'The Minister of Public Works, Herr von Thielen, left Berlin this morning for the scene of the disaster.' [411]

---

[408] Ibid, 7th December 1900
[409] *The Times*, London, Illinois, 9th October 1900
[410] *Eisenbahnkatastrophen in Deutschland 1* by Hans Joachim Ritzau
[411] *The Times*, London, England, 10th November 1900

CHAPTER 5

SIDELONG, HEAD-ON, DOUBLE AND FOULING COLLISIONS

Trains can come into collision with one another in a variety of ways. Rear collisions, that is ones involving one train running into the back of another, trains meeting one another head-on on a single track and trains colliding on rail/rail level crossings are detailed in other chapters. This chapter will deal with other types of collision such as those involving trains running into one another at junctions, meeting head-on on multiple tracks and colliding with vehicles left in sidings but which ended up foul of running lines.

A mistake by a signalman was the cause of a collision at Desford West Junction, England, on the Midland Railway, on 22nd October 1881. Desford station was on the Burton – Leicester line. East of Desford station were Desford East and West Junction signal boxes with extensive sidings on the south side of the line between them. The East Junction was with the single track West Bridge branch while the West Junction was a non-geographical junction, with a facing connection between the up main line and an up loop siding. By this date the layout here was fully interlocked, so that normally it would have been impossible to clear the up main home signal with the points set for the up loop siding. Unfortunately, on the 14th and 15th of October England had been affected by a very severe gale, which had caused widespread damage to the country's railway network, bringing down 88 signals at various points on the Midland Railway system alone. The company rule book had provisions, of course, to cope with such an eventuality, and Rule No. 149 directed that, 'In the event of a home or starting signal becoming defective, or not working efficiently, a competent person must be placed outside such signal with hand signal and detonators, and act under the instructions of the signalman.' This was done and a handsignalman was stationed at the signals, about 80 yards short of the facing points. The up main home signal, which had simply canted over somewhat, was soon repaired but the up main to loop home signal alongside of it required a new post. The signal inspector recommended that the up main signal be brought back into use, with a handsignalman retained for trains going into the loop siding, but the Desford stationmaster disagreed, considering that this would only confuse the train drivers, and as he was supported by the district traffic superintendent, the handsignalman alone was still directing all trains there on the 22nd.[412]

At 8.52 a.m. on the day of the accident Signalman Thomas Butler at Desford West Junction had turned the 4.45 a.m. coal train from Burton into the up loop siding. He then gave "Line Clear" back to Desford Station box. At 9.1 a.m. he had received the "Be Ready" signal from the station box for the 8.25 a.m. express train from Burton to Leicester, and the "Train on Line" signal at 9.4. The express consisted of an engine and tender, brake van, bogie composite, bogie third-class, and ordinary composite carriages and a brake van, in charge of Guard William Barrett, with Driver John Whitfield and Fireman William Cholerton. Running two minutes late, the express was probably travelling at 50 miles an hour as it approached the facing points, Driver Whitfield having had a clear distant signal and being shown a white flag by the handsignalman, Platelayer James Grimes. Despite the clear signals the facing points had been left set for the loop siding and Driver Whitfield, who must have had very little warning that anything was wrong, was unable to reduce speed very much before his engine was diverted into the loop and collided violently with the brake van of the Burton coal train. The impact destroyed the brake van of the coal train and the waggon in front of it. The express engine was overturned, the two rear compartments of the bogie composite were torn away, and this vehicle telescoped three compartments of the following bogie third-class carriage. Three passengers, Mrs Whetstone of Leicester, Miss Wainwright from Yorkshire (a guest of Mrs Whetstone) and Mrs Orton, of Coalville, were killed outright and a fourth passenger, Frederick Astle, a foreman coppersmith, of Burton, and Driver Whitfield, died a few days later from their injuries.

Five other passengers were injured, three seriously, and Fireman Cholerton was also very badly hurt.

Major F.A. Marindin, the Board of Trade inspecting officer, noted that the accident was unquestionably due to the carelessness of Signalman Butler, who had failed to restore No. 5 lever, working the facing points, after admitting the coal train to the siding. The frame in his box had only nine levers, he had been under no pressure, the points were within full view of his box and the accident had happened in broad daylight. The only explanation was that he had been chatting to Platelayer James Bennett, who according to the company rules should not have been in the box.

As regards the pressure put upon the signal department by the blowing down of so many signals, the inspecting officer noted that only six of the signals affected protected facing junctions and he considered that these should have been given priority. He also was of the opinion that the provisions of Rule No. 149 were quite inadequate for the proper security of the public at facing junctions (Platelayer Grimes had told the inquiry that his showing of a red or white flag to the train drivers had depended entirely on what colour flag the signalman had shown to him, and that he had had no instructions to actually check the lie of the facing points, and had not done so). A man, Major Marindin stated, should be stationed actually at the facing points to inform the signalman of their position, and trains should be brought almost to a stand before being called forward by hand signal. The express should also have been fitted with continuous brakes, which

412 Board of Trade report

would not have averted the collision altogether but would have reduced the speed so as to prevent the fatal telescoping. The time should not, he said, be far off when all passenger trains were fitted with continuous brakes and any company which continued to run express trains without them could not be said to show due regard either for the safety of their passengers or their own interests.

In conclusion Major Marindin noted that three of the most fatal accidents which had occurred in Great Britain, at Locomotive Junction, Nine Elms; Penilee near Glasgow, Scotland and at Desford had been due to almost inexplicable signalman's errors, which showed that railway companies could not be too careful to impress upon their signalmen that telegraph block working and the interlocking of points and signals were only aids to safe working, and it was just as necessary for them to exercise the greatest care and vigilance. There was also a need for the electrical interlocking of the block telegraph with the levers working the points and signals and he urged the railway companies to employ one of the inventions which now made this possible.[413]

Two inquests were held on the victims of the accident as they had died at different times and in different places. Both returned verdicts of manslaughter against Signalman Butler.[414] In the usual, though by no means inevitable sequel, the jury at Leicester on Tuesday 24th January 1882 found that he had been guilty of neglect but that that neglect had not been of a criminal nature, and he was discharged.[415]

Six men died in a collision near what was then called Charleston Junction, two and a half miles from Savannah, Georgia, United States, on Thanksgiving Day (Thursday 24th November) 1881. A pusher engine named *Pennsylvania*, attached to a single car, was backing down to the Savannah Florida and Western Railroad yard to make up a freight train. At the same time a Central of Georgia Railroad locomotive, *Vulcan*, was backing towards the junction, drawing a lumber train. The tenders of the two locomotives struck one another and telescoped. The *Vulcan* engine was 'almost entirely ruined' and the *Pennsylvania* engine as driven violently into the Green Line freight car it had been pulling.

Engineer Robert Meldrim and Fireman Byrd Wade of the *Vulcan* and Engineer Julius Cummings and Fireman John Brannan of the *Pennsylvania* attempted to save themselves by jumping. In doing so Wade's skull was crushed and his left arm and leg were badly mangled by flying timber, but he was apparently not killed. Less fortunate were a wood passer named Klunken on the lumber train and some negroes who were riding on the car of the lumber train adjoining the engine. Klunken and four of the negroes were killed outright and four more were seriously injured, one of whom died en route to St Joseph's Infirmary. A white man named Taylor, a hand on the lumber train, was also badly hurt. Whether the negroes, whose names were not given, were train hands or trespassers the newspaper report does not make clear.

As to the cause of the accident, a local newspaper reporter wrote, 'At this time it is impossible to tell where the fault for this sad catastrophe lies. All that is known is that the lumber train was on schedule time and had been flagged by the regular passenger which

---

413 Ibid.

414 *The Times*, London, England, 24th, 25th and 28th October and 1st, 8th, 9th, 11th, 12th and 21st November 1881

415 *The Magnet*, London, England, 28th January 1882

preceded it. The engineer of the *Pennsylvania*, however, it is thought, failed to notice the flag, and supposing that the way was open, started off at his regular time to the railroad yard to attend to his duties there. The entire affair will be fully and thoroughly investigated.' [416]

The day after the Savannah accident there was a serious collision at Tayport, Scotland, on the North British Railway. Again, it resulted from a signalman's error. With the fall of the Tay Bridge at the end of 1879 Tayport had gone back to its previous status as a terminus, with a ferry service across the Tay estuary to Broughty Ferry, but there was also a single line which continued to Newport. There was a goods yard on the east side of the station, which was controlled by south and north cabins. The line was not worked on the block system.

On the evening of the 25th November 1881 a goods train which was booked to run "as required" left Ladybank Junction at 5.40 p.m., Leuchars Junction at 6.38 p.m. and Leuchars Old station at 6.42.

Behind it the 5.5 p.m. passenger train from Burntisland, which had left Edinburgh at 4 p.m., left Ladybank Junction at 6.20 p.m., Leuchars Junction at 6.52 p.m. and Leuchars Old station at 6.54. The goods train arrived at Tayport at about the same time as the passenger train was leaving Leuchars Old. The goods engine was detached, ran forward to Tayport North cabin, was crossed over there to the up line, came back along the up line to another mains crossover at the south cabin and was there crossed over to the tail of the train and coupled up to it. It then started forward with the intention of its being taken over the crossover at the Tayport South cabin to the up line and thence into the goods yard. Instead Driver Andrew Cairns found his train was coming back along the down line and before he could quite pull up his engine had collided head-on with that of the passenger train, which was running into the station at about 15 or 20 miles an hour. Driver Brodie Crawford of the passenger train told the Board of Trade inquiry into the accident that he had found all the Tayport signals off for him except the Tayport North down distant, and the collision had taken him completely by surprise. The collision drove the goods train back about an engine's length but apart from its engine and tender it was little damaged. In the passenger train, on the other hand, the two third-class carriages behind the engine were telescoped into one another. Three passengers, Mrs Parker, of St Andrews, Captain McKechnie, of the *S.S. Curfew* and John Sorley, of Tayport, were killed outright and passengers Francis Hannah, of Dairsie and David Carmichael, of Springfield, were fatally injured, as was Fireman James Pearson of the goods train. Nine passengers, Driver Cairns of the goods train, and Driver Crawford and Fireman Maxwell of the passenger train were injured.[417]

Major F.A. Marindin, the Board of Trade inspector, stated that it was perfectly clear that Thomas Dipple, the Tayport South signalman, had forgotten all about the goods train when he had heard the passenger train whistle for the signals and had at once reversed his crossover and cleared his signals. At first sight it was difficult to understand how the signalman could have forgotten a train which had only just arrived, the engine of which he had just let back onto its tail and which invariably ran forward through the south

[416] *Columbus Daily Enquirer Sun*, Columbus, Georgia, United States, 27th November 1881
[417] Board of Trade report

crossover en route to the yard. Leaving aside the fact that the engine of the goods train was showing no lights towards the south cabin or passenger train and was concealing the lights of the brake van, the explanation for Dipple's mistake was doubtless to be found in the evidence of Alexander Thompson, the Tayport stationmaster, who said that Dipple's hours were from 6.20 or 6.40 a.m. till after the arrival of the 10 p.m. train on weekdays, plus twice on Sundays. Thus, at the time of the accident he had been on duty for almost 13 hours and his daily hours, assuming that the last train was on time, were over 15½ hours, not including walking time to and from his home. The inspecting officer went on to say that it was hardly too much to say that it was a scandal that a man upon whose vigilance the safety of the public depended should be expected to work such long hours. As usual other factors had also contributed to the accident. The lights of the goods engine should, according to rule, have been reversed. Driver Cairns stated that he had directed the deceased Fireman Pearson to do so, but whether the fireman had done so or not it was the driver's duty to ensure that it had been done. It was also a grave error of judgement, though not an actual breach of the rules, not to have shunted the goods train out the way of the passenger train at Leuchars Old.

The best safeguard against such accidents, Major Marindin stated, was the introduction of the absolute block system. The North British Company had during the past year been introducing block telegraph working on their main line between Edinburgh and Berwick, but on the whole of the Fifeshire railways with their very heavy traffic there were only a few short stretches of recently opened line equipped with the block system. It was to be hoped that such a state of things would soon cease to exist but in the meantime the strictest attention should be paid to keeping a proper interval between trains and signalmen and others should not be practically incapacitated by overwork. A disc signal should also be provided to prevent goods trains having to be hand-signalled back through the crossover at Tayport South.[418]

Both signalman Dipple and, a couple of days later Guard George Barnett of the goods train were apprehended on charges of culpable homicide.[419] The case against Guard Barnett appears to have been dropped. Signalman Dipple stood trial at Cupar, Fife, and after a prolonged hearing was acquitted on Friday 3rd February 1882. [420]

Germany suffered a major disaster at about 1 a.m. on the morning of Tuesday 30th May 1882 at Wieblingen, near Friedrichsfeld station, on the Heidelberg – Mannheim line. Owing to a mistake by a pointsman a train from Heidelberg was shunted onto the line on which the other train was approaching from Mannheim. The engine and most of the carriages of one of the trains were completely destroyed and 12 passengers and an engine driver were killed, and 60 other passengers injured. Of 30 of the injured passengers lying in the Heidelberg hospital half were reportedly fatally hurt. The pointsman responsible fled after the accident but was found hiding in a barn and placed under arrest. *The Times* in London added that, 'Great grief and excitement prevail in the neighbourhood. Heidelberg is much frequented by young Englishmen, but none are likely to have been in either of the shattered trains.' [421] [422] The accident, it was said in another British

---

[418] Ibid.
[419] *The Times*, London, England, 26th, 28th and 29th November and 3th and 5th December 1881
[420] *Lloyds Weekly Newspaper*, London, England, 5th February 1882
[421] *The Times*, London, England, 30th and 31st May 1882

newspaper, had attracted very little attention in the Berlin Press. 'It is not that the Press are unwilling to publish full accounts of such occurrences, but that the railway authorities in Germany manage to hush up affairs of this nature so cleverly that the reality seldom becomes known.' [423]

A collision on the railway between Nefiche and Mahsameh in Egypt on or about 26th September 1882 was very briefly reported in the British press. Eight or ten 'natives' were killed, and a large number injured but, 'A company of the Commissariat Corps {part of a British Army sent to Egypt in 1882 for reasons debated by historians}[424] had a narrow escape. They were travelling by the train, but, seeing that a collision was inevitable, leaped off in time to escape the shock. The line was not damaged.' [425] [426]

At Salem switch, Kansas, United States, on the Atchison Topeka and Santa Fe Railroad nine miles west of Hutchinson, on the evening of 3rd October 1882, passenger train No. 6 was waiting in the siding for the westbound Cannonball express, due to leave Hutchinson at 5.20 p.m. The points at the east end of the siding were unfortunately in the wrong position and the express was diverted into a head-on collision with the standing train. Both engines were badly damaged but the carriages of the two trains more or less escaped. The two engineers, Westlake and Fisher, their unnamed firemen and Brakeman Holliday were killed outright and Road Carpenter Shafer, who had been riding in the baggage car, died later from his injuries. Several passengers were badly hurt. All the dead men except Holliday were from Topeka, Kansas. The collision was blamed on a green brakeman named Ed Campbell who had reportedly disappeared after the accident but had given himself up at Nickerson, the next station west of Salem, three hours later.[427]

At an investigation a few days later at Hutchinson Brakeman Campbell denied having interfered with the switch and said that he had gone to Hutchinson for assistance after the accident. He was said to be a very good railwayman and a number of witnesses stated that he had been nowhere near the switch. It was also stated that Salem was a common place for throwing tramps off trains and that switch keys were often found on non-railroad employees, while a local farmer spoke of hearing one tramp say to another 'that he would get even with the Santa Fe soon.' Public sentiment was much in favour of the brakeman and at the end of the investigation it was decided that it would be folly to send him to trial.[428] Clearly there was at least a possibility that the accident was the result of an act of sabotage.

The Caledonian Railway main line which had seen the Kirtlebridge disaster in 1872 witnessed another bad accident just two stations further north at Lockerbie, Scotland, on 14th May 1883. Lockerbie was the junction for a single track branch to Dumfries, which

---

[422] *Lloyd's Weekly Newspaper*, London, England, 4th June 1882
[423] *The Evening News*, London, England, 2nd June 1882
[424] Wikipedia – Anglo-Egyptian War
[425] *North London News and Finsbury Gazette*, London, England, 30th September 1882
[426] *The Times*, London, England, 27th September 1882
[427] *The Deseret News*, Salt Lake City, Utah, United States, 11th October 1882
[428] *Belleville Telescope*, Belleville, Kansas, United States, 26th October 1882

diverged to the west at the north end of the station and had a double line junction with both up and down main lines. The junction was controlled from Lockerbie North cabin. There was absolute block working on both main and branch lines, combined with staff and ticket working for the branch. The up branch line was provided with home and distant signals, together with an intermediate home signal, 178 yards in the rear of the home signal protecting the junction. Late in the evening three up trains were due at Lockerbie close together; the 9.15 p.m. up mail from Glasgow (which was combined at Carstairs with the 9.25 p.m. mail train from Edinburgh) which ran through Lockerbie without stopping, the 7.30 p.m. up mail train from Perth and the 8.40 p.m. up passenger train Stranraer (10.57 p.m. from Dumfries), which was booked to connect into the Perth train, though were either train late they worked forward separately to Carlisle. The Glasgow mail was due to pass at 11.20, the Stranraer train to arrive at 11.25 and the Perth mail to arrive at 11.31. Up branch trains normally ran to the up main platform but shortly after the service of Stranraer trains had begun in July 1881 it had been found that should either the Stranraer or Glasgow mail trains be running late and the Perth mail train be on time that delay was caused if the Stranraer train was run into the up platform because the Perth train had to be held back at Nethercleugh cabin in accordance with the block regulations. The Lockerbie stationmaster, James Stothhart, had therefore adopted the procedure of having the up Stranraer train work into the down main platform, its through coaches being detached and added to the front of the Perth mail train after the latter had arrived. No signal was provided for such a movement, and the up branch points being set for the down line the up branch home signal was locked at danger, so trains had to be hand signalled past it.[429]

On the night of the accident Signalman Joseph Ross in Lockerbie North cabin had the 9.10 p.m. down goods train from Carlisle to Edinburgh arrive at 11.7. It did some shunting at the south end of the station and he then lowered the down starting signal for it to proceed at about 11.22. At 11.12 p.m. he received the "Preparatory" signal for the Stranraer train from Lochmaben cabin on the branch and the "Block" or "Train on Line" signal for it at 11.18½. Nethercleugh cabin on the main line gave him the "Preparatory" signal for the up Glasgow mail train at 11.20 when he lowered his signals for it. He kept his signals on for the Stranraer train until he heard it whistle, when he lowered the up intermediate home (No. 14), intending to bring the train past the home signal at "Danger" into the down platform as soon as the down goods had gone, set the branch points for the down platform and waved with a hand lamp to Driver John Robb of the down goods to hurry up and get his train clear of the platform. He saw the up mail approaching and then realised that the Stranraer train was not going to stop at the up branch home signal. Running down the steps of his cabin he waved a red light at the engine of the Stranraer train, but it did not pull up, and its engine struck the third or fourth waggon of the goods train. There was no time for him to stop the up mail. His telegraph boy, Robert Ross, had thrown the up home to danger, but the mail train was probably then too close for its driver to see the signal being thrown back in his face.

The initial collision was only a slight one, but it threw one or more waggons of the goods train foul of the up line, where they were struck by the up mail train with disastrous results. The pilot engine of the mail train was thrown against the up platform and overturned, and the leading vehicles of the train were dragged along the side of the

[429] Board of Trade report

goods train and greatly damaged. Driver William Mingis and Fireman George Sowerby of the pilot engine and five passengers were killed, and a large number of passengers were injured, some seriously. The passengers who lost their lives were Edward Annan of Glasgow, described as a writer; James Wynn of Glasgow; James Foster, a Carlisle clerk; Hugh McKenzie, a Newcastle fitter and a boy named William Blakemuir. All those killed had been riding in the mail train.

Major F.A. Marindin investigated the accident on behalf of the Board of Trade. He found that the second collision was purely accidental and that the first had arisen from a number of causes, the principal one of which was the introduction of the stationmaster without proper authority of a most dangerous method of working. The District Inspector, William Martin, was also to blame for not having stopped this method of working when he had accidentally become aware of it and having instead effectively endorsed it. Thirdly Driver George Easton of the Stranraer train had approached the junction at a dangerous rate of speed, without keeping a proper lookout for signals, and had overrun the home signal. Driver Easton contended that he had always regarded the lowering of the intermediate home signal as authority to run into the down platform if the home signal was not lowered for the up platform without a further hand signal, but other branch line drivers denied this. Signalman Ross, supported by his colleagues, stated that the lowering of the intermediate signal was only meant to allow trains to come as far as the branch home. Major Marindin did not consider that Signalman Ross was at fault in lowering the intermediate signal (which, the inspecting officer characterised as 'worse than useless') as he had only been following the normal practice, though he had shown an error of judgement in not throwing his up main line signals back before leaving his cabin to signal to the Stranraer train. The system of signalling branch trains had, however, become very loose in many respects, for branch trains had been brought forward past the home signal by hand signals before they had been brought to a stand at the signal, merely to give up the train staff or ticket, and such a habitual use of hand signals was sure sooner or later to lead to an accident.

As usual in Board of Trade reports at this time Major Marindin also considered the brake power of the mail train. The pilot engine had only been provided with a hand brake on its tender, whereas the train engine had been fitted with a Westinghouse brake and its driver had had control of a chain brake on the second, third and fourth vehicles; but having had less warning than the deceased driver of the pilot engine had had no time to apply either. There was also the absurd anomaly that the Caledonian Company had adopted the Westinghouse brake, while their West Coast partners, the London and North Western, would have nothing to do with it, so that Caledonian trains were being run without continuous brakes because the London and North Western was experimenting with another system of continuous brakes and had yet to adopt any.

Continuing, the inspecting officer stated that the intermediate stop signal should be either removed or interlocked with the home signal and interlocked safety points should be provided to protect the main line from branch trains running past signals. Finally, he gave praise to the rear guard of the mail train, Charles Owen, for running back to protect his train although injured and Telegraph Boy Ross (the son of the signalman?) for his promptitude in replacing the up home signal.[430] [431]

---

[430] Ibid.

Eight people died, and many were injured in a collision on the morning of Sunday 24th June 1883 on the Havana – Matanzas Railway in Cuba. The injured were taken to Matanzas.[432]

Disaster struck the Rome Watertown and Ogdensburg Railroad at Carlyon, New York, United States, on the evening of Friday 27th July 1883. The train involved in the accident was Excursion Train No. 53, known as the steamboat express, which was a regular working from Niagara Falls to Cape Vincent. It consisted of locomotives Nos. 51 and 61, a baggage car, a smoking car, regular coach and eight sleeping cars in charge of Conductor E. Garrison with Engineer L.D. Boynton and Fireman George McCathra on the lead (pilot) engine and Engineer James McCarthy and Fireman Lucius France on the train engine. It had left Niagara Falls at 6.30 p.m. and its passengers included a party of 200 excursionists from Michigan bound for the Thousand Islands, a popular summer resort. It was a dark night and there was a severe gale blowing accompanied by torrential rain. By the time that the train reached Carlyon, at about 9.30 p.m., it was running half an hour late.

When Track Walker Albert Parry had passed the scene only 30 minutes before the collision there had been nothing amiss, but during that interval the strong winds had blown a freight waggon from the station siding 750 feet down the track and onto the main line, where it had become derailed. Station Agent William Barry and the two men who had parked the car in the siding had gone off duty at 7 p.m. and there was no one left to warn the express. The train was travelling at 35 or 40 miles an hour when Engineer Boynton spotted the obstruction on the line ahead. He had only time to close his throttle and whistle for brakes when the collision occurred. The impact of the collision was heard by local residents above the noise of the storm and they rushed out to see what was wrong. The pilot engine was thrown to the north and the train engine to the south, making a somersault and ending up overturned in the ditch. The leading vehicles of the train piled up behind the second engine and were badly wrecked. At least 19 passengers, Engineer McCarthy, Fireman France and coloured porter Sill lost their lives and about 50 passengers, Engineer Boynton and Fireman McCathra were injured. Trainmaster W.H. Chauncey was also injured but nevertheless supervised the rescue work. Wrecking gangs were despatched from Oswego and Lewiston and surgeons sent from Oswego and Rochester.[433] [434]

The accident was the first one of importance to be investigated by the New York Railroad Commission, which decided that the car could not have been properly secured, and that Station Agent Barry was therefore mainly to blame. Barry insisted that it had been secured. The railroad company was also criticised for running a heavy double-headed train rather than two separate lighter trains, even though this would have introduced the alternative risk of a collision between the two sections. The coroner's jury made the unreasonable suggestion that a watchman should remain on duty at every

---

431 *The Times*, London, England, 16th, 17th, 18th and 19th May 1883

432 *The Philadelphia Times*, Philadelphia, Pennsylvania, United States, 26th June 1883

433 *Brooklyn Daily Eagle*, New York, United States, 29th July 1883

434 *New York Times*, New York, New York, United States, 29th July 1883

station until the passing of the last train, an idea not likely to appeal to the railway company's accountants given the expense in wages that it would incur. No one suggested the installation of "throw off switches" to protect the main line from stray vehicles, even though these were already in use at drawbridges, rail/rail level crossings and other such dangerous locations.[435]

A major disaster in France on 29th November 1883 probably received less publicity because it did not involve an important express or even a commuter train but only a lowly workman's train. The line from La Brohinière to Ploërmel was still under construction, and still owned by the State pending its leasing to the Ouest Company, when on this date there was a collision near Saint-Méen, France, between a workman's train and a ballast train. Eighteen workmen were killed and 15 seriously injured.[436 437]

Fifteen people died and 17 were seriously injured in a head-on collision between a mail train from Goalunda to Calcutta and a goods train at Arrunghatta, India, 50 miles from Calcutta on the Eastern Bengal Railway, on Thursday 2nd October 1884. Those killed included the European driver of one of the trains, a 'native' fireman and a European child. The others who lost their lives were 'native' passengers. The goods train had been due to shunt out of the way of the mail train at Arrunghatta. The cause of the collision was not entirely clear. A report on the accident blamed the dead European driver for running past signals but there were calls for a full public inquiry. The guard of the mail train, who was unhurt, stated that the Arrunghatta distant signal had been at clear for his train but he had then seen the home signal at "Danger", whereupon he had applied the brakes, but too late to prevent the accident.[438 439 440]

The Colonies and India commented that, 'A careful examination of the available evidence would, however, seem to point to confusion and blundering on the part of the Aranghatty station officials; and it is clear that these men, all of whom are natives, were found quite useless from terror immediately after the accident. The Indian newspapers comment severely on the reticence observed by the railway authorities; while their action in burying the dead in one large grave a few hours after the collision, and without awaiting the arrival of their friends, calls forth general reprobation. It is also stated that since the Government took over the railway, in July, dangerous economies have been introduced; but the truth of this assertion is doubtful.'[441]

On the afternoon of Thursday 1st July 1886 an express train and a mail trail collided head-on between Würzburg and Rottendorf in Germany. Sixteen people lost their lives and over 70 were injured. The accident gave rise to strong protests, especially from the

---

435 *A History of Railroad Accidents, Safety Precautions and Operating Practices* by Robert B. Shaw, pgs. 308-9
436 *The Times*, London, England, 1st December 1883
437 *Le Figaro*, Paris, France, 1st December 1883
438 *The Rimes*, London, England, 3rd October 1884
439 *The Guardian*, London, England, 8th October 1884
440 *The British Mail*, London, 3rd November 1884
441 *The Colonies and India*, London, 17th October 1884

sensationalist *Würzburger Damenwelt* newspaper. Between Würzburg and Rottendorf the Würzburg – Bamberg and Würzburg – Nürnberg lines ran alongside one another and each of the two tracks was used as required by trains travelling in either direction. On the day of the accident the signalman was instructed to switch the mail train coming from Bamberg onto the Nürnberg line and the Nürnberg express onto the Bamberg line, so that they were both running on what was the left-hand track relative to the direction in which they were going. The Würzburg pointsman realised too late that the Nürnberg express, running 16 minutes late, had left his station on its own, the right-hand, track, while coming towards it on the same line was the mail train, itself running 20 minutes late, and travelling fast down a falling gradient. The observant driver of the Nürnberg express, which was running slowly uphill, had time to bring his train to a stand but a disastrous collision could not be avoided. The wreckage of the accident formed a heap about 10 metres high. Those killed in the accident included the drivers of both trains and the fireman of the mail train. Apart from the driver of the express all those killed were from the mail train.[442] [443]

At 9 p.m. on Sunday evening, 29th August 1886 a local train from Baden entered the station at Mödling, near Vienna, Austria, on the Südbahn line and ran into a slow passenger train standing at the platform which was 'crammed with Sunday excursionists'. Three carriages of the local train were smashed to pieces and one report spoke of five dead and 20 others very seriously injured.

Another report stated that there were seven dead. Among those who lost their lives was M. Wieniawaki, a noted pianoforte player who was a professor at the Vienna Conservatory of Music. A correspondent in Vienna added that, 'The disaster was entirely due to negligence on the part of the engine driver of the local train.' [444] [445] [446]

Very brief reports appeared in the British and United States press of a collision at Charkow, Russia, on 18th December 1886, in which both engines and seven carriages were wrecked, 13 passengers were killed and 30 injured.[447] [448]

At about 1 a.m. on 26th July 1888 a freight train on the Omaha and Republican Valley Railroad broke in two as it was going over the brow of a hill near Yutan, Nebraska. As it was descending from the summit the two portions came together with great force, derailing and wrecking four or five cars. About an hour after the accident, while railwaymen were transferring shelled corn from a derailed Wabash Railroad car to another car, they discovered the bodies of six tramps who had been stealing a ride in the car, and who had been buried in the corn and suffocated when the car had overturned. Three of the bodies were identified as a John C. Taylor, a cigarmaker of Louisville,

---

442 *The Times*, London, London, 2nd July 1886
443 *Eisenbahn-Katastrophen in Deutschland* by Hans Joachim Ritzau, pgs. 19-21
444 *The Times*, London, England, 30th August 1886
445 *The Standard*, London, England, 31st August 1886
446 *Daily Times*, Dubuque, Iowa, United States, 31st August 1886
447 *Boston Post.*, Boston, Massachusetts, United States, 21st December 1886
448 *The Magnet*, London, 27th December 1886

Kentucky; Charles Sparkes, who had a brother at Wymor, Nebraska and Carl R. Talbot, who had letters from Columbus, Nebraska.[449 450]

Just after 8 a.m. on the morning of 16th October 1888 a collision occurred at Tamanend siding, Pennsylvania, United States, on what was then a new line on the Hazleton and Pottsville division of the Lehigh Valley Railroad. A Pennsylvania Railroad freight train (the Pennsylvania Company had trackage rights – equivalent to British running powers – over the line) with Engineer John Peck left Delano with orders to run to Lofty regardless of all other trains. Meanwhile a Lehigh Valley ballast train had left Lofty with Engineer John Messersmith with orders to shunt at Tamanend siding out of the way of the freight and passenger trains. Engineer Messersmith's locomotive was pulling ballast waggons but also propelling a caboose and a gondola car, the latter containing about 50 workmen. As the ballast train was just about to go into the siding the freight train came thundering downhill and collided with it. Engineer Messersmith claimed that he had sent out flagmen in both directions when he arrived at the siding, Engineer Peck stated on the other hand that he had not seen any flagman. Six labourers and the brakemen of both trains were killed and about 20 labourers were injured. They were mostly Hungarian immigrants. An investigation blamed the flagman of the ballast train for being too slow in protecting it.[451 452]

On 7th July 1889 fast trains No. 4 from Nürnberg and No. 8 from Würzberg were combined at Treucchtlingen, but despite being double-headed from there the train was already 23 minutes late at Ingolstadt. The booked crossing with fast train No. 1 at Dachau was therefore altered to Röhrmoos and the engine crews of both trains were so informed. Shortly before the arrival of the two trains at Röhrmoos a waggon was brought out onto the main line and then pushed into a siding, but the points were not reset for the main line. When the train from Ingolstadt arrived it was therefore diverted into the dead-end siding, where it collided with the waggon which had been just been stabled there. The driver of the pilot engine saw the points standing open for the siding, whistled and applied the Westinghouse brake but despite his quick reactions and the train's considerably reduced speed the collision was still violent enough to badly damage several carriages of the train. In particular there was a telescoping of one of the carriages and the adjoining luggage van; 10 passengers in that carriage were killed and some of its remaining nine occupants were seriously injured. Some passengers in other carriages were slightly hurt.

A court sentenced one of the Röhrmoos pointsmen to a year and three month's imprisonment, a second pointsman to a year's imprisonment and a signalman to six months' gaol for their parts in causing the accident.[453]

Gross carelessness by an engineer and possibly his intoxication were the alleged

---

[449] *Las Vegas Daily Optic*, Las Vegas, Nevada, 28th July 1888
[450] *Madison Daily Herald*, Madison, Indiana, 27th July 1888
[451] *The Times*, Philadelphia, Pennsylvania, United States, 17th October 1888
[452] *Train Wrecks of Eastern Pennsylvania* by Charles J. Adams III and David J. Seibold, pg. 72
[453] *Eisenbahn-Katastrophen in Deutschland* by Hans Joachim Ritzau, pgs. 21-2

causes of a collision at the 87th Street station of the Chicago Rock Island and Pacific Railroad on the outskirts of Chicago, Illinois, United States, on 24th September 1889. In accordance with the normal working a passenger car known as the Washington Heights car was attached to suburban train No. 57, which left the Rock Island depot in Chicago at 5.35. On arrival at 87th Street station the car was detached from the suburban train and left on the main line pending the attachment of a dummy engine to take it forward to its destination. While the car was still standing on the main line freight train No. 97, in charge of Conductor Charles Buford, with Engineer Seth Twombly and Fireman Henry La Cloche, appeared on the scene, running 10 minutes early. Failing to stop at the semaphore signal protecting the passenger car the freight train struck the coach with terrific force, sending it forward into the main suburban train. It then rebounded and was struck by the freight train engine a second time, completely telescoping it.

The detached car had contained about 30 passengers who had been reading and talking while they waited complete their journeys. The first shock sent them all forward to the front of the coach, where some attempted to escape through the door, while others tried to jump through the windows. None of the passengers was actually much hurt by either of the collisions, but damage to the steam chest of the freight engine led to the coach being instantly enveloped in scalding steam. Passengers Mrs Captain Brown, Mrs Ward, Fred Hipner and James Mckenzie, all of Washington Heights, and Miss Saddie Kelly of Dummy Junction were killed more or less outright and passenger Gus Mulcahy died of his injuries the next day. Several other passengers were more or less injured, one possibly fatally.

Conductor E.C. Lewis blamed the accident entirely on Engineer Twombley, whom he said was not only running 10 minutes early but was going too fast. The son of the master mechanic of the road, Twombley was said to be addicted to excessive drinking and to have been under the influence of alcohol at the time of the accident, and not in a fit state to run a train. Engineer Twombley and Fireman La Cloche were arrested, as were Conductor Buford and Brakemen Whitted and Parker of the freight train, required as witnesses. [454]

A coroner's inquest found Driver Twombley and Fireman La Cloche responsible for the accident. The fireman subsequently admitted that despite his testimony at the inquest that his driver had been sober he had in fact been very drunk; they had had several drinks together that evening. The inquest also censured Conductor Lewis for failing to exhibit red lights at the rear of his car and the railroad company for failing to enforce its rules and regulations.[455]

A seventh passenger, Edward Smith Jr, died of his injuries on 9th November and Engineer Twombley, Fireman La Cloche and Conductor Buford were all indicted for his murder.[456] The case against the Conductor Buford, who had testified earlier that he and his two brakesmen had done their best to stop the freight train, was doubtless dropped, and on Saturday 22nd February 1890, 'after a long and bitter debate', a jury acquitted Engineer Twombley and Fireman La Cloche.[457]

---

[454] *South Haven Sentinel*, South Haven, Michigan, United States, 28th September 1889
[455] *Hawarden Independent*, Hawarden, Iowa, United States, 3rd October 1889
[456] *Ironwood Times*, Ironwood, Michigan, United States, 18th January 1890
[457] *Logansport Journal*, Logansport, Indiana, United States, 23rd February 1890

What was presumably a head-on collision, possibly on a single line though the brief descriptions of the accident do not make this clear, occurred at Wildpark station, Germany, on the line from Stuttgart to Waihingen, at about noon on Tuesday 1st October 1889. A goods train proceeding to Waihingen stalled on an incline and its driver asked for an assisting engine, but before the latter had arrived the goods train was able to get going again and reached Waihingen. A passenger train, going in the opposite direction, was then allowed to leave Waihingen and collided with the assisting engine, which was running at full speed. The two locomotives were destroyed, and the luggage van and three carriages of the passenger train were 'smashed to splinters'. Ten people were killed outright in the accident and about 40 were injured, 11 most severely. Those killed included one of the drivers and one of the firemen, who was 'burned to cinders'. Minister Herr Schmidt, a high judicial official and others went to the scene of the disaster, 'which presents a most distressing spectacle'. The line was expected to be closed for some time, with trains being worked to and from the point of obstruction.[458]

The British London and North Western Railway seemed to have more than its fair share of bad accidents, including one which occurred at Longsight, England, on Friday 4th October 1889. One passenger was killed outright, five more were fatally injured and 37 were hurt, four seriously, when the 7 p.m. up local passenger train from Manchester London Road collided with two engines and a brake van which were standing on the line near Longsight No. 4 (Hyde-road) signal box. One railwayman was also injured. The passenger train consisted of a tank engine running bunker first, a third-class carriage, a brake van, two composite, a third class, a composite and a brake composite carriage and was in charge of Guard William Haystead, with Driver John Burgess and Fireman Thomas Brocklehurst.

The accident arose as a result of what the Board of Trade inspecting officer, Major-General C.S. Hutchinson, called an extraordinary delusion on the part of Signalman James Swift in Longsight No. 4 box. At about 7 p.m. an engine standing on the down loop line at No. 4 box and an engine and van standing in sidings on the up side had to be sent to Ardwick Junction box, the next one on the down (Manchester) side of Longsight No. 4. Signalman Swift set the road for the engine and van to cross from the up sidings to the down loop line, telling Driver Edwin Clarke to couple up to the engine there and then come back out onto the up fast line. This Driver Clarke did and as he was crossing back to the up fast line in accordance with a hand signal from Signalman Swift the latter received the "Be ready" signal from Ardwick Junction for the local passenger train. Signalman Swift then gave the drivers of the engines and van a hand signal for them to cross from the up to the down fast line and proceed towards Ardwick Junction by means of a crossover which did not exist. Then thinking that the engines and van had moved onto to the down line by the non-existent crossover Signalman Swift accepted the local train from Ardwick Junction and instructed Porter James Woolrich, who was learning the box, to clear the signals for it, whereby the latter lowered the up home signal, No. 59, without looking to see where the engines and van were. Signalman Swift then belatedly realised his error and threw back the home signal again, but it was too late. Both Driver

---

[458] The Standard, London, England, 2nd and 3rd October 1889

Clarke and Driver William Sheldon on the other engine stated that they had crossed over to the up fast line in obedience to the signalman's hand signal, unaware of his intentions, but they had refused to move when he signalled for them to proceed to Ardwick and had instead shouted to the signalman, "We are on the wrong road!" [459]

Driver Burgess of the passenger train stated that the Longsight No. 4 distant signal had been on but when he had seen the home signal through the slight mist it had been off, with the Longsight No. 1 up fast to up slow distant signal beneath it also off, and he had given his engine steam. He had then seen a red light at a distance of about 50 yards and had applied the automatic vacuum brake (which operated only on the vehicles of the train and not on the engine) and told Fireman Brocklehurst to apply the engine hand brake. There had been no time to either reverse his engine or to put down sand.

The inspecting officer did not blame any of the engine crews and placed the entire responsibility for the accident on Signalman Swift, stating that it was impossible to say how the signalman had acquired the delusion that he could cross the engines and van via a connection which did not exist. Signalman Swift himself could not explain it, adding that he had had a good night's rest prior to the day of the accident and was a total abstainer. Major-General Hutchinson commented that quite apart from his strange mistake Signalman Swift had been gravely negligent in not assuring himself of the position of the engines and van before accepting the passenger train. He did, however, deserve credit for taking all the blame for the accident and not attempting to implicate anyone else.

Turning to other factors the inspecting officer stated that there had been no necessity for the engine of the passenger train to have been running bunker first as there was turntables at Longsight depot, whence it had gone to London Road station to be attached to its train. Again the brake power of the passenger train had been inadequate. The engine weighed about 33½ tons and the vehicles about 63 tons, but the engine was only equipped with a hand brake, which the fireman had had no time to apply. The six drivers and fireman would also have been on duty, had there been no accident, for periods ranging from 12½ to 14¾ hours, lengths of duty for which, except in emergency, there could be no justification.

To prevent a recurrence of the accident the inspecting officer suggested the installation of a disc signal from up fast to down loop lines and a repositioning of signals so as to render hand signals unnecessary. He added that, 'If a simple and trustworthy means could be devised for automatically preventing a signalman from accepting a block signal for a train before it is right for him to do so, it would be a great safeguard against such a mistake as was the immediate cause of the present collision.' [460] [461]

Signalman Swift appeared in court on 15th October charged with causing the death of the six passengers, but the magistrates decided that a jury would not convict him, and he was discharged.[462]

[459] Board of Trade report
[460] Ibid.
[461] *The Standard*, London, England, 5th October 1889
[462] *St James Gazette*, London, England, 16th October 1889

The Lake Shore and Michigan Southern Railroad Cleveland and Chicago express, due at Buffalo at 8.55 p.m., broke in two at Bay View station, about six miles from Buffalo, on the evening of 6th March 1890. The train had been running at high speed to make up time when the division occurred. The front portion of the train, consisting of the engine and tender, a smoking car and two day coaches was quickly brought up, whereupon the rear portion, consisting of five heavy Pullman cars, came down the bank at full speed and collided with the rearmost day coach. The second day coach climbed up onto the top of the Pullman car and the leading day coach was telescoped. The day coaches were full of passengers. A passenger told a reporter that a coupling was defective, the train having become divided at exactly the same place previously as it was leaving Cleveland. Conductor John Houghtaling, in charge of the train, had most unwisely pulled the bell rope, thus getting the engineer to apply the brakes, when he found that the train had parted. Most of the passengers in the rear day coach were warned of their danger in time to escape but those in the leading Pullman had no warning before it was buried beneath one coach and telescoped another. Six or seven passengers were killed including Mr and Mrs Stewart, of Chester, New York, whose 18-month-old baby was unhurt, and negro porter J. Swain of Cincinnati. Another coloured porter named Waldron was the hero of the day when he seized a lamp which was burning in the debris and threw it out of a window, thus preventing a fire adding to the accident. Pullman Conductor Fest also distinguished himself by his presence of mind and bravery.[463]

On Tuesday 18th March a coroner's court in Buffalo found the management of the Lake Shore Company guilty of gross negligence, and that Conductor Houghtaling was primarily responsible for the accident and culpably negligent, while Brakeman David Beckwith was also censured for not notifying passengers that the train had parted. Paradoxically the jury also decided that the brakes of the first portion had been applied by some person or persons unknown. Engineer Mooney was freed of blame.[464]

Norton Fitzwarren, England, on the Great Western Railway main line from Bristol to Exeter and the junction for branches to Barnstaple and Minehead, was the scene of a major disaster in the early hours of 11th November 1890. The Great Western at this time was still two years away from finally discarding its broad gauge and the main line was equipped with both broad and narrow (standard) gauge tracks. At 12.36 a.m. on the morning of 11th November a narrow gauge down goods train consisting of an engine, 38 waggons and a brake van arrived at the station, assisted by a broad gauge pilot engine. It had to do some shunting here and before this was done the Norton Fitzwarren signalman, George Rice, first put the pilot engine out of the way onto the Minehead branch. (The Board of Trade report states that the Barnstaple branch was used but at the inquest Signalman Rice stated that it had in fact been the Minehead branch.) The shunting having been completed Signalman Rice shunted the goods train onto the up main line at 1.10 a.m. to leave the down main line clear for a fast down goods train, which passed at 1.17 a.m. Immediately it had done so the signalman brought the pilot engine back onto the down main line. Upon seeing the pilot engine regaining the down line Driver Charles Noble of the goods train assumed that the signalman was about to let them proceed and told his fireman, Alfred Dowling, to change the engine's headlight back from red to

463 *Bloomington Daily Leader*, Bloomington, Illinois, United States, 7th March 1890
464 *Dunkirk Evening Observer*, Dunkirk, New York, United States, 20th March 1890

green, which the latter did. They then waited to be crossed back over onto the down line, but nothing happened, and a minute or so later they saw the headlights of an up train coming towards them on their line at high speed. There was no time to pop the whistle or do anything but jump for their lives.

Shortly after he had put the broad-gauge pilot engine back onto the down line Signalman Rice had received the "Is Line Clear" signal from Victory Siding, the next box towards Exeter, for an up special boat train carrying passengers from the Donald Currie Company's Cape mail steamer *Norham Castle* from Plymouth Millbay to London Paddington. Without giving a moment's thought to the fact that he had placed the goods train on the up main line he had accepted the special, obtained the road for it from Silkmill Crossing, the next box on the Taunton side, and cleared his signals. Victory box gave him "Train on Line" at 1.23 and the train passed his box and the collision occurred a minute later, through Signalman Rice said that he did not hear the noise of it.[465]

With the Norton Fitzwarren signals all off for him and the engine of the goods train displaying a green light, Driver Scott of the boat train had no warning of anything being wrong. His broad-gauge train had just descended Wellington bank and was travelling at about 50 miles an hour when it collided head-on with the standing goods train. Driver Scott and his fireman survived the accident, though seriously injured, but 10 of the 47 passengers on the train were killed and nine more seriously injured. All the fatalities were in the leading carriage of the special, which was completely smashed. The two other vehicles of the special were damaged, the two engines were badly damaged, and three waggons of the goods train were destroyed and nine damaged. Most of those killed in the accident were skilled miners from the north of England, who had been returning home after working in South Africa, no doubt in the gold and diamond mines. Another of those who lost their lives was a Titus Baylis, a negro missionary who had been on his way from South Africa to the United States.

Colonel F.H. Rich conducted the Board of Trade inquiry into the accident. His report was not a long one, as it was obvious that Signalman Rice was primarily responsible for the accident; he ought, the inspecting officer stated, to have shunted the goods train onto one of the branches instead of the up main line. Driver Noble was also partially responsible for the accident for having prematurely instructed his fireman to change the engine headlight back to green. Had it been red it might have reminded the signalman of the presence of the goods train and would have served to warn the driver of the up special. Furthermore, the regulations stated that the guard, William Lowe, should have gone to the box to remind the signalman that the goods train was there, or else have sent his brakeman. Guard Lowe had pleaded that he had been taking the waggon numbers and that he had been on his way to the box when the collision occurred.

Colonel Rich recommended that refuge sidings should be provided at all stations where much shunting of trains was undertaken and suggested the installation of lifting bars which locked signals on lines where trains were likely to have to stand to await "Line Clear". He also noted that many signalmen put one of their flag sticks in the spring catch of a signal lever which was not to be pulled and suggested the introduction of a slide bar, wedge, loop or other contrivance marked "Train Waiting" to serve the same

[465] Board of Trade report

purpose.[466] This recommendation was adopted, and lever collars of various designs introduced.

Ignored by the inspecting officer, though mentioned by Signalman Rice in his evidence, was the fact that the latter had suffered a serious head injury the previous January, when he had been run down by a light engine. He had sustained a fractured skull as well as three broken ribs and had been off sick for 14 weeks. On the night of the accident he said he had been bad in the head for the whole of his shift, worse than usual. Signalman Rice said the same thing at the inquest at Taunton on the victims of the disaster, and here the jury were sympathetic, and very reluctant to find a verdict of manslaughter against him until effectively told by the coroner that they must do so. His trial at the Taunton Assizes took place on 22nd November, only 11 days after the accident. After an absence of almost an hour the jury found him 'Not Guilty'. [467]

Saturday 18th April 1891 saw a bad accident at Kipton, Ohio, United States on the Lake Shore and Michigan Southern Railroad. Eastbound fast mail train No. 14 collided head-on with westbound passenger train No. 21, the Toledo express, almost in front of the depot. The main line was single track but there were sidings at Kipton station where trains could cross one another. The siding opposite the depot was occupied by a freight train and train No. 21 was attempting to reach another siding just west of the depot when the collision occurred. Both trains were running at some speed when they met and as might be expected the collision was a violent one. The two engines, three postal cars, a baggage and an express car were telescoped together in one mass of wreckage. Engineer Ed Bacon, of train No. 21, Engineer Charles Topliff, of the mail train, and postal clerks J. Bowerfield, Charles Hammel, J.L. Clement, Charles McDowell, Frank Nugent and J.C. McKinley were killed. Fireman Curtis Starkey of the mail train, Express Messenger J.N. Klinefelter, one passenger on train No. 21 named Abram Dereamer and a four-year-old boy named Charlie Dazey, who had been standing on the platform at Kipton, were more or less seriously injured.[468]

General Superintendent P.P. Wright quickly opened an investigation into the accident, which found that the responsibility for it lay with the crew of train No. 21, which should have waited for the mail train at Oberlin, the previous station. Train No. 21 was booked to leave Oberlin, 4½ miles east of Kipton, at 4.38 p.m., but did not do so until 4.46, eight minutes late, then ran fast to Kipton, where the collision had occurred at 4.52 p.m. Kipton was the regular meeting point of the two trains when they were on time, but as No. 21 was late it should have waited at Oberlin, or if its crew had discovered en route that they could not reach Kipton on time they should have stopped their train and sent a man forward. The newspaper report noted that Engineer Bacon had paid for his mistake with his life, while Conductor Tim Daley would likely lose his position after many years' service.[469]

An inquest on the victims followed and was held at Elyria. It heard much conflicting

---

466 Ibid.

467 *The Times*, London, England, 12th, 13th, 14th, 15th, 17th, 18th, 24th and 25th November and 4th December 1890

468 *Norwalk Daily Reflector*, Norwalk, Ohio, United States, 20th April 1891

469 Ibid, 21st April

testimony. There was no jury, the evidence being assessed solely by the coroner, R.E. Braman, whose verdict was that the railroad company were to blame for the accident for allowing insufficient time for train No. 21 to be in clear of train No. 14 at Kipton, for overworking the Kipton agent, for not removing an old water tank which obstructed the view of the Kipton operator and for not introducing the automatic block system. While he considered that those in charge of *both* trains had been guilty of carelessness, he did not consider that there had been criminal negligence, but that the accident had been due to the faulty operating system.[470]

The railroad investigation having determined that the watch of Engineer Bacon was possibly four minutes slow, Superintendent Wright appointed Cleveland Ohio jeweller Webster Ball to look into time-keeping issues on the line. As a result watch performance and inspection standards were introduced in 1893.[471]

Six people died and a number were injured in a collision between a passenger train and a goods train near Warsaw, Russian Poland, on about 8th July 1891.[472]

In the summer of 1891 an accident occurred on the Colorado Midland Railroad at Aspen Junction Colorado, United States (renamed Basalt 1901). It involved a returning excursion train from Glenwood Springs to Aspen. These excursion trains were nicknamed laundry trains because they carried excursionists to the public hot springs at Glenwood, promoted locally and nationally as a health resort. [473] There was a wye at Aspen Junction, with the passenger platform situated east of the junction with the branch to Aspen, so that passenger trains from Glenwood Springs bound for the branch had to be set back to the junction after calling at the depot. At about 10.55 p.m. on the evening of Saturday 11th July 1891 a laundry train in charge of Conductor Riley Miller, with Engineer William Switzer, arrived at Aspen Junction from Glenwood Springs. The train consisted of an engine, a baggage car and a compartment car, the latter containing 30 passengers. It pulled up at the junction platform, where the passengers for the junction alighted, drew forward along the main line to the water tank, took water and then began to set back towards the junction points preparatory to continuing its journey, Conductor Miller having given Engineer Switzer his orders, which instructed him to meet train No. 64 at Watson and an extra train at Rathbone, these being adjoining stations on the branch.[474]

Meanwhile at the roundhouse situated within the wye Engineer J. Sheppard with engine No. 22 was ready to work an extra train to New Castle, and his conductor, C.F. Rickard, told him, "They are all in, we will go and hitch on." With that, Brakeman Thomas Mahoney had removed the track block from in front of the engine and had walked forward to set the switches (points) while he (the engineer) had let his locomotive drift slowly to the main line, keeping 40 of 50 feet behind the brakeman. The view of the

[470] *Norwalk Daily Reflector*, Norwalk, Ohio, United States, 28th and 29th April and 5th and 26th May 1891
[471] Webpage – The Great Kipton Train Wreck by Nancy Pope
[472] *Decatur Morning Review*, Decatur, Illinois, United States, 10th July 1891
[473] Webpage – Laundry trains brought valley residents to Glenwood Springs
[474] *Leadville Herald Democrat*, Leadville, Colorado, United States, 14th July 1891

light engine from the main line and of the main line from the light engine was obstructed by coal bunkers, a tool house and a sand house. Engineer Sheppard testified at the subsequent inquest that his engine had been 25 feet from the junction switch when Brakeman Mahoney had suddenly signalled for him to stop his engine and back up. Before he could stop his engine was on the points, foul of the main line, and before he could reverse it had been struck by the compartment coach of the excursion train. What made the accident much more serious was that the coach tore off the check on feed pipe of the injector, leaving a two-inch hole in the side of the boiler. Steam and hot water poured from the damaged boiler through the broken coach windows and scalded the passengers inside. Engineer Sheppard said that after the collision he had attempted to back his engine away but could not get it to move even after engine No. 10 had been attached to it to give additional traction.[475] All that he could do was to open the blow-off cock on the opposite side of the boiler to allow the steam to escape faster that way, lessening the amount which was pouring into the passenger car. None of the passengers was killed in the collision but by the following Thursday 10 had died of scalding and suffocation and several others were seriously hurt.[476] Trainmaster R.C. Bowdish told the inquest that the excursion train had the right to make passenger speed on the main line, and that each side of the Y (wye) was considered a main line. It was the duty of anyone to see that the main line was clear before allowing the engine to go onto it. It was the brakeman's responsibility to stop the engine until he was sure that the main line was clear rather than the engineer's responsibility to stop his engine without any signal from the brakeman.[477] The inquest jury found a verdict that Brakeman Maloney and Conductor Rickard were guilty of criminal carelessness 'and they will be prosecuted.'[478] The conductor and brakeman reportedly escaped from the country after the trial.[479] The jury also criticised the location of the sand house and other buildings which had obstructed the view of the two crews and declared that 'the local tracks at Aspen Junction are such as to render the handling of trains at that point dangerous to public travel.'[480]

A bad accident on the morning of Thursday 24th September 1891 involved a work train of the Pittsburgh and Western Railroad. The accident occurred at McKim's Siding, near Zelienople, Pennsylvania, United States. A gang of 50 Italian labourers was engaged in laying a new track at that point. At about 8 a.m. the work train went into the siding out of the way of a westbound freight train, but its crew failed to realise that the freight train was running that day in two sections, with the second section following five minutes behind the first. As soon as the first section had passed the work train re-entered the main line and the labourers were busy throwing off dirt from the waggons when the second section of the freight crashed violently into them. The cars of the work train were piled into a shapeless mass and steam and as at Aspen Junction boiling water from the wrecked freight engine poured onto the poor workers trapped in the wreckage. Nine people died in the accident, eight of whom were Italian labourers whose names the newspaper reporter

---

475 *Aspen Evening Chronicle*, Aspen, Colorado, United States, 14th July 1891
476 *Centralia Daily Sentinel*, Centralia, Illinois, United States, 16th July 1891
477 *Aspen Evening Chronicle*, Aspen, Colorado, United States, 14th July 1891
478 *Castle Rock Journal*, Castle Rock, Colorado, United States, 22nd July 1891
479 *Centralia Daily Sentinel*, Centralia, Illinois, United States, 16th July 1891
480 *Galveston Daily News*, Galveston, Texas, United States, 17th July 1891

had not been able to obtain. The ninth death was that of Engineer John Houghton of the freight train, who had stuck bravely to his engine in an effort to minimise the effects of the collision. Twenty other labourers were injured, some it was feared fatally.[481]

Intoxication, or at any rate the consumption of alcohol, was a feature of a collision around midnight on 18th October 1891 at Kohlfurt, Germany (now Węgliniec in Poland). A fast train which had left Breslau at 10.32 p.m. for Berlin with a heavy compliment of passengers was close to Kohlfurt station when it was run into from the side by a shunting engine. A second-class carriage was first struck, its side was smashed in and it was overturned. The funnel of the shunting engine then forced itself into the leading compartment of the following third-class carriage. Although the train was travelling at reduced speed the force of the collision was also sufficient to derail its two locomotives. *The Times* in London reported that five passengers had died, and another passenger and two railwaymen had been seriously injured. A German report stated that that there were eight deaths and 20 more people injured. Among those who lost their lives were a Herr von Kardorff, 'a son of the well-known member of the Reichstag.' [482] [483] [484]

Aladdin, Illinois, United States, on the Wabash Railroad, was the scene of a bad accident in the early hours of Tuesday 5th January 1892. The station there was the regular meeting point of eastbound passenger train No. 41 and westbound passenger train No. 46. At 1.47 a.m. train No. 46 was stopped in the station waiting for train No. 41 to arrive when the latter train came in on the wrong track and collided head-on with the stationary one. The violent collision demolished both engines and baggage, smoking and a chair car of train No. 46. Engineer Bouchard and Fireman Warberry of train No. 41, Engineer Wilson of train No. 46 and four Italian immigrants who had been riding in the smoker were killed outright and a female Italian immigrant was said to be so badly injured that she could not live. Fireman D.E. Conwill of train No. 46 and Express Messengers E.P. Clough and D. Dunham were injured. There was an impenetrable fog at the time of the collision, making it likely that the accident was due to Engineer Bouchard losing his whereabouts and not realising that he had reached the station.[485]

More than 50 persons were killed, and a large number injured in a collision between two trains near Cuyambuca station, Brazil, on the evening of 23rd February 1892 according to reports published in United States newspapers in March. The accident occurred on the Estrada de Ferro Recife ao São Francisco.[486]

Somewhat unusual circumstances surrounded an accident on the St Louis Southwestern (Cotton Belt) Railroad at about 7 p.m. on 20th May 1892 at Crooked

[481] *Alton Daily Telegraph*, Alton, Illinois, United States, 25th September 1891
[482] *The Times*, London, England, 20th October 1891
[483] *Evening News and Post*, London, England, 20th October 1891
[484] *Eisenbahn-Katastrophen in Deutschland Bd 1* by Hans Joachim Ritzau, pg. 123
[485] *The Indiana State Sentinel*, Indianapolis, Indiana, United States, 13th January 1892
[486] *Madison Daily Democrat*, Madison, Indiana, United States, 3rd March 1892

Bayou, between Humphreys and Golden, Arkansas, United States. Reports of the circumstances of the accident were scanty due to a storm having brought the telegraph wires down, but it appears that the progress of a southbound passenger train in charge of Conductor Field was interrupted due to a bridge over the Arkansas River having been washed away. Finding that they could go no further the crew of the train elected to back their train to Humphreys, but in the progress of doing so it was run into by a freight train which had followed them southwards. All the cars of the passenger train were thrown from the track and upset in the collision. Reports of the casualties were conflicting, especially with regards to the names of those who had lost their lives, but the dead appear to have included Fireman Nichols of the freight train, the coloured porter of the sleeping car and five or six passengers. About a dozen other passengers were seriously injured. The accident was blamed on the crew of the passenger train having failed to obey orders, presumably in having set back without protecting the rear of their train against a following movement.[487] [488]

Esholt Junction, England, on the Midland Railway, was the point where branches from the Leeds and Bradford directions united to continue roughly northwards to the town of Ilkley. On the afternoon of 9th June 1892, the 3.10 p.m. passenger train from Ilkley to Bradford was run into as it was passing through the junction by the 3.7 p.m. passenger train from Leeds to Ilkley with serious consequences. The last six vehicles of the Ilkley train were damaged, the sixth, a third-class carriage, being completely broken up. The engine, tender and leading third-class brake carriage of the Leeds train were also damaged. One passenger in the Ilkley train was killed outright, four more died of their injuries, and 26 were more or less injured, as were one of the guards of the Ilkley train and the driver, fireman and guard of the Leeds train.

Although the lines were worked, like all the passenger railways in Great Britain by this date, on the absolute block system, the signalling regulations permitted two trains to approach the junction simultaneously under the 'Warning Arrangement', provided that the driver of the second train had been duly cautioned by the signalman at the box in the rear. Thus, Signalman Harry I'Anson at Esholt Junction had accepted the Ilkley train from Rawdon Junction cabin at 3.24 p.m. under full line clear. A minute later he received the "Is Line Clear?" signals from Apperley Junction for the 3.7 p.m. Leeds train and also from Baildon cabin for the 3.15 p.m. North Eastern Railway train from Bradford. He accepted the Leeds train at once under the warning arrangement and the North Eastern train likewise two minutes later. The North Eastern train appeared first, and he cleared No. 7 home signal for it, the train having already passed his distant signal. The North Eastern train passed his cabin at 3.33 as the train from Ilkley was approaching. At this point he had seen nothing of the train from Leeds, but as he replaced No. 7 signal behind the North Eastern train and the train from Ilkley to Bradford was passing his cabin at 3.33½ he saw the Leeds train about to overrun the junction, but told the Board of Trade inquiry into the accident that he had been reluctant to show its driver a red flag for fear that the driver of the Ilkley train would see it. He had hoped that the Ilkley train might just clear the junction in time, but it did not.

---

[487] *The Daily Huronite*, Huron, South Dakota, United States, 23rd May 1892

[488] *The Hawarden Independent*, Hawarden, Iowa, United States, 26th May 1892

Signalman Thomas Aubrey at Apperley Junction remembered his colleague at Esholt accepting the 3.7 Leeds train under regulation 16 and said that he had kept his home signal at danger until the train had almost reached it, then had lowered the signal and exhibited a green flag from his cabin. As he had been about to shout, "Junction blocked," the engine driver nodded and held up his hand. Signalman Aubrey said that he had then lowered his starting signal and the train had whistled and proceeded. He had not caught sight of the guard. According to a strict interpretation of the regulations he should have stopped the train dead before cautioning its driver, but it was not customary to do so at Apperley Junction with branch trains because of the 1 in 60 incline towards Esholt which began 200 yards beyond his cabin. Main line trains were stopped because the gradient on the main line was less severe.[489]

Driver Archibald McLay, in charge of engine No. 179 pulling the Leeds train, admitted to seeing the green flag shown by Signalman Aubrey from Apperley Junction cabin and understanding that he was proceeding to Esholt Junction under the Warning Arrangement. He supported the signalman's statement that branch trains were not normally brought to a stand to be cautioned because of the steep incline. The Esholt Junction distant signal he had found to be at danger, and he had realised that he might have to stop at the junction. Emerging from the short tunnel near the junction he had seen two signals off, the Esholt Junction stop signal and the distant arm for Rawdon Junction below it, and had assumed that he had the road. It was only when he had looked again when his engine was close to the junction that he had realised that he had been mistaken, and that the signals which were off were those for the Bradford line, while his own signals were at danger. He had immediately shut off steam and applied the brake but the next moment the collision occurred. What his fireman, Walter Bolton, was doing at that moment he did not know but before entering the tunnel his fireman had been fiddling with the lubricator, which had been giving trouble all day. The only explanation that he could give for his mistake was that he had been dazzled by the bright light after leaving the tunnel.

Major-General C.S. Hutchinson, who conducted the Board of Trade inquiry into the collision, remarked that the immediate cause of the accident was clearly a want of reasonable care on the part of Driver McLay, who had frankly confessed to his mistake and not attempted to implicate anyone else. The inspecting officer, nevertheless, also placed some of the blame on Fireman Bolton, whose duty it had been to watch for signals when approaching a junction, and on Guard Arthur Miller for the same reason. Signalman I'Anson could not be blamed for the collision and while Signalman Aubrey was wrong in not stopping the train before cautioning its driver this had made no difference in the present case and given that the practice must have been known to the district inspector, he could not much be blamed.

Major-General Hutchinson pointed out that this was the second serious collision at a junction involving the Warning Arrangement in less than a fortnight (following one at Derby Junction, Birmingham) and that it was time that the use of the regulation at junctions was abandoned, as otherwise railway companies could not be regarded as complying with the Regulation of Railways Act of 1889 as regards absolute block working. He also recommended a repositioning of the Esholt Junction home signals to

[489] Board of Trade report

make a mistake by drivers less likely.[490]

At the inquest on the victims of the collision the coroner's jury found that Driver McLay had not exercised the necessary vigilance but returned a verdict of "Accidental death". [491]

An accident at Flat Rock Tunnel, near West Manayunk, Pennsylvania, United States, on the Philadelphia and Reading Railroad, illustrated the difference in the manner of operating a double-track railway between Great Britain and the United States. In Great Britain each of the two tracks was normally only used by trains running in a specified direction; up trains ran on the up line and down trains on the down line unless one line was blocked by accident or engineering work and single-line working had to be introduced. Even if there were half a dozen trains following one another on for example the up line with no train expected on the down line for two hours an up train would not be permitted to use the down line to overtake a slower up train, and there would no fixed signals to control such a move. In the United States on the other hand trains in either direction were commonly permitted to use either of the two tracks provided they had been authorised to do so by an order issued by the dispatcher. Thus, on the morning of Monday 24th October 1892 Conductor George Hamilton of southbound passenger train No. 2 received an order at Phoenixville from Reading Division Superintendent Whitcomb authorising his train to use the northbound track from West Conshohocken to West Falls 'regardless of all trains', the southbound track being occupied by one or more coal trains. Before the order had been issued to the passenger train an order had been sent to Yardmaster John Rupp at West Falls at 7.46 a.m. to hold all northbound trains after engine No. 1019, on train No. 27, until the arrival of southbound train No. 2 and an engine, the *Ariel*.[492]

Train No. 2, known as the *Shamokin Express*, with Conductor Hamilton, Engineer Thomas Fitch and Fireman Thomas Walsh, had duly left Phoenixville at 8.45 a.m. and was proceeding along the west bank of the Schuykill River, a few miles north of Philadelphia, at about 25 miles an hour, when Engineer Fitch caught sight of a coal train coming towards him on the same track. The engineer of the coal train saw the danger at about the same time, and both reversed their engines and applied the brakes but there was no time to stop their trains which were coming together at a closing speed of about 40 miles an hour. Most of the crew of the freight train jumped to safety before the impact but the enginemen of the express did not, nor did Fireman Bernard Gallagher of the freight train, who was taken unawares by the collision, but he somehow escaped with slight injuries. Both engines were badly smashed in the collision, the baggage car of the passenger train was smashed to splinters and the leading passenger car driven through the smoker. Four of the empty coal cars were thrown from the track into some trees lining the riverbank and six other cars, four of them gondolas filled with coal parked on the southbound track, were wrecked. Seven passengers, Fireman Walsh of the express and Brakeman Jacob Kilrain of the coal train were killed outright and about a dozen passengers and Engineer Fitch of the express were injured, some seriously. A fire, started

[490] Ibid.

[491] *The Times*, London, England, 10th, 11th, 16th and 18th June 1892

[492] *Philadelphia Inquirer*, Philadelphia, Pennsylvania, United States, 28th October 1892

by the contents of one of the engine fireboxes, also broke out but the wreckage was well scattered about and Conductor Hamilton, assisted by the uninjured passengers, was able to extricate all the injured and dead before the flames reached them.[493]

Train No. 27 had passed the station at Pencoyd, a mile on the West Falls side of West Manayunk, at 7.36 a.m. while train No. 2 was not due there until almost 9 so Yardmaster Rupp decided that he had ample time to move a local freight train out to Pencoyd. Engineer Charles Billick of engine No. 538 was thus given an order at West Falls at 8.35 instructing him to pick up some coal cars at Pencoyd and at Campbells Siding. The engineer understood that these orders gave him authority to run from West Falls to Bridgeport (beyond West Conshohocken). Yardmaster Rupp stated that he had not intended the freight train to proceed beyond Pencoyd and had instructed the Pencoyd operator to hold it there. He considered that he had authority to permit the freight train to run to Pencoyd as Pencoyd was within yard limits.

At the resumed inquest into the victims of the accident Engineer Billig (as his surname was now spelt) denied knowing of any southbound train or of having been instructed to wait for any opposing trains at Pencoyd. His conductor, William Gerlatcher, stated that he had received verbal orders from Trainmaster Rupp to pick up cars at Campbells Siding and at Pencoyd but had had no orders to wait there, but rather orders to run from West Falls to Bridgeport. Contradicting this, Trainmaster Rupp insisted that he had told the conductor to lay over at Pencoyd for trains Nos. 32 and 2. Operator C.M. Lesher and Agent Robinson Body spoke of going out with a red flag to stop the coal train after having received an order from West Falls to hold it there, but Engineer Billig denied having been given any stop signals, though his train had pulled up there anyway to pick up some empty coal cars. He added that he was almost entirely ignorant of the road south of Reading, but that in executing his orders he was subject to the instructions of his conductor. Operators L.K. Altendorfer and C.H. Turten, at West Manayunk and the north end of Flat Rock tunnel respectively, had not been informed that train No. 2 was running on the southbound track and hence took no steps to stop freight train No. 538.[494]

At the conclusion to the inquest Philadelphia and Reading General Manager Bonzano stated that Trainmaster Rupp was always subject to orders from Reading, notwithstanding Rupp's testimony to the contrary. Rupp's sending out of engine No. 538 had been contrary to the rules of the road. The verdict of the jury was that, 'The said collision was due to the carelessness of John R. Rupp, yardmaster at West Falls, in disobeying order No. 54 from Reading, which order he received at 7.46 a.m. on the morning of the accident, and we hereby censure the said railway company for not having a check on the said yardmaster, in case he would either disobey or misconstrue the said order, such check to be in the form of an order from Reading, issued to all stations south of Conshohocken, as follows, "Trains numbers 2 and 23 and engine *Ariel* will run on northbound track from Conshohocken; hold all trains and engines after engine No. 1019 on train No. 27 passes north," and we further censure the company for running green crews, as was the case with engine 538.' [495]

---

[493] Ibid, 25th October 1892
[494] Ibid, 3rd November 1892
[495] Ibid.

On about 28th November 1892 a collision occurred near a location given in a newspaper report as Szelna, Austria, but which was probably Zsolna in neighbouring Hungary. Owing to the engine driver of one train missing his signals two trains met and one full of passengers was wrecked. Several persons were killed outright, and others had to endure long agony before they could be released. One woman passenger was impaled by a long splinter which went right through her body and suffered for some time before she died, it being impossible to remove the splinter until later. Others were crushed beneath the carriage wheels. 'Many persons were permanently crippled, and the number injured is very great.' [496]

On Wednesday 4th January 1893 a sidelong collision with features similar to that at Esholt Junction, England, the year before occurred at the junction of two lines at Kunizsa, Hungary. The driver of a passenger train from Vienna misread the signals protecting the junction and at moderate speed entered the junction in front of another passenger train from Stuhlweissenburg, which was travelling at 20 miles an hour. The Stuhlweissenburg train ploughed through three carriages of the Vienna train which were filled with passengers and the vehicles were more or less completely wrecked while the colliding locomotive was 'twisted out of shape.' Eight passengers in the Vienna train were killed instantly, five sustained probably fatal injuries and 10 more were seriously but not fatally hurt. The passengers in the Stuhlweissenburg train were thrown from their seats and almost all of them were injured, but none fatally.[497]

The so-called Congo Free State, which was basically a colony established by King Leopold II of Belgium to extract resources from the area, was the scene of a disastrous accident at an unclear date, but probably about 10th January 1893. A train carrying both Europeans and 'natives' left Matadi, on the Congo River, for Emposi and then returned to Matadi. On the return journey the driver was instructed to attach a waggon loaded with explosives at a point only two and a half miles from Matadi and bring it to the latter point. The waggon contained 54 boxes of dynamite and 82 barrels of gunpowder used for blasting, probably in connection with mineral extraction. The waggon was to be attached to the rear of the train and presumably the driver expected to find it left in a siding. At any rate he approached the point where the waggon was waiting for him at a speed too high to be able to stop when he found that it was on the main line, collided with it violently and a fearful explosion ensued, with a report which was heard in Matadi and for many miles along the banks of the Congo. The waggon disappeared in the explosion, which left a huge crater, the rails were twisted in all directions, the engine turned completely round and utterly destroyed, with the carriages containing most of the passengers piled on top of it. The officials at Matadi, guessing the cause of the tremendous noise, had hurried to the scene, as did Captain Murray of the British and African Steam Navigation Company's steamer *Lualaba*, which had been lying at Matadi, together with his officers and the ship's doctor. They found an appalling sight, with some of the bodies blown to pieces and others with limbs blown off. The ship's doctor and its

---

[496] *Silver Standard*, Silver Plume, Colorado, United States, 3rd December 1892

[497] *Boston Daily Globe*, Boston, Massachusetts, United States, 5th January 1893

crew did all that they could to assist the victims and the injured and dying were taken as soon as possible to a temporary hospital. The only doctor at Matadi had been on the train and was himself seriously injured. Three of the Europeans who had been on the train were killed outright and a fourth – M. Legene, the head of the Société Anonyme Belge – died of his injuries the next day. From the number of bodies and parts of bodies recovered it was initially calculated that about 20 Africans had been killed outright but the following day it was found that another 26 were missing, and it was feared that they had been blown into the Congo River. The newspaper report concluded that, 'The people of Matadi were most profuse in their thanks for the humane efforts of the English crew, the doctor of the steamer being particularly successful and energetic in rendering medical aid to the sufferers. The bodies were buried in the little cemetery at Matadi, but for the most part they were unrecognizable.' [498] [499]

Only brief reports appeared in several American newspapers of a collision at South Ripley (Dover), Kentucky, United States, on the Chesapeake and Ohio Railroad, on 12th January 1893. A westbound Maysville accommodation train collided with a freight train, killing the engineer, fireman and six passengers on the accommodation. 'The railroad officers are close-mouthed, and particulars are meagre.' [500]

The Southwestern Limited, a fast train of the 'Big Four' Railroad (the Cleveland Cincinnati Chicago and St Louis) was involved in a bad accident at Alton Junction (or Wann), two miles east of the city of Alton, Illinois, United States, on Saturday 23rd January 1893. Alton Junction was served by both the Big Four and Chicago Burlington and Quincy Railroads, for whom it was a station used for the watering, transfer and feeding of livestock. The accident was due to a set of facing points wrongly being left open for a siding, that is, it was of what was known in the United States as the 'open switch' type. The Southwestern Limited due to leave St Louis at 8.05 a.m. in charge of Conductor Howard Clelland was 13 minutes late and was running at an estimated 47 miles an hour, attempting to make up time, as it approached Alton Junction. Engineer Webb Ross saw that the points were open and applied the air brakes as his train was diverted into the siding, which was occupied by a long freight train consisting mostly of loaded oil tanks. He did not jump but stuck to his post, was buried in the wreckage and burned to a crisp. His fireman, Richard White, jumped and escaped unhurt. After the accident he stood by the body of his engineer, 'moaning and weeping as if his heart would break', and exclaimed, "He saved the passengers but lost his life in doing it." [501] The train passengers were badly shaken up but none of them were seriously hurt.

The shock of the collision split two of the oil tanks wide open according to an eye-witness and the oil immediately ignited, with flames shooting 50 feet into the air. The locomotive and baggage cars were wrecked but the remainder of the train was pulled back clear of the fire. In a very short time the freight train was a mass of flames. A wrecking train arrived from East St Louis at 11 a.m. but could do little. A switch engine

---

498 *The Times*, London, England, 16th January 1893
499 *Mid Surrey Times and Advertiser*, London, England, 21st January 1893
500 *New North*, Rhinelander, Wisconsin United States, 19th January 1893
501 *Telegraph*, Alton, Illinois, United States, 26th January 1893

was used to clear the yards of waggons. Traffic over the Big Four and Chicago Burlington and Quincy was entirely suspended, passengers for St Louis being carried over the Chicago and Alton line.

At this point the accident had claimed only one victim, Engineer Ross, but hundreds of local people soon gathered at the scene to view the spectacle. A few minutes at 12 noon disaster struck. One of the tank waggons exploded, scattering debris all around and setting fire to the stockyard enclosures. Failing to realise that this explosion was only a preliminary one the spectators rushed to the stockyards to save them from destruction. A minute or so there was a deafening report that shook the earth for half a minute and spread burning oil in all directions. There was no escape for anyone within a hundred yards. Their clothing was burned and fell from their bodies and those who could ran to and fro screaming for help. Panic reigned for a short time, then several doctors who chanced to be in the area began to treat the casualties, using the contents of two barrels of linseed oil taken from a grocery store, while every house in the vicinity was used to take in the victims. A special train also took 20 sufferers to St Joseph's Hospital in Alton. All that could be done was done, but even so 41 people eventually died and another 75 were injured in what became known as the Wann Disaster. Eighteen cars were burnt to ashes, as were two nearby houses. The property loss was placed at $100,000.[502 503]

Blame for the accident was laid upon a pointsman named R. Grattan, who combined his railroad duties with working as a barber. He fled the scene in the confusion after the accident and officers were said to be searching for him.[504]

In the autumn of 1893 Kingsbury, Indiana, United States, was a small station on the Wabash Railroad on a line opened only in May of that year in anticipation of traffic to the Chicago World's Fair. West of the passenger depot there were two sidings on each side of the single main track, and early on the morning of 22nd September 1893 the second south siding was occupied by an eastbound freight train, No. 92, with about 30 cars of meat, in charge of Conductor H. Parks, with Engineer Peter Whitman and Fireman John Barber and Head Brakeman Herbert Thompson. The train had been waiting at the siding since 4.27 a.m. for two sections of the Niagara Falls express, train No. 55 from Buffalo to Chicago. This train was running in two sections because of the World's Fair traffic. The first section was due to pass at 4.40 a.m. with the second section 12 minutes behind it, but the first section did not in fact pass until 5.10 and the second section, did not arrive until 5.40. The second section was in charge of Conductor John Coulter, with Engineer John Green and Fireman John Volking. It approached Kingsbury running at 50 miles an hour. Just as the passenger train crossed the trestle approaching the station Head Brakeman Thompson, in a moment of thoughtlessness, seized the switch lever of the points which led to the south sidings, and the express was diverted into a head-on collision with engine No. 63 of the freight train. The two engines met with a thundering crash and were left welded together. The freight engine was also forced backwards through a heavy car of dressed meat and at the same moment the boiler of engine No. 473 of the passenger train exploded. Two of the passenger coaches were smashed to pieces and the third ended up

---

502 *Illinois Weekly Courier*, Jacksonville, Illinois, United States, 25th January 1893

503 *East Alton* by Jason D. Bricker and Judith M. Richie, pg. 27

504 *The Belleville Telescope*, Belleville, Kansas, 27th January 1893

sideways across the line at the switch and crushed the first half of the following sleeping car *Kansas*. Fortunately, there was no outbreak of fire. Eight passengers died in the accident, together with Engineer Green, Conductor Coulter and baggagemaster W.N. Lyon of the passenger train. A large number of passengers and the crew of the freight train were injured.[505]

Before the arrival of the first section of the express Engineer Whitman, knowing that he had 20 minutes in hand, had taken his locomotive out onto the main line to get to the water tank.

Brakeman Thompson had operated the switch to let him out and closed it again when he returned to the siding. That he had properly closed it at that point there could be no doubt for the first section had passed through safely and had given the usual whistle code – one long and two short blasts – to indicate that a second section was following. Why Thompson had then reopened the switch could not be known. After the accident Engineer Whitman, his right arm broken and with other injuries, had extricated himself from the wreckage and had found the brakeman standing by the track 'as pale as a ghost and trembling in every fibre'. "What have you done?" Whitman demanded. "I must have left that switch open," Thompson had replied. With that he had left the scene and he could not subsequently be traced.[506]

Two reports from Milan, Italy, on 29th and 30th November detailed a disastrous collision which had occurred at Limitro station, on the line between Milan and Treviglo, on the evening of 28th November 1893. A heavy goods train from Verona was being shunted out of the way of a Milan express (the reports do not make it clear whether the express was going to or from Milan). The trains collided sidelong and engine to engine, both locomotives rearing up in the air and then falling onto their sides. One waggon of the goods train and the two leading carriages of the express were smashed to pieces, the remainder of the passenger train was not derailed. Burning coals from the fireboxes of the engines set fire to the wreckage. By 8 a.m. the next morning 13 bodies, mostly badly mutilated and charred, had been recovered and 22 injured passengers had been taken to hospitals in Milan. It had been very foggy at the time of the accident; the fog had delayed the goods train, which was 40 minutes late, and supposedly prevented the driver of the express from seeing the signals protecting the shunting movements.[507]

By the time of the second report the number of dead had climbed to 43 and there were 183 people injured. Nearly every passenger on the train was more or less hurt. Most of the victims were emigrants, on their way to Venice and thence to the United States. Following the accident telegrams were sent to Milan for assistance and a relief train with doctors soon arrived, while railwaymen, local inhabitants and a company of soldiers engaged in the rescue work. The burning woodwork of the carriages was cut away with axes and streams of water were poured upon the wreckage, in which a number of passengers were trapped. The second report stated that some of the imprisoned passengers had died horrible deaths, being cremated before the eyes of the rescuers who were powerless to save them. Others were finally extricated only to die in the arms of

---

[505] *Westville Indicator*, Westville, Indiana, United States, 28th September 1893
[506] Ibid.
[507] *The Times*, London, England, 30th November 1893

their rescuers. All of the injured had suffered burns and many of them also had broken limbs, some requiring immediate amputation. The passengers who had been lucky enough to escape with minor injuries were cared for at the station and afterwards most of them proceeded on their journeys.[508]

On Thursday 14th December 1893, 11 persons were killed in a collision between a passenger train and a freight train near Soshowica, Russian Poland.[509]

In a type of accident which was to become much more common in the following century, a passenger train collided with a freight train near Samara, Russia on about 22nd January 1894. Five naptha trucks were set ablaze as a result of the collision and 15 persons lost their lives.[510]

Russia apparently suffered another disaster in the same area on 16th February 1894. Two trains collided near Jelan, between Samara and Slavansk, in the east of the country. Thirteen people died and several others were badly injured.[511]

On the morning of Wednesday 31st October 1894, the inhabitants of Sydney, in the Australian colony of New South Wales, must have been shocked to learn of a serious accident which had happened at the Redfern station. At 9.30 a.m. that day a six-coach passenger train from Strathfield was approaching the station and two passenger trains were timetabled to depart; at 9.30 a.m. for Goulburn and three minutes later a departure to Parramatta. On the engine of the Strathfield train, F class tank locomotive No, 366, were Driver William Pearce and Fireman Charles Orr, while the outgoing Goulburn train was hauled by a larger P class engine, No. 509, with Driver William Williams and Fireman Thomas Maddison. The station approaches were controlled by a busy signal box which during the day was double-manned; on duty at the time of the accident were Signalman George Lawrence and Assistant Signalman Luke Carter.

A couple of minutes before the departure time of the Goulburn train Signalman Lawrence should have set points No. 26 for it to pass from Platform No. 6 onto the down line and then cleared the platform starting signal, No. 3. He forgot to do so but at, he said, 9.32 he had cleared the starting signal, No. 4, for the Paramatta train to depart from the adjoining Platform 5. Meanwhile Guard Wilson Stanley of the Goulburn train had checked that all the doors of his train were closed, and everything was in order and then given his driver the right-away, at, he testified later, exactly 9.30 by the station clock. Possibly he actually got his train under way a minute or so later, or perhaps Signalman Lawrence had pulled off for the Parramatta train a minute or so earlier than he imagined. At any rate it seems certain that Fireman Maddison, seeing the guard's flag, told his driver that they were right to go, and Driver Williams had looked up, seen No. 4 signal

---

[508] *Madison Daily Democrat*, Madison, Indiana, 30th November 1893
[509] *Monticello Express*, Monticello, Iowa, United States, 21st December 1893
[510] *Lowell Sun*, Lowell, Massachusetts, United States, 25th January 1894
[511] *Hermann Advertiser and Advertiser-Courier*, Hermann, Missouri, United States, 21st February 1894

off for the Parramatta train, which he knew should have followed their train out, and took it for his own. The Goulburn train thus started away against its signal, and Signalman Lawrence not having set up the road to take it onto its proper course, it was instead diverted into a head-on collision with the incoming Strathfield train. Neither of the trains was travelling very fast at all, probably not even 10 miles an hour, but the effects of the collision were disastrous. The heavier engine of the Goulburn train forced the lighter engine of the incoming Strathfield train 10 or 12 feet back into the leading first-class smoking car while the following car telescoped the smoking car at its other end, and there was some telescoping of the remaining carriages as well. What really made the accident much more serious though was the breaking of the engine steam pipes, which as in a number of accidents in the United States and elsewhere filled the smoking car with scalding steam as well as hot cinders, while the passengers found their attempts to escape frustrated by the barred windows of the carriage.

Several passengers on the ill-fated train described their experiences to a newspaper reporter: Dr C.J. Weekes, who was taken to the Sydney Hospital, was not much hurt, and was able to get about all right after a few minutes. In a short description he said: "I come through from Croydon and was sitting in the first car two seats from the engine. We were all right till we got to within 100 yards of the platform, when the train collided suddenly with a country train just coming out. The first thing I knew was a terrific shock, and the carriage was full of steam and cinders from the engine. My glasses were covered with the steam immediately, and I could see nothing. The hot coals flying about fell on my clothes and burnt them through in one or two places, and the steam scalded me slightly, but considering that I was so close to the end of the car which telescoped on to the engine I escaped wonderfully. The scene when we got out of the carriage was all confusion – engine and cars all piled up on one another, and the injured passengers being brought out."

Mr E.A. Pyman, the Enfield council clerk, who was on board the suburban passenger train, and whose right hand was scalded, made the following statement: - "I was in the first car next the engine, at the end furthest from the locomotive. Dean M'Cartney was sitting opposite to me. I was reading a paper when I heard a loud crash, breaking of wood and smashing of glass. It was the seconds car telescoping our carriage. I saw the end of the car burst in just where I was. The carriage soon became filled with steam. I put my coat over my head, as it was stifling, and put my head out of the window. I could have got out through the window but for the iron bars outside. I think after this experience that the bars should be removed. There was a general rush for the windows. I did not say anything to Dean M'Cartney, and he did not say anything to me. There was much screaming and shrieking. There were many passengers in my car. I also heard the engine end of the carriage being broken and smashed up. When at last I did get through the window I saw Mr Alfred Muddle, the solicitor, helping Mr Quodling. I at once went to his aid. Numbers of the passengers were hurt, being cut and bleeding. It was a perfect panic in the carriage for fully three minutes. I thought it was all up. The scene was a very exciting one. There were soon several doctors present. Dr Long came to me and attended to my hand, which is very painful. I know that Mr and Mrs Osborne were badly hurt. I also saw that Mr Brierly, of Messrs Kent and Brierly, and Mr Bradley, who is engaged in Mr Sydney Burdekin's office, were injured."

The initial report in the *Sydney Australian Star* newspaper the same evening was able to reassure its readers that although a serious accident had happened no passengers had lost their lives, though unfortunately Fireman Orr had succumbed to his injuries during the afternoon.[512] Tragically although no passengers had been killed outright over the next few days no less than 12 of them, along with Driver Pearce, were to die of their injuries, and almost 30 more in the Strathfield train were injured. None of the passengers in the Goulburn train were hurt and the accident caused little disruption to train services.

An inquest was opened on Fireman Orr the day after the accident. Driver Williams insisted that signal No. 3 had been cleared for him and there was even a suggestion that Signalman Lawrence had initially pulled off for the Goulburn train and then changed his mind and thrown the signal back and reversed the points, but Superintendent Parry stated that the signalman would not have had time to do so. In the end the inquest jury found Driver Williams guilty of manslaughter. At his trial at the end of the month his defence counsel, Mr Edmunds, suggested that there was no more evidence that the starting signal had been at danger than that it had been clear. The jury apparently believed him and took only a short time to acquit Driver Williams. Possibly he was lucky.[513]

The day after his acquittal the railway company dismissed Driver Williams but a month later, he was re-employed at a gland packer, at a considerably lower wage than he had been earning as a driver.[514]

The danger posed by shunting operations being carried out during high winds was well demonstrated by a disaster on the London and North Western Railway at Chelford, England, on Saturday 22nd December 1894. The 4.15 p.m. up express from Manchester London Road to London Euston consisted of engine No. 418, *Zygia*, and tender, Engine No. 520, *Express*, and tender, a brake van, third-class, bogie composite and third-class carriages, these four vehicles all belonging to the Great Western Railway; and a composite, third-class, and two composite carriages, a milk van, a brake composite, composite, third-class, two composite and third-class carriages and a brake van; these vehicles all belonging to the North Western Company. In charge of the train and riding in the rear brake van was Head Guard Arthur Jessop with Guard Samuel Scott in the Great Western brake van, Driver Thomas Kerr and Fireman James Rowlands on the pilot engine and Driver Joseph Capper and Fireman Henry Cheshire on the train engine. On duty in Chelford signal box was Signalman Frank Fearn.

A down pick up goods train arrived at Chelford at 3.55 p.m. in charge of Brakesman John Roughsedge, who was relieved there at 4.25 by Brakesman Joseph Whalley, with Driver Samuel Needham and Fireman William Hopwood. It was a very windy afternoon. The train did a little shunting and then Signalman Fearn put it into the down loop out of the way of the 3.45 p.m. down passenger train from Crewe. That train having passed, Signalman Fearn let the pick-up goods back out of the down loop and shunting re-commenced. At 4.38 p.m. he accepted the Manchester express from Alderley Edge and received the "Train Entering Section" signal two minutes later. He obtained "Line Clear" for it from Chelford Loop box and cleared his up main signals. He did not hear anyone calling out or see any red lights but as the express neared his box, he saw a tremendous number of sparks coming from the engine wheels and heard

[512] *The Australian Star*, Sydney, New South Wales, 31st October 1894
[513] Ibid, 31st October and 1st, 2nd, 3rd, 26th and 27th November 1894
[514] *Australian Railway Disasters* by Kenn Pearce, pgs. 24-6

a rumbling as if the train was off the line. The next moment the front of his box was demolished by the train. He tried to reverse his signals and sent the "Obstruction Danger" signal in either direction on the block bells but got no reply. He then tried to send a warning on the speaking telegraph and found that that was still working. Ironically his box had been demolished by a derailed train once before, in July 1885. It had been getting dark at the time of the accident and with the gale-force wind blowing he doubted if he would have been able to hear the noise of any collision in the goods yard unless it was very loud.[515]

Stationmaster John Hyde was supervising the shunting operations. When they restarted after the departure of the Crewe train a part of the goods was left on the down main line while the engine pushed the front waggons back into the goods yard or back onto the main line portion of the train as directed by the stationmaster, while the brakesman uncoupled the waggons. A waggon of potatoes for Stalybridge was pushed back towards those on the main line, the next waggons were placed in the yard and a high-sided North Western waggon was pushed back along the main line in turn. The engine then went forward again to push six more waggons back into the goods yard. As it was pushing these trucks back the high-sided waggon was blown forward towards them along the main line and the waggons met at the V crossing of the down main line and the connection to the goods yard. The blow was sufficient to derail the empty high-sided waggon and it overturned fouling the up main line. Stationmaster Hyde had seen the danger and was running towards the express waving a red light, but the latter was close at hand, and although both drivers saw the red light the collision occurred only a couple of seconds later.[516]

The express was travelling at about 60 miles an hour. Both engines were derailed and after running some distance the leading engine overturned across the up line and the remainder of the train piled up against it, all the vehicles of the train except the last three leaving the rails. The Great Western bogie composite carriage marshalled third in the train was the one which struck the signal box, while the fifth to tenth vehicles were badly damaged by striking the rear waggons of the pick-up goods train standing on the down main. Fourteen passengers died in the accident and 79 were injured. Driver Kerr, Fireman Rowlands and Guard Jessop were seriously injured (Fireman Rowlands was initially reported as having been killed) and Driver Capper, Fireman Cheshire and Guard Scott were slightly hurt.

Major F.A. Marindin investigated the accident on behalf of the Board of Trade. He noted that the second collision was unavoidable but was of the opinion that the same could not be said for the initial mishap. Although he could not see that Stationmaster Hyde had breached any rules or regulations, he did not think that he could be entirely acquitted of any culpable negligence, as he should have realised that the force of the wind was sufficient to move the waggon and applied its brake.

As far as recommendations were concerned Major Marindin pointed out that while accidents of this type could be prevented by suspending all shunting operations whenever a passenger train was passing, such a course of action would be practically unworkable except on very lightly used lines. He did, however, recommend the addition of a

---

[515] Board of Trade report
[516] Ibid.

regulation to the effect that during high winds all waggons being set back onto their trains should be kept attached to their engines, or failing that, the brakes of any loose-shunted waggons should be immediately applied. To make this operation easier all waggons should have brake handles on both sides, instead of one just one side as was the case at the time.[517]

At the two inquests on the victims of the disaster (two had to be held because the last victim did not die of his injuries until 4th February 1895) the juries returned verdicts of "Accidental Death", with a recommendation that shunting operations should cease while express trains were passing.[518]

A brief report from Reuters news agency stated that a telegram from San Paolo, in the province of Corrientes, Argentina, gave news of a collision between two trains carrying immigrants on about 10th July 1895. Fifteen persons had lost their lives and 30 were injured.[519]

An excursion organised by the Ancient Order of United Workmen of Chicago, which took 1,500 excursionists over the Wisconsin Central Railroad to Schiller Park for a picnic on Sunday 12th July 1896 seems to have been jinxed from start to finish. On the outward journey in the morning one section of the train was boarded by roughs who assaulted and robbed the conductor and engineer. Worse was to come on the return trip in the evening, for the first section of the train consisting of an engine and 13 coaches collided with a train of empty coaches at Altonheim station, Illinois, United States, and was wrecked near the Madison Street crossing. Five passengers were killed outright, five seriously injured and many more slightly hurt.

Shortly before the arrival of the southbound excursion train the station agent had set the road for a northbound freight train to depart, and had forgotten to reset the points, so that the excursion train was diverted onto the wrong line. Switchman or Station Agent J.V. Kniskern and the train crew were arrested. The passengers who lost their lives were Mary Arnold, aged 11; Lena Herbert, 24; Frank Koch, Fred Kertell and Charles Samuel. The accident happened at 8.40 p.m. The train was only travelling at about 10 miles an hour and the engine was little damaged. The passengers who lost their lives were standing on the platform of the first coach, which was partially telescoped by the baggage car.[520] [521] [522]

The narrow gauge (1,000mm) Rajputana – Malwa Railway, by this date under the control of the Bombay Baroda and Central India Railway Company, suffered a major disaster in the Rajputana – Malwa yard in Delhi, India on Friday 25th July 1896, when

517 Ibid.

518 *The Times*, London, 24th, 25th and 27th December 1894, 2nd, 5th and 12th January and 5th, 6th and 14th February 1895

519 Ibid, 12th July 1895

520 *The Harvard Herald*, Harvard, Illinois, United States, 17th July 1896

521 *Cedar Rapids Standard*, Cedar Rapids, Iowa, United States, 23rd July 1896

522 *Crawfordsville Weekly Journal*, Crawfordsville, Indiana, United States, 17th July 1896

down mail train No. 2 from Malwa was diverted onto the wrong line and collided with a shunting engine. Twelve 'native' third-class passengers and the second guard were killed, and 21 passengers injured. The driver and head guard suffered contusions. [523 524 525]

In another accident of the same type as that at Malwa in 1896 an up mail train which was passing through Gholvad station, India, on the Bombay Baroda and Central India Railway on 9th January 1897, was diverted through a mistake by a pointsman onto the platform line on which a down mixed train was standing. Realising his mistake after the engine and four leading vehicles had passed through the connection, the pointsman tried to put matters right by reversing the points beneath the train, splitting and derailing part of it. The front portion of the train nevertheless collided with the mixed train, causing serious damage. Four passengers and a railway servant were killed, and 22 passengers and two railway servants injured. The deputy stationmaster and the pointsman were both prosecuted, the former being sentenced to 10 days' simple imprisonment and a fine of 10 rupees and the latter to two years' rigorous imprisonment. (Some newspaper reports spelt the station name as Golwood.) [526 527]

In Germany a troop train travelling from Westphalia to Merz and carrying five officers and 1,124 reserve men of the 98th, 130th, 135th and 143rd regiments, came to grief as it was descending a steep incline near Gerolstein at 11.30 p.m. on Tuesday 18th May 1897. Seven carriages were wrecked and nine soldiers and a brakesman were killed on the spot and 35 soldiers and two railwaymen were injured, many of them it was believed fatally. 'Among the dead and wounded are many reserve men who are heads of families.' Thirty of the most seriously injured were taken by train to Trèves (Trier) by train and distributed among the hospitals of the town. The accident was believed to have originated in the fracture of a coupling in the middle of the train, followed by a collision between its two portions when the front portion reduced speed.[528]

On the morning of 30th November 1897, a serious collision occurred at the Warsaw terminus of the Warsaw – Pushto line in Russian Poland. Due, it was supposed, to an error by a pointsman a goods train entering the station was diverted onto the line on which a passenger train had previously arrived. A violent collision ensued which completely telescoped the rear carriages of the passenger train. Eleven passengers were killed outright and 22 injured, four seriously. By chance four doctors were on the train and were able to assist the station staff in administering first-aid before the casualties were removed to Warsaw Hospital.

'The judicial authorities at one proceeded to hold an inquiry into the matter and have already inspected the scene of the disaster. The pointsman, the staff of the goods train, and the traffic superintendent have been suspended from their functions pending the

---

523 *The Daily Mail*, London, United States, 28th July 1896
524 *The Standard*, London, England, 28th July 1896
525 Administration Report on the Railways in India 1896-7
526 *The Times*, London, United States, 11th January 1897
527 Administration Report on the Railways in India 1897-8
528 *The Times*, London, England, 20th May 1897

result of the inquiry.' [529]

An accident very reminiscent of that at Chelford four years earlier occurred at Wrawby Junction, England, on what had now become the Great Central Railway, on 17th October 1898. At 4.20 p.m. that day a goods train left Grimsby West Marsh (Armstrong Street) timber sidings for Hexthorpe, in charge of Guard James Starkey with Driver William Simpson and Fireman George Green. It consisted of an engine and tender, 43 waggons and a van. Nine of the waggons were carrying 72 42-foot-long poles, with 37 poles chained to two waggons and 35 to two more, with a runner waggon at each end and another runner waggon between the two groups of four waggons. The train proceeded to Barnetby, where it was looked over by Guard Starkey and Waggon Examiner T. Fish, who saw nothing wrong with it. From Barnetby the train was routed via the down goods loop to Wrawby Junction. Here nine waggons at the rear off the train had to come off and the guards van was detached and left on the down loop while the engine pulled the train onto the down Lincoln branch and then set back into the cattle siding to detach the nine waggons. This having been done the engine pulled forward clear of the siding points and then began to set back onto the guard's van.

The passenger train involved in the accident was the 4.45 p.m. express from Cleethorpes to Manchester, consisting of an engine and tender, brake van, third-class carriage, two composite carriages, third-class carriage and rear brake van. It was in charge of Guard George Ogden with Driver William Hulse and Fireman James Osborne. The train approached Wrawby Junction at 5.24 p.m. just as the front part of the goods train was being set back onto its brake van. As it was nearing Wrawby Junction box and passing the goods train in the down loop, Fireman Osborne saw a waggon fouling their line and shouted to his driver to pull up. At about the same moment Signalman John Kidney heard what he described as an unusual noise and threw back his down main signals in the face of the express.[530]

As the waggons of the goods train were being set back one or more of those loaded with the poles became derailed and fouled the down main line just as the express was passing. Part of one of the poles wedged itself in the express engine and other poles swept away the bodies of the carriages in the centre of the train, whereby six passengers were killed outright, two fatally injured and died the next day and 26 injured, some very seriously. No vehicle of the express was derailed.

After the accident help quickly arrived from Barnetby, Brigg and Grimsby. The injured were taken in a special train to Grimsby, the bodies of the six who had died at once being left at Barnetby.

At the adjourned inquest on the victims of the accident, held at Barnetby, the jury decided it there was insufficient evidence to show how the waggons had become derailed and like the jury at the inquest on the victims of the Chelford accident they condemned the practice of carrying out shunting operations on lines adjoining those on which express trains were running. They commended the promptitude of Signalman Kidney, Fireman Osborne and Driver Hulse in their efforts to avert the accident.

---

[529] *The Times*, London, England, 1st December 1897
[530] Board of Trade report

Lt-Col G.W. Addison, the Board of Trade inspecting officer, suggested that the initial derailment had been caused by the buffering-up of the waggons as they were being pushed back. As to the suggestion of stopping all shunting operations while passenger trains were passing, he concurred with the opinion of his colleagues that such a blanket restriction was out of the question. He could only suggest that the distance between sidings and running lines be increased and that poles would be better carried on specially constructed bogie waggons. [531] [532]

In late 1898 the Grand Trunk Railway main line between Montreal, Quebec and Toronto, Ontario, in Canada, was a mixture of double and single track. West of Trenton, Ontario, station the line was single track, but it became double after two miles at Murray Hill, where westbound trains took the left-hand (southern) track. There was no station at Murray Hill, just a signal box to control the single to double line connection. Signalling at the junction was rudimentary; at night a white light showed in one direction and a red lamp in the other. The engineer of a westbound train seeing a white light would know that the route should be set for him to continue to Toronto, while a red light would indicate that he must bring his train to a stand.

On duty at Murray Hill in the early hours of Tuesday 15th November 1898 was Pointsman John Murphy, working a normal 12-hour shift, 7 p.m. to 7 a.m. At 1.52 a.m. a westbound freight train approached his box and Murphy directed it to the southbound track, as usual. He then reset the points so that they were right for an eastbound movement. Westbound trains would thus be confronted with a red light at the signal, and eastbound trains a white light. An eastbound freight train with 24 loaded boxcars, No. 96, arrived at Murray Hill at about 4.45 a.m. but despite the clear signal Engineer Thomas Ivens brought it to a stand clear of the junction, presumably because either the working timetable or orders received by him had directed him to wait there for westbound express No. 5. Switchman Murphy should have been aware that the westbound express was approaching him as well and should therefore have reset the road to let it cross to the westbound track, but perhaps overcome by tiredness in the early hours, he did not do so. There was, however, a second line of defence against an accident, for his signal was showing a red light eastwards against the oncoming express.

Westbound express No. 5 consisted of an engine and tender, express, mail and baggage cars, a second-class coach, first-class coach and two Pullman sleeping cars at its rear. It had climbed the hill out of Trenton and then gathered speed as it ran downhill towards the junction and was probably travelling at 50 miles an hour when it reached there. For some reason Engineer William Brady and his fireman, John Macdonald, failed to reduce speed and thundered past Murray Hill box, where their train was diverted onto the eastbound track by the wrongly set points. Not only had he ignored the red signal, Engineer Brady had failed to react to the fact that his engine had not given the usual jerk as it went through the crossover points, nor did he and his fireman apparently notice that there was no eastbound track to their right. Engineer Ivens and his fireman, Alex Toppin, did notice in time to jump to safety before the inevitable collision and it would seem that Engineer Brady also realised his fatal mistake at the very last instant.

---

531 Ibid.

532 *The Times*, London, England, 18th, 19th, 20th and 27th October 1898

The force of the collision destroyed both engines and the baggage car completely telescoped the following second-class coach, which contained about 25 passengers, including a number of cattlemen and a party of German-speaking immigrants from Russia. Hardly any of them escaped being killed or injured. The mail car was driven right on top of the baggage car, and the express car was partially wrecked. The first-class coach and sleepers escaped. Nine passengers died in the accident; six of them were immigrants (four of them from one family), two were cattle dealers from Toronto and the last an off-duty engineer travelling on the cushions. Engineer Brady and Fireman Macdonald also lost their lives. Fourteen passengers were seriously injured.[533]

Newspaper reports on the inquest into the victims of the disaster described its result as a 'Peculiar Verdict.' The cross-examination of Pointsman Murphy was especially harsh, though in fact his failure to reset the junction would not have of course in itself have caused an accident. Engineer Brady, and to a lesser extent his fireman, were directly responsible for the disaster but various witnesses at the inquest attempted to defend the actions of all three railwaymen. Engineer Ivens suggested the possibility of a sudden problem on the locomotive, such as a broken steam gauge, though there was no evidence to support such a theory. The more obvious possibility, that the engineer had dozed off, was not, it would seem, mentioned. The jury returned a verdict that the deaths were the result of Engineer Brady's running past a red signal and so ending up on the wrong track, then added that he had been 'as faithful and trustworthy a driver as was in the service of the company' despite his mistake which had resulted in 11 deaths proving just the opposite. The jury did not consider that Brady had been overworked or that the junction at Murray Hill was unsatisfactory. They added a rider that 'the Grand Trunk company deserved censure for the actions of the present management', whatever that meant.[534] [535]

An accident occurred in India on the Assam-Bengal Railway on 28th April 1899. A light engine going to collect the waggons of a ballast train stabled in at siding at Fajilpur station collided with them with considerable violence. Six coolies were killed outright, five seriously and one slightly hurt. Two of the injured men died in hospital. Four of the waggons were badly damaged.[536]

On about 30th March 1900 a serious accident occurred at a location whose name was variously given as Alkan Tilpust and Alkhan Tehurt in Russia. A mail train collided with a freight train, petroleum tanks were set ablaze and the drivers, postal clerks 'and many of the passengers' perished in the flames.[537]

A brief report on an accident sent from Strassburg in Austria described a collision between an express train from Bamberg in Germany to Riscingew and a freight train on

---

[533] *Wellsville Alleghany County Reporter*, Wellsville, New York, New York, United States, 18th November 1898

[534] *Fort MacLeod Gazette*, Fort Macleod, Alberta, Canada, 25th November 1898

[535] *Wreck!* By Hugh H. Halliday, pgs. 95-100

[536] Administration Report on the Railways in India for 1899-1900

[537] *Albert Lea Freeborn County Standard*, Albert Lea, Minnesota, United States, 4th April 1900

6th June 1900. The author has not been able to identify any place named Riscingew. The location of the accident, in which 30 people reportedly lost their lives, was given in another report as Staffelbach, Russia, but it would appear more likely to have been a place of that name not far from Bamberg in Bavaria, Germany.[538 539]

A northbound Chicago and North Western Railroad passenger train carrying excursionists to the Saengerfest (singer festival) at Green Bay, Wisconsin, United States, collided with a freight train at Depere, five miles south of Green Bay, at 10.45 a.m. on Sunday 24th June 1900. The passenger train was pulling into Depere station and the freight train was being backed into a siding out of its way but had not cleared the main line. The excursion train had been made up at Fond du Lac and was carrying passengers from that place, Oshkosh and Neenah. Its two leading coaches, a combination baggage and passenger car and a smoker, were filled almost exclusively with Fond du Lac passengers. The freight train had come from Green Bay and had orders to cross the excursion train at Depere. It was too long to entirely fit in the crossing loop and part of it having been left in the loop the remainder was being set back into a siding at the back of the station when the accident occurred. A flagman went back to warn the excursion train, but the air brakes of the latter allegedly did not work properly, and it collided with the freight train. The two leading coaches of the excursion train were telescoped and demolished in the collision and almost all the passengers in them were killed or injured. Eight passengers were killed outright and over 50 were injured, one of whom apparently died later. Having stated that the air brakes of the passenger train had failed to work properly the newspaper report confusingly stated that the accident had been caused by the failure of the freight train to keep out of the way of the excursion.[540]

Twenty people were killed and injured in a 'railway smash-up' near Plevna, Bulgaria, on about 20th August 1900. Among those who lost their lives was a M. Petroff, 'brother of the former minister of war.'[541]

On Sunday 11th November 1900 the Chemin de fer de Paris à Orléans slow train No 215 which had departed from Paris Quai d'Orsay station at 10.37 a.m. arrived at Choisy-le-Roi station, its destination, and its passengers alighted. Its empty coaches were still standing at the platform, waiting to be disposed of, while train No. 214, which had left Brétigny at 10.22, for Paris, was approaching Choisy le Roi, 20 minutes late. Due to a signalman's error train No. 215 was then run into by express train No. 9 to Nantes and Le Croisic, which should have run through Choisy-le-Roi without stopping. The engine of the express overturned onto its left side after striking the van at the rear of the local train, killing its driver and fireman who were buried beneath it, but at least falling clear of the opposite running line, on which train No. 214 was passing at that moment. The luggage van and third-class carriage at the front of the express were smashed, and six passengers were killed and 23 injured, some seriously. When news of the accident reached Paris the

538 *Sedalia Sentinel*, Sedalia, Missouri, United States, 6th June 1900
539 *Sycamore True Republican*, Sycamore, Illinois, United States, 13th June 1900
540 *The Nebraska State Journal*, Lincoln, Nebraska, United States, 25th June 1900
541 *Winchester Journal*, Winchester, Indiana, United States, 29th August 1900

Minister of Public Works, the Procurer of the Republic and the Prefect of Police set off at once for Choisy-le-Roi. The dead and injured were taken to Paris, where a number of beds had been improvised at the station for the injured, who were attended by two doctors from the Salpetrière Hospital.[542] [543] [544]

Six people reportedly died and a large number were injured in a collision between two ore trains in Nattavara station, Sweden on Tuesday 13th November 1900.[545] Agnes, the 22-year-old widow of Driver Fritiof Torstensson, one of those killed, was left to bring up three small children on her own, but she subsequently received sufficient compensation from the state to allow her to buy a house in Luleå, where she lived until her death in 1962.[546]

542 *The Times*, London, England, 12th November 1900
543 *The Standard*, London, 12th November 1900
544 *Le Petit Parisien*, Paris, France, 12th November 1900
545 *The Evening Democrat*, Warren, Pennsylvania, United States, 14th November 1900
546 Webpage – Agnes: om en lokförarhustru och tågolyckan I Nattavaara.

CHAPTER 6

SINGLE LINE COLLISIONS

The operation of trains in opposite directions over a single line of rails carries the obvious danger of head-on collisions between them. By 1880 there were many hundreds of miles of single-track railways throughout the world, especially in places where the density of traffic did not justify the provision of a second line. Most of them were worked with the aid of the electric telegraph, with decisions as to the meetings between trains being either made locally or, as in United States, by a central dispatcher. In Great Britain some lines were worked telegraph supplemented by a staff and paper tickets, while an improved method of single-line control had been brought into in 1880 after the invention by Edward Tyer of the electric train tablet system (described below in more detail in the account of the 1888 Hoylake accident) but this system was not used extensively outside of the British Empire. The invention of a failsafe track circuit in 1872 by the American William Robinson[547] led to the introduction in the United States of automatic block signals, by which train movements on a single line could be safeguarded by signals warning of the approach of an opposing train, but the cost of installing such signals was high and as late as 1906 the Interstate Commerce Commission was stating that out of 194,727 miles of railroad there were still only 41,916 miles protected by them.[548] With the protection of trains running over a single line left solely to the exchange of telegraph messages, rather than by the more secure electric train tablet or its later equivalents, and not supplemented by some form of automatic signalling. collisions due to human error in the transmission or interpretation of telegraph messages became all too common.

Two collisions in the United States on 3rd and 5th March 1881 demonstrated how chance factors could determine the degree of publicity that an accident received. The first accident, which received little publicity, was in fact the more serious of the two in terms of fatalities. A freight train collided head-on with a materials train near Caldwell Station, West Virginia, on the Chesapeake and Ohio Railroad in what was presumably a single-line collision, though the brief reports on the accident do not make this entirely clear. The

547 Wikipedia – Track Circuit
548 Wikipedia – Automatic Block Signalling

locomotives were 'shattered to the tenders' and the two engineers and two firemen were killed, and a Dr Caldwell of White House Springs also lost his life. What the doctor was doing on one of the trains was not stated. The wreckage of the cars was piled up four feet deep and the line was blocked until the evening.[549]

The second accident resulted in only two deaths but received much more publicity because a passenger on one of the trains involved was ex-President Rutherford B. Hayes, who was returning from Washington, D.C. to his home in Ohio the day after the inauguration of the new President, James A. Garfield. The accident happened just south of Severn station on the Baltimore and Potomac Railroad at 2.5 p.m. on Saturday 5th March 1881. The passenger train carrying ex-President Hayes was running as the second section of a northbound fast train due to leave Washington at 10 a.m. but because of heavy traffic, it did not start from the capital until just before 1 p.m. It was in charge of Conductor Armacost, with Engineer John Unglaub and Fireman B.L. McKeever and consisted of an engine No. 283 and tender, four Pullman sleeping cars, the first of which, the *Syria*, was chartered by J. Wyman Young of Shamokin, Pennsylvania and contained his friends and citizens of Shamokin and the area around there. The next three cars, the *Brockton*, the *Arcola* and another, contained the Cleveland City Troop, who had formed a special escort on the train carrying President Garfield to Washington and were returning as a special escort to ex-President Rutherford. The latter was riding with his family and friends in a special car fifth from the engine, which was followed by another special car carrying counsel James Mason and officer A.C. Armstong of the Lake Shore and Michigan Southern Railroad and their families and friends. Next came a baggage car, a special car carrying Daniel P. Eells, President of the Ohio Central Railroad, and finally a car containing 12 horses of the Cleveland City Troop. [550]

Coming in the opposite direction were a pair of locomotives coupled together; engine No. 1, with Engineer Henry Freeburn, Fireman G. Frere, Conductor Roy and baggagemaster John Oliver; and engine No. 147 with Engineer Jacob Rider, Fireman Robert Eberhart, and Conductor Robert Oliver. The engines and their crews were running as an extra to Washington D.C. to work northbound trains and had an order to wait at Severn for the train carrying ex-President Hayes, but they ran past the siding there. The northbound train had just rounded the curve south of Severn station when the two met, with each running at about 30 miles an hour. Fireman McKeever of the northbound train and Conductor Roy of the engines jumped and escaped injury; Fireman Everhart jumped and was slightly hurt. The others on the three engines stayed put. baggagemaster Oliver lived but a short time after the collision, and the others who had remained on the engines were all more or less hurt, Engineer Freeburn suffering a fractured skull. The engines were thrown from the track and badly damaged and the *Syria* Pullman car was driven into the tender of engine No. 283, while the *Brockton* Pullman car was driven partly into the *Syria*. Mr Young in the *Syria* car was the other fatality in the accident; he was killed outright while his 12-year-old son, who was also riding in the car, escaped unhurt. Ten railroad employees and 18 passengers were injured. Ex-President Hayes and his party, situated well back the train, escaped harm.

---

549 *Boston Daily Globe*, Boston, Massachusetts, United States, 5th March 1881
550 *The Sun*, Baltimore, Maryland, United States, 7th March 1881

After the accident Engineer Unglaub was asked why he had not jumped from his engine and he replied, "If I had jumped off, I did not know that many passengers might be killed, and the train wrecked through me not standing at my post." He was taken at first to the station house at Severn. 'He said that while he was in the station house ex-President Hayes and Mrs Hayes entered, showing their sympathy, and the wife of the ex-President took him by the hand, all dirt-stained and grimy as it was, and spoke a few kind words in praise of those who had done their duty in a time of danger. He also said she burst into tears at the distress and suffering caused by the accident and furnished bandages for the wounded from some of her own apparel.' The engineer was later taken first to Baltimore station and then to his own house. 'Mrs Unglaub said when her husband reached home Saturday night about the first thing he asked for was something to eat, having been a long time without food; that, referring to his really very narrow escape from death, he said; "In the name of God, old woman, I don't know how I got out of my engine." Unglaub was born in Germany but came to this country as a boy.'[551]

News of the accident was telegraphed to Superintendent George Wilkins at Baltimore and a breakdown train was sent to the scene, accompanied by Drs Atkinson, Bosley and Latimer, while Dr Thomson remained at the Union Depot in Baltimore to await the arrival of the injured there. Trainmaster Kapp, who accompanied the relief train, decided that it would be quicker to build a bypass line around the wreckage than it would be to remove it, ordered rails and ties (sleepers) and in only three hours after the accident a diversion had been constructed, over which 15 trains were worked the same evening. President Hayes and his party and the Cleveland City Troop continued their journey west on the 11.30 p.m. departure from Baltimore, which also carried the body of Mr Young and his unharmed and doubtless much traumatised son.

Superintendent Wilkins stated soon after the accident that a thorough investigation into it would be launched but that it was evident that it had been due to the disobedience of orders by the engineer and conductor of the southbound extra, who had clearly forgotten that they should have waited at Severn. The conductor and engineer of the second engine also had the order but the crew of the leading engine were regarded as the ones to blame. The northbound train had been obeying orders. Conductor Roy stated that he had tried to call the attention of Engineer Freeburn to the need for him to stop so that he (the conductor) could work the switch to put the train into the siding, but the train had passed the switch before it could be checked. Engineer Rider on engine No. 147 stated that he had also realised that Engineer Freeburn was going too fast but had been unable to do anything about it.[552] [553]

Seven people died and 12 were injured in a collision on 4th July 1881 at a location 12 miles south of Covington, Kentucky, United States, on the Kentucky Central Railroad. Those killed, who were stated to have all been residents of Covington, were Fireman John Pye, passengers Michael Dixon, William Callapy, William Weichoug, John Cotter and two boys who had been stealing a ride and whose names were not known. Both engines were demolished in the collision, which as caused by the negligence of the

[551] Ibid.
[552] *Times*, Philadelphia, Pennsylvania, United States, 6th March 1881
[553] *The Sun*, Baltimore, Maryland, United States, 7th March 1881

conductor of an express train in not obeying an order to side-track his train for an opposing excursion train.[554]

The rashness of a conductor and the inexplicable behaviour of an engineer were the features of a collision on 29th September 1881 between Aylmer and Orwell, Ontario, Canada, on the Great Western Railway. Once again, one of the trains involved was an excursion train, heading for the exhibition grounds at London, Ontario; the other was a freight train. The trains were due to cross one another at Aylmer but Conductor Maxwell of the excursion decided that as it was running 10 minutes early, he could advance his train to the siding at Orwell, two and a half miles further on. The engineer of the freight train saw the excursion approaching and had reduced speed almost to a standstill and was blowing his whistle continuously. The line was dead straight, and Engineer Walmsley should have been able to see the freight train over two miles away, but he continued to forge ahead as if the line was clear and did not whistle. As he was killed in the collision, along with his son, the fireman, no explanation was forthcoming for his seemingly suicidal behaviour. Five of the 10 coaches of the excursion train were telescoped by the force of the collision and at least four passengers were killed in addition to the Walmsleys. The passengers who lost their lives were William Cooke; J. Hines of Jarvis, Ontario; E Stanley, of Copenhagen, Ontario and an unknown man from Delhi, Ontario. About 20 passengers were severely injured and a large number of others were slightly hurt. The casualties would have been higher had not most of the excursionists been seated in the rear of the train. The accident left the two locomotives, six passenger coaches and several freight cars piled up in a heap. Conductor Maxwell took all the blame for the collision upon himself and disappeared after the accident.[555] [556]

A collision on the Lake Shore and Michigan Southern Railroad near South Bend, Indiana, United States, on Tuesday 1st August 1882 began with the derailment of eastbound through freight train No. 60 at Terre Coupee, Indiana, some miles further west. The derailed train consisted of engine No. 35 and 40 loaded waggons, with Engineer Charles Stanton. It was got back on the road and was hurrying eastwards again when it was in collision with local freight train No. 73, consisting of engine No. 109 and 28 loaded waggons, between Palmer and South Bend. Both trains were travelling at about 40 miles an hour when they met in a thick fog. The two locomotives were thrown from the track in the collision and the leading six cars and their loads piled up about them, completely burying engine No. 109 and almost burying No. 35. Burning coals from the engines then set fire to the wreckage. Engineer Stanton was severely bruised, had a badly cut forehead and suffered possibly fatal internal injuries. His opposite number, Engineer William Harrington (or Herrington, or Henry) of the local freight suffered severe bruising and a broken leg. Both had to be dug out of the wreckage and Engineer Harrington was only just released before the advancing flames would have become unbearable to the rescuers. The other railroad men escaped unhurt, but it was believed that there had been eight tramps riding on one or both of the trains; their voices could be heard moaning and

554 *La Plata Home Press*, La Plata, Missouri, United States, 9th July 1881
555 *Weekly New Mexican*, Santa Fe, New Mexico, United States, 3rd October 1881
556 *Terre Haute Weekly Gazette*, Terre Haute, Indiana, United States, 6th October 1881

calling to be released but nothing could be done to help them. Only two of them managed to get out. One of the tramps, presumably one of the two who escaped, was identified as a J. Reynolds of Cleveland, Ohio. The track was expected to be blocked for two days. Some trains were cancelled, and others diverted. The loss was estimated at $150,000.[557] [558]

The collision was blamed on 'criminal carelessness'. Early reports stated that it had been due to a failure to deliver an order to the eastbound train to take the siding at Palmer for the westbound one.[559] A later report stated that responsibility had been placed on the conductor of the eastbound train, who had been dismissed, and a despatcher, who had been suspended. The newspaper commented, 'If these men were really to blame for the terrible wreck and its sad results, they should be made to suffer some more severe penalty than a mere dismissal or suspension from the employ of the company. It is frequently the case, however, that a subordinate is made to suffer for the mistakes of his superiors. It is so convenient to have a scape-goat in an emergency like this.'[560]

In early August 1882 in The Netherlands work needed to be carried out on bridge No. 24 on the Hollandsche IJzeren Spoorweg Maatschappij (HIJSM) main line near Vogelenzang. A notice issued on 3rd August indicated that in connection with the bridgework there would be single line working with a pilotman between km 24.717 and km 24.886. The mark the point where the single line working commenced red stop boards were erected at km 24.632 and km 24.971, with green warning boards 300 metres in their rear.

At 9.20 a.m. on Wednesday 9th August train No. 8, consisting of the engine *Pegasus* and 25 axles, slowly approached the stop board at km 24.971 and Pilotman C. Vermeulen boarded its engine to conduct it over the single line. It started off again and was negotiating the single line when train No. 9, which consisted of the engine *Cerberus* and 22 axles, came into view coming in the opposite direction. Pilotman Vermeulen saw that train No. 9 was not pulling up and shouted to his driver, J.W. Muller, who waved a red flag and whistled for brakes, which were applied at once from the two brake vans. Pointsman A. de Vink at the single to double connection at km 24.717 also saw the danger and waved a red flag, but the two trains collided at the points.

The two locomotives were only slightly damaged. The leading luggage van of train No. 8 was pushed into the following goods waggon (No. 79) and two compartments of composite carriage No. 66 were telescoped. In train No. 9 third-class carriages Nos. 51 and 74 and mail van No. 4 were damaged. Nine passengers were badly hurt and two of them later died from their injuries. None of the railwaymen was hurt.[561]

Driver L.H. Bink of train No. 9 stated that he had seen the notice regarding the temporary single line working in the engine shed at Westerdok on 5th August. He had been at Vogelenzang on Sunday 6th August but had not noticed that the two double to single connections had been installed over the previous night, then from Monday 7th to

[557] *New Albany Ledger*, New Albany, Indiana, 2nd August 1882
[558] *Elkhart Monitor*, Elkhart, Indiana, United States, 4th August 1882
[559] *Daily Evening Democrat*, Shelbyville, Indiana, United States, 2nd August 1882
[560] *Elkhart Monitor*, Elkhart, Indiana, United States, 18th August 1882
[561] *Spoorwegongevallen in Nederland 1839-1993* by R.T. Jongerius, pgs. 76-9

Thursday 10th he had not been there, so that the train he had worked on the day of the accident was the first over the line since the single line working had come into operation. He said that he had misunderstood the notice and had been under the impression that the pilotman was on the far side of Vogelenzang station; when he saw the green warning board, he had shut off steam, thinking that he must reduce speed because of work on the bridge. It was only when he had seen the red board that he had realised that he must stop, and had applied the brakes and reversed his engine, but by then it was too late.

The official report on the accident stated that the red and green boards had not been placed a sufficient distance from the beginning of the single line to allow a train safely to pull up. In November 1882 the HIJSM had Driver Bink examined by Professor F.C. Donders, a famous ophthalmologist, who declared that the driver's colour and far vision were perfect. The professor was then asked to examine the visibility of the warning boards themselves and in October 1883 he reported that the green board was hard to make out not only against a green background but also against a yellow or brown background. The red board could be distinguished more easily, a freshly painted one being visible twice the distance of a faded one. Against a sky background the boards were hard to distinguish from one another. In order to make them more visible against a green background the boards should have a white border and against a sky background a black border. The driver's assertion that the boards were only visible when he had come close to them was thus borne out.[562]

On Monday 4th December 1882 a Melbourne, Victoria newspaper reported that the city had been thrown into a state of excitement on Saturday night, 2nd August, by a deplorable railway accident which had happened on the Hawthorn line, on the Melbourne and Hobson's Bay Railway, near Picnic station at 6.23 p.m. Two long trains, both filled with passengers and travelling in opposite directions, had collided with such violence that four carriages were completely wrecked. The drivers of the two trains had only seen one another at a short distance, when onlookers in a park and neighbouring streets had heard them whistle and seen them shutting off steam and trying to reverse their engines. Then came a loud crash, a moment's silence and then the uninjured passengers went to the aid of their less fortunate fellow travellers, aided by the onlookers and 70 more or less injured passengers were extricated from the wreckage. No one was killed outright, but a number of passengers were seriously injured, and one of them, Mr William Cozens Thomas, a caretaker at the Wesleyan church in Lonsdale Street East., Melbourne, died in hospital the morning after the accident.[563]

The line on which the accident occurred was double track but at the time of the accident one line was out of use, coincidentally, like the Dutch accident at Vogelenzang in August, because of bridgework, this time on a bridge over the Yarra River. The bridge had been built for double track but initially only a single line had been provided. A second line had been brought into use on 1st December but at the same time the original line had been taken out of use for repairs to be made to the bridge. The trains involved were the 6.7 p.m. departure from Melbourne Flinders Street and the return working of a special train from Box Hill run in connection with a land auction. The return special

---

[562] Ibid.

[563] *The Argus*, Melbourne, Victoria, 4th December 1882

consisted of an engine (No. 191) and tender, seven first-class carriages and a van, and was in charge of Guard William Hunt with Driver Thomas Kitchen and Fireman R. Gleeson. The special was due to leave Box Hill at 5.55 p.m. and arrive at Hawthorn at 6.21, the same time as the down train from Melbourne. As it approached Hawthorn station at 6.15½ or 6.16 p.m. the special slowed down but the stationmaster, James Thompson, gave Driver Kitchen an "All right" signal and it continued its journey, gaining speed. After about a quarter mile it crossed the Yarra River on an iron bridge, and it then passed through Picnic station, shortly beyond which Driver Kitchen saw the smoke of the Melbourne train, but assumed at first that the other line had been brought into use. It was only when it got closer that he realised it was on the same track.

The train from Melbourne consisted of a tank engine (No. 32), nine carriages and two vans in charge of Guard Donaldson, with Driver William Statter and Fireman Thomas Phillips. It had been well filled when it started out, but many passengers had got off at Richmond. Seeing the special approaching Stetter exclaimed, "Good God Almighty! Tom, put your brakes on, lad!" as he reversed his engine and sounded the danger whistle.

Screwing down the brake the fireman cried, "What is it?" and then, "What shall we do?"

"Its life or death," his driver had replied, adding, "Stick to her." They wished one another goodbye and held on as hard as they could as the collision occurred. Both survived the collision, though they were injured to some degree.[564]

The accident was the subject not only of an inquest but also of a very lengthy inquiry, which began on 23rd January 1883 and finally submitted its report on 11th July of that year. The inquiry found that the direct cause of the accident was that the return special had arrived at Hawthorn station at 6.16., five minutes before its booked departure time, having run fast from Camberwell. The inquiry concluded that Stationmaster Thompson, seeing that special was before time, had then taken a chance and given it an "All clear" signal in the hope that it would be able to get to Burnley Street station, where the double track recommenced, before the 6.7 p.m. train from Melbourne left there, and that Guard Hunt and Driver Kitchen, seeing that the stationmaster was not stopping their train, had concurred in taking the risk.

The inquiry pointed to a number of other factors which had increased the risk of an accident.

1. No semaphore signal was provided at Hawthorn to control the departure of trains.

2. The opening of a double line had been unnecessarily delayed in order to strengthen the bridge, a job which was not urgent, and which could in any case have been carried out at night without interfering with the traffic or needing single-line working.

3. Had the two trains been fitted with an automatic or even a continuous non-automatic brake the collision would have been prevented.

4. Had either the block system or the staff system been in use the accident would not have occurred.

5. Evidence showed that the time shown by the clocks at various stations often differed and there was inexcusable carelessness and neglect in their regulation.

[564] Ibid.

6. The railway department had no permanent head. The secretary considered himself as the head, but the professional heads of the different branches refused to recognise him as such.

7. There was no recognised means of communication between the officers of each branch or between the different branches.

8. A deficiency in organisation, an absence of effective supervision and a want of discipline pervaded the department and were the cause of the accident.[565]

The inquest on the death of William Thomas found a verdict of manslaughter against Guard Hunt and Stationmaster Thompson. At their trial on 27th February 1883 they were both acquitted. [566]

Doubtless because all but two of the people who lost their lives were Chinese labourers, a collision at the west end of the Missoula division of the Northern Pacific Railroad on Saturday 23rd June 1883 did not receive a great amount of coverage in the press. It happened at Elk Creek, 1½ miles east of Heron Siding, Montana, United States. A gravel train of five or six flat cars carrying about 50 'Chinamen' was backing down the line at a reported 35 miles an hour when it was in collision with what was variously described as a wood train and an engine, wrecking both locomotives. Eighteen labourers were killed outright, and 20 or 25 others were injured, several fatally. Engineer George Pike of the gravel train was also killed, and his fireman jumped but was seriously injured; though the crew of the other engine jumped and escaped unhurt.

A newspaper reporter wrote, 'The engines were entirely demolished, the cars were wrecked, and the helpless wounded, piled in with splinters, scraps of iron, and the dead, made up a horrible scene. As soon as possible the wounded were conveyed to Heron siding, and a force of men set to clear the track. The scene of the accident is indescribable, some having arms cut off, others with legs severed from their bodies, and many of the killed ground into jelly. A painter named Hayes who was riding on the engine of the gravel train started to jump but was too late and both legs were cut off; death occurred during the night. The Chinamen refused to touch their dead countrymen.' [567]

The accident happened at 6.30 p.m.; the trains meeting on a curve so that those involved had little warning of what was going to happen, this being probably especially true of the labourers who were riding on a flat car behind the engine. The dead were buried at Heron Siding. [568] Another report spoke of the death of 23 labourers and stated that the accident had been the result of gross carelessness.[569]

The crews of both trains were at fault for the accident inasmuch as they both had orders to work under flag protection. In the event neither train had been flagged, its being supposed that the crews of each had depended on the other train being so protected. Engineer Pike was considered to have been most to blame, however, for quite apart from

---

565 Ibid, 14th July 1883

566 Ibid, 28th February 1883

567 *The Dakota Leader*, Huron, Dakota Territory, 17th July 1883

568 *Davenport Daily Gazette*, Davenport, Iowa, United States, 27th June 1883

569 *Terre Haute Weekly Gazette*, Terre Haute, Indiana, United States, 28th June 1883

the question of flag protection the speed of trains was restricted by order to 10 miles an hour, so he had clearly been driving the gravel train much too fast as well.

Six months after the Elk Creek accident a major disaster occurred in Canada. The accident on 2nd January 1884 occurred on the Grand Trunk Railway at Toronto High Park, Ontario, Canada. Engineer Richard Jeffery had begun work the night before with a light freight train and in the early hours of the 2nd he arrived at Hamilton. He had asked for a pilotman as he was unfamiliar with the line to Toronto but was told that none was available; instead he was given an experienced conductor named George Barber. At Hamilton new orders were delivered for the train to run to Queen's Wharf, Toronto, avoiding regular trains. The freight train, which consisted of only an engine, its tender, a baggage car and a caboose, left Hamilton at 5.20 a.m. One of the regular trains which it had to keep out of the way of was a local passenger train, due to leave Toronto Union station at 6.40 a.m. and consisting of a small combined engine and tender, known as a "dummy", and two coaches. On the day of the accident this train actually left nine minutes late at 6.49, in charge of Conductor James Carter with Engineer John Kennedy and Fireman James Gasken. It was carrying 43 workmen bound for the Toronto Bolt and Iron Works foundry between Toronto High Park and Mimico stations. This workmen's train featured in the working timetable only as a footnote and its presence in the timetable was not noticed at all by Engineer Jeffery, while Conductor Barber was aware that such a train was running, but simply forgot about it.[570]

The last point where the freight train could have been held until the passenger train arrived was Mimico but when the freight train passed through there at about 6.50 a.m. the station was unstaffed, as it was usually until 7.30 a.m. The trains met on a curve just on the Toronto side of the bolt works where the workmen would have disembarked. Engineer Kennedy saw the freight train running fast down the grade, reversed his engine, blew his whistle and jumped, as did his fireman. Hearing the warning Conductor Carter shouted, "Jump, boy, for your lives!" and sprang into the snow. Engineer Jefferey of the freight train also jumped but Fireman Charles Thomas did not and was killed instantly. In the collision the heavier freight engine forced the lighter dummy engine and tender through the front coach and the boiler of the dummy engine exploded. Fire then broke out in the wreckage to add to the horror of the scene.

'Shrieks and heart-rending cries for help from the agonized sufferers rent the air. Several poor fellows suffering untold agonies with limbs and bodies burnt to a crisp, piteously imploring those near them to pour water upon their scalded limbs or put an end to their sufferings. Strong men looked on appalled and many an eye was dim with tears. Every possible phase of disfigurement was to be seen. Limbs cut, bruised, mangled and half eaten away by the fire and swollen to enormous size by the steam and water. No possible horror was left to the imagination. No more awful scene could be witnessed. The first passenger car was a terrible wreck, and several hours elapsed before the poor fellows buried beneath were extricated. A relief party of thirty or forty men were soon on the scene and at once set to work, slowly and tenderly, to rescue the sufferers, and about 10 o'clock a carload of dead and injured was on the way to the city. On arrival at the Union station wagons, and several cabs, and buses were in waiting. A large crowd had gathered,

[570] *Wreck!* By Hugh A. Halliday, pgs. 57-63

and as one after another of the dead and wounded were carried from the car to the conveyance many turned sick at the sight. The wounded men bore their sufferings with fortitude and patience, a few groans being the only indication of their extreme agony. The scene at the morgue baffles description. There were fifteen bodies laid side by side and in rows on the opposite side of the room. Mothers, sisters, fathers and brothers were to be seen passing from body to body and with trembling hands lifting the coverlet to gaze on the faces of the dead. Now and again cries of anguish tell too plainly the discovery of some dear one carried off in the prime of manhood.'[571]

There could have been another collision as news of the collision did not reach Toronto Union station before the departure at 7.35 a.m. of passenger train No. 1, but the latter was stopped by men who saw it approaching and ran back down the line shouting and waving their arms. It eventually set back to Queens Wharf station with some of the casualties and brought the first news of the disaster to the Toronto headquarters.

By the next day the papers were reporting 23 dead and with some of the badly injured later succumbing the final death toll was 29; all of them were workmen apart from Fireman Thomas. About 10 other passengers were injured.

Conductor Barber was arrested after the disaster. The subsequent inquest returned a verdict of manslaughter against both Conductor Barber and Engineer Jefferey, while stating that they considered that the men's hours of work had been too long, that the engineer should have been provided with a pilot, that the orders issued were not as clear as they might have been, that Mimico station should have been staffed and that the workman's train was late. The jury recommended that given the amount of traffic the line between Toronto and Mimico should be doubled. The Grand Jury returned a true bill on the manslaughter charge against Conductor Barber but at his trial on 25th January he was found "Not Guilty." [572]

The ordinary passenger and mail train which had left Geelong at 9.15 p.m. on the evening of Wednesday 2nd April 1884 collided head-on with a special train of empty sheep trucks which had left Melbourne at 8.20 p.m. about a mile from Werribee station, and between there and Little River, Victoria. The line on which the accident occurred was single track and was worked by the staff and ticket system in conjunction with the electric telegraph. The two engines were not derailed in the collision, but the leading vehicle of the passenger train, a guard's van, and the following second-class carriage were thrown down a slight embankment and the first-class carriage third in the passenger train was badly damaged. The tender of the goods train reared up in the air and three of the empty sheep waggons piled up in a heap. A young woman passenger, later identified as Mrs Ellen Johnson, screamed in agony for an hour and a half after the accident before she died in the arms of mail guard M'Mahon, and Driver Thomas Kitchen of the goods train, who was partly buried in the coal from the tender of his engine, died shortly afterwards. Mrs Johnson had been returning to Melbourne from a visit to her brother at Horsham. Thomas Kitchen was the driver involved in the Hawthorn accident exactly 16 months earlier. He had been lucky then, but this time his luck ran out. Driver James Craike of the passenger train was also fatally injured in the accident and died just over a

---

571 *Manitoba Daily Free Press*, Winnipeg, Manitoba, Canada, 3rd January 1884
572 *Wreck!* By Hugh A. Halliday, pgs. 57-63

day later. About eight passengers were injured, some seriously. [573]

Guard Samuel Best, the guard of the goods train, told the jury at the inquest on the deaths of Mrs Johnson and Driver Kitchen that his train had been booked to meet the passenger train at Little River and was not booked to stop at Werribee. It had arrived at Werribee at 9.23 p.m. He saw Mrs Biddle, the stationmaster's wife, and her daughter on the platform, along with Porter McAree. His driver, Thomas Kitchen, was given the staff for the Werribee – Little River section and indicated to him, the guard, that he had received it. The porter then gave him a light to proceed, he signalled to his driver and the train set off. Before they reached Little River, he heard his driver sound the whistle, applied the brakes and the collision occurred.

Charles Coles told the inquest that he was stationmaster at Little River. He said that at 9.40 p.m. he had received a telegraph message from Werribee. "Please send on the 7.10 train. I have staff and will keep line clear." The passenger train arrived at 9.45. He gave Driver Craike a copy of the telegraph message and the train proceeded. He considered that the telegram cancelled the staff arrangements and that he had been justified in permitting the passenger train to proceed. After the train left, he had returned to his office and had had his attention drawn by his daughter to light and steam visible up the line. He had dismissed these as simply the engine firing with the fire-box door open but when a nearby gatekeeper had also drawn his attention to them and the gatekeeper had said, like his daughter, that he feared that there had been an accident, the two of them had gone to investigate and had met one of the mail guards of the passenger train, who confirmed that there had been indeed been an accident. Continuing his evidence Stationmaster Coles stated that he had the power to suspend the staff arrangements in an emergency, and he considered a delay to a passenger train to be an emergency. He had frequently suspended the staff arrangements before to avoid delaying passenger trains.

Stationmaster Coles' daughter Frances had actually taken the telegraph. She testified that it had read, "Sixteen W.R.O.S. To S.M. L.i Please send on 7.10, I have staff and will keep the line clear until its arrival (sig) Thomas Biddle W.R." She had sent, "G.A. Time," and Werribee had replied, "Time 9.40." She explained that 'Sixteen W.R' meant sixteen words; 'O.S.' meant on Her Majesty's time; S.M. was stationmaster; L.i was Little River and G.A. meant go ahead. She had sent G.A. Time because the Werribee operator had forgotten to add the time.[574]

Porter M'Aree testified that Stationmaster Biddle had left the station at 8.30 p.m. He (M'Aree) had handed the staff to the driver of the goods train, whom he did not know.

Annie Biddle testified that she was the daughter of the stationmaster. Her father had left home at 8 p.m. and had left her in charge of the station while he went to the church choir practice. He told her that if the Melbourne goods train arrived before 9.30 p.m. to let it go on to Little River, but if not to put it into the siding and to send "Line clear" to Little River not later than 9.40. The goods train had arrived at 9.25. Its driver had thrown out the Geelong Junction staff; the porter had given him the Werribee – Little River staff and the train had proceeded. She had waited until 9.40 p.m. and then had sent a "Line clear" message to Little River. She knew that the goods train was in the section and could not explain why she had sent the "Line clear" message. There had been no one in the

---

573 *Geelong Advertiser*, Geelong Victoria, 4th April 1884

574 Ibid. 9th April 1884

office when she had sent it. It took a train 20 minutes to run from Werribee to Little River. She was 17 years of age and had never been officially employed on the railway but had been using the telegraph for two or three years. There was a lot of telegraphing to be done at Werribee, up to 50 messages a day. She had not known what effect her message would have – she had never anticipated an accident.

Thomas Biddle testified that he had been stationmaster at Werribee for six years. He had left the station at 8 p.m. after the departure of the 3.25 p.m. up train from Ballarat and 7 p.m. down train from Melbourne. Before he had left for church (where he was the organist) he had told the porter to give the goods train the staff and told his daughter that it was too late for her to accompany him to church and that if the 8.20 goods came in before 9.30 it was to proceed to Little River, otherwise to put it into the siding, and "Line clear" to be sent to Little River not later than 9.40, but that he would be back before then. He could hear every train passing from the church, a quarter mile away, and had the goods train arrived after 9.30 he would have returned at once. He had heard it pass at 9.25 and had assumed all was all right. He had not given his daughter any instructions to use the telegraph. The porter had been in charge of the station in his absence and there had been no authorised person at the station capable of working the telegraph. He had returned to the station at 9.45 but had not realised there was anything wrong until Little River had asked where the goods train was at 10.5 when he had telegraphed to Little River that the goods had left him at 9.25. Little River had then telegraphed, "Why did you send line clear?" He had replied either, "I did not send line clear," or "Line clear not sent." Little River had answered, "You did send line clear." He had then called his daughter and asked her if she had sent any message to Little River. She admitted that she had and had looked terrified but could not explain why she had done so. (Here he wept.) He added that he was on duty daily from 6.45 a.m. until 10.15 p.m. (Stationmaster Coles had testified that his hours were 3.45 a.m. until 9.45 p.m. but that he frequently had to work all night.) Stationmaster Biddle accepted that he was entirely to blame for the accident.

Rather surprisingly Assistant Traffic Manager Henry Barter stated that he was aware that the staff system was at times suspended and trains worked just using the telegraph and agreed with Stationmaster Coles that a delay of 20 minutes to a passenger train constituted an emergency.

After half an hour's deliberation the inquest jury returned a verdict that the deaths of Mrs Johnson and Driver Kitchen were due to the sending of a telegram by the stationmaster's daughter while he had been absent from the station without permission. The coroner, Mr Heron, interpreted the verdict as one of manslaughter and committed Stationmaster Biddle for trial.[575]

After the inquest the details of the accident were gone over again when it was the subject of an official court of investigation and here Assistant Traffic Manager Barter stated that staff working was also suspended where necessary in New South Wales and England. He might have been correct as regards New South Wales, it seems very doubtful if he was as regards English practice. The Chairman of the Railway Board, Mr Richard Speight, stated on the other hand that the regulations should have been strictly adhered to, which presumably did not condone suspending the staff working merely to

---

[575] Ibid.

avoid delay, but Mr John Anderson, the General Traffic Manager, did not think that the staff system pure and simple would answer on the Victorian lines.[576]

Thomas Biddle stood trial for manslaughter on 18th July. It took the jury an hour and a half to find him "Not guilty". [577]

Between 9 and 10 a.m. on Wednesday 14th May 1884 a disastrous collision occurred a mile east of Connellsville, Pennsylvania, United States, on the Baltimore and Ohio Railroad. The trains involved were a westbound freight train, heavily laden with ore, and an eastbound construction train. The latter consisted of an engine and tender and six camp cars carrying 50 labourers who had been transferred from the Pittsburgh Southern Railroad and were on their way to work on grading for double track at Ohio Pyle Falls. The trains met on a curve. Both were travelling at speed, the engines telescoped one another, and their tenders jumped into the air and then fell, the freight tender onto the locomotive and the gravel train tender onto its first car. Six men in that car were crushed to death and four others, trapped by the timbers, burnt to death when a fire broke out, started by an overturned stove. Three other men escaped from the car and one of them said that the fire could easily have been extinguished at the start had water been applied to it.

By noon a large body of men were at work putting out the fire and the wreck train was busy pulling the debris apart and rescuing the charred remains. 'The burned trunks of four of the victims have been taken out. Across the breast of one the white unsinged flesh contrasts strongly with the blackened surroundings. That spot was where the heavy timbers caught him and held him in the grip of fate until the flames left nothing but a headless, armless, legless trunk. The remains were reverently laid on a board, carried aside and laid down upon the green grass spotted with purple forest flowers.' [578]

When the rescue work was finished it was found that there were 12 persons injured and another 14 were missing. The bodies of only 10 victims were recovered, it was supposed that those of the other four victims had been entirely consumed by the fire. Only two of the bodies could be positively identified. One of the injured, J.T. O'Shea, died on the way to the West Pennsylvania Hospital in Pittsburgh, increasing the number of dead to 15.

Coroner Britton held an inquest on the bodies of the victims the same afternoon. It found that the accident had been due to a mistake made by train dispatcher Gilbert Lewis at Connellsville, who had issued an order authorising the camp train to run wild to Ohio Pyle, 18 miles distant, forgetting that the second section of westbound freight train No. 80 had not passed. He had realised his mistake before the camp train had passed out of sight and tried to stop one or other train but failed. The inquest jury duly returned a verdict that the deaths had been caused by the carelessness of despatcher Lewis. District Attorney Cochran attended the inquest but at the time of the newspaper report had not decided whether or not to prosecute Lewis.[579]

[576] Ibid. 19th and 21st April 1884
[577] Ibid. 19th July 1884
[578] *The Philadelphia Inquirer*, Philadelphia, Pennsylvania, United States, 15th May 1884
[579] *Norwalk Daily Reflector*, Norwalk, Ohio, United States, 15th May 1884

Exactly a month after the Connellsville disaster, on Saturday 14th June 1884, a bad accident occurred on the Camden and Amboy Railroad between Haddonfield and Ashland, New Jersey, United States. This time the fault lay with one of the engineers rather than with the despatcher.

Two passenger trains were involved in the Haddonfield accident; the regular accommodation train which left Atlantic City at 7.5 a.m., and a special excursion train, organised by Mr Frank Fenton, a civil engineer and surveyor of the line who was also a trustee of the Second Presbyterian Church of Camden. The excursion train was carrying Sunday School children of that church and of the Wenonah Presbyterian Church on an outing to the resort of Lakeside Park. The Camden portion of this train had left Cooper's Point at about the same time that the accommodation train had left Atlantic City. At Tenth Street Junction connection was made with the train carrying the Wenonah excursionists and it was almost 8 a.m. before the combined train left there. Earlier that morning the prospect of bad weather had caused Mr Fenton to telegraph the despatcher to cancel the train, but the weather had presumably cleared up as later Mr Fenton subsequently rescinded that request.

From Tenth Street Junction the excursion train had made a quick run to Haddonfield, which it had reached three minutes late, and its crew were informed that the accommodation train had just passed through Kirkwood, five miles distant. At Berlin the accommodation train had received an order reading, 'Train No. 30 at Berlin. To conductor and engineer: Meet train No. 25 at Lakeside, then meet train No. 9 at Ashland.' Train No. 9 was the excursion train, which was thus cleared to run to the siding at Ashland to cross the accommodation train, No 30. Meanwhile at Haddonfield, the excursion train received an order reading, 'Meet train No. 30 at Ashland, No 32. at Lakeside. Follow train No. 25 past block signal and return to Haddonfield.' Conductor Albert Smith reportedly came out of the telegraph office at Haddonfield and told Engineer George Baxter, "The regular will lay over at Ashland." The train departed at 8.37, hurrying so as not to delay the accommodation train. Only 25 members of the Camden party were in the five cars of the excursion train as the remainder were coming down at 11, while there were 150 members of the Wenotah party.[580]

The trains collided violently in a cutting a mile and three-quarters from Haddonfield, near Pike's Peak Crossing. The collision cost the lives of Engineer Baxter, Conductor Smith and Brakeman Augustus Edwards of the excursion train, Engineer Sylvester Palmer and Fireman Nicholas Barber of the accommodation train, mail agent Winfield Scott Hyles of the accommodation train, and on the excursion train Mr Fenton and ten-year-old Willie Caskey of Camden, who survived long enough to be taken home but died the same evening. Mr Fenton's daughter was on the train with him, but news of his death was kept from her until she was back home in Camden. Her mother had died a year earlier, so that she was now an orphan. Four passengers, Conductor A.T. Glenn of the accommodation train, Brakeman John Layre, baggagemaster James Rosenbaum of the accommodation train and Fireman Louis McLain of the excursion train were seriously injured, and a handful of passengers were slightly hurt. Supervisor Dale, who had been on the accommodation train, ran to Haddonfield after the collision to fetch help.

580 *The Philadelphia Times*, Philadelphia, Pennsylvania, United States, 15th, 20th and 24th June 1884

Following the accident Superintendent Bannard declared that the company was blameless for it and that clear orders had been issued to the two trains by himself, but the engineer and conductor of the accommodation train had disregarded them and failed to wait at Ashland, despite the fact that Engineer Palmer was the oldest engineer on the line and known to be both trustworthy and cautious. Conductor Glenn spoke from his sick bed, "I rang the bell to stop twice. I knew we had to lay off at Ashland and so did Palmer, and when he didn't pull up, I did my best to stop the train."

baggagemaster Rosenbaum stated, "I was at work in the baggage car just before the collision when I heard the scream of the whistle to put down brakes. I knew it was a danger signal and I attempted to get to the door to see what it was. I never did, for from that time my mind was a blank until I came to my senses lying on the grass and a lot of people standing around me. I never knew what struck me." He probably survived because he was thrown from the shattered baggage car into a ditch.[581]

At the inquiry into the accident Conductor Glenn testified that he had signalled to his engineer to move down to the far end of the siding at Ashland and had assumed that the engineer would wait there for the excursion train. He had then passed through the last two cars of his train and issued two ferry tickets when he realised that they had left the siding. He had pulled the bell cord twice with no response, the third time the brakes had been applied. He insisted that he had delivered the order to Engineer Palmer and that his train had almost stopped. He surmised that his engineer had forgotten the order, as he had done two or three years previously.[582] Other evidence, for example that of baggagemaster Rosenbaum and Jesse Taylor, the rear brakeman of the accommodation train, cast doubt on Conductor Glenn's story of his having attempted to stop his train before the collision and the inquiry blamed both him and Engineer Palmer for the accident.[583]

A single-line collision between two passenger trains in the south of France on 10th March 1886 was probably unique inasmuch as several hundred people could see that it was about to happen, but none could do anything to prevent it. It occurred between Roquebrune, France and Monte Carlo, Monaco, apparently just on the French side of the frontier, at about 5 p.m. The platforms of Monte Carlo station were crowded with passengers, some of whom had alighted from train No. 483 from Nice and others who were waiting to go to Nice on the train coming from Mentone. Train No. 483 then departed towards Roquebrune and Mentone, to the apparent consternation of the Monte Carlo stationmaster, who seems to have been already aware that a train had left Roquebrune on the single track to Monte Carlo. He quizzed first one and then another porter as to whether they had despatched the train but each in turn denied having done so.

The railway between Monte Carlo and Roquebrune Cap Martin curved around the coast along the cliffside so that the whole of it was visible from Monte Carlo station and the people there could watch the accident unfold. An eyewitness described what happened. 'No idea can be formed of the anguish of the scene. At first everybody hoped that the collision would not take place. The coast was all open. "They will see each other in time," said people on all sides. The station-master ordered all the bells to be rung, in

581 Ibid, 15th June 1884
582 Ibid, 18th June 1884
583 Ibid, 25th June 1884

order to make as much noise as possible. Four hundred men rent the air with their cries. The two columns of smoke, however, continued to get closer and closer. Those in the two trains did not see each other. Everybody screamed and gesticulated. Such impotence, in presence of such a danger, was enough to drive one mad. At last the driver of the train coming from Mentone saw the danger and reversed his engine. But the other, in consequence of the curve, noticed nothing, and continued to advance. They were within 200 metres of each other – 100 metres – 10 metres. A great shudder ran through the crowd. The women turned away their heads, not to see what happened. A sharp cry was heard as a girl went into hysterics. The collision took place. The two trains were soon tilted up one against the other, and then there was a horrible incident – two or three carriages, falling over the parapet, were precipitated on to the rocks on the coast. A thick smoke spread all round. Driven to despair, the station-master no longer knew what he was doing. There were cries on all sides of "The doctors! The doctors!" Everyone rushed forward. Twenty-five or 30 persons went up the line and reached the spot where the accident had happened. It is impossible to describe the scene. No one can form an idea of it but those who saw it.'

It was very fortunate that the two carriages which fell 30 metres down the cliffside were almost empty, containing only three passengers, one of whom, M. Ferraro, a Ventimiglia builder, was killed and the other two injured. The other three people who lost their lives were Driver Domergue, Fireman Laurent and Guard Ferraud, all on the Mentone train. Twenty-three passengers, all French and Italians, were injured. The accident was blamed on a mistake made by the head porter or stationmaster at Roquebrune who had not realised that train No. 483 was running in two parts and had allowed the train from Mentone to depart for Monte Carlo after the first part of train No. 483 had arrived.[584]

Rather more than two years passed after the Haddonfield accident before there was another disastrous single line collision but on 15th September 1886 the *Boston Daily Globe* reported 'Sixteen Killed in a Railroad Accident in New York.' The accident it was referring to had happened the day before, 14th September, near Silver Creek, New York, United States. At 6 a.m. on the 14th a special excursion train of 14 coaches left Ashtabula for Niagara Falls, picking up passengers at Erie, Dunkirk and other points along the line. The eastbound excursion had orders to pass a westbound freight train at Silver Creek, east of Dunkirk. When it arrived there the freight train, due there at 10.10 a.m., was not there, and at 10.15 the excursion train, crowded with 600 passengers, set off eastwards once more in defiance of the train order. It passed slowly over a bridge just east of the depot and then accelerated as it entered a deep cutting. It had not travelled much further when its engineer saw the freight train advancing towards him, coming downhill at 25 miles an hour. Just before the impact the two engineers and the two firemen jumped or were thrown from their engines and thus saved their lives. The two engines were crushed into a shapeless mass, two freight cars were 'shattered into fragments' but more to the point the engine, baggage car and smoking car of the excursion were telescoped into one another, the smoking car being completely telescoped. The cars behind the smoker were undamaged. Fourteen passengers were killed outright and possibly three more died of their injuries in the hours afterwards. About 15 others were injured. (The number of dead

---

[584] *The Times*, London, England, 12th and 13th March 1886

is not very clear, one newspaper stating that 14 had died and then proceeding to list 17. Another newspaper stated that there had been 15 deaths while Robert Shaw gives the number as 20.) All of the killed and injured were men; ladies would not think of riding in smoking cars in the 1880s. One young man riding in the smoker had the presence of mind to drop to the floor when the collision occurred and escaped with a few bruises.[585]

How the accident had occurred was fully explained by the evidence of witness at the inquest on its victims, opened at Dunkirk by Coroner Booth on 17th September. The first witness was William Bryant, train dispatcher at Conneaut, Ohio, who testified that his fellow dispatcher, James Dale, had issued an order to Conductor William Harrison and Engineer Lewis Brewer of the excursion (train second 36) reading as follows: "Extra west, engine No. 152, has until 8.45 a.m. to make Brocton for you. Meet and pass No. 29, engine No. 6, at Summit." No. 29 was the local freight train and Summit was 1¼ west of Silver Creek. A second order was issued to the conductor and engineer of the excursion train at Dunkirk: "To conductor and engineer of No. 36 at Dunkirk, N.Y. No. 37, Engine No. 65, has until 9.25 a.m. to make Sheridan and No. 41, Engine No. 33, until 9.30 a.m. to make Summit for you. Meet and pass No. 29 at Silver Creek instead of Summit." Harrison and Brewer had signed the order at 9.15. According to it they should have stopped their train west of the switch at Silver Creek until No. 29 had taken the siding. The excursion train had arrived at the siding at 9.45. An order had been delivered to freight train No. 29 at Angola: "Meet and pass No. 36, Engine 159, at Silver Creek," and signed by Conductor Cooper and Engineer William Harris at 8.54 a.m.

Engineer Harris testified to receiving the aforementioned order at Angola. He said they had overtaken freight train No. 41 and pushed it into the siding at Irving. He and his conductor knew nothing as to any change in the order to meet train No. 36 (the excursion) at Silver Creek.

Conductor Harrison testified that he had met the flagman of the failed freight train, No. 41, at Summit, west of Silver Creek, and the latter had told him that his train (No. 41) had broken down east of Irving siding, and that the excursion train could therefore advance to Irving. On the basis of the flagman's information he had told Engineer Brewer to continue beyond Silver Creek. Conductor Harrison maintained that in certain cases the flagman had the right to change the orders of the dispatcher and that the accident was entirely the flagman's fault. The inquest was then adjourned.[586] At the resumed inquest Superintendent George Kimball stated that the telegraph stations were Angola and Silver Creek. There were no telegraph facilities at the intermediate locations of Farnham or Irving. Freight train No. 41 having broken down east of Farnham it was the duty of the flagman to go west of Irving and hold an eastbound train until No. 41 had reached Irving, and there was nothing wrong with the flagman going beyond Silver Creek to Summit, but his meeting train No. 36 there did not justify that train going beyond Silver Creek. It was right for train No. 29 to push train No. 41 to Irving and then continue to Silver Creek as per its original order. The heights of the baggage car and the smoker of the excursion had been the same and had not contributed to the telescoping (as had been suggested). Superintendent Kimball stated that he considered Conductor Harrison and Engineer Brewer to be mainly responsible for the collision and they would be dismissed but the flagman, named Reed, would be retained. A juror read part of Conductor Harrison's

585 *Boston Daily Globe*, Boston, Massachusetts, United States, 15th September 1886
586 *Warren Daily Mirror*, Warren, Pennsylvania, United States, 18th September 1886

testimony and asked if the flagman's information had not excused the excursion train being taken beyond Silver Creek, but the superintendent replied that Harrison and Brewer had no right to go against the dispatcher's order, and had they reached Irving safely there would have still been an investigation and censure of them.

Engineer Brewer testified to meeting train No. 37 at Summit and to picking up the flagman, who told him that train No. 41 had broken down in the "hole" east of Farnham. The flagman told him that train No. 41 had had to pull up suddenly to avoid running into the rear of train No. 37, whose flagman was not out far enough. The engine of train No. 41 had then broken its eccentric strap and become disabled. The flagman said that train No. 29 would push train No. 41 to the siding at Irving and then wait there for the excursion train and assured him that there was no danger in the latter going on to Irving. Engineer Brewer stated that his conductor would not have given him a proceed signal at Silver Creek had he not thought it right to do so.

Conductor William Mahony stated that he had sent a flagman to hold the excursion train at Irving till his train arrived there. His answer as to whether or not Conductor Harrison should have proceeded beyond Silver Creek was that he considered that the conductor would have been justified in proceeding provided the flagman had assured him that trains 41 and 29 would be held at Irving.

Conductor Herman Haight of train No. 37 stated that the flagman had ridden on his train to summit and implied that the flagman could have alighted at Irving, though his train had been travelling at 25 miles an hour at that point.[587]

The inquest jury found Conductor Harrison and Engineer Brewer guilty of manslaughter and they were arrested at the close of the inquest pending the action of the grand jury. The engineer had disappeared between the accident and the inquest, hiding in Buffalo and saying afterwards that from the garbled reports of the accident appearing in the newspapers that he had feared that he would be lynched.[588] This seems to have been quite a common fear of railwaymen involved in accidents in the United States and perhaps elsewhere. There were certainly plenty of lynchings in the United States at this date, but the author has yet to come across a case of a railwayman having been a victim of such for causing an accident.

A report on the accident made by the Railroad Commissioners came to the same conclusion as the coroner's jury. The commission concluded that the conductor and engineer had assumed that train No. 29 would wait behind train No. 41, forgetting Rule 100, which stated that, "A train overtaking a train of the same or inferior class, unable to proceed from any cause, will run around it if practicable, and proceed on its own rights." The conductor and engineer could have contacted the dispatcher at Silver Creek to advise him of their situation or have sent a flag out ahead before proceeding.

Brakeman Reed was cleared of responsibility. The commissioners theorised that in the collision the forward end of the empty baggage car might have been depressed and its rear end lifted up sufficiently to cause the telescoping with the loaded sleeping car. There was no evidence that the sills of the two cars had been of different heights.[589]

---

[587] *Dunkirk Observer Journal*, Dunkirk, New York, United States, 22nd September 1886

[588] *New Castle Weekly News*, New Castle, Pennsylvania, United States, 29th September 1886

[589] *Dunkirk Observer Journal*, Dunkirk, New York, United States, 8th October 1886

In mid-January 1887 it was surmised that the forthcoming trial of Harrison and Brewer would result in an acquittal. No report has been traced on the trial itself; possibly it was eventually decided not to proceed with it.[590]

Shortly after 2 a.m. on Tuesday 4th January 1887 Baltimore and Ohio Railroad westbound express No. 5 collided with a freight train about eight miles from Tiffin and a mile west of Republic, Ohio, United States. The express consisted of an engine No. 726 and tender, a baggage car, an express and mail car, two passenger coaches and two sleeping cars and was in charge of Conductor Tom Haskell with Engineer Len Eastman and Fireman William Fredericks. It was running late and travelling at about 60 miles an hour. The eastbound freight train consisted of Engine No. 925 and its tender, 16 loaded and a few empty gondola cars and was in charge of Conductor Lyman Fletcher with Engineer Edward Kilen and Fireman W.J. Gullison. It was trying to make the siding at Republic. The two locomotives buried themselves in one another and several coaches were telescoped. Fire broke out in the smoking car following the collision and soon spread to the other cars and in an all too familiar scenario some of the passengers who had survived the initial impact were trapped in the wreckage and slowly burned to death. The shock of the collision was severe enough to make neighbouring farmers think that there had been an earthquake. Bitterly cold weather added to the misery of the survivors. Locals hastened to the scene and help came from the village of Republic and the city of Tiffin. There had been 65 passengers on the train, and it was initially reported that 22 of them had lost their lives. Some of the bodies of those killed were said to have been entirely consumed by the flames along with the whole structure of the smoking car. The contents of the mail and express car were destroyed. Three of the four enginemen survived; both engineers reversed their engines and all four jumped but Fireman Fredericks of the express in his panic fell too close to his engine and was crushed to death by it. A later and probably more accurate report spoke of 17 deaths, comprising 15 passengers, Fireman Fredericks and baggagemaster W.S. Price.

The newspapers contained moving reports of the plight of the surviving passengers. 'Those who were saved from the wreck remaining in Republic were a mother with three little ones hardly out of their swaddling clothes, the husband and father, with two sons, having been devoured by the flames of the burning cars. In some way that even this mother was unable to relate, she got out of the car in which she had been travelling, and wandering away from the burning train, carrying the two youngest children in her arms, entered the first house in which she found a light. She was almost crazed, and with difficulty it was ascertained that her name was Mrs Mary Postlethwaite. The family were on their way to Missouri, and expected to find a home in Chillicothe, in that State. They have been living near the town of Belton, Wetzel county, W. Va. Her husband, Wm Postlethwaite, was 58 years of age, and the sons killed with him were 18 and 11. The agents of the Baltimore and Ohio endeavoured to take care of the mother and children by sending them to Chicago Junction, where the company have a hotel. On advice, however, of her friends, who were telegraphed, she consented to remain in Republic until she should hear from her son, who remained in West Virginia. It was a pitiful sight to see the poor woman, now breaking forth in the wailing tomes of her awful grief, and then to see her little boy, not over four years of age, endeavouring to comfort her with the words,

[590] Ibid, 21st January 1886

"Oh, mamma, don't cry; mamma, don't cry, mamma; I'll get papa." Postlethwaite had on his person $600 in money, a draft for $500, government check for $18, and a note for $350, besides tickets and money for immediate use. With one awful stroke all that the woman had in the world of support and property was swept away, and she left with her fatherless children among strangers.' [591]

The conductor told the inquest held on the victims of the disaster that Engineer Kilen had had two drinks of whisky at Bloomdale and two more at Fostoria, places about 18 and 10 miles west of Tiffin. He did not know whether or not the engineer was in the habit of getting drunk or whether he had been drunk when they left Tiffin, or if he had had anything to drink there. There was no drink on board the engine. The train had been stopped 25 minutes at Tiffin. He had not learnt where No. 5, the westbound express, was at Tiffin and had not enquired. They had left Tiffin at 1 a.m. with no orders regarding train No. 5. His train had been side-tracked at Seneca three and a half miles from Tiffin, to let an eastbound express pass. (The name of the place is mistakenly given in many contemporary newspapers as Scipio, which was the next station to the east of Republic.) After this train had passed, they had agreed that they could make it to Republic for train No. 5. It was four miles from Seneca to Republic, up-hill, and it was snowing and blowing. When they had covered a mile and a half, he had gone to the engine and told his engineer to hurry up and get to Republic out of the way of the express, as it was then 1.30 a.m. and the express was due at Republic at 2 a.m. His engineer had taken no notice of him and had said nothing. The steam pressure had fallen very low and the train had been making very little progress. It was the engineer's duty to send out a flagman to warn the passenger train or he could have backed his train back down the grade towards Tiffin, but he had done neither. The engineer had a time-card and watch. His engineer had acted like a man who had taken too much whisky and had given no notice that the steam had run out. When they had left Garrett, Indiana (their starting point) at 9.40 a.m. on the 3rd the engineer had been lively but at Seneca he had been dull and indifferent. Conductor Fletcher said that when his train had stalled from lack of steam he had gone forward with red and white lights, saying nothing to the engineer. He had got about 10 car-lengths when he had seen the express coming 100 yards away, when he had run forward as fast as he could and signalled to it.[592]

Brakeman Charles Snyder, who was also riding on the engine, corroborated the conductor's testimony.

Fireman Gullison denied seeing his engineer consume any whisky and said that he had not drunk any himself. They had had 140 lbs of steam when they left Seneca, but it had then dropped rapidly and was down to 60 lbs a quarter mile short of the scene of the collision. He agreed with his conductor in stating that Engineer Kilen had paid no attention to the latter when he came up to the engine and he added that the engineer had seemed to be perfectly indifferent to the consequences and had permitted the water to run down in the boiler to one gauge just before the collision, when he should have had three gauges. Furthermore, the engineer had shut off steam but not reversed before he had jumped off; had he reversed his engine there had been sufficient steam to start the engine

[591] *The Sun*, Baltimore, Maryland, United States, 5th January 1887
[592] *Goshen Weekly News*, Goshen, Indiana, United States, 14th January 1887

backwards down the bank and give the air brakes on the passenger train time to work.[593]

Engineer Kilen did not attend the first day of the inquest, having stated that he was damned if he would attend.

Conductor Haskell of the express stated that they had been three minutes late at Republic. He had had 60 passengers, 14 or 15 of whom had been in the smoker. After the collision he had seen five men who had told him they had been in the sleeper; two of them were unhurt. His engine had whistled at Republic, a sound which could be heard a mile away.[594]

The inquest did not come to an end for over a month. When it did the conclusions of Coroner Lepper were not what might have been expected from the evidence of the witnesses given above. To begin with he found that the engine of the freight train had been in an unsafe and unserviceable condition and that the officials of the railroad had been aware of this but had allowed it to remain in use regardless. Engineer Kilen had not been intoxicated, as was alleged, but rather had been exhausted after having been on duty continuously for 16 hours and 50 minutes with a green fireman and a poor engine. There should have been plenty of time for the train to get from Seneca siding to Republic, but the engine had been in such poor condition that its boiler had leaked, the fire had been put out and it had died on the track. Conductor Fletcher had failed in his duty to flag the express although he had had 10 minutes to do so after the freight train had stalled. The brakes of the express had been inferior and ineffective and had failed to halve its speed in the thousand feet up to the point of collision. The manner of heating of the cars of the express had been inferior, unsafe, dangerous, and contrary to the laws of Ohio. Had the lighting and heating been proper the cars would not have caught fire and the loss of life would have been very little.

Coroner Lepper found that the accident had resulted from the gross negligence of Conductor Fletcher in failing to signal the express train and the gross negligence of the officials and managers of the Chicago division of the Baltimore and Ohio Railroad for running an express with inferior and ineffective brakes and with an improper, dangerous and unlawful manner of lighting and heating.[595] It will be noted that the conductor had been working excessive hours like his engineer so his failure to take action more quickly to flag the express was probably due to his tiredness. No report has been traced of his having stood trial for manslaughter and it seems likely that the matter was dropped after the inquest.

The last day of 1887 saw a collision one mile south of Greenwood, Kentucky, United States, on the Cincinnati Southern Railroad, between express trains No. 1 southbound and No. 2 northbound, the two expresses being advertised as the fastest trains in the South, covering the 336 miles between Cincinnati, Ohio and Chattanooga, Tennessee, in nine hours and 42 minutes at an average speed of about 35 miles an hour. The trains met at about 12 noon on a 60-foot-high embankment midway between tunnels 7 and 8. At the moment of the collision train No. 1 was travelling at 35 miles an hour and train No. 2 at 60 miles an hour so the collision was a very violent one. The two locomotives were

[593] Ibid.

[594] *Hillsdale Standard*, Hillsdale, Michigan, United States, 11th January 1887

[595] *Sandusky Daily Register*, Sandusky, Ohio, United States, 21st February 1887

jammed together, and their boilers shattered, discharging their contents in a few minutes. The baggage and mail cars of train No. 1 immediately caught fire and the baggage car and smoker of train No. 2 rolled down the embankment, and the ladies' car was thrown upon the engine, whereby its occupants were scalded by the steam escaping from the wrecked engine.

The details of those who had lost their lives in the accident varied somewhat between different reports, but the papers generally agreed that Fireman T. Candee of train No. 2, baggagemasters Lee Withrow and Lawrence Callan, Express Messenger Byron Powell (or Howell), Postal Clerk James Severance (or Stephens) and passenger Miss Jessie Green had been killed outright or had died of their injuries. It was also reported that Engineer Pat Taylor of train No. 1 and a Mr Avey (or Avery) had died of their injuries. Mr Avey had thrown his wife through a window and jumped after her, both rolling down the embankment. There was also a possibility that the bodies of other victims had been consumed by the fire, for example no trace could be found of a man seen to enter the toilet room of the ladies' coach an instant before the crash.[596] Fireman Pat Murphy, Mail Agent W.P. Roberts and Express Messenger W.E. Jackson of train No. 1; Engineer William Mitchell, Conductor Sam Schrumm and coloured Porter Joe Neland (or Nolan) of train No. 2, Brakeman Dick Patton (train unclear) and several passengers were injured, some seriously.[597] [598]

Sharkey Withers, a passenger on train No. 1, described his experiences to a newspaper reporter. "We left Somerset about half an hour late, and were running about thirty miles an hour up grade. I was in the coach next in front the sleeper, close to the front door, when a shrill whistle alarmed me. I ran out on the platform and was pitched down an embankment of fifty feet. I jumped up and went to work. We helped to extricate the victims. I helped take out three alive, two dead; one of them., Pat Murphy, I helped to dig out of the mud in which he was jammed up to his arm-pits. The next man we got out was an express messenger who was crushed under the trucks; next a baggageman who was under the tender of No. 1. We could hear his groans and dug him out. He died in an hour. Then we got the injured man out of the mail-car of No 1. Withrow died in a few minutes after he was put on the relief train. Postal Agent James Severns, who was appointed one month ago, lay at one end of the car, blood streaming from his mouth. We reached Somerset at 10 p.m. Withrow's body was laid in the baggage room and the injured on our train were placed in the hotel." [599]

The cause of the accident was a simple mistake. The trains normally crossed one another at Flat Rock, a location apparently also referred to as 'Summit'. After the collision Conductor Bennett of train No. 1 ran up to Conductor Shrumm of train No. 2 and exclaimed, "I'm not to blame for this. Read your orders and see." Conductor Shrumm had then taken out his orders (which he had received at Winfield, Tennessee, about 23 miles south of Greenwood) and studied them carefully. Then he exclaimed, "Oh, my God, I have made a mistake!" The order had read Summit and he had mis-read the name as Somerset, a place 22 miles further north. It was reported that he had become a raving maniac after discovering his mistake; in modern language he obviously suffered

[596] *Muscatine Evening Journal*, Muscatine, Iowa, United States, 3rd January 1888
[597] *Columbus Enquirer*, Columbus, Georgia, United States, 2nd January 1888
[598] *Indiana State Sentinel*, Indianapolis, Indiana, United States, 4th January 1888
[599] Ibid.

a complete mental breakdown.

The very same day as the Greenwood accident, 31st December 1887, there was another serious single line collision near Meadville, Pennsylvania, United States on the New York Pennsylvania and Ohio Railroad. The first rumours of the accident were, as was and is usually the case, wildly exaggerated, speaking of 40 deaths and many injured. In fact, the accident was nowhere near as bad as that, but it was serious enough all the same. The trains involved were eastbound limited express No. 8, known as the Thunderbolt, a through train from Chicago to New York, and westbound freight train No. 23, which was double-headed. They collided two miles west of Meadville, at a location known as Trace's Cut, at 9.17 a.m. At the moment of the collision the express, which was running two hours and 40 minutes late, was travelling at 40 to 45 miles an hour, while the freight train was moving at about 15 miles an hour. Heavy snow was falling at the time, limiting visibility. Engine No. 57 of the passenger train and leading engine No. 84 of the freight train were lifted up into the air by the force of the collision, standing on their furnaces with their driving wheels locked together, and hours elapsed before the bodies of the engine crews could be released.[600]

The express consisted of an engine and tender, an express car, a baggage car, a smoker, a passenger coach and two sleeping cars, and was in charge of Conductor F.E. Buffington, with Engineer E.P Swan and Fireman Arthur Irwin. In charge of the leading engine of the freight train were Engineer William Googe and Fireman Edwin Humes, while the train engine was manned by Engineer Alfred McFarland and Fireman C.A. Crumm. The latter two enginemen escaped with slight injuries, the other four were killed outright. The express car of train No. 8 was but little damaged, but the floor of the baggage car struck the smoker about a foot above its platform and telescoped it. Strangely, if not miraculously, only one passenger in the smoker was killed, a 'commercial man' named variously as William Stevens, of Detroit, Michigan and William Stephenson, of Toledo, Ohio.[601] baggagemaster Joe Boynton and about a dozen passengers were injured. Shortly before the collision it was said that Mr Stevens or Stephenson had been talking to a young lady in the palace car about railway accidents, and how lucky he had always been in avoiding them. Another passenger, 'a Hebrew', had been sat at the front end of the smoking car and was thrown from one end of it to the other by the impact but escaped with a sprained ankle. Overjoyed by his good fortune, he had exclaimed to a reporter, "Shesus Christ! Shesus Christ! Wasn't I lucky?" Philip Foulk, of San Francisco, who had been lying asleep on the third seat from the front of the smoker, was also very fortunate. The stove in front of him was shot through the side of the car like a cannonball, while two-thirds of the baggage car passed over his head, but his sole injury was a slight wound to his left arm. As in so many accidents of this period fire also broke out in the express and baggage cars, but the passengers were able to extinguish the blaze by throwing snow onto it.

Local farmers George Smith and Edward Snowden were the first on the scene. Smith said that the shock of the collision had shaken his house and had sounded like dynamite. Snowden remarked that, "I never beheld such a sight before, nor do I want to again as

[600] *Oil City Derrick*, Oil City, Pennsylvania, United States, 2nd January 1888
[601] *The Daily Gazette*, Burlington, Iowa, United States, 2nd January 1888

long as I live." Coroner Squire Hannan visited the accident scene, empanelled a jury and then postponed the inquest until Monday 2nd January.

An experienced local railwayman told a newspaper reporter how he thought the accident had occurred. Freight train No. 23 had been waiting in the lower yard at Meadville for westbound train No. 5 to overtake it and for eastbound train No. 8, which had orders to meet train No. 5 at Geneva, the next station west of Meadville. Train No. 5, running 40 minutes late, had passed, and Engineer Googe, supposing that train No. 8 had arrived (he could not see it through the blizzard, only hear it), had then instructed the head brakeman to open the switch to let his train out onto the main line. Trains Nos. 5 and 8 duly crossed at Geneva as per their train orders, train No. 8 left Geneva for Meadville and the collision occurred. Train Dispatcher Mansfield only became aware that train No. 23 had left when its passing was reported by the operator at the junction five minutes before the collision. He then knew that there was nothing that could be done to avert the collision. Conductor Murry of the freight train, who was unhurt, told the Meadville operator shortly after the accident that he had known that train No. 8 had not arrived, but had then forgotten about it. He had remembered it again before shortly after his train had started away and had called out to his brakemen, "For God's sake stop the train!" The brakes had been applied on every car but with two locomotives pulling a light train they could not hold it, nor could they signal to the engine crews through the blinding snowstorm.[602]

Various American newspapers reported an accident which had occurred in Cuba on Saturday night, 21st January 1888 on the Ferrocaril del Oeste line from Havana to Vuelta. A passenger train and a freight train collided head-on at a location given in the reports as having been 17 miles east of La Herradura, which would imply somewhere near San Cristóbal. The drivers and firemen of both locomotives were killed instantly and their bodies terribly mangled, the locomotives themselves were wrecked, and 40 passengers were more or less injured. According to the report it was over seven hours before news of the accident was received, the conductor of the passenger train having had to walk 12 miles to telegraph the accident. It seems strange that it took the conductor so long to walk 12 miles, especially as he would presumably be hurrying, but perhaps he was badly injured, the report does not elaborate. As a result of the delay in getting aid an unspecified number of the passengers died before medical aid reached them. The accident was blamed on 'a mistake by a signal station'. [603]

On Tuesday 7th February 1888 a collision occurred near Hoylake station, England, on the Seacombe Hoylake and Deeside Railway. Signalling Engineer Edward Tyer had invented the electric train tablet system in 1878 and it had come into use in 1880, initially on the Caledonian Railway in Scotland, but its spread was, as usual with such things, slow. It involved the issue to the driver of every train of a metal tablet engraved with the names of the stations between which it applied. The tablets were housed in a pair of instruments which were electrically interlocked so that once a tablet had been withdrawn from one instrument another tablet could not be obtained until the first tablet had been

[602] *Oil City Derrick*, Oil City, Pennsylvania, United States, 2nd January 1888
[603] *Burlington Daily Gazette*, Burlington, Iowa, United States, 27th January 1888

replaced in the machine, thus ensuring, at any rate in theory, that no two trains could occupy a stretch of single line. As each machine contained a number of tablets, the system gave greater flexibility in working than the train staff and ticket system. However, the Seacombe Hoylake and Deeside line was still worked by staff and ticket, but without the block system.

On the day of the accident the 6.48 p.m. down passenger train from Birkenhead Docks arrived at Hoylake at 7.3 p.m. It was carrying a ticket because a down special was to follow it. Fireman Montague Goodwin of this train handed the ticket to Porter James Glass, who was working the signals, and the latter went looking for his stationmaster, but could not find him. Meanwhile the 6.59 p.m. up train from West Kirby was standing in the opposite platform, and Stationmaster Frederick Wilmott, unaware of the down special, handed Driver Frederick Weston of the up train a ticket for the Hoylake – Moreton section which he had most irregularly written out in advance. The stationmaster alleged that he had asked the driver and fireman of the West Kirby train if they had seen the staff for the section and one of them had replied, "Yes." Driver Weston on the other hand testified that he had asked the stationmaster that question and the latter had replied in the affirmative. None of them had really seen the staff of course, because it was not there. The up train, which consisted of a tank engine running bunker first and seven coaches, then departed.[604]

The down special consisted of another tank engine and six carriages. It left Moreton under clear signals, with the Hoylake – Moreton staff. Eighty yards inside the down Hoylake distant signal the two trains collided head-on. The surrounding countryside was flat, and the collision took place on a gentle curve, so the engine crews of the two trains should have seen one another some distance away had they been keeping a good lookout. Both engines were badly damaged in the collision, the leading carriages of each train were telescoped and the following carriages considerably damaged. Five passengers and six railwaymen were injured, Fireman William Stretch of the up train fatally so.

Investigating the collision on behalf of the Board of Trade, Colonel Frederick Rich found that it had been caused by a total disregard by the Company and several of its servants of the regulations under which it was supposed to be worked. To begin with the railway was not worked with the aid of the block system as agreed in a letter from the secretary of the Company to the Board of Trade dated 1st April 1878. Had it been, the collision would not have occurred. Furthermore, Stationmaster Wilmott admitted that for the last two years instead of keeping the single-line tickets in the box locked by the train staff he had left the train ticket book outside of the box and that he normally wrote out the ticket for the 7.4 p.m. up train (which usually carried a ticket) before the arrival of the down train, so as to save time. Porter Glass had not replaced the down signals before attending to the down train, pleading that the stationmaster had told him to attend to the train first. The signals had thus still been off for the following special. Driver Weston of the up train had failed to check that the staff had arrived with the down one and Stationmaster Michael Mullen at Birkenhead Dock had failed to place a second tail lamp on the down passenger train to denote that a special was following. His excuse that there was nowhere to hang a second tail lamp was not valid as an ordinary hand lamp could have been hung alongside the ordinary tail lamp. Finally, Traffic Inspector George Hill had neglected to send out notices informing staff of the working of the special train.

---

[604] Board of Trade report

Colonel Rich stated that a simple remedy for the state of affairs was to work the staff and ticket with the block system, as it had been stated it would be in the letter of 1878, and to require the men to work strictly in accordance with the rules. Proper signal cabins, with interlocking, should also be provided at Hoylake and other stations.[605] Interestingly he made no suggestion as to the introduction of the electric train tablet system, mentioned at the beginning of this chapter and now in use elsewhere for eight years. 1888 also saw the first introduction by the London and North Western Railway of the rival electric train staff system, with the electric train token system coming into use in 1892. Within a few years one or other of these systems had largely replaced the train staff and ticket method of working in Great Britain, and they were used on British colonial railways, French colonies in the Middle East, Russia and China,[606] but did not find favour in other countries, probably because of the necessary delay in replacing and withdrawing tablets when crossing trains, plus the expense of the equipment.[607]

At the inquest on the death of Fireman Stretch the jury found Stationmasters Wilmott and Mullen, Driver Weston and Porter Glass all guilty of manslaughter. A month later, on 11th March, after a two-day trial at Chester Assizes the jury found the driver and porter not guilty but the two stationmasters guilty, while condemning the lax management of the line and recommending the prisoners to mercy. Mr Justice Wills sentenced Wilmott to six months' imprisonment without hard labour and bound over Mullen to the sum of £50.[608]

On Saturday 19th May 1888 a passenger train and a goods train collided San Pol de Mar and Canet de Mar stations, Spain, just east of Barcelona. Eight people died and over 200 were injured according to a list of accidents given by Wikipedia.[609] A contemporary newspaper report on the accident does not specifically state that it was a single-line collision though the implication would seem to be that it was. The accident occurred in the early morning, the trains involved being No. 42, the first freight train out of Barcelona, and train No. 45, the first inward passenger train to the city. One girl on the passenger train saw the approaching goods train and instinctively threw herself from the carriage immediately before the collision. The first report on the accident spoke of only one death but by the evening there were known to be five dead and two of the seriously injured passengers had succumbed after emergency surgery.[610]

Eight lives were lost in another accident on the New York Pennsylvania and Ohio Railroad between Kent and Tallmadge, Ohio, United States, on 14th January 1889. The events leading up to the accident began with the breaking in two of the third section of westbound freight train No. 71 between the two places. The decision was made to take the front portion of the freight train forward to the siding at Tallmadge and leave it there, the engine then returning to collect the remainder of the train. The front part of the freight train was duly brought to Tallmadge as planned and the engine set off back for the rest of

---

605 Ibid.
606 *Eisenbahnunfälle i Europa* by Erich Preuß, pg. 114
607 *The Railway Magazine* 1959, pgs. 746-8
608 *Lloyd's Weekly Newspaper*, London, England, 11th March 1888
609 Wikipedia – Anexo: Accidentes ferroviarios en España
610 *La Vanguardia*, Barcelona, Spain, 21st May 1888

the train. while a flagman was left at Tallmadge to protect the obstruction against eastbound passenger train No. 8, which was shortly due. It was an inexplicable mistake on the part of this flagman, an old and experienced employee named James Bradley, which led to the accident. He heard a whistle, which he got into his head was the engine of the freight train whistling to recall him, and without further thought set off towards it, forgetting all about flagging the passenger train, which was now close at hand. The latter consisted of an engine and tender, a combination baggage and express car, a smoking car, an ordinary coach and two sleeping cars. Running at about 50 miles an hour down an incline it collided head-on with the engine of the freight train, which was just starting off with the rear portion of its train. The collision occurred on a curve and in a dense fog about a mile and a half east of Tallmadge at 2.30 a.m. The passenger engine was 'crushed to bits', the freight engine reared up on end and the baggage and smoking cars were telescoped and shortly caught fire.

Those killed in the accident were Engineer Robert Huntington and Fireman William Walters of the passenger train, Brakeman J.F. Rushman, Wells Fargo & Co express messenger William Lundy, three un-named Chinamen and Mary Ann Lyon, of Idaho, aged six. baggagemaster David Thomas and possibly a dozen passengers were injured. The engineer and fireman of the freight train jumped, the engineer suffering cuts. No one in the sleeping cars of the passenger train was hurt and they were uncoupled and pushed clear of the spreading fire by hand. Five companions of the three Chinamen who lost their lives were rescued from the burning smoker, half-dead with fright.[611] [612]

A reporter for the New Albany Ledger described a particular tragedy associated with the accident. 'A most pathetic scene attended the death of little Mary Lyon. She was an orphan and was being sent through to relatives in Cherry Creek, N.Y. Edward Peltzer, a passenger in one of the sleepers, took much interest in the little girl, and when the crash came his first thought was of her. He found her wedged by a seat, the flames already surrounding her. This so unnerved Mr Peltzer that he threw himself upon the ground and sobbed. Another eyewitness says that the little girl, whose shrieks were heartrending, released herself just before the fire got to her, a for a moment groped wildly about, then fell over choked by smoke, the flames quickly coming up to complete the dreadful work.' [613]

The line was cleared at 1 p.m.

The earliest railway accident of any magnitude to have occurred in China was a head-on collision between two trains at Chung Liang Cheng on 25th March 1889. Like the first accident in Japan in October 1877 it was a single-line collision, again caused the negligence of a European engine driver. In this case Driver M. Jarvis had waited a considerable time at a crossing place for an opposing train and becoming impatient had decided that he would wait no longer and had taken his train forward onto the single line. Driver Jarvis, a fireman and seven passengers were killed outright in the accident. Driver Dawson of the opposing 'innocent' train survived. Four carriages caught fire after the collision and several passengers were burned alive. More passengers died from burns and injuries over the following days, the final number of dead being variously stated as 10

[611] *The Fort Wayne Sentinel*, Fort Wayne, Indiana, United States, 14th January 1889

[612] *The Belleville Telescope*, Belleville, Kansas, United States, 17th January 1889

[613] *The New Albany Ledger*, New Albany, Indiana, United States, 15th January 1889

and 20. Junior staff alleged that Davis had been drinking heavily before the crash.[614]

The directors of the railway were heavily punished for the accident. By order of Li Hung Chang, a prominent politician, the managing director of the railway had three black marks, or marks of gross error, put down against his name in the official records, barring his chances of promotion or of obtaining any lucrative new position for some time. Two other directors received two black marks. The dead were buried at public expense, and their relatives each awarded a sum equivalent to about £50. Meanwhile villagers living near the scene of the disaster reported hearing in the dead of night the piteous wailings of the spirits of the victims who had been so suddenly forced from their earthly habitations. There were also reports of people saying that they had seen a ghostly train travelling at full speed in the night sky, only to suddenly stop and disappear. The accident clearly had a profound effect on the minds of the local populace.[615]

As well as seven persons who were killed outright 15 valuable racehorses en route from Latonia, Kentucky, to Kansas City were lost in a collision on the Wabash Railroad near Warrenton, Missouri, United States, at 1.25 a.m. on Monday 9th June 1890. The trains involved were the second section of westbound train No. 5, consisting of engine No. 341 and eight palace horse cars and eastbound freight train No. 26, hauled by engine No. 388 and in charge of Conductor Holland, with Engineer Shivel. Both engines, two horse cars and 16 cars of the freight train were badly wrecked.

Killed outright in the collision were William Thomas, trainer Henty Stone, Frank Kellar, James Kellar, hostler Ed Simmons, Morris Green and jockey Samuel Davis. Engineer Lee Parsons, Brakeman E. Norton and five passengers were seriously injured. Frank Kellar, who was killed along with his young son, was the owner of a stable at Cynthiana, Kentucky. His five horses were also killed. The stable of Jack Chinn, whose horses had occupied the second palace car, was also almost wiped out; the horses *Ganges, Little Prince, Leo, Oklahoma* and *St Bridget* being killed. *Josie M,* a fast filly, and *Mary McGowan*, survived and were taken by the stableman to a livery stable at Warrenton, but were so badly hurt that it was doubtful if they would ever race again. The third car contained the horses of the Lamasney Brothers. Its front end was staved in, but the horses escaped except for *Sea Fog*, which was so badly affected by the shock that it had to be left behind.

The accident had been caused by a simple mistake in the delivery of a train order. The order issued to train No. 5 was for it to run two hours five minutes late on its schedule, whereas the order issued to train No. 26 by the operator at Montgomery, Missouri read for No. 5 to run two hours *fifty* minutes late.[616] The crew of No. 26 thus believed that they had 45 minutes more than they actually had on train second No. 5 and could progress further along the line. The accident clearly demonstrated, like, for example the collision near Greenwood at the end of 1887, the fragility of the train order system, where a simple clerical error could lead to disaster.

---

614 Webpage – Kaiping Tramway and Imperial Railways of North China

615 *Newcastle Morning Herald and Miner's Advocate*, Newcastle, New South Wales, 10th August 1889

616 *The Democrat*, Defiance, Ohio, United States, 12th June 1890

The same morning as the Warrenton accident an error by Train Dispatcher Davidson, on the St Louise, Keokuk and North Western Railroad, led to a single-line collision between a freight train and a construction train near Busch, Missouri, which killed Fireman Arthur Thayer outright and fatally scalded Fireman Berry Nelson.[617]

Extremely brief reports appeared in a couple of American newspapers and *The Times* in London, England, of a collision between two trains near Rinconada, Mexico, on 19th September 1890 in which 10 persons were killed and several injured. It has not been possible to trace any further information on the accident or even to confirm that it was a single-line collision.[618]

A couple of Iowa newspapers reported a collision near Girard, Iowa, United States, on the Iowa and Minnesota division of the Chicago Milwaukee and St Paul Railroad on 16th January 1891 in which nine persons were stated to have been killed and three more fatally injured. The trains involved in the collision were a westbound passenger train and a freight train. The accident was supposedly due to the failure of an operator to deliver a train order.[619]

Another newspaper cast doubt on whether the accident had actually taken place at all, stating that no notice of it had been received in the Milwaukee Company dispatcher's office in Chicago by 2 a.m. the next morning.[620] There seems to have been no further mention of the accident, if one did indeed occur.

A collision near Irving, Kansas, United States, on 26th January 1891, was a sequel to the 'Battle' of Wounded Knee a month earlier, where up to 300 Lakota Indians died together with 25 soldiers in what was effectively the last engagement of the Indian Wars.[621] The trains involved were a northbound passenger train and a southbound special, the latter consisting of two engines, seven coaches and 25 stock cars carrying the men of the Seventh Cavalry, their horses and three batteries of artillery. The accident occurred a mile north of Irving on the Manhattan Blue Valley division of the Union Pacific Railroad. The cavalry were returning from the Pine Ridge Indian reservation, the site of the engagement, to Fort Riley.

All the coaches of the troop train fell into the ditch as a result of the collision and two men, named as St. Schortbat, of Battery E, and a Private Meil, were killed outright and another 12 injured, six fatally. A later report stated that a Captain Godfrey and six men had died.[622] The engine crews jumped and escaped and none of the passengers on the passenger train were injured, though a baggagemaster on one or other of the trains was somewhat hurt. A number of the cavalry horses were also killed and others so badly injured that they had to be shot. The accident was blamed on the conductor of the troop

617 *The Morning Constitution*, Chillicothe, Missouri, United States, 11th June 1890
618 *The Jackson Sentinel*, Maquoketa, Iowa, United States, 25th September 1890
619 *Burlington Hawkeye*, Burlington, Iowa, United States, 17th January 1891
620 *Waterloo Daily Courier*, Waterloo, Iowa, United States, 17th January 1891
621 Wikipedia – Wounded Knee Massacre
622 *The Waukesha Journal*, Waukesha, Wisconsin, United States, 31st January 1891

special, whose train was running in the time of the passenger train. He was missing after the accident. Upon hearing of the collision Superintendent Brinkerhoff of Kansas City proceeded to the scene with a special train and physicians were sent from Fort Riley, Frankfort and Concordia.[623 624]

Two passenger trains travelling at full speed between Ostrowo and Loswitz in Prussian Poland on 18th August 1891 met in what was probably a single-line collision, though again the brief newspaper reports do not specifically say so. As might be expected if it was a high-speed head-on collision 'the trains were terribly wrecked' and 10 people were killed outright and many injured. Burning coals from the locomotives set fire to the wreckage, killing some of the trapped victims before rescuers could extricate them.[625]

A month after the accident in Prussian Poland, on 20th September 1891, what was probably another single-line collision occurred between Wolbrom and Golaczowy, this time in what was appears to have been Russian Poland. Again, two passenger trains were involved, and seven passengers and four railwaymen died, and many were hurt, 20 seriously. Both locomotives were reportedly under full steam and were reduced to shapeless masses of metal, while eight carriages were wrecked.[626] The similarities between this accident and that in Prussian Poland a month earlier make it possible that the various newspaper reports relate only to one accident, but the two events were apparently four weeks apart in time and in different locations, so this seems unlikely.

Spain suffered a major disaster late in the evening of Wednesday 23rd September 1891 involving an Irun-Madrid express and a mixed train from Madrid, which collided between Burgos and Quintanilla stations on the Norte line. The express was travelling at about 60 kilometres an hour and the mixed train at perhaps 40 kilometres an hour when they met about six kilometres from Burgos. The engines, tenders and leading vehicles of each train were smashed, but as there were six goods waggons at the front of the mixed train it was in the express that most of the casualties were found. First reports stated that there were 24 dead, but it was later determined that there were 14 dead and about a hundred injured. Those killed included the driver and fireman of the express and the driver of the mixed train. The fireman of the mixed train survived but reportedly lost his mind. The express had been carrying a number of politicians and members of the Spanish nobility and those who lost their lives included the six-year-old daughter of the Marquis de la Camarinas, while the injured included the Marquise de Castroserna and the Marquise de Montesiou. Also killed was an Englishman, Mr Maurice Long, British Vice Consul at Malaga, while Mr William Cotton, a London barrister, an English artist, Mr Seymour Lucas, A.R.A. and Mr Fletcher, son of Professor Banister Fletcher, an architect and Liberal M.P., were injured. Mr Seymour had been on his way to El Escurial (a Spanish monastery), to obtain materials for his next Academy picture. Fears were entertained for his life and a telegram was sent to London to his wife, who immediately

---

623 *The Atchison Daily Globe*, Atchison, Kansas, United States, 27th January 1891
624 *Muncie Daily Herald*, Muncie, Indiana, United States, 28th January 1891
625 *Jacksonville Daily Illinois Courier*, Jacksonville, Illinois, United States, 19th August 1891
626 *Times*, Philadelphia, Pennsylvania, United States, 23rd September 1891

left for Spain by the night mail train.[627] Mr Seymour did in fact survive the accident, but several of the injured passengers died in subsequent days, including Mr Cotton on 30th September, following the amputation of one of his legs in a last vain attempt to save his life. The final death toll appears to have been 25.[628]

Señor Canalejas, a former Spanish Minister of Justice, who had been travelling in the express with his wife, described his experiences in the accident. "Some time after leaving Burgos station, most of the passengers being fast asleep, we were suddenly awakened by a terrible shock, accompanied by the noise of splintered glass and the crash of breaking timbers. Deep silence followed, but it was almost immediately broken by the heart-rending cries of the injured. Both my wife and myself had been thrown to the floor of the carriage but finding ourselves unhurt we opened the door and got out. A terrible sight met our eyes. The front of both trains, but especially the express, had been completely wrecked, one carriage being scarcely distinguishable from another. Most of the killed were horribly mutilated, some being scarcely recognisable as those of human beings. My wife and myself remained on the scene of the disaster, rendering such help as we could, until all the injured had been recovered and taken to Burgos. I wish to mention that the driver of the mixed train showed the courage of a hero. He saw the express approaching and might have attempted to save his life by jumping from his engine. He remained, however, at his post to the last, doing all he could in the little time he had to lessen the force of the collision. He was among the killed." [629]

Señor Canovas de Castillo, the Spanish Premier, ordered a searching inquiry into the causes of the accident, and expressed his determination to hold the railway company responsible should it appear that the accident had been due to the neglect of any necessary precautions on their part. A Reuters report stated that following the disaster the Government had resolved upon several drastic amendments to the railway regulations. The details of the reforms would be considered by a committee of engineering experts, but it was decided that certain modifications were absolutely necessary, 'For example, the telegraphic instruments generally employed on the Spanish railways are those of the Breguet system, which admits of all sorts of misunderstandings arising, there being no printed tape or other record of the message transmitted. It is also considered necessary to increase the number of signalmen and other employés at certain stations. It came out in the discussion upon the accident that many of the railway employés have of late been working 16 hours a day.' [630]

One report stated that the accident had resulted from a mistake by a clerk who had telegraphed that the express had left Quintanilla, having mistaken that name for Quintanapalla and that this confusion of names had caused the catastrophe.[631] Exactly who was to blame for the accident is not clear from the different reports. Some stated that the departure of the express from Burgos was telegraphed to Quintanilla but that the telegraph clerk there mistakenly told his stationmaster that the line was clear for the local train to proceed. Another account had it that the mistake was made by a young man of 18,

---

[627] *The Standard*, London, England, 25th September 1891

[628] *The Times*, London, England, 25th, 26th, 28th, 29th and 30th September and 1st, 2nd, 3rd and 5th October 1891

[629] *St James Gazette*, London, England, 26th September 1891

[630] *The Times*, London, England, 26th September 1891

[631] *The Evening News and Post*, London, England, 25th September 1891

an apprentice telegraphist at Burgos, who received no salary, and was employed in miscellaneous services at the station there. An official, presumably this telegraph clerk, was arrested after the accident while the Burgos stationmaster, upon learning of the disaster, attempted to commit suicide. [632 633 634]

Several weeks after the accident Mr Lucas, now in Paris and finally well enough to endure the long journey back to England, gave his account of the accident. "We had spent a very pleasant day at Burgos, and returned to the hotel to dress for dinner. Mr Long, now dead, spent some time over his toilet, so that we missed the train we intended to go by. This made a difference of two hours. The later train started at 10 o'clock. The night was bitterly cold, so we put on out overcoats, and, lighting our cigars, took out seats in one of the carriages nearest to the engine, the only available one, in fact. My friend Cotton remarked that we were unpleasantly close to the engine, but as we had a long journey before us, we made ourselves comfortable for the night, and Cotton, seated on the arm of the carriage, was studying his 'Murray' with a view to our future movements. Suddenly there came a crash, after which all was a total blank.

"I might have been unconscious ten or 15 minutes, but the next thing I remember was seeing what appeared to me like a transformation scene at a theatre, but gradually I came to myself. It was a beautiful moonlight night and the stars were shining brightly. There were two engines on my right jammed into one another and piled up behind them lay a heap of carriages, one above the other. On the top of these I distinguished a lady screaming for help and unable to stir. She was extricated afterwards, but I think she died. During all this time I was half buried in debris, with my leg jammed, and I was suffering most intense pain. Intuitively my eyes were busy. I soon saw my friend Mr Cotton on my right, and it was easy to see that he was in a critical state. He said in a broken voice-'Kiss my mother and my sister for me, for I know my time has come.'

"It was fully an hour before the rescue party arrived from Burgos, and after the rescuers had lighted signal fires all along the road I was at last extricated and laid on the bank some little distance from the scene. The immense excitement saved me then from a good deal of the pain I have since gone through. The scene is not an easy one to describe. The awful wreck, the human suffering, and the Spanish rescuers – some of whom were bull fighters, with their dark faces and coloured cloaks thrown back upon their shoulders – gave the scene a most weird appearance. I was kept waiting some little time, and at last a doctor came. He perceived that I was in great pain and badly wounded, so he cut up my clothes, and put on a splint. I was suffering from a compound fracture of the right leg, and he said that my case was not a hopeless one. At last a crowd of helpers arrived, and, suffering untold agony, I was conveyed on a stretcher down to a train going to Burgos specially chartered to convey the wounded there. At that place I was on the point of being carried off to a hospital when a Spanish lady, who spoke a little English, asked me if I would not prefer to go to a hotel, to which I replied 'Yes.'

"Here I was left from 4 o'clock in the morning until 4 in the afternoon before the doctors could get near me. Understand that I do not blame them for this. They could not have done otherwise. At last they came and set my leg. The clergymen who were sent to

---

[632] *The Standard*, London, England, 25th September
[633] *The Evening News and Post*, 25th September 1891
[634] *The Times*, London, England, 26th September 1891

attend me were very good and kind; but, unfortunately, I could not make myself understood. Almost daily I had visits from the Governor of the province, the alcalde, or mayor, the Judge and the general of the district, who came to make inquiries about me."[635] At Burgos Assize Court on 23rd May 1894 Assistant Stationmaster Miguel Setien was sentenced to 16 months' imprisonment and Telegraph Clerk Endaldo Pujo, who had been only 17 years old at the time of the accident, to six months imprisonment. They were also ordered to pay half of the legal costs.[636]

A major disaster, probably a single-line collision, occurred in what was then India on Saturday 5th December 1891 when up and down mail trains collided between Mooltan (now Multan) and Lahore in the Punjab, on the North Western Railway. The accident occurred at a point 70 miles from Lahore, near Sahiwal in what is now Pakistan. Thirty-four or thirty-five people died and many more were injured but an English newspaper was able to reassure its readers that 'All the Europeans on the train escaped.' The accident was blamed on a mistake on the part of a stationmaster.[637] [638]

The Pittsburgh Fort Wayne and Chicago Railroad experienced a head-on collision near Wooster, Ohio, United States, on Wednesday 21st September 1892. Eastbound express passenger train No. 8 collided with westbound freight train first No. 75 in Brown's Cut, one and a half miles west of Wooster, on a sharp curve in a cutting. The two engines met with such force that the mail and smoking cars of the express were thrown onto the passenger engine and a freight car loaded with printing presses was in a similar manner thrown on top of the freight engine. Jacob Weltman, a passenger on the express, was half asleep when he was awoken by the shock of the collision. He told his story to a reporter. 'He said there was a terrible rumbling noise and almost instantly the air was filled with cries and screams for help. He was pinned to the floor of the car by wreckage but with great difficulty he finally freed himself and went with others to the aid of a lady who with her child was held by heavy timbers. They worked with might and main to rescue the two but were finally driven back by the flames and compelled to leave the woman and child to their fate. Mr Weltman said that the cries that came from the imprisoned people were such that he can never forget. The fireman on the freight train must have had a horrible fate, as he was caught in the cab and his frightfully burned body dangled in the air in the view of the hundreds who visited the scene, the wreckers being unable to get at it because of the heat. The postal clerks imprisoned in the mail cars, which were thrown upon the passenger engine, had no chance to escape, and they were doubtless all burned to death.'[639]

Killed in the accident were Fireman George Smith of the passenger train, Fireman N. Hammond and Front Brakeman A.D. Glenn of the freight train, Express Messenger Samuel Jackson, postal clerks D.E. Reese, H.S. Allen, G.C. Mann and J.D. Patterson, and four passengers, among them the mother and child who Jacob Weltman and others had

---

[635] Ibid, 18th October 1891

[636] *The Times,* London, England, 25th May 1894

[637] *St James' Gazette*, London, England, 8th December1891

[638] *The Times*, London, England, 8th December 1891

[639] *The Sun*, Baltimore, Maryland, United States, 22nd September 1892

been unable to save, and whose burnt bodies were unidentified at the time of the first reports. Engineer Frank Burt and half a dozen passengers were seriously injured. The injured passengers were treated at hastily improvised hospitals at Wooster and Shreve, while the uninjured passengers were brought to Alliance for breakfast and arrived at Pittsburgh on a special train at 2 p.m.

The fire, which had originated in the fire-box of the passenger train locomotive, destroyed the postal car, two express cars, a baggage car, and a passenger coach, while five freight cars were broken up. Two Pullman cars towards the end of the passenger train and a pay car at its rear escaped the fire; the passengers in the Pullman car were only shaken. By the time that wreck trains arrived at the accident site an hour afterwards the locomotives and burnt cars were a mass of red-hot iron and embers, but rescuers were able to recover $50,000 in silver bullion from the debris, some of it melted. A temporary track was laid around the wreckage and the local coroner opened an inquest.

In its cause the accident resembled the collision at Mast Hope, 23 years earlier. The crew of the freight train, which had been waiting for the express on a siding, had fallen asleep, and hearing a freight train going past had assumed that it was train No. 8 and without further ado had pulled out onto the main line.[640] At the inquest on the victims of the accident Conductor Conklin of the freight train admitted to having assumed that the passenger train, which was an hour and a half late, had gone by, and giving Engineer Bradley the order to start. The freight train had been waiting on the siding for three hours.[641]

Two freight trains on the New London and Northern Railroad (by this date leased by the Central Vermont) came together on the single track 1,000 feet south of the flag station at Harrison's Landing, two miles above New London, Connecticut, United States, at 5.45 a.m. on Friday 14th October 1892. The trains were the northbound Boston express freight and the southbound Brattleboro freight. Both engineers and both firemen jumped from their engines and escaped more or less unhurt, but the leading car of the Brattleboro train included was loaded with four racehorses bound for the races at Croton (another source said for a fair at Poquonnock) and the five men accompanying the horses were killed outright or fatally injured, two of them being named as Charles Reeney and William Gillen, of Boston Springs, New York. Three horses were also killed, the engines badly damaged and three cars were described as having been smashed into kindling wood. Blame for the accident was placed on the train dispatcher at Norwalk, who was placed under arrest after the accident.[642] [643]

For six months from 1st May 1893 the American city of Chicago had hosted a World's Fair to mark the 400th anniversary of the arrival of Christopher Columbus in the New World [644] and many special trains were run in connection with the event. It was one of these trains which was involved in an accident near Colehour, Indiana, United States, at

640 *The Philadelphia Times*, Philadelphia, Pennsylvania, United States, 22nd September 1892
641 *The Connersville Times*, Connersville, Indiana, United States, 28th September 1892
642 *The Daily Journal*, Logansport, Indiana, United States, 15th October 1892
643 *The Gazette*, Lanark, Illinois, United States,19th October 1892
644 Wikipedia – World's Colombian Exposition

8.40 a.m. on Thursday 7th September 1893. The outgoing special train had just passed through Colehour, where the train order signal was showing "Clear" when it collided head-on with an incoming Valparaiso accommodation train. Both trains were running at full speed when they met on a curve, the engineers just having time to reverse their locomotives and apply the brakes. In the collision the baggage car was lifted from its trucks and its floor raised above the platform of the following smoking car and driven through it from one end to the other. It was in this car that all the deaths and most of the injuries occurred, the passengers in the accommodation train escaping with slight injuries. Ten passengers and a tramp were killed more or less outright and baggagemaster William Brewer and about 15 passengers were injured to a greater or lesser degree; one of them died later. The engineers and firemen jumped and escaped.

After the accident Engineer Maurice Hart of the Valparaiso train said, "I have my orders in my pocket and I am all right. We had the right of way. I didn't see the Pan-Handle train till it was too late."

Conductor J.W. Earley of the World's Fair train said that his train had been running on schedule time. "When we passed Colehour the white signal was displayed, and we went on. The white signal meant that the track was clear."

In the absence of train orders, the World's Fair train, as an eastbound working, had the right of way. Neither of the train crews was in fact at fault, the responsibility for the accident rested with Pennsylvania Railroad Train Dispatcher F.B. O'Connor at Fort Wayne, who had sent a "time order" to Conductor Studor and Engineer Hart of the northbound Valparaiso "milk train", due at Colehour at 8.45 a.m., to take the siding there and cross the World's Fair special, due to leave there at 8.40 a.m. What he forgot to do was to also send the order to the special, which had therefore received clear white signals at every point.[645]

Another special train being run in connection with the Chicago World's Fair was involved in an accident at Battle Creek, Michigan, United States, on the Chicago and Grand Trunk Railroad, on 20th October 1893. The trains involved in the accident were a special train run on behalf of the Raymond and Whitcomb travel agency and the westbound Pacific express. The special was being run as the first section of the eastbound night express (train No. 6) and had left the 60th Street depot in Chicago at 8.15 p.m., crowded with returning excursionists from the eastern United States who had visited the World's Fair and were now on their way home. The train was in charge of Conductor Burt N. Scott and Engineer Harry Wooley. It did not reach Battle Creek until 3.35 a.m. Here there was a mile and a half of double track before the line became single once more, and at the depot the night operator handed Conductor Scott two copies of an order instructing the train to run to the end of double track, east of Main Street and about half a mile distant, and there to wait for the westbound train.

The Pacific express (train No. 9), which was running nearly three hours late, consisted of 13 day coaches and two baggage cars. Its coaches were old ones, poorly equipped to withstand the forces of a collision, and each was packed with excursionists from the East who were taking advantage of the cheap fares being offered to visit the exposition in its

645 *Iowa State Reporter*, Waterloo, Iowa, United States, 14th September 1893

last week. It was in charge of Conductor John Bird and Engineer Gil Cranshaw, who had received orders at Lansing, Michigan, to look out for the excursion on the mile and a half of double track at Battle Creek.[646]

Having collected his orders Engineer Wooley on eastbound first No. 6 proceeded up the double track but instead of waiting at the end of it he continued onto the single line. He had barely covered more than 200 yards when the Pacific express, running at 20 miles an hour, came into view coming around the slight curve behind the telegraph office of the Battle Creek railroad yards. Neither engine crew had much warning of the accident though the brakes of the Pacific express had already been applied for a level crossing. The two engine crews jumped or attempted to do so and a moment or so later the two locomotives came together with a tremendous crash. The engine of the special ploughed itself halfway through that of the Pacific express, driving it back into the leading baggage car, and the latter into the day coaches behind; the four leading day coaches were completely telescoped. Fire, caused by either the stoves or the oil lamps in the coaches, then broke out in the wreckage, and there were the all too familiar scenes of passengers trapped in the wreckage burning to death in front of rescuers who were powerless to save them, in this case the rescuers being the passengers on the eastbound special, who had escaped with a shaking. 'No pen can describe the last moments of Mrs Charles van Dusen,' a newspaper reporter wrote, and he then attempted to do anyway. 'She had succeeded in getting halfway out of the window, but her limbs were pinned by a heavy framework of the seat she had been occupying with her husband. Thus held, roasted from the feet up, she begged and pleaded for the help that those on the outside were helpless to give. Despite her terrible agony she retained consciousness to the last and as the flames crept up and surrounded her, she called out her name and address and that of her friends to be notified. "I am a teacher in the Methodist Sunday school at Sprout Brook, N.Y.," she cried. "Say I died like a Christian." Then the car gave way and she fell back into the flames. Her husband had meanwhile been rescued from the next car, into which he had been driven by the force of the collision. Both his limbs were fractured, and he had also received internal injuries. He retained consciousness until his death but was kept in ignorance of the fate of his wife. In his last breath he asked the Rev Dr B Kulp to send his love to the children he had left at home and also to give them his watch. He was 47 years old and a deacon in the Methodist church of his native village.' [647]

The accident was witnessed by Henry Canfield, one of the night clerks in the nearby Grand Trunk office, who pulled the fire alarm box and telephoned to the engine house for assistance. Delay occurred at this point due to a mislaid key and so it was several minutes before the alarm bell was sounded to rouse the citizens. There was a quick response from the fire brigade department, but the nearest fire hydrant was nearly two thousand feet away, and when a line of hose was laid the pressure was insufficient to throw a stream. A proper stream was eventually secured from another hydrant but by the time that a supply of water was available the telescoped coaches had been reduced to pieces of charred wood, with little more than the trucks remaining. The firemen then set about recovering what remained of the victims. Almost the entire population of the town turned out to assist in response to the alarm and cared for the injured passengers, who were taken to the Nichols Home in buggies and carts, where they were attended by a corps of physicians

646 *The Warren Mail*, Warren, Pennsylvania, United States, 25th October 1893
647 Ibid.

and nurses. At about 6 a.m. the first body was recovered, and in a few minutes a dozen were discovered around the stove in the second car. Twenty bodies or parts of bodies were eventually taken out. 'As the work went on portions of satchels and valises, several gold and silver watches, pocket books containing currency and a quantity of silver money were brought to light and turned over to the officers. Tonight, after the wreckers went through, a heap of ashes alone remains to tell the story of one of the most awful holocausts in the history of railroading.' [648]

A day or two after the accident it was stated that 28 passengers had lost their lives in the accident, 18 of whose bodies were so badly burnt and charred as to be unrecognisable. About 25 other passengers were injured.

Conductor Scott and Engineer Wooley were arrested after the accident, and before the inquest, and charged with manslaughter. Each made a statement blaming the other. Engineer Wooley stated, 'After I got into the yards, I saw a headlight. I thought at first it was the switch engine, but in a moment, I saw she was coming too fast. I shut the throttle and reversed my engine. My train came to a stop and was standing still when No. 9 struck my engine. I went to get off, but before I could do so I was thrown against the tank and my side hurt. I asked my fireman after the wreck if he remembered what Conductor Scott had said to him. He said he did. He told me that No. 9 had gone through. I saw Conductor Scott after the wreck. He was much excited. I asked him in regard to the order again. He muttered out something and said he thought that No 9. had gone.'

Conductor Scott said, 'I got my orders at Battle Creek to meet No. 9 at the double track, and I understood them and did not forget that. I compared my order with the engineer's, and we agreed. I had no conversation with Engineer Wooley after our train had started. I did not tell him that No. 9 had gone by us. I went into the baggage car soon after we left Battle Creek, taking it for granted that the engineer knew where he was going. I did not discover that he had run by the double track until we were struck. In fact, I had no time to find out if such was the case, as I judge it was not more than half a minute after we left the double track till we met the other train.'

At the first day of the inquest Train Dispatcher Seth Cornell testified to issuing the order for the trains to meet on the double track between Battle Creek and Nichols, adding that he had told Conductor Scott that train No. 21 was at Lancing, to leave as soon as possible, and to look out for the train order board at Belleville, where he would probably be stopped for orders. Head Brakeman William Teat testified that the special had been travelling at ten or fifteen miles an hour through the yards, He heard the conductor tell the engineer to hurry to Charlotte and that the order board might be out at Belleville. Nothing was said about No. 9. After the collision he heard Wooley ask Scott, "What in hell was that?" and Scott replying, "You dammed fool, it's No. 9." [649]

At the conclusion of the inquest the jury duly found that the accident was due to the gross negligence of the engineer and conductor. Conductor Scott was tried at the end of December and acquitted.[650]

Following the conductor's acquittal it was decided that there was no chance of obtaining a conviction against Engineer Wooley and the case against him was dropped in

648 *The Philadelphia Inquirer*, Philadelphia, Pennsylvania, United States, 21st October 1893
649 *The Marshall Statesman*, Marshall, Michigan, United States, 27th October 1893
650 Ibid, 29th December 1893

mid-January 1894.[651]

On 4th April 1895 No. 151 up mela (fair or Hindi festival) passenger train collided with No. 46 down goods train between Jaunpur City and Mihrawan, India, on the Oudh and Rohilkhand Railway. Seven passengers were killed, nine injured and there was considerable damage to rolling stock. The accident was judged to have been the fault of the assistant stationmaster at Mihrawan, who was subsequently prosecuted and sentenced to two years' rigorous imprisonment.[652]

A major single-line disaster occurred in Belgium on the evening of Sunday 6th October 1895 near Céroux-Mousty, on the line between Wavre and Ottignies. A passenger train had just passed through Céroux Mousty station when it collided with a light engine coming in the opposite direction. Even from first reports it was clear that the accident was a bad one, for these spoke of 10 deaths and 40 injured. Later reports increased the number of deaths to 18 and of injured to over 100, including 30 very seriously hurt, some of whom were not expected to survive. The high number of casualties was due to three of the passenger carriages being piled on top of one another and two of them telescoped. The passengers on the train included M. Boernaert, President of the Chamber of Representatives, and seven members of his family. The President was unharmed, his wife suffered a slight chest injury and his sister-in-law, Madame Mourlon, wife of a well-known engineer, was killed on the spot. A doctor, a priest, and several local businessmen were also killed and a Mademoiselle Crets, of Paris, was among the injured, suffering a head injury. M. Vandenpeereboom, the Minister of Railways, who had been attending the fêtes at Nivelles, had also been a passenger on the train, but had changed trains at Baulers to reach the Anderlecht quarter of Brussels. Relief gangs and doctors were sent to the scene from Ottignies and Brussels and the dead and injured were extricated as quickly as possible.

The accident had its origin in the unusual method of working what should have been a double line. In a manner reminiscent of the situation at South Shields, England, back in 1844, the line was worked between Ottignies and Court St Etienne as a pair of single tracks, one belonging to the State railway and the other to the Grand Central Company. It being a Sunday the regular signalman had the day off and his place was taken by a foreman platelayer. Initially it was suggested that it was the employment of the inexperienced platelayer which had been the cause of the accident [653] but a letter to *The Times* in London from M. Duhot, a representative of the Belgian State Railway, denied that this was the case. He stated that a Board of Inquiry had investigated the accident and had determined that the relief pointsman had been standing in for the regular man for many years, was thoroughly familiar with the job and had always previously performed it in an irreproachable manner. The inquiry had found nothing wrong with the Sunday working arrangements, that the staff had not been overworked and that the accident had been due to an act of absent-mindedness and could just as easily have occurred during the

[651] Ibid, 19th January 1894
[652] Administration Report on the Railways in India for 1895-6
[653] *The Standard*, London, England, 8th October 1895

week.[654] Another report stated that M. Dewez, an assistant stationmaster, who had been regarded as having been partly to blame for the accident, had been completely exonerated by the inquiry, which placed the blame on the driver of the light engine from Ottignies, who was amongst those killed, and who had passed signals at danger.[655]

The *Courrier,* described apparently correctly by a Brussels correspondent to a London newspaper as a fanatical publication, attributed the accident to Divine wrath directed against Sabbath breakers.[656]

In 1896 the French were in the process of colonising the island of Madagascar, and to supplement their troops there sent reinforcements from Algeria. On 10th May 1896 a troop train conveying 140 Algerian riflemen from Mostaganem to Algiers, en route to Madagascar, was in collision with a passenger train coming from Algiers. The collision occurred at the entrance to a tunnel near Adelia station, Algeria, on a dark and wet night. The troop train was descending an incline and the collision was a violent one. The leading carriage of the military train was a first-class one occupied by 10 officers, six of whom were killed outright and three injured. Forty privates and the driver and fireman of the troop train were also injured, while the passengers in the other train were only shaken. The casualties were looked after by local French colonists prior to their being taken to a military hospital. A seventh officer later died of his injuries.[657]

The Adelia stationmaster was to blame for the accident. He had been dozing when the troop train had arrived at his station and imagining that the ordinary train had also arrived from the opposite direction, Adelia being the place at which they were booked to cross one another, he had let the troop train proceed on its journey. Waking up fully and realising what he had done, he had been going to shoot himself, but his porters persuaded him that the two engine drivers might be able to avert a collision.[658]

Tsar Nicolas II of Russia and Kaiser Wilhelm II of Germany sent their condolences to Félix Faure, the French President, in connection with the accident.[659]

A Chicago and North Western Railroad excursion train returning to Omaha, Nebraska, collided with an eastbound fast freight train near Logan, Iowa, United States, at about 6.45 p.m. on the evening of Saturday 11th July 1896. The excursion train consisted of an engine and tender and 15 well-filled coaches, carrying members of the Union Pacific Pioneers' Association and the friends, from Omaha, Nebraska and Council Bluffs, Missouri Valley and Woodbine, Iowa, and was in charge of Conductor A.L. Reed with Engineer J.D. Montgomery and Fireman D.H. Lamh. Fast freight train No. 38 was in charge of Conductor Modlin with Engineer Cheeseborough and Fireman Rowe.

The excursion train was not far past Logan station when it met the freight train on a curve, there being barely time to whistle for brakes before both engine crews jumped. As

654 *The Times*, London, England, 12th October1895
655 *The Standard*, London, England, 10th October 1895
656 *The Times*, London, England, 11th October 1895
657 *The Times*, London, England, 11th and 12th May 1896
658 Ibid, 14th May 1896
659 Ibid, 15th May 1896

was so often the case in accidents of this type at this time there was severe telescoping of the baggage car and the leading passenger coach of the excursion train and it was here that most of the deaths and injuries occurred. The engine of the freight train was forced through and over the engine of the excursion train and lifted itself almost clear of the rails. The left-hand side of the baggage car passed outside the left side of the passenger coach and its right-hand side cut through the passenger coach in line with the right row of seats. Fortunately for once there was no fire. The cries of the injured passengers could be heard by the people of Logan, who turned out en masse to assist the victims. The skating ring, the opera house and other buildings were converted into emergency hospitals and other victims were looked after in nearby farmhouses. A reporter stated that the citizens had been well organised and were to be highly complimented for their herculean efforts on behalf of the sufferers. The local Odd Fellows hall was thrown open and the bodies of those killed were laid in there side by side on settees. Thirty-one passengers were reportedly killed or died of their injuries and about 50 were injured. [660]

Conductor Modlin stated that his freight train had been travelling at 24 miles an hour and had been running under orders.[661] Another report agreed that the accident was the fault of the crew of the excursion train, stating that Engineer Montgomery should have waited for fast mail train No. 2 and the fast freight train (No. 38), but had started off again after only the passenger train had passed.[662] Conductor Reed of the excursion train allegedly gave his engineer the tip to proceed and the excursion train departed from Logan at 6.40 p.m., the fast freight train being due there at 6.43. Having pulled out onto the single line the excursion train stopped to reset the switch and had only just restarted again when the collision occurred. Although Conductor Reed had given the signal to depart Engineer Montgomery shared the blame for the accident. He was said to have been crazed with grief and there was a rumour, denied by his friends, that he had attempted to commit suicide afterwards.[663] Conductor Reed was said to be similarly affected. The two men had reportedly both forgotten their orders. The station agent realised almost at once that the excursion train had departed and ran after it but was unable to overtake it or make anyone on board understand what was wrong.

As might be expected the coroner's jury found the conductor and engineer guilty of criminal negligence.[664] Both men, who were among the oldest and most trusted employees of the company, admitted full responsibility for the accident and the Harrison County grand jury, after a full investigation, adjourned without indicting them, leaving them free men. The accident cost the Chicago and North Western Company an estimated $300,000 in damages.[665]

At 9.44 a.m. on the morning of Sunday 25th October 1896 two St Louis San Francisco Railroad passenger trains collided head-on in Big Bend cut, two miles west of Spring Park, Missouri, United States. One of the trains was the regular eastbound accommodation train, No. 12, and the other was the second section of an excursion train

---

660 *Lincoln Semi-Weekly State Journal*, Lincoln, Nebraska, United States, 14th July 1896
661 *The Des Moines Daily News*, Des Moines, Iowa, United States, 13th July 1896
662 *The Breeze*, Monroeville, Indiana, United States, 16th July 1896
663 *Lincoln Semi-Weekly State Journal*, Lincoln, Nebraska, United States, 14th July 1896
664 *The Ames Times*, Ames, Iowa, United States, 23rd July 1896
665 *Cedar Rapids Evening Gazette*, Cedar Rapids, Iowa, United States, 14th October 1896

of 11 crowded coaches which had left St Louis at 9.1 a.m. for St James, Missouri, where the excursionists were to attend the dedication of the Missouri Home for Aged Veterans, a Grand Army of the Republic (Union) veterans' home. In charge of the excursion train was Conductor George Atwood, with Engineer Joseph Dryden and possibly Fireman Frederick Miller; the accommodation train was in charge of Conductor George Wolff with Engineer Adolph Hohl and possibly Fireman Frank Halser (The newspaper report is contradictory as to which fireman was on which train.)

The excursion train, had it left on time, should have crossed the accommodation train at Valley Park, but as it was running 34 minutes late it should instead have met the latter train at Spring Park, six miles further east. The accommodation train was not at Spring Park when the excursion train reached there, and instead of stopping there for orders as he should have done Engineer Dryden, who appeared to have forgotten the accommodation train, even though it was a train which he himself had driven previously, continued towards Valley Park. Conductor Atwood, preoccupied with collecting tickets, failed to notice that they had passed Spring Park. Engineer Dryden was said to have been one of the best engineers on the road.

The accommodation train consisted only of an engine and a combination baggage car and coach. It was running fast downhill. The trains collided violently, both engines being demolished, and a number of cars telescoped. The two engine crews jumped, but Engineer Hohl was killed, and his fireman seriously injured. In the excursion train those killed included Barney McKenna and his 14-year-old daughter Maud, in charge of refreshments in the leading commissary car. Passengers Charles Mobine, Conrad Kuntz, C.C. Blevins, H Thall and John Cartwright also lost their lives and 21 were more or less injured.

Immediately after the accident thousands of people gathered at the scene from all around, coming in buggies, waggons and on foot and the dead and injured were soon released and attended by doctors from nearby towns. A relief train carrying the survivors of the accident arrived back at the Union station in St Louis at 2.40 p.m. and was at once besieged by anxious relatives and friends.

Those injured were taken to their homes or to hospital; it was feared that three were fatally hurt.

Station Agent J.D. Dishman told a reporter that he had been in his office when the excursion train had passed at high speed, presumably despite the fact that his train order board had been displayed. He then knew that the trains would collide but had been powerless to do anything to prevent it.[666]

The Atchison Topeka and Santa Fe Railroad suffered a major disaster between Emporia and Lang, Kansas, United States at 7.30 p.m. on 8th September 1897. Two passenger trains were involved; the double-headed westbound Mexico and California express and the eastbound fast mail. These trains were said to be the fastest two on the Santa Fe system. The boilers of all three locomotives exploded as a result of the collision and reportedly tore a hole in the ground so large that the smoking car of the westbound express went in on top of the three locomotives and the mail and baggage

[666] *The Salt Lake Tribune*, Salt Lake City, Utah, United States, 26th October 1896

and express cars and balanced there without turning over. The passengers in the sleeping cars escaped through its windows. The front end of the car was enveloped in smoke and steam belching from the wreck below and its rear door was jammed in the following vehicle. Fire from the engines set the wreckage alight and in a short time the two mail cars and the smoking car were reduced to ashes. In attempting to climb from the smoking car several men fell into the wreckage and it could not be determined whether they had got out or been trapped and burned to death. The coach following the smoking car was badly damaged.

The westbound train of eight cars was carrying many excursionists who had been to hear William Jennings Bryan, the noted orator and Democratic Party politician, speak at the fair at Burlingame, Kansas. Bryan himself was also a passenger on the train, but he was riding in the rear Pullman car, well away from those affected. He stated that the passengers in his car had experienced nothing beyond a severe jolt. After the accident he went to help the injured passengers, one of whom said to him, "I went to hear you today. I am dying now and want to shake your hand and say God bless you. If you possibly can, Mr Bryan, get me a drink of water." Mr Bryan went into one of the mail cars, the other end of which was burning, and got some water for the dying passenger. He also brought out some cushions for others of the injured and was said to be everywhere present to minister to the needs of the suffering.[667] Also prominent in the rescue work was T.M. Potter, a Kansas senator.

The names of the dead quoted in different newspapers did not entirely agree but it would appear that 13 persons died in all, some of them succumbing to their injuries in the days after the accident. They comprised Engineers James Brennan and Ed Gunsaully (or Gonzales) on the westbound express, their two fireman, Nate Hollister and John Shirley; Engineer William Frishbey and Fireman Ben Walters on the mail train; Brakeman van Cleve; Wells Fargo Express Messenger Julius Saure, Postal Clerks R.O. McGee (or Michael McGlade?) and R.A. Doran; Daniel McKennon and Harvey Fowler and an unknown tramp. Express Messengers Claude Holliday and D.C. Erter and half a dozen passengers were seriously injured and a number of other passengers less badly hurt.

There were only about half a dozen passengers in the mail train, all situated in one coach, and although they experienced a terrible shaking up none of them was seriously injured. Every seat in the coach was torn from the floor. One passenger, named John Sweeney, was thrown over three seats and through a window, but sustained only cuts and bruises. The baggage and mail cars of the fast mail were totally wrecked.

The westbound express was running at 40 miles an hour or more with orders to meet the mail train at Emporia, while the eastbound fast mail had received orders at Emporia to meet the express at Lang. It was stated that a failure to deliver an order to the westbound express to wait at Lang for the mail train was the cause of the accident. A further report in a Kansas paper threw more light on how the disaster had occurred. It said that Conductor Dave Ferguson of train No. 1 (the westbound express) had expected to receive orders at Lang regarding the mail train as they were running on its time but as none were received there, he had assumed that the mail must be late, like his own train which was an hour late and was running fast to try and make up time. According to this version of how the collision had occurred the agent at Lang, finding that the westbound express was late,

[667] *The Herald Democrat*, Leadville, Colorado, 9th September 1897

had decided that he had time to go and eat his supper, and was not at the station when it passed. He had displayed the red train order signal, to indicate that the express must stop to receive orders, but its light had gone out.[668]

A local newspaper reported some unsavoury events in the aftermath of the disaster. 'Human ghouls delved in the burning wreckage and plundered the baggage and mail sacks which strewed the ground. One tried to snatch a diamond from the breast of an Emporia doctor, who, weak and nervous, was creeping slowly out of the debris. He had strength enough left to hit the brute a blow in the face which made him turn with a curse and sneak away. Mail sacks were dragged into the corn field and rifled. The wreck was smoking this morning. The burning corpses of five men were dug from the cinders. One was a mere handful of bones. Thousands of people stood and watched the wreckers. It will not be cleared away before tonight. It was such a complete smash up that everything of iron is a tangle. All mail on both trains was burned. A little man dressed in a summer suit of linen, which was covered and smeared with mud, stood in the firelight and talked loudly of his loss. He said he was a representative of Eastern investors in Mexico. "All I've got on earth is lost in there and burning up," he said excitedly. "What is it, some kin?" asked a man. "No, but it's all my papers and patents, and contracts and concessions from the Mexican government and—" "Shut up!" shouted a man in overalls, whose left hand was mangled and bloody. "Shut up, young feller, and catch hold of this beard. You're lucky you're alive." A boy, prodding around in the wreckage late last night, found a $20 silver piece. A man saw him find it and knocked him down and took it from him."'[669]

The coroner's jury put the blame for the disaster on Train Dispatcher King of Topeka.[670] However, in October it was reported that Assistant Superintendent Avery Turner of the Santa Fe had placed the blame for the accident on the crew of train No. 1, the westbound express, all of whom were killed except Conductor Ferguson, who was still working on the railroad and whom it was not indicated would be dismissed. The assistant superintendent added that Train Dispatcher King, blamed by the coroner's jury, and Agent Larson at Lang were blameless, casting doubt on the story that Larson had been away having his meal and had let the train order signal go out.[671]

Only two days after the Emporia wreck, what was accurately stated to have been the worst railroad accident in the history of the state of Colorado, United States, occurred at 12.25 a.m. on the morning of Friday 10th September 1897 a mile and a half west of Newcastle, on a part of the Denver and Rio Grande Railroad over which the Colorado Midland Railroad had running powers. Westbound Denver and Rio Grande passenger train No. 1 collided with an eastbound Colorado Midland stock train. The trains came together on a curve as the line wound around a mountainside, with very limited visibility. Estimates as to their speeds varied between 20 and 40 miles an hour for the passenger train and 10 and 30 miles an hour for the stock train. At any rate the collision was powerful enough to wreck both engines, baggage and express cars, smoker and day coaches and two stock cars. What made the accident much worse was the explosion of a

668 *Republic City News*, Republic, Kansas, United States, 17th September 1897
669 Ibid.
670 *Bureau County Tribune*, Princeton, Illinois, United States, 24th September 1897
671 *The Belleville Telescope*, Belleville, Kansas, United States, 22nd October 1897

Pintsch gas tank which started a fire which quickly spread through the wreckage, burning to death a number of passengers trapped in the wreckage.

Fred G. Rockwell, a stock agent for the Chicago Burlington and Quincy Railroad, was a passenger on the Denver and Rio Grande train, and described his experience to a newspaper reporter. "I had just left the sleeper, in company with a gentleman, and passed through the day coach into the next car. We had barely seated ourselves in the smoking compartment when we were hurled violently from the seats by the shock of the collision. We got out safely. A moment later a terrific explosion of gas rent the air, and after it came a burst of flames. After getting my bearings I was horrified to discover that the smoking car had ridden over the day coach, cutting its way almost to the seats and encasing the car in such a manner to shut off all hope of escape. It was in this day coach where the horrible harvest of death took place. I procured an axe and with others managed to chop through a window and got out two women and three men who were badly hurt, and another man who was struggling to get his foot loose. He was self-possessed and told me he was not injured otherwise. He could not free himself and I was compelled to stand helplessly by and see him consumed by the flames. The five that we rescued were the only ones who could possibly have escaped from the car. It was crowded with men, women and children. I should say that fully thirty adults were cremated there aside from the children." Conductor William Hawthorne of the passenger train stated that he thought that only eight had been killed.[672]

Two express messengers on the Rio Grande train, named James Foley and William Messmer, both of Denver, had piled up the through baggage at one end of the car and were busy with the egg cases at the other end at the moment of the collision. They were pitched headlong about the car, and when it toppled over were stunned and bruised by the loose baggage flying about. It was several minutes before they were able to pull themselves together. Then the car began to fill with smoke, and they faced the prospect of death by burning or suffocation. Fortunately, they had an axe with them to aid their escape. Hacking away with it, they had just started to see daylight through the hole they had made when the car burst into flame, but with the help of some men outside they were able to pull themselves out through the hole that they had made, bruised and bleeding.

As at Emporia not everyone was a hero. Two Italians attempted to break open trunks taken from the baggage car. Marshall Pearson shot at one of them and thought that he had hit him while the other, Nick Finck, was arrested. The express car of the passenger train was entirely consumed with its contents.

Another passenger on the Rio Grande train was Frank Mannix, editor of the *Victor Record*, who said, "I was in the smoker when the collision happened and was jammed down in the seat. I saw daylight on one side and managed to pull myself out, and with the help of Brakeman Daniels helped to pull three people from the wreck. At the time of the collision the tank under the smoker exploded and set fire to the train. The scene was awful. The mail, baggage, smoker, a day coach and a tourist sleeper were burned. The rear Pullman and a private car remained on the track." Mr Mannix thought that at least 40 passengers had died and burned.

Mr and Mrs E.H. Strouse, who lived a quarter of a mile from the scene of the crash, and on whose ranch it occurred, reported that the shock of the collision had been so

---

672 *Salt Lake Tribune*, Salt Lake, Iowa, United States, 11th September 1897

severe as to literally throw them out of their bed. The noise was allegedly heard, and the shock felt in Newcastle itself. 'It is thought that between 25 and 30 are dead. The bodies that have been found are ten in one car and four in another. The trunks of two ladies were found linked together in each other's arms, heads and lower extremities entirely gone, but each had a gold watch in good condition. These people will eventually be identified by their watches, and these are the only bodies which can possibly be identified.' [673]

The passenger train was due at Newcastle at 11.05 p.m. and should have passed the accident site at 11.10 but was over an hour late. Early reports surmised that the blame for the collision rested with Conductor Burbank of the stock train, who had made a mistake of 10 minutes in calculating the time that the passenger train would reach Newcastle. Engineer Ostrander of the stock train was missing, believed buried under the wreckage, and his fireman, named Hines, was probably fatally injured, being shockingly burned. Engineer Gordon of the passenger train was also badly hurt, having been thrown over a barbed wire fence, and his fireman, Robert Holland, died of his injuries at 5 a.m.[674]

A quickly held inquest cast the blame for the disaster on Conductor Burbank but no further action seems to have been taken against him.

The Lehigh Valley Railroad in the vicinity of Bound Brook, New Jersey, United States, was double track but on the morning of Monday 9th January 1899 the derailment of a freight train due to a broken axle had completely blocked the eastbound track near Bound Brook and trains heading to New York were being worked using the westbound track over the six miles between Bound Brook and Newmarket, while westbound trains were being held at South Plainfield, the next station east of Newmarket. Train No. 20, which had left Shamokin, Pa, at 7 a.m. was a very heavy one on this day, and as a result was run in three sections. The first two sections arrived at Bound Brook, crossed over to the westbound line, ran over it to Newmarket, and crossing back to the eastbound track continued their journeys to New York. The third section was running over an hour late on the schedule of train No. 20. It was carrying 400 excursionists in seven cars, mostly from the Pennsylvania towns of Shamokin, Mount Carmel, Mahoney City, Hazleton and Pottsville, travelling under the auspices of an annual businessmen's trip, with a three-day stay in New York. At Bound Brook it was crossed over to the westbound track and followed the two previous sections towards Newmarket.

Meanwhile waiting at Newmarket was westbound local train No. 71, due at Bound Brook at 11.59 a.m. but the single line working had meant that it had had to wait for a succession of eastbound trains, and it was thus also almost three-quarters of an hour late. Finally, the operator at South Plainfield gave the local permission to proceed. At the stop at West Dunellen Engineer Rick slowed down his train in case there were any passengers to pick up, but the signalman waved his arms to indicate that there were none. Engineer Rick put on steam and the train gathered speed again as it rounded a curve and had probably attained a speed of 25 miles an hour when disaster struck.

Engineer James Prendergast and Fireman George Cheshire on the excursion train saw the approaching local train come around the corner and did their best to stop before they, like the engine crew of the local train, jumped for their lives. Alarmed by the continued

[673] Ibid.
[674] *Boston Daily Globe*, Boston, Massachusetts, United States, 10th September 1897

whistling of the engine the passengers in the excursion train opened windows, women grabbed their children and men began to get up from their seats; then the crash came. The excursion train was travelling at about 15 miles an hour. The engine of the local train was turned a complete somersault by the collision and landed beside the engine of the excursion train, the tender of which was pushed back through almost the whole length of the leading car of the train. It was in this car that all the deaths and most of the injuries occurred, the passengers in the remainder of the train escaping with a severe shaking. The two engines, the leading coach of the excursion train and the baggage car of the local train were the only engines and vehicles so badly damaged that they could not be removed on their own wheels after the accident. There should have been a baggage car marshalled at the head of the excursion train, but it had been detached at Mauch Chunk after it had been discovered to have a hot axle box. Had it not been missing the number of deaths would probably have been minimal. Probably once it had been removed the passengers in the leading coach should have been moved back to another part of the train and the leading coach locked out of use, but that is being wise after the event and also would doubtless have been an unpopular decision if the remainder of the train had then became seriously overcrowded.[675]

Doctors were soon summoned after the collision from Bound Brook, Dunellen and other local places and they were able to reach the accident site over the tracks of the Central of New Jersey Railroad, which at this point ran parallel to and about a quarter mile distant from the Lehigh Valley route. Relief and wrecking trains were also soon on the scene and the bodies removed from the wreckage of the demolished car were loaded into the rear cars of the excursion train and taken back to Bound Brook. Twenty-one seriously injured victims were taken to Plainfield Hospital; the less badly injured were taken to Jersey City and New York.

Engineer Prendergast was a big man, weighing almost 300 lbs, and he delayed jumping from his engine until it was too late, with the result that he was pinned beneath it and lay there in great pain for three hours before a crane could be brought to the scene to lift the locomotive. It was not thought likely at the time of the report that he would survive. Engineer Rick was likewise found badly injured after the accident and bleeding from the mouth but managed to say, "Someone gave wrong orders. I know how it happened but can't speak now."

Superintendent W.G. Sprigg, in charge of the Eastern and Amboy division of the Lehigh Valley, agreed, saying, "The accident is evidently due to a blunder in orders."

Continuing with its explanation of how the accident might have happened a local newspaper the next day wrote that the train dispatcher at Easton was said to have wired the South Plainfield operator to let train No. 71 through, but then appeared to have changed his mind and sent a further order to the operator to hold all westbound traffic for the third section of train No. 20. Harry Rutland, the operator at South Plainfield, denied having received any order from the dispatcher to hold train No. 71. Martin Brennen at West Dunellen, although described by the newspaper as a signalman, had no access to the telegraph and nothing to do with the dispatching of the trains.

The Lehigh Valley general office issued a statement regarding the disaster. 'The third section of train No. 20, which is due in New York at 1.03 p.m., collided with train No.

675 *New Brunswick Daily Times*, New Brunswick, New Jersey, United States, 10th January 1899

71, a local accommodation train, which left New York at 10.50 a.m. near West Dunellen Station, shortly after 12 o'clock today. The eastbound train was occupied by an excursion party from Mahoney City, Shamokin and other points in the coal regions destined to New York. Both trains were using the west bound track on account of the east bound track being occupied by freight trains which had temporarily blocked the road. The engines telescoped, likewise the first car of the excursion train, resulting in the injury of 15 or more persons and the death of 12 persons.'

In a footnote to its report the same newspaper wrote, 'An instance of real pluck and heroism was the work of Father Lawrence, of Bound Brook, at the railroad wreck yesterday afternoon. Engineer Prendergast was caught under his locomotive, with blazing coals from the furnace about him, steam hissing from the boiler and the iron work of the wrecked locomotive penning him in. When Father Lawrence arrived, he crawled on his hands and knees under the locomotive, in imminent danger of being crushed, and asked the engineer if he was a Catholic. Receiving an affirmative reply, he calmly proceeded to administer to him the last sacraments. It was a bit of heroism as sublime as that of the doctor who ministers to his patients under fire on the battlefield. Father Lawrence is a brother of Joseph Lawrence, of this city {New Brunswick}, and formerly pastor of St Joseph's Church, at Metuchen.'[676]

Two days after the accident it was reported that the number of dead had risen to 18 with the death of George Joseph of Shamokin at the Muhlenberg Hospital in Plainfield. He had been one half of a couple who had been eloping to New York to get married there the same afternoon. His wife-to-be, Mary Soherimkey, also of Shamokin, was killed outright in the accident, bringing their elopement to a tragic conclusion.[677]

The investigation into the accident disclosed a series of errors and misunderstandings. When dispatcher J.A. Sullivan at Easton had issued the order to South Plainfield at 11 a.m. to hold westbound trains the operator there had been an employee named Mehaffey. At 12 noon Mehaffey had been relieved by Rutland as operator but had remained in the tower as leverman, so there was no clear transfer of responsibility between the two men. When Rutland came on duty there were two westbound trains waiting at South Plainfield; train No. 71 and a shunting engine going a short distance down the line to do some work. The conductors of the local train and the engine came into the tower and Operator Rutland gave them some other orders not relating to the single line working. Without checking Mehaffey, who had neither delivered the holding orders to the two conductors nor told his mate of the single-line working, assumed that they were clearances and operated the train order board, whereby the local train set off, 36 minutes late. Rutland, sat with his back to the track, did not notice the local train departing, or did not react.

Shaw suggests, and it seems very plausible, that the two operators were at loggerheads, as there does not seem to have been much communication between them. No record of the holding order was written on the train sheet and, perhaps because they were barely speaking to one another, Rutland did not ask Mehaffey what the local train and the engine were waiting for and was not told. Train Dispatcher Sullivan was also considered to have been at fault for not checking that Mehaffey had issued the holding orders to the two conductors, and he was suspended. The crew of train No. 71 might also

---

[676] Ibid.

[677] Ibid, 11th January 1899

have been considered to be at fault for they were aware of the derailment and the disruption to traffic, also from the green flag carried by the second section of train No. 20, that another section was to follow; however, they had not received a holding order and the order board had been cleared for them. Finally, although it was a busy metropolitan line, the route still lacked automatic block signals at this rather late date.[678]

At 2 p.m. on the afternoon of Saturday 23rd September 1899 a Judson excursion train from Chicago to San Francisco left Pueblo, Colorado on the Denver and Rio Grande Railroad. Hauled by a large Mogul engine, it consisted of an empty passenger coach and two sleeping cars filled with passengers, most of them bound for their homes in the West. About two hours later the excursion train reached Florence, Colorado, United States, where it received orders to meet an eastbound freight train at Reno siding, four miles further on. The same orders were delivered to the freight train at Canon City, but for some reason it ran past the siding and continued towards Florence, running downhill at considerable speed. The line here ran along the Arkansas River valley with many twists and turns. The excursion train was speeding towards Reno with a full head of steam while its passengers admired the scenery. A group of children were stood on the platform of the forward sleeping car. The crews of the two engines caught sight of one another at close range and had just time to whistle, reverse their engines and jump. The two engines were locked together by the collision, 'the boiler of the big passenger engine forming a coat for that of the smaller engine.' The passenger coach telescoped half of the forward sleeping car and waggons of ore were piled up in a heap of debris.

The children who had been standing on the sleeping car platform were the main victims of the accident, six of them being killed outright. They were Eva Walker, 12, of Westford, Massachusetts; Eugenie Judkins, 12, of Spokane, Washington; Addie Judkins, 10, Spokane, Washington; George Judkins, 6, Spokane Washington; Merville Wills, 10, of Los Angeles, California and F. Salter, of Boston, Massachusetts, the manager of the excursion. Two other passengers were probably fatally injured, one of them Mrs Ellie Wills, the mother of Merville; eight passengers were less badly hurt.[679]

The Grand Rapids and Indiana Railroad was not noted at the beginning of 1900 for having had any major disasters but on the 15th August of that year it had a bad accident between Sand Lake and Pierson, Michigan, United States, the worst in its history up to that date. The northbound Northland express (train No. 5) which had left Grand Rapids at 4.05 a.m., collided head-on with southbound train No. 2, due in Grand Rapids at 6 a.m., at about 5 a.m. between Pierson and Sand Lake. Dawn was just breaking at the time of the collision and a fog reduced visibility to 100 yards.

The Northland express was considered the finest train on the line and was made up largely of Pullman cars. It was made up of baggage cars, a Grand Rapids coach, a day coach, several sleeping cars from Cincinnati and the South and a buffet breakfast car. According to the timetable it was to cross train No. 2 at Sand Lake, two miles south of Pierson, at 4.52 a.m.

---

678 *A History of Railroad Accidents, Safety Precautions and Operating Practices* by Robert B. Shaw, pg. 144

679 *New Castle Nonpareil*, New Castle, Colorado, United States, 28th September 1899

The trains collided at speed and both locomotives and both baggage cars were demolished. Killed in the accident were Conductor Charles Letts, Engineer Gilbert Groetveld and Fireman Edward Woodhouse of train No. 5; Engineer William Fish and Fireman Louis Boyle of train No. 2; Ralph Levan, who was riding with his father Baggageman Levan, C. Pierson and Mark Blossom, a newsagent who was fatally injured. Baggageman David Powers of train No. 5; Coloured Porter C.M. Ford, Coloured Waiters W.M. Graves and Harvey Taylor, and Dining Car Conductor William Barnes, all of train No. 5; trainman Frank Roroff and passengers H.A. Dennis and W.G. Hartshaw were all more or less injured. Baggage, mail and dining cars at the head of train No. 5 and three freight waggons at the head of train No. 2 prevented a heavier death toll.

Initial reports were that either the engineers had blundered their orders or that they had been unable to make out the signals in the fog.[680] It soon emerged that in fact neither was the case, the real cause was a little more complicated. Train No. 2 was running late and the dispatcher therefore issued orders advancing train No. 5 northwards from its normal meeting point at Sand Lake to Woodstock, four miles north of Pierson, and this order was delivered to trains Nos 2 and 5. Subsequently the dispatcher learned that train No. 2 had made up some time, and decided to move the crossing place southwards again to Pierson. Before he sent a further order, he asked Operator Wells at Mill Creek, four miles north of Grand Rapids, if train No. 5 had passed there yet. Assured that it had not he issued the order to train No. 2 at a point north of Woodstock and to Operator Wells at Mill Creek for train No. 5. Operator Wells, who had nodded off in the early hours, was mistaken; train No. 5 *had* passed him. Train No. 5 therefore did not receive the order and the crews of both trains assumed that they had the right to the track between Pierson and Woodstock, with the collision as the result.[681]

The *Grand Rapids Evening Press* wrote, 'Operator Wells, of Mill Creek, was considered one of the best operators on the system and was also believed to be one of the most trustworthy of the road's employees. He was awake all night, but just at the sleepiest time of his watch his faculties failed him. After he received the order for No. 5 to stop at Sand Lake, he waited at the door for the train. He became anxious after a time and was about to report the matter when he received word from farther up the line that No. 5 had already passed him. Half crazed by the situation, he wired Pierson in an attempt to hold the train, but his message got there a fraction of a minute too late. Train No. 5 had just gone through there. The operator at Pierson rushed to the door when the message came, in time to see the tail of the express vanishing into the fog. A few moments later he heard the crash of the collision and at once notified headquarters. Wells had been stationed at Mill Creek about two weeks. He was an experienced operator and had been in the employ of the Pere Marquette and Michigan Central before coming to the Grand Rapids and Indiana. Before being engaged for the Mill Creek office Wells was in the employ of the Michigan Central at Grand Rapids. He was an abstract clerk at the freight office.' [682] The newspaper's statement that Wells was an experienced operator seems doubtful given that he had only been doing the job a fortnight and had previously been a clerk, unless of course he had worked as an operator some time earlier, perhaps on

680 *The Republican Atlas*, Monmouth, Illinois, United States, 17th August 1900

681 *The Jasper Herald*, Jasper, Indiana, United States, 17th August 1900

682 *The Fort Wayne Weekly Sentinel*, Fort Wayne, Indiana, United States, 22nd August 1900, reproduced from the *Grand Rapids Evening Press*

the Pere Marquette.

A man named Edward Risch arrived in San Antonio, Texas, United States, at the beginning of December 1900 with information that a terrible train wreck had occurred the previous Thursday, 29th November 1900 on the Ferrocarril Central Mexicano (Mexican Central Railway) midway between Tanamachi and Symon, Mexico, in a valley between two great hills, 50 miles south of Jimuico. A construction train, carrying 150 men, and a freight train of 33 empty cars, each travelling at 30 miles an hour, had collided. Three engines and 40 cars had ended up forming a mass of wreckage 30 feet high, beneath which the dead and injured were buried. Two of the railwaymen had been Americans, and they had immediately fled the scene for fear of their lives, 'as in Mexico employees of a train must suffer death if fatalities attach to a train wreck'. This statement is presumably meant to refer to the possibility of surviving railwaymen being lynched rather than an official mandatory death penalty for causing an accident. At any rate it was supposed that the Americans had made it back across the border to their own country. Mr Risch had left the scene before the names of the dead and injured had been obtained and could furnish no more details of the accident, which was stated to be the worst in the history of the country [683] (not very likely given the 1881 bridge catastrophe at Cuartla).

---

683 *Portland Morning Oregonian*, Portland, Oregon, United States, 3rd December

CHAPTER 7

RUNAWAY TRAINS AND VEHICLES

A variety of situations could lead to either a complete train or part of a train running away out of control. If there were railway staff in attendance the train or engine could only run away out of control if there was a failure of the brakes or if the brakes were incapable of holding it on a steep gradient. In the case of unattended vehicles, the action of gravity alone would be sufficient to precipitate a runaway if the train or vehicles were left standing on an incline. In contrast to the situation on a road a train or vehicles can sometimes run for many miles without mishap, an accident occurring only if the runaway meets with another train, comes to the end of the line or gets to a curve too sharp for the runaway for it to negotiate at the speed at which it is travelling.

Ten people died when a train ran away and derailed on the British owned Taltal Railway in Chile on or about 15th June 1882. A report on the accident reached New York, United States, via Panama only in mid-July. The narrow gauge (3ft 6in) railway was not opened to traffic until 20th October of that year[684] so it was presumably a construction train which was involved in the runaway. The railway was built to carry saltpetre (sodium nitrate) from mines in the Atacama Desert down to the port of Taltal (referred to simply as 'Port' in the newspaper account of the accident) on the Pacific Ocean coast, from where it could be exported. Saltpetre was used in the production of both fertilisers and explosives and in 1883 Chile controlled 80% of the world's supply. The single-track railway fell steeply from the mines to the coast. According to the report the control of descending trains was usually assisted by reversing the engine of the train and then giving it full steam but on the day in question this was either forgotten or overlooked. The result was that the train got out of control, dashed through the station at the end of the line and was 'finally brought up in fragments on the beach'.[685]

On 4th August 1882 a heavy coal train got out of control on an incline between

[684] Webpage – Grace's Guide to British Industrial History: Taltal Railway.
[685] *Milwaukee Weekly Wisconsin*, Milwaukee, Wisconsin, United States, 26th July 1882

Cranberry Mines and Oil City, Pennsylvania, United States, on the Buffalo Pittsburgh and Western Railroad. It consisted of 17 flat cars laden with coal and 15 or 18 miners were riding on it using it as a means of getting into town, intending to ride back on a train of returning empties. The train ran at increasing speed until the rear cars were hurled from the track and demolished, killing six men more or less outright, including Conductor Charles McGinley, and injuring about 13 others, three of whom later died.[686] At the end of its run the engine of the train collided with some stationary cars.[687]

An American circus train was involved in a runaway accident early on Sunday 24th September 1882 near Paint Lick, Kentucky. At about 3.30 a.m. that morning the first of three special trains carrying the performers, animals and equipment of the circus became unmanageable as it was descending a '72 feet to the mile' (1 in 73) gradient and it ran at increasing speed until it left the track on a curve 300 yards from Paint Lick station. The fourth car from the engine was the first to leave the rails and roll down an embankment, followed by 15 more of the 21 cars of the train. The cars were principally loaded with equipment; tableaux, waggons and an electric light machine, but there was also a cage containing a tiger. The train was also carrying a number of circus employees, scattered along the train, two of whom were killed outright, three probably fatally injured, and seven or eight less badly hurt. A third immediate fatality was young William Underwood, from Mt Vernon, Kentucky, who had been stealing a ride. The cage containing the tiger was burst open and the big cat was thrown out, with the result that no one wished to go anywhere near the cage until daylight. The door of the cage having been left open the tiger, which had been crouching nearby, then crept meekly back into it and was re-secured. The wrecked cars were piled up in a confused heap and the tableaux and electric light cars destroyed, while 100 yards of track was torn up. The locomotive and leading three cars of the train were undamaged and were immediately despatched to Lancaster for surgeons and railwaymen.

A gang of workmen cleared the track by 5 p.m. but Mr Sells declined to go on to London where the circus was due to perform the next day. He decided instead to rest at Silver Creek Monday and continue to a planned stop at Lebanon on Tuesday. The cause of the accident was stated to be unknown and it was hinted that an official investigation would be necessary but earlier in the report it had been mentioned that the train had had only four brakemen and, probably more significantly, that Mr Sells had removed the brakes from the cars to facilitate their loading.[688]

The Paint Lick accident was not the only wreck of a circus train in the United States which was followed by the escape of big cats. Eleven years later Walter L. Main's circus train ran away on the Pennsylvania Railroad and was derailed at McCann's Crossing, near Tyrone, Pennsylvania, on 30th May 1893. Four circus labourers and Brakeman William Heverley died in the wreck and a number of other people were injured, including one who was bitten by a lion. This time lots of animals escaped including two Bengal tigers, three lions, a silver tip panther and 'a ferocious ape'. One of the lions, doubtless hungry, found its way to the farm of a Mr Alfred Thomas and surprised Mrs Thomas

[686] *The Democrat*, Olean, New York, New York, United States, 8th August 1882
[687] *Chester Times*, Chester, Pennsylvania, United States, 5th August 1882
[688] *The Atlanta Constitution*, Atlanta, Georgia, United States, 27th September 1882

while she was milking the cows. Her husband, alerted by her screams, appeared with a gun and shot the beast dead. One of the lions and the panther were also shot, the other large animals being eventually recaptured after their escape had 'caused great consternation' in the neighbourhood.[689]

The overland express of the Southern Pacific Railroad left San Francisco, United States, at 9.20 a.m. on Saturday 19th January 1883. At Sumner an assisting engine was attached to help the train to the summit of the Tehachapi Incline, by which the railroad negotiated a pass between the Sierra Nevada and Tehachapi mountain ranges. Shortly after midnight on Sunday 20th the train stopped at Tehachapi Summit, 47 miles from Sumner, and both engines were detached and moved forward. One of the two brakemen had gone forward to operate the points to let the engines into a siding and the other had escorted a lady to the station and then stopped to relight his handlamp. Apart from the engines the train consisted of express, mail and baggage cars, two sleeping cars a day coach and a smoker. These seven vehicles then all began to roll down the 1 in 44 bank, slowly at first but then at increasing speed. Conductor Reed had remained on the train and tried in vain to halt the runaways by screwing down the hand brake harder. The runaways covered about four miles before the hindmost sleeping car left the track on a curve, taking with it the other sleeping car and the mail, baggage and express cars. All went over an embankment about 15 feet high and piled up in a heap of wreckage which then caught fire and was consumed. According to newspaper reports some victims were trapped in the wreckage and roasted to death before the eyes of rescuers powerless to save them.[690] The day coach and sleeper remained on the rails and after running another two miles were brought to a halt, no one in them being hurt.

Seventeen or possibly eighteen people died in the accident, mostly passengers but also Sleeping Car Porters Wright and Lawrence and two tramps. Those killed included Mrs John Downey, wife of the ex-Governor of California and ex-Congressman Charles H. Larrabee of Wisconsin. Downey himself, who was first president of the Los Angeles & San Pedro Railroad, was injured. The uninjured passengers included Porter Ashe and his wife, who was the daughter of railroad magnate Judge E.B. Crocker. After the accident a special train was sent to the scene from Los Angeles with physicians and an undertaker and the dead and injured were taken there. A breakdown train was also despatched from Sumner. The train had been carrying a pouch of registered letters for St Louis and letters and printed matter from Lost Angeles and San Bernardino and all post offices south of there, Arizona, New Mexico, Texas, part of Colorado and Sonora state, Mexico. The mail and express matter was variously reported as having been burnt and saved.[691]

The runaway was at first attributed to the air brakes having leaked off and the hand brakes either having not been applied or not having been strong enough to hold the train. Later it was theorised that the runaway had been caused deliberately. The two tramps killed in the accident had been seen loitering at Tehachapi and two more tramps were found on the cars which escaped the derailment. It was postulated that the tramps had boarded the train at Tehachapi and deliberately released the brakes so as to run the

---

689 *Tyrone Daily Herald*, Tyrone, Pennsylvania, United States, 1st June 1893.
690 *The Daily Gazette*, Colorado Springs, Colorado, United States, 21st January 1883
691 Ibid.

vehicles down to a point where they could easily rob the train but had instead lost control of it.

Ex-Governor Downey did not accept the tramps sabotage theory but put the accident down to the carelessness of the railroad employees and announced that he would bring a suit for heavy damages.[692] The coroner's jury apparently did not believe it either and found Rear Brakesman Patton guilty of manslaughter through culpable negligence, but at his trial at Bakersfield on 23rd March he was acquitted.[693]

Early on Sunday morning, 1st July 1883, an accident occurred near Rasselas, Pennsylvania, United States, on the Rochester and Pittsburgh Railroad. Since the consummation of a traffic agreement with the Erie Railroad and the opening of a new branch to coal mines at DuBois the Rochester Company had run a lot of coal trains, most of which operated at night so as not to interfere with the completion of the work on the new line. Thus on the evening of Saturday 30th June the regular coal train out of DuBois was run in four sections, to the first of which was attached a passenger coach, carrying about 15 late-night travellers. This section was in charge of Conductor S.N. Toles, with Engineer Mike Vaughn. Three-quarters of a mile from Rasselas the engine of the train stopped at a tank situated on a steep incline to take water, at about 3 a.m. on Sunday morning. As Engineer Vaughn was on the ground filling the tank, he was surprised to see part of his train running back down the bank. Assuming that the brakemen would soon pull up the runaway cars up he jumped back onto his locomotive and began to set back after them at a moderate pace. The cars were not pulled up, however. They had divided twice, with the passenger coach and four loaded coal cars being followed by another two cars. Engineer Pat Downs on the locomotive of the second section of the train was in the act of taking a chew of tobacco from Fireman Prosser when he glimpsed the light of the runaway coach coming around a bend. He just had time to reverse his engine and both jumped. The next moment there was a terrific crash as the runaways, travelling at 40 miles an hour, collided with the engine of the coal train, which had been struggling up the hill at about six miles an hour. The engine split the coach like a wedge, and the wreckage was then hit by the other two runaway cars. To add even further to the horror the passengers then found themselves scalded by steam from broken steam pipes, though one woman, Mrs W.H. McCurdy, had the presence of mind to jump out of the coach at the last moment and saved herself and her son.

Conductor Toles, Brakeman Michael Downs and passengers Robert Clement, George Quinn, Adamo Angelo and Samuel McKee were killed outright, and passenger John Collins was fatally injured and died the next day. All the other passengers were more or less seriously injured, but Mrs McCurdy and her four-year-old son suffered only cuts and bruises.

The accident and subsequent inquest were reported upon at length in a local newspaper but the report of the conclusion of the inquest has not been traced. It would appear, though, that the cars had become divided and run away because the brakemen had failed to apply their brakes when the train stopped, possibly because they had fallen asleep, and the inevitable uncontrolled sagging back of the heavy coal cars at the front of

[692] *Galveston Daily News*, Galveston, Texas, United States, 1st February 1883
[693] *Weekly Nevada State Journal*, Reno, Nevada, United States, 31st March 1883

the train had proved too much for the couplings. No attempt had been made to apply the brakes on the runaways at any point.[694]

An American circus train again featured in a runaway accident on 4th October 1885, this time on the Fergus Falls branch of the Northern Pacific Railroad. Two sections of a train conveying the equipment and performers of John Robertson's circus left Walperton, Dakota Territory, for Fergus Falls, Minnesota. When near French, Minnesota, seven miles from Fergus Falls, the first section became divided while climbing a steep grade, and 10 or 12 cars ran back down the hill. The second section included three sleeping cars carrying over 200 workmen, who were all sound asleep. The runaway cars struck the engine of the second section with tremendous force and threw the three sleeping cars off the track, 'smashing them beyond recognition'. Over a hundred men were buried in the wreckage, five of whom were found to be dead and many others injured.[695] The runaway cars probably covered a mile before they met the second section. The brakesmen on the first section tried to pin down their brakes but their progress from car to car was impeded by the circus waggons loaded onto them. The engineer of the second section saw the runaways approaching but only had time to bring his train to a stand before the collision.[696] The Northern Pacific hospital car took the injured to hospital in Brainerd, Minnesota.[697]

The colony of New South Wales had a bad accident on 21st June 1887 at Hawkesbury River (Peat's Ferry). A fully loaded excursion train left Redfern station, Sydney, at 10.29 a.m. on the morning of that day for Hawkesbury River, four minutes late. All went well until the train reached a steep incline at Beecroft, which it failed to climb. After two failed attempts the train was divided, and the two parts were worked forward separately to Hornsby, where the train was reunited. With no further steep banks to climb no further problems were anticipated. Instead, the train had to descend the 5¼ mile Cowan Bank to Hawkesbury River and it was here that disaster arose. The gradient on this bank averaged 1 in 40 and the train got out of control, owing, it was stated initially, to a problem with the Westinghouse brakes. Some of the passengers apparently thought that the increased speed and oscillation were normal, a newspaper report likening their reaction to that of the Paris theatre-goers at the disastrous Opera Comiqe fire a month earlier, who had thought the crackling and flames were part of the play. They were soon to find how wrong they were.

Two loaded passenger trains were standing at the platform of Hawkesbury River station, but Stationmaster Thomas Cavanagh, seeing the runaway train, told Pointsman Patrick Proctor to 'stick to his points'. The pointsman did so and the runaway train was diverted onto a siding leading to a bridge being built over the river.[698] The train lost some speed on the siding and then collided with some contractor's waggons. Driver Thomas Wilson stuck to his engine and was drowned when it rolled down the embankment into

[694] *Bradford Era*, Bradford, Pennsylvania, United States, 2nd, 3rd and 4th July 1883
[695] *Reno Evening Gazette*, Reno, Nevada, United States, 5th October 1885
[696] *Stevens Point Journal*, Stevens Point, Wisconsin, 10th October 1885
[697] *Bismark Daily Tribune*, Bismark, Dakota Territory, United States, 6th October 1885
[698] *The Daily Telegraph*, Sydney, New South Wales, 1st July 1887

the river, and six passengers also lost their lives. Fireman John Pye also ended up in the river but was rescued; his driver was trapped beneath the engine.[699]

A few days after the accident the theory was put forward that the apparent failure of the Westinghouse brakes was due to the fact that the pipe connection had had to be closed when the train was divided to climb the bank at Beecroft and that it was possible that the taps had not been properly re-opened when the train was re-joined at Hornsby. This theory was reportedly reinforced by the observation that the wheels of the Redfern carriage and the American car, those immediately in the rear of the locomotive, were extremely hot, while those of the remaining carriages were cold.[700] It also emerged that Guard Alfred Clissold was making his first trip on the line, that Driver Wilson was making only his second trip, that four carriages added to the end of the train at Redfern were not fitted with brakes, that there was no brake van at the end of the train as none was available, and that the train had not had a proper brake examination before it left Redfern because the brake examiner, Andrew Weirick, had had insufficient time to do so.[701] Guard Clissold admitted that he had not tested the air brakes at Hornsby after re-joining the train as per the regulations, because he said been very much hurried and jeered by the excursionists, fed up with the endless delays to the train.

The accident was investigated by the railway department and by a coroner's jury, who came to opposite conclusions as to its cause. The railway department inquiry decided that the accident was due to the mishandling of the Westinghouse brakes by Driver Wilson on the long descent to the terminus, with the result that the air was exhausted to the point where he could not apply the brakes fully when they were most needed. The fact of a brake tap at the front of the third carriage having been found closed after the accident was explained by its having been knocked closed by the impact of the collision itself. The board relied upon the evidence of Guard Clissold and the Hornsby stationmaster, George Stead, that the brake connections had been properly made there.

The coroner's jury passed a verdict that the accident was due to 'the great want of supervision on the part of the officials in the Railway department in not carrying out the instructions laid down for their guidance' but did not specifically charge any individual with neglect, and passed a rider recommending the Government to appoint a commission to inquire into the workings of the carious branches of the department.[702]

The local press was not very happy with the railway department attempting to blame the accident on the dead driver and thought that the coroner's jury had discovered its real cause.[703]

A major disaster in Russia on Sunday 13th May 1888 was briefly reported in the British press. It occurred near Galitsyno station on the Moscow-Kursk railway. Several carriages of a passenger train bound for Moscow became detached and running back down an incline were struck by a following goods train. Eleven passengers died and 30

[699] *Illustrated Sydney News*, Sydney, New South Wales, 15th July 1887
[700] *The South Australian Advertiser*, Adelaide, South Australia, 25th June 1887
[701] *The Daily Telegraph*, Sydney, New South Wales, 29th June 1887
[702] *The Australasian*, Melbourne, Victoria, 23rd July 1887
[703] *The Daily Telegraph*, Sydney, New South Wales, 6th September 1887

were injured. The victims were all workmen.[704] [705]

The Great North of Scotland Railway suffered a runaway accident on 11th September 1888 at Kintore, Scotland, where the Alford branch left the main line. A down goods train had left Kittybrewster (Aberdeen) for Alford at 5.30 p.m. and having left Kintore stopped to shunt at Ratchhill Siding, about 1½ miles down the Alford branch, at 7.36 p.m. The waggons to be left at the siding were in the middle of the train and the train was divided on either side of them, after which they were tail-roped by the engine into the siding, an operation which took nine or ten minutes. The twelve waggons and brake van at the rear of the train had been left on a gradient of 1 in 226, which a little nearer Kintore steepened to one of 1 in 137 and then to 1 in 86, all falling towards Kintore.

When the guard returned to the rear waggons, he found that they had run back down the incline, with the lights of the brake van just disappearing under an overbridge 330 yards away. All that the staff of the goods train could then usefully do was for the engine to whistle to alert the people at Kintore station.

The engine whistle was heard by the stationmaster and an inspector on the platform at Kintore but unfortunately not by the signalman, who could otherwise have reversed the road to let the runaways out onto the main line, which was clear. When he heard the runaways approaching it was too late to do anything and they ran into a brake van standing on the Alford dock line and pushed it into the buffers. The two brake vans were broken up and seven of the waggons, all of which had been loaded, were more or less damaged. Two young men who had been riding as passengers in the brake van were killed.

Major F.A. Marindin investigated the accident for the Board of Trade and concluded, having experimented at the siding, that the brakes in the van must have been initially applied as the guard had stated for otherwise the waggons and van would have run away at once. He could only conclude that the young men in the van had released the brakes by turning the brake wheel, either accidentally or thoughtlessly, and that when they had realised in the darkness that they were running away had probably turned the wheel the wrong way in an attempt to reapply the brakes. The accident proved once again, he said, the objection to letting members of the public ride in brake vans. Unlike the Hereford Hay and Brecon accident of October 1872 (detailed in volume on 1811-1880 accidents) they had been legitimately there, the Great North of Scotland Company permitting passengers to travel in brake vans at first-class fares! The two men had been drinking but were not sufficiently drunk to prevent their being carried as passengers. Finally, the inspecting officer pointed out that by the date of the accident no such siding as Ratch-hill would be sanctioned to by the Board of Trade, unless all shunting operations could be conducted clear of the main line.[706]

The worst railway accident in Irish railway history, not only in the 19th century but up to the present time (July 2019) occurred near Armagh on 12th June 1889. On that day the

[704] *The Times*, London, England, c17th May 1888
[705] *London Evening News*, London. England, 15th May 1888
[706] Board of Trade report

Armagh Sunday schools had arranged a day trip for children and adults from Armagh to the seaside resort of Warrenpoint, over the Great Northern (Ireland) Railway line between the two places. To carry the estimated 800 excursionists the Dundalk locomotive department provided 15 vehicles and a 2-4-0 locomotive, though the assigned driver, Thomas McGrath, had only had instructions to take 13 vehicles, the line having severe gradients, including a mile climbing at 1 in 82 followed by 2½ miles climbing at 1 in 75 shortly after leaving Armagh.

John Foster, the Armagh stationmaster, considered that the engine, No. 86, was well capable of taking 15 vehicles up Armagh Bank, and finding that the actual number of excursionists was not 800 but about 940, wanted to send 16 or even 17 vehicles to accommodate them. An argument then ensued between the driver and the stationmaster. Driver McGrath told the stationmaster that his instructions were only to take 13 vehicles, to which the stationmaster replied that other drivers did not grumble about taking excursion trains. Driver McGrath asked the stationmaster why he had not sent word to Dundalk so that he could have been provided with a more powerful six-wheel-coupled engine but left it at that and walked down the platform, confident that the engine that he had could manage 15 vehicles. Like Driver Wilson at Peat's Ferry Driver McGrath did not know the line very well, having only ever been over it as a fireman with ballast and excursion trains.[707]

The train was thus started at about 10.15 a.m., a quarter-hour late, with 15 vehicles; 13 carriages and a brake van at each end. In charge of the train was James Elliott, chief clerk in the traffic manager's office at Dundalk. Before starting, on seeing the size of the train and the number of passengers, Elliott had mentioned to the stationmaster that it would be better if the train was assisted up the Armagh Bank by the engine of the ordinary train in charge of Driver Patrick Murphy, and this would have been done but Driver McGrath, who had obviously taken the huff, told Elliott that he did not need assistance and could manage perfectly well with just engine No. 86. Because of the number of passengers Porter Guard William Moorehead had 10 people with him in the brake van behind the tender and Porter Guard Thomas Henry had 14 or 15 passengers with him in the van at the rear of the train.

Steam pressure was 130 lb on starting from Armagh and the train made fair progress over the first two miles but then its speed gradually diminished over the next mile and a quarter until it came to a stand within a couple of hundred yards of the end of the 1 in 75 climb, and for no obvious reason, as the steam pressure was still 125 lb. It stopped at 10.33 a.m. having covered 3¼ miles in 18 minutes at an average speed of 11 miles an hour.

It was at this point that Chief Clerk Elliott, in consultation with Driver McGrath, made the fateful decision to divide the train and work it forward in two parts to Hamilton's Bawn station, a mile and a half distant, where there was a siding which could hold half of the train, instead of sending the rear guard back to protect the stalled train and waiting for the ordinary train (which was due to depart 20 minutes after the excursion) to arrive and push the excursion to the top of the bank. Even had no accident occurred, no time would have been saved by working the train forward in two parts as intended.

Having made this most unwise decision Elliott went back to the rear van and told

[707] Ibid.

Porter Guard Henry to put his brake on tightly and he was assured that it was on. Elliott asked if Henry had any sprags but, because it was a passenger train, Henry had none, so Elliott directed him to put some stones under the wheels. Learning that the siding at Hamilton's Bawn could probably hold five vehicles Elliott then decided to divide the train between the fifth and sixth carriages. Porter Guard Moorehead uncoupled the vacuum brake pipes forthwith and the train only being fitted with the simple vacuum brake, whereby the creation of a vacuum was needed to apply the brakes, the rear carriages were now entirely dependent on the brake in the rear van. Moorehead next unhooked the side chains, put a stone beneath the wheel of the sixth carriage and unscrewed the screw coupling. So far, so good, both portions of the train remained at rest. Moorehead then jumped in his van, released its brake, and told Driver McGrath to proceed. McGrath had to set back his engine slightly to get a start and this movement was enough to set the rear vehicles moving back down the bank. Seeing what was happening, Moorehead shouted to McGrath to gently set back so that he could recouple the vehicles. Twice he almost succeeded in doing so, but both times he tripped over some rails which were lying beside the line at the crucial moment. After that the runaways were going too fast to make a third attempt.[708]

Elliott meanwhile had vainly tried to stop the runaways by placing more stones beneath the wheels, but these were just crushed to powder. He then jumped onto the rear van and told Henry to try and screw the brake down harder, but Henry and another man were already doing all they could.

Porter Guard Henry described Elliott jumping on the step of his van and telling him to try and put the brake on harder. Elliott, Henry said, had then cried, "Oh, my God! We'll all be killed," and had jumped off. The speed had then increased until it was difficult to see the hedges as they passed but he had remained in the van until the crash when he was knocked unconscious, coming round in the infirmary. He said that he had intended to go back to protect his train once he had finished putting stones under its wheels.

Driver Murphy of the ordinary train said that they had been travelling at between 20 and 30 miles an hour and had been near the 19 milepost (it must have actually been the 19½ milepost) when his fireman, William Herd, who had a better view at that point from his side of the engine, had shouted, "Hold, hold, hold!" whereupon he had whistled, shut off steam, and looked out to see the runaways coming towards them. He had then applied the vacuum brake with full force while Fireman Herd had reversed the engine. The speed of the ordinary train was thus reduced to about five miles an hour at the moment of the collision. Fireman Herd jumped from the engine before the collision, but Driver Murphy clung on to the tender, which ran away down the bank with a horse-box attached, and which he was able to stop with the tender brake short of the remainder of the ordinary train, which had also run away after the collision but was stopped by its guard, Daniel Graham.[709]

The results of the collision were catastrophic; the last three of the runaway carriages were completely destroyed and their debris thrown down a high embankment, while the engine of the ordinary train was also thrown over the embankment. The Board of Trade report speaks of 78 passengers killed and 260 more or less injured. Several the injured

---

[708] Ibid.
[709] Ibid.

passengers succumbed during the following months, the last probably being John Armstrong, former deputy governor of Armagh Prison, who survived until 15th November, his death bringing the probable final number of fatalities to 83.[710]

Major-General C.S. Hutchinson had the task of investigating the catastrophe on behalf of the Board of Trade. He stated that the immediate cause of the accident was the want of sufficient brake power to hold the 10 detached carriages on the incline. Experiment showed that the brake of a brake van identical to the one in the accident was sufficient to hold 10 similar carriages and if, as Porter Guard Henry stated, the brake had been applied, the only conclusion was that the brake had been released by one or more of the passengers in the van, while Henry was on the ground placing stones under the wheels. It was Elliott's duty to ensure that the brake in the rear van was in good order before the train was divided and then to have ensured the rear guard remained in the van to prevent anyone meddling with the brake, instead of calling him to the ground. Elliott said that he wished to divide the train in order to save time but here again his judgement was faulty, as it would have taken at least 10 minutes to take the first five vehicles to Hamilton's Bawn and another 10 at least to take the remainder of the train there and recouple the two portions, by which time the ordinary train was due there anyway. In failing to send Porter Guard Henry back at once to protect the rear of the stalled train Elliott had also failed to comply with Rules 223 and 226(c) relating to the protection of stalled trains. Finally, Elliott should have insisted before the journey began that the excursion train be assisted up the bank from Armagh and not let himself be overruled by Driver McGrath.[711]

Going on to consider the less immediate causes of the collision, the inspecting officer criticised the Dundalk locomotive department for the provision of an engine, No. 86, 'with less haulage power than was judicious' and for using a driver who had never before worked between Armagh and Goragh Wood as such, especially as the line had gradients as steep as 1 in 70 on curves of 33 and 30 chains radius. Another factor was Driver McGrath's over-confidence in being able to work the train with just his engine, though he must also have failed to manage his engine properly, as an experienced driver was able to take an experimental train of the same composition up the bank with the same engine with no difficulty.

Major-General Hutchinson echoed the comments of his colleague Major Marindin in his report on the Kintore accident by stating that it had been a grave error to have allowed passengers to ride in the brake van and that while, with an excursion train all of whose carriages were full it was doubtless difficult to prevent such an eventuality, it was a practice which should be strictly prohibited. It was also wrong that most if not all the doors of the excursion train had been locked, a common practice on excursion trains carrying children but one which prevented some of the passengers from jumping out before the collision. Also, given the weight of the train, both brake vans should have been marshalled at its rear. Mr Elliott appeared to have been responsible for allowing passengers to ride in the brake vans and for the positions of the vans in the train and Mr Foster, the Armagh stationmaster, for the locking of the carriage doors.

Major-General Hutchinson then considered what measures were needed to prevent a repetition of the disaster. In the first place, he wrote, it would in all probability have been

---

[710] *The Times*, London, England, 16th November 1889

[711] Board of Trade report

prevented had the excursion train been fitted with an automatic continuous brake instead of a non-automatic one. The Board of Trade was in the process of asking Parliament to introduce a bill making the use of continuous automatic brakes compulsory on all passenger trains, so he did not think it necessary to comment further upon the subject.

Secondly the effects of the collision would have been somewhat mitigated had the line been worked on the block system, so that the ordinary passenger train would not have left Armagh, and the speed of the runaway vehicles would have been reduced on the level stretch before the station was reached.

Further recommendations in the inspecting officer's report were that the length of excursion trains on lines with heavy gradients should be limited to about 10 vehicles and that the vans of all passenger trains should carry wooden sprags for use if necessary.[712]

The inquest on most of the victims of the accident was quickly held and on 21st June its jury found verdicts of culpable negligence against Locomotive Superintendent James Park and Running Shed Foreman William Fenton of Dundalk, for their selection of the engine, Chief Clerk Elliott, Driver McGrath and Porter Guards Henry and Moorehead. The coroner pointed out to them that that meant verdicts of manslaughter against all six, but the jury persisted in their verdicts except in the case of Fenton, as Park was his superior. The jury recommended the introduction of continuous automatic brakes, unspecified special precautions in the running of excursion trains, an overhaul of the organisation at Dundalk and a clarification of who exercised authority over what.[713]

On 6th July True Bills for manslaughter were returned against Elliott, McGrath and Moorehead (who Major-General Hutchinson had ruled in his report was 'in no way to be blamed') Elliott was tried first, first in August when the jury could not agree on a verdict and again in October when he was acquitted. The charges against McGrath and Moorehead were then dropped.[714]

Within two months of the accident the British Parliament had passed the 1889 Regulation of Railways Act forcing railway companies to provide all passenger trains (but not goods trains) with automatic continuous brakes, the introduction of the automatic block system on all passenger lines and the interlocking of all points and signals, all measures which the Board of Trade inspecting officers had been recommending for many years.[715] It had taken a major catastrophe to force the British Government to finally take action on the matter.

Some boys looking for a ride on some coal cars caused an accident near Shamokin, Pennsylvania, United States, on 17th July 1889. Four coal cars were standing on a siding below the Excelsior colliery and it was thought that the boys, intending to ride on two of them, uncoupled them from the other two and released their brakes. When the two cars began to run away too fast, they attempted to reapply the brakes but were unable to tighten them up enough. They then became frightened and ran off into the woods and the two runaway cars entered the Pennsylvania Railroad main line at Coal Run switch.

---

[712] Ibid.
[713] *The Runaway Train Armagh 1889* by J.R.L. Currie, pgs. 75-6
[714] Ibid, 77-91
[715] *The Times*, London, England, 13th, 14th, 15th, 17th, 18th, 20th, 21st, 22nd, 24th, 25th, and 28th June 1889

Continuing at high speed along the main line the runaways overtook a passenger train carrying 200 miners back home from their day's work at Hickory Ridge colliery to Shamokin. A miner on the rear platform of the passenger train saw the runaways coming, shouted a warning and jumped and he was followed by Flagman John Metz, who tried to put some sleepers across the track to stop the runaways, but these obstacles were just swept aside. Some of the other miners also had time to jump off and escape but others were not so lucky and in the inevitable collision two were killed outright and some others badly hurt, at least one possibly fatally.[716] [717]

The 8 p.m. London and North Western Railway express from London Euston to the North was in serious trouble early on the morning of Tuesday 4th March 1890. It got out of the control on the 28-mile descent from Shap Summit (ruling gradient 1 in 125), entered Carlisle, England station at considerable speed and collided with a Caledonian Railway engine. Four passengers were killed and 11 injured, as well as the drivers and firemen of both engines. The accident occurred at the period when the London and North Western Company were in the process of converting all their rolling stock to the automatic vacuum brake, but they still had stock fitted with only the simple vacuum brake, whereby the creation of a vacuum was needed to apply the brakes, and loss of vacuum released them, in direct contrast to the mode of operation of the automatic vacuum brake. While the transition to automatic brakes was still in progress some of the stock was fitted with a five-way cock whereby by means of a lever either brake could be used.[718]

A problem arose before the train left Euston due to ice in a brake pipe which was resolved by heating the area with some burning oily waste. The Northchurch signalman then sent the 'Stop and Examine' emergency signal to Tring, the train was stopped there, and it was found that the brakes were binding badly on the wheels of the Gourock van, which were red-hot. Colonel F.H. Rich, who investigated the accident for the Board of Trade, surmised that this was due to the five-way cock of the van having been left at 'Simple' so that the vacuum applied to the rest of the train to release them was applying the brakes of the van, as the van brakes had released as soon as it was uncoupled from the other vehicles of the train. To avoid having to detach the van the remainder of the vehicles of the train were then switched to 'Simple' operation as far as Rugby, where they were all switched back to 'Automatic'.

The automatic brake had worked well between Rugby and the approaches to Carlisle. Driver Thomas Rumney said that he had shut off steam at Wood Bank box down distant signal and tested the vacuum brakes. but they did not appear to be working properly. Further attempts to get the vacuum brakes to work were equally unsuccessful and he eventually told his fireman to apply the tender hand brake and put down sand. Giving evidence at the inquest he stated, "When I was trying to create a vacuum, I took the pin out of the slide and pulled the lever right back, so as to keep the steam brake on the engine and tender and at the same time create a vacuum." This statement, Colonel Rich commented, explained the mystery as to why the train had not stopped, for by doing so

[716] *Daily Gazette and Bulletin*, Williamsport, Pennsylvania, United States, 18th July 1889
[717] *Edwardsville Intelligencer*, Edwardsville, Illinois, United States, 24th July 1889
[718] Board of Trade report

Driver Rumney had inadvertently switched the brakes from 'Automatic' to 'Simple' operation, thus releasing them throughout the train except the screw brakes applied by the guards in the front and rear vans.

Surmising the course of events which led to the accident Colonel Rich supposed that Driver Rumney had shut off his small ejector at Shap Summit, where the train had stopped for its pilot engine to be detached and had then forgotten to re-open it. The brakes had thus gradually leaked on as the train descended the bank, but not sufficiently to especially retard the train on the falling gradient with greasy rails and the engine steaming hard. When the driver then tested the vacuum at Wood Bank there was little left and at that point he had panicked, and instead of whistling for the guard's brakes as per the rules, he had taken out the pin and altered the lever, converting the brakes to simple operation.[719]

Colonel Rich commented that Driver Rumney's confusion was understandable and the sooner the London and North Western Company did away with the simple vacuum altogether the better.

In contrast to the conclusions of Colonel Rich the coroner's jury at the inquest (which was held before the Board of Trade report was published) found a verdict of "Accidental death" and acquitted Driver Rumney of all blame, while passing the opinion that the London and North Western Company 'incurred a very grave responsibility in using a brake of such an uncertain and unreliable character.'[720]

A more recent writer had concluded that Colonel Rich was wrong and that in fact Driver Rumney had made no mistake, but the brake pipe had been blocked by a lump of ice, a possibility dismissed by the inspecting officer.[721] It has been suggested, doubtless accurately, that the continued use by the London and North Western Railway of the combined Smith's simple and Sander's automatic brake system had much to do with the fact that Francis W. Webb, the Chief Mechanical Engineer of the Company, received royalties for every vehicle fitted with it, as he had before with the Clark-Webb system. It was only after the Armagh catastrophe of June 1889 (see above) that the London and North Western and other companies were forced by Parliament to use continuous automatic brake systems on all passenger trains.[722]

At 5.30 p.m. on 25th April 1890 train No. 26, a mixed goods and passenger working, left Bathurst, New South Wales, for Sydney. It had passed Kelso station and was climbing the incline to Raglan when the pointsman at Kelso heard a train coming from Sydney and 16 vehicles rushed past him and onwards towards Bathurst. Meanwhile stationmaster George Farquhar at Bathurst had despatched No. 28 goods after the mixed train and there was nothing that he could do when Kelso wired him to say: "No. 26 broken away." The runaways and the goods train collided at the Russell-street crossing in Bathurst.[723] Three passengers were killed outright and a fourth died later; two of the dead were schoolchildren returning home from their lessons.

---

[719] Ibid.
[720] *The Times*, London, England, 27th March and 10th and 16th April 1890
[721] *The Railway Magazine* 1961, pgs. 200-5
[722] *Railwaymen Politics and Money* by Adrian Vaughan, pgs. 210-1
[723] Wagga Wagga Advertiser, Wagga Wagga, New South Wales, 29th April 1890

Guard Edwin Davis stated that when the vehicles had begun to move backwards, he had caught hold of the brake in his van and tried to put it on tighter. He had then tried to sprag a truck through the angle irons but the sprag had been thrown back and hit him in the chest and knocked him down. In all he had made three attempts to use the sprag. He had made another attempt at the brake but could not get it on any harder. He had then run to the station and told them to wire Kelso and Bathurst. He thought it would have been better if he had had an assistant.[724]

Driver James Palmer of train No. 26 told an inquest at Bathurst that the draw bar of the third truck behind his engine had broken and 16 vehicles had run away as his train had arrived at Raglan. He said that his train had stopped without any jerk.

Stationmaster Farquhar at Bathurst gave evidence as to receiving a wire that there had been a break away and said that it had then been too late to do anything. There was no necessity for two guards, and he did not think side chains would have helped. He was challenged as to his opinion but stuck to it.[725]

At the resumption of the adjourned inquest Traffic Manager David Kirkcaldie was asked about the lack of an assistant guard on the train and the non-use of side chains. He replied that assistant guards were only employed to reduce delays and for the exceptional conditions of the Penrith-Katoomba line, they were certainly not required on train No. 26. In answer to other questions Traffic Manager Kirkcaldie said that as regards to side chains they would not have held the runaways on the 1 in 48 gradient at Raglan. Stronger draw gear was being introduced as quickly as possible. There was, he thought, not a railway company in the world that had continuous brakes on all its goods stock. It would cost £200,000 to supply automatic brakes to all the trains of the colony.

Going back to the subject of side chains Mr Kirkcaldie said that if they had been hooked up on the train involved in the accident, he did not suppose for a moment that they would have held. Side chains had been dispensed with in England 18 years previously. He attached no blame whatever to the drivers or guards. He acknowledged that it was dangerous for passengers to ride in goods trains, but it had been done for so long that there would be great complaints were it stopped.

New South Wales Engineer William Thow also gave evidence at the inquest and stated that 430 or 450 drawbars had been strengthened and the work was being done as quickly as possible, but all the alterations could not be effected in a day. To withdraw all the trucks which required attention would paralyse traffic completely.

The jury at the inquest on one of the victims, Mrs Rebecca Franklin, returned a verdict recommending the use of side chains until all the rolling stock in the colony had been fitted with stronger gear and the re-introduction of a second guard.[726]

The Railway Commissioner's inquiry rejected the evidence of the Raglan stationmaster and others that there had been no recoil when the train had stopped there and instead stated that there must have been a violent recoil sufficient to break the draw-bar when the driver applied the brake on getting a red light from the guard. They doubted

[724] *Newcastle Morning Herald and Miner's Advocate*, Newcastle, New South Wales, 29th April 1890
[725] *Evening News*, Sydney, New South Wales, 29th April 1890
[726] *The Sydney Mail and New South Wales Advertiser*, Sydney, New South Wales, 3rd May 1890

if either side chains or an additional guard would have prevented the runaway, nor was the use of sprags or the application of the brake in the guard's van likely to be effective in stopping the runaway on the steep incline. Sounding very much as if they were contemporary inspectors for the British Board of Trade, they stated that stations and sidings should not be provided on lines steeper than 1 in 280 unless runaway catch points could be installed.[727]

The Mount Penn Gravity Railroad was opened in March 1890 to provide transportation to the top of the mountain outside of Reading, Pennsylvania, United States. It provided sight-seekers and tourists with spectacular views of the mountainside as they were hauled three miles to its summit by a steam locomotive. Their carriage was then released to descend by gravity for five miles to the starting point by a different route while the steam locomotive returned the way it had come. The railroad had only been running for five months when there was a bad accident. On 22nd August 1890 car No. 9 ran away, left the track on the 'Hairpin Bend' and fell over a 70ft embankment, killing five people and injuring 20. One of those killed was the conductor, Charles Rettew; the others were all passengers.

Car No. 9 had an air brake and a hand brake, situated next to one another. One theory as to how the accident occurred was that the vacuum chamber was partially exhausted so that the power simply pressed the brake blocks against the wheels with insufficient force to stop them, while at the same time destroying the leverage needed to apply the hand brakes. Following the accident, the company provided two brakemen with each car and also equipped the cars with for cone-shaped iron wedges, accessible by trap doors through the floor, acting directly on the brake shoes. Another accident occurred nevertheless in 1891, killing a conductor and a brakeman and attributed to wet leaves.[728] [729]

Facit station, England, on the Lancashire and Yorkshire Railway branch from Rochdale to Bacup, was the scene of the next serious runaway accident on 29th August 1891. The 7.25 a.m. up passenger train from Bacup to Rochdale was waiting at the Facit up starting signal for a down passenger train from Rochdale to clear the single line when it was run into from behind by a stone train, the driver of which had lost control of it on the descent from just beyond Bacup station. The engine of the runaway train was badly damaged in the collision and a composite carriage and a third-class of the passenger train were completely smashed up; other rolling stock was less badly damaged. Seven passengers were injured, three of whom later died. Two of the dead were one of two sisters on their way to Liverpool and thence to the Isle of Man for a holiday and a 60-year-old missionary for the New and Latter House of Israel, residing in Lincoln.[730]

The Bacup branch fell very sharply for 1¾ miles from a summit just on the Rochdale side of Britannia station, the gradients including half mile stretches at 1 in 39 and 1 in 42 and even a few chains at 1 in 36, and there was a written order that no goods or passenger

---

[727] *The Daily Telegraph*, Sydney, New South Wales, 15th May 1890
[728] Webpage – Pagoda Skyline Inc. – The Gravity Railroad
[729] *The Marion Star*, Marion, Ohio, United States, 23rd August 1890
[730] *The Times*, London, England, 31st August 1891

train should travel at more than 10 miles an hour over this stretch, that no goods train must exceed 30 waggons and that the guards of goods trains must be careful to pin down a sufficient number of brakes.[731]

Goods Guard Benjamin Ferriday told the Board of Trade inquiry into the accident that he had started from Britannia at 7.40 a.m. with a train of 25 waggons, one empty next to the engine and 24 loaded with stones. Before starting his driver, William Barratt, had told him that the engine brake was not to be depended on, and he had replied that they would have to pin down sufficient waggon brakes, and gave the driver a brake stick so that he could assist in doing so. The train did not stop at the summit, but proceeded at a slow walking pace. He had pinned down seven waggons on the six-foot side and considered that with those that his driver had pinned down that would be sufficient. He had not objected to his driver not stopping dead at the summit. Nothing was done to increase the train's speed, but it seemed to accelerate as soon as the whole of it was over the summit, reaching 20 miles an hour between Shawforth and Facit. As soon as the speed had increased, he had become alarmed and feared they would not be able to stop at the Shawforth signals. He had got back into his van, screwed down the brake as hard as he could and applied sand, and could do no more.

Driver William Barratt said that he had been in charge of a bogie tank engine, No. 226. He had left Britannia at about 6.35 a.m. with one empty and 24 loaded waggons. He knew the gradients of the branch and that they would have to pin down brakes at the summit, and that he ought properly to have stopped completely to do so, but on the day of the accident they had been just creeping along. He had pinned some waggons down and when he returned to the footplate there had still been some waggons on the Bacup side of the summit. Steam had been shut off before this, and the hand brake was applied, and sand put down. The train's speed had increased to 15 miles an hour before they reached Shawforth and he had realised that they were running away, had applied and released the engine brake several times, reversed the engine and given the alarm whistle. There was a drizzling rain and the rails had been greasy. The highest speed they had reached was about 20 miles an hour at the Shawforth down distant signal.[732]

Signalman John Dawe at Shawforth said that he had had his signals against the stone train as Facit Station cabin had not cleared out the passenger train, but the stone train had passed him at 18 to 20 miles an hour, running past his signals and its engine giving alarm whistles. He had at once told Facit Station cabin on the telephone that the stone train was running away and then sent the 4-5-5 'Train running away on right line' bell signal. He had never known a train to run away on the bank before.

Signalman Charles Mansfield at Facit Station cabin said that as soon as he heard from Shawforth that the stone train was running away he had shouted to the stationmaster, Robert Jackson, that the pilot was running away. The stationmaster had not understood what he said the first time and he had to repeat it. He had also sent the 'train running away' bell signal on to Facit Goods Yard cabin and advised the signalman there by phone.

Stationmaster Jackson said that as soon as he had comprehended what Signalman Mansfield was trying to tell him he had run along the platform and told the driver of the

[731] Board of Trade report
[732] Ibid.

up-passenger train to cautiously draw forward as the stone train was running away. He had not ordered the passengers out of the train as he had hoped to get it away and avoid a collision. The down passenger train did actually clear the single line shortly before the collision, but the driver of the up-passenger train had only had time to just about get his train moving when it was run into.

Major-General C.S. Hutchinson found that the accident was the result of a serious error of judgement on the part of Guard Ferriday in not pinning down sufficient brakes and that given the nature of the engine, the weight of the train and the greasy rails a more judicious guard would have pinned down 19 or 20. Driver Barratt had only once before worked a stone train over the line and on that occasion had had a first-class tender engine. He could not be blamed for insufficient brakes having been pinned down as he was acting under the direction of his guard but he should not have left his engine in charge of the fireman (who was actually only a passed cleaner) to help with the pinning down and it was doubtless because the engine was in charge of the fireman that through his inexperience the train speed was allowed to climb too high. The signalmen at Shawforth and the two Facit cabins were to be commended for their promptitude in giving warning of the runaway. Stationmaster Jackson on the other hand would have displayed more judgement had he ordered the passengers out of the train on learning of the runaway.[733]

As to recommendations the inspecting officer stated that only engines with the most brake power and the most experienced drivers should be used to work stone trains over the branch and that trains of more than 20 waggons should be supplied with two brake vans and two guards.

At 4 a.m. on the morning of Friday 18th September 1891 Union Pacific Railroad eastbound passenger train No. 2 was approaching Port Neuf, Idaho, United States, six miles beyond Pocatello, where it had received orders to cross westbound passenger train No. 1, which being westbound had the right to the main track. Half a mile outside of Port Neuf Engineer Charles Hoyt went to apply his air brakes, having seen that train No. 1 was just pulling into the station. He found to his consternation that they would not work and immediately whistled for brakes. Conductor Hogan 'tugged in vain at the air valve in the car', and the next moment the trains collided head-on, the points at the west end of the station not having been reversed to send train No. 2 into the loop. Both drivers and both firemen jumped before the collision, and Engineer Hoyt was the only one of the four who was hurt. Others were less fortunate, for Mail Clerk L. Weidermyer and four Indians from Fort Hall reservation were killed, and a man named Lyons, Mail Clerk P.P. Gunderson, Christopher MacKenzie, a squaw and two tramps were injured. Lyons was badly scalded and received internal injuries, but it was hoped that he might pull through. After the collision engines and mail cars were piled high in the air and the groans of the injured filled the air. The status of the Indians killed is not clear, for the report stated that none of the passengers had been injured, implying that they were 'tramps', though it did not explicitly say so. An inquest was held on the body of Mail Clerk Weidemeyer the next day and decided that no blame would be attached to the railroad company as the

733 Ibid.

brakes had been functioning correctly when the train left Pocatello.[734] The cause of the apparent brake failure was not stated. Did perhaps one of the tramps accidentally close off the brake pipe, preventing the air being exhausted from the engine?

On 16th September 1892 as a work train consisting of an engine and four waggons loaded with earth and stones was descending a steep incline, which included a sharp curve, on the railway between Alcanena and Torres Novas in Portugal it suffered the reported failure of its brakes near the beginning of the descent. The train soon became uncontrollable and rushed down the bank at an ever increasing speed until eventually it left the rails, its final speed being estimated by survivors as over 60 miles an hour. The engine and waggons were stated to have been completely wrecked and the track much damaged. Riding on the waggons were a ganger and 16 labourers, and several men ended up buried beneath the overturned trucks and their contents. Six people were stated to have been killed outright and most of the remainder injured, most of them seriously, while the final death toll was nine.[735 736]

The Ferrocarril Nacional de México suffered a runaway accident on the afternoon of Sunday 13th November 1892. Where the railway passed through the Sierra Madre mountains near Saltillo, Mexico, the gradients were very steep and freight trains were routinely double-headed. On the day in question such a train was climbing the incline 12 miles south of Saltillo when a coupling snapped, and the caboose and six waggons loaded with 3,000 bushels of corn separated from the front of the train. Riding in the caboose were Conductor James Martin and Brakeman Frank Harwood. They attempted to set the brakes but were powerless to check the momentum of the runaway trucks, the speed of which was said to have reached 90 miles an hour before most of them left the rails, killing both men. The caboose and one truck continued their wild flight and passed through Saltillo yards, reportedly at 120 miles an hour. The train dispatcher there knew that another freight train was coming up the hill, and was at that moment 20 miles below Saltillo, but as there was no telegraph office within that distance, he could do nothing but order out the breakdown train. Engineer Frank Davis on the leading locomotive of the second train saw the runaways coming but did not have time to save himself and was buried beneath a pile of corn and wreckage; both engines being demolished. His fireman and the engineer and fireman of the train engine jumped off and escaped with slight injuries. The reports implied, though did not actually state, that Engineer Davis lost his life.[737 738]

Train No. 12 of the Cleveland Cincinnati Chicago and St Louis (Big Four) Railroad, the overnight vestibule express from Chicago to Indianapolis, was wrecked at Lafayette, Indiana, United States, at 1.14 a.m. on the morning of Sunday 7th May 1893. It had left

---

734 *Brooklyn Daily Eagle*, Brooklyn, New York, United States, 19th September 1891, reproduced from the *Idaho Statesman*, Pocatello

735 *The Times*, London, England, 17th September 1892

736 Wikipedia – Accidentes ferroviarios en Portugal

737 *Dubuque Evening Times*, Dubuque, Iowa, United States, 18th November 1892

738 *Winnipeg Free Press*, Winnipeg, Manitoba, Canada, 18th November 1892

Chicago 10 minutes late on Saturday night and consisted of an engine and tender, two postal cars, a baggage car, a combination car, smoking car, ladies' car, chair car and two Wagner sleeping cars in charge of Conductor Tyner with Engineer Michael D. Welsh and Fireman Sterling McInnis. The sleeping cars were well filled, there were six passengers in the chair car, a number of men women and children in the ladies' car and two men in the smoker. No problem was found in making the relatively few stops timetabled before Lafayette, the last of which was at Templeton, where several tramps riding 'blind baggage' (at the end of the baggage car next to the tender where there was no door) were thrown off the train. The brakes were working perfectly at this point. Approaching Lafayette from Templeton the line fell steeply for four miles and close to the station there were bridges over the Wabash River and a canal, followed by a sharp left-hand curve and then a right-hand curve.[739]

As the train neared Lafayette the surviving train men heard the usual whistle for the station, followed by a wild shriek for brakes. Conductor Tyner opened the brake valve in the ladies' car, but the brakes did not respond. The brakeman in the rear sleeping car applied the hand brake there and was hurrying through to set the brake in the next car when the crash came. The crews of a freight train and a yard engine shunted out of the way of the express at the station watched it coming in, clearly out of control, with its engine reversed. At the canal bridge Engineer Welsh could be seen to climb down to the step of his engine as if ready to jump but after hesitating he climbed back into the cab. Travelling at 35 miles an hour the engine then left the track on the reverse curve, demolished some sheds adjoining the line, collided with a Lake Erie and Western box car and finally overturned on its side. The two postal cars, baggage car and combination car piled up on top of the engine and the first three vehicles were badly damaged. Four times the wreckage caught fire, but each time the blaze was extinguished before much damage had been done.

Engineer Welsh and Fireman McInnis died in the wreck, together with Postal Clerks E.D. Myers, A. R. Chadwick and Jesse Long; Express Messenger Thomas M'Mahan; and Otto Gesselson of Montreal, a theatrical agent; and Charles Cahill of Lafayette, a moulder, these two being passengers who had been waiting at Lafayette to board the train. Also killed by the crashing train while they were awaiting its arrival at the station were mail carrier John Leman and bus driver Charles Meyers, making 10 fatalities in all. About 10 other people were injured.

There were several theories as to why the brakes had failed. One was that tramps had closed the stop cock behind the engine, but in this case the brakes should have been applicable from further back the train. Another was that Engineer Welsh had accidentally exhausted the air and not discovered his error until it was too late to replenish it. Or some obstruction thrown up from the track might have blocked the pipe. Yet another theory put forward by an old railroad employee was that the brakes had not failed at all, but that Engineer Welsh had fallen asleep for a minute or two at the crucial point in the journey at the top of the bank, and that when he awoke and belatedly applied the brakes it was too late.[740]

---

[739] *The Indianapolis News*, Indianapolis, Indiana, United States, 8th May 1893
[740] Ibid.

The short narrow-gauge line from Lezama to Bilbao in Spain suffered a disastrous runaway accident on the morning of Saturday 7th July 1894. A train consisting of an engine and two carriages left Lezama at 8.30 a.m. and at Dean Roque three waggons, heavily laden with stone and bricks, were attached. According to a survivor of the accident the train was then carrying 39 passengers and six railwaymen (other reports stated that there were only 38 people on board the train in all). The train left Dean Roque and proceeded without incident until it emerged from Zurbaran Tunnel, where the line traversed a curve on a steep embankment. Here the driver realised that the train was getting out of his control as it descended a steep incline and he whistled for brakes, which were applied, but were insufficient to check the train. The driver told his fireman to jump and save his life, but the latter decided it was safer to remain on the engine and escaped unhurt. The driver was badly injured and subsequently died. Also riding on the engine was a contractor who in attempting to jump caught his foot in a wheel and was dragged along the line and badly mutilated. One woman, seeing the danger, threw her baby out of a window. The infant survived, its mother was amongst the dead. The train then left the track and both carriages were wrecked. All seven passengers in the leading carriage lost their lives and five in the second. Nineteen people were injured, eight seriously and about a dozen escaped unhurt or with slight injuries. Most of the passengers were peasants going to Bilbao.

After the accident help came from the peasants living in the cottages and farms in the vicinity. 'One poor villager offered all the bedding, sheets and linen in her cottage for an ambulance when the dead bodies, six men and six women, were extricated from the debris.' A relief train was sent from Bilbao, just a kilometre away, with railway officials and medical staff of the Red Cross Society. The line had only been opened the previous May.[741] [742] [743]

An 1895 runaway train accident in Paris, France, was hardly a major disaster, resulting as it did in only one fatality, but a photograph of the wrecked engine taken afterwards has become so iconic that a brief account of it seems appropriate. The photograph is reproduced on the cover of this book.

On 22nd October 1895 the Chemins de fer de l'Ouest express from Granville to Paris Montparnasse departed from Granville on time. It consisted of 2-4-0 engine No. 721, three luggage vans, a postal van and six carriages. By the time that it reached Paris it was running several minutes late, and it entered the terminus at too high a speed, reportedly 40-60 kilometres an hour. Failing to stop in time the engine rammed the buffers and the momentum of the train carried it across the station concourse, which was almost 30 metres wide, and through the end wall beyond. It then fell 10 metres into the Place de Rennes below, the locomotive ending up on its nose. A woman named Marie-Augustine Aguilard, who was selling newspapers while her husband was away, was unfortunately killed by falling masonry. Two of the 131 passengers on the train, the fireman, two guards and a passer-by in the street were injured. The carriages and tender were drawn back into the station and an effort to move the engine with 14 horses having failed it was

741 *The Times*, London, England, 9th July 1894
742 *The Standard*, London, England, 9th July 1894
743 Webpage – Anexo: Accidentes Ferroviarios en España

eventually lowered to the ground with a 250 tonne winch and 10 men.

Driver Guillaume-Marie Pellerin was fined 50 francs for entering the station too fast and Conductor Albert Mariette 25 francs for not applying the handbrake because he was preoccupied with paperwork.[744] [745]

The Caribbean island of Jamaica suffered a major disaster on its branch from Bog Walk to Port Antonio after the line had only been opened a few months. From its junction with the main line at Bog Walk the branch climbed for 18 miles at a ruling gradient of 1 in 173 to Highgate, from whence it fell at a ruling gradient of 1 in 63 for 9 miles to sea level at Wag Water Bridge before continuing more or less on the level for the remaining 27 miles to Port Antonio.[746] The runaway train involved in the accident of 11th July 1896 must have begun to get out of control somewhere on the descent from Highgate to Wag Water Bridge. It reached the bottom of the hill without accident but evidently was still out of control and some miles further on, on the level stretch, it collided with an engine near Annotto Bay station. The train consisted of six flat cars carrying male and female labourers, 17 of whom were killed and 22 injured. The air brakes of its locomotive had allegedly failed.[747]

The Columbia and Donegal Electric Railway of Pennsylvania, United States, bore similarities to the Mount Penn Gravity Railroad mentioned earlier. It had opened in 1893 and climbed from Columbia at a ruling gradient of 6% (or about 1 in 17) up Chickies Ridge to an amusement park on the top of the ridge overlooking the Susquehanna River.[748] Unlike the Mount Penn line it was worked by electric traction, but like the Mount Penn line, it had a bad runaway accident, in this case on 9th August 1896.

The accident occurred at about 10 p.m. on a Sunday evening. A crowded car, with about 90 people on board, was descending from the summit when its brakes failed to work, and its speed gradually increased until it left the rails on a bend, crossed a turnpike road, struck a tree and fell down a steep embankment. Seven people were killed outright or died from their injuries and about 50 were hurt. The dead included the motorman, Adam Foehlinger.[749] [750] [751]

A fortnight or so later a coroner's jury found the railway company to be largely to blame for the accident for permitting the cars to become overcrowded, employing inexperienced hands and not properly inspecting the cars before use.[752]

Another runaway accident happened in Pennsylvania, United States, on 30th

---

744 Wikipedia – Montparnasse derailment
745 Webpage – The Infamous Montparnasse Derailment
746 Wikipedia – Bog Walk to Port Antonio railway
747 *Evening Bulletin*, Decatur, Illinois, United States, 21st July 1896
748 Wikipedia – Chickies Ridge
749 *North Adams Transcript*, North Adams, Massachusetts, United States, 10th August 1896
750 *Gettysburg Compiler*, Gettysburg, Pennsylvania, United States, 11th August
751 *Fort Wayne News*, Fort Wayne, Indiana, United States, 10th August 1896
752 *Gettysburg Compiler*, Gettysburg, Pennsylvania, United States, 25th August 1896

September 1896 but this time it was on the Baltimore and Ohio Railroad at Philson. Fast eastbound freight train No. 74 in charge of Engineer John Seesa and Conductor William Pickless was scheduled to meet westbound freight No. 95 at Philson at about 10.30 p.m. Train No. 95 arrived first and stopped on the main track. Train No. 74 passed Meyersdale on time and stopped at Sand Patch for inspection, at the summit of the Sand Patch grade, the steepest and longest on the Baltimore and Ohio system with 18 miles at a ruling gradient of about 1 in 50. The operator at the east end of the 1,592-yard-long Sand Patch tunnel said that the train had been running at a terrific speed when it had passed him, so its crew must have lost control of it shortly after reaching the start of the descent.[753] They decided to take their chances staying on the runaway rather than jumping. The two trains met in a violent collision which wrecked both engines, all the cars of the eastbound train and most of them of the westbound, the wreckage being piled up as high as the telegraph poles.[754] Both engineers and both firemen were injured in the collision and possibly six tramps were killed. A dozen or more tramps had been riding on the train, some of them returning to their homes in Myersdale, Pennsylvania, after having gone to listen to William Jennings Bryan, the orator and politician who would be involved in the Emporia collision the following year. It was thought that some of the tramps riding on the buffers might have inadvertently caused the train to run away by blocking the brake pipe.[755]

Reports of the accident were even more conflicting than such reports usually were. Some gave the surname of the engineer of train No. 74 as Reane, some stated that all both engineers and firemen had been killed in the collision, the time of the accident was variously given as 10.30 p.m. on 30th September and 12.15 a.m. on 1st October and so on.

On 18th May 1897 a serious accident occurred near Vega Alta in Santa Clara province, Cuba. A freight train of 28 waggons heading for Placetas was ascending an incline when unbeknown to its driver the four rear waggons became detached. Running back down the hill they collided with a following passenger train on a curve. The engine of the passenger train mounted the runaway vehicles in the collision, its boiler exploded, and fire broke out as well. The engine driver, a policeman and five passengers were killed, and 27 passengers were injured.[756]

An accident which was almost a repetition of that at Philson, Pennsylvania occurred at American Falls, Idaho, United States on 27th May 1897. A westbound passenger train on the Oregon Short Line Railroad was stopped in front of the station buildings at American Falls awaiting a meet with an eastbound freight train at about 4.30 a.m. The freight train, which was carrying mostly timber, ran away coming down the incline leading to the station and crashed head-on into the waiting passenger. The runaway train was reportedly travelling at 50 miles an hour when it struck the engine of the passenger train, which was reversing to try and get out of the way. Both engines were severely damaged, the station buildings were wrecked, and 20 freight cars were piled up in a heap. Eight tramps riding in a box car, who had been working as sheep shearers and were beating their way back

753 *Fort Wayne News*, Fort Wayne, Indiana, United States, 1st October 1896
754 *Fort Wayne Gazette*, Fort Wayne, Indiana, United States, 2nd October 1896
755 *Tyrone Daily Herald*, Tyrone, Pennsylvania, United States, 2nd October 1896
756 *The Daily Herald*, Delphos, Ohio, United States, 19th May 1897

east, were killed. A man named John Cooper who had been standing on the station platform and was about to board the passenger train was also killed and another passenger probably fatally injured. Cooper had made the fateful decision to run into the station building when he saw the runaway freight train and was killed there. Engineer George Moore of the freight train was seriously injured and his fireman, Dick Cosgrave, who had climbed up onto the freight cars to help Conductor Cook and brakemen Davis and Westron set the brakes, suffered a broken leg. Engineer E.C. Heckman of the passenger train was slightly hurt, and several other tramps were injured. It was thought that the air brakes had been tampered with. Engineer Heckman probably saved the lives of his passengers and of the family of the station agent by sticking to his post and reversing his engine. Communication having been broken by the accident bringing down the telegraph wires a man set off on horseback to Pocatello, only to find when he got there that communication had been restored and a relief train was about to set off for the scene.[757] [758]

At 6.30 a.m. on the morning of Wednesday 24th November 1897 a mixed train from Montréjeau came to a stand in Tournay station,[759] France, on the Chemins de fer du Midi. It had been stopped there three minutes when it was run into from behind by a following train carrying workmen, which had got out of control coming down the 33/1000 (thus about 1 in 30) Capvern bank and was unable to pull up. The rails were slippery, and it was foggy. The mixed train was marshalled with its passenger carriages at the rear and these were crushed in the collision. Fourteen people were killed and eight seriously injured. The casualties were mostly in the colliding train.[760]

A runaway train accident in Italy on 11th August 1898 had its origin in the collapse of an engine crew overcome by fumes in a tunnel and is described in the Tunnel Accidents chapter.

A major disaster occurred in the Cape Colony on 16th August 1898 though it was only briefly reported in *The Times*, presumably because most of those killed and injured were only 'natives'.

The Cape Government Railway line from Cape Town to Matjesfontein had very severe gradients east of Matjesfontein where to avoid difficult engineering works it took a course which instead involved gradients of 1 in 40 and 1 in 60 on the climb from Bavians (now Baviaan) to Mostertshoek. On the evening of Tuesday 16th August, No. 48 up goods train in charge of Guard Thomas Marsh was brought to a stand north of the crossing loop at Mostertshoek and on a gradient falling at 1 in 60 towards Bavians at 6.37 p.m. It was really a mixed train, for in addition to 13 goods waggons there was a bogie carriage at its rear, which had been attached at Fraserburg Road and was carrying 48 black passengers, this probably being the only train they were permitted to travel in. The

---

757 *Idaho Statesman*, Boise City, Idaho, United States, 28th May 1897

758 *Fort Wayne Gazette*, Fort Wayne, Indiana, United States, 28th May 1897

759 *The Times*, London, England, 25th November 1897

760 Wikipedia – Liste des accidents ferroviaires en France au XIX siècle

train was not fitted with continuous automatic brakes. Guard Thomas then ordered his driver, James Simpson, to detach his engine and pick up a wagonload of bricks from the siding at the station. Driver Simpson stated later that he had objected, telling the guard that it was not safe to begin shunting there with the bulk of the train left on a falling gradient and with fast passenger and mail train No. 4 following them. Driver Simpson also said as much to Station Foreman Albert Milner but nevertheless eased his engine back so that the guard could uncouple the waggons behind it, telling the foreman that he, the driver, would not be responsible if there was an accident. Guard Marsh later denied having heard his driver make any objection to the proposed shunt move and said that he had screwed down the chain-brake in his van and pinned down the brakes of seven waggons. He was probably anxious to complete the shunt move quickly to avoid delaying the passenger train or alternatively delaying his own train by postponing the shunting until after the passenger train had gone and meanwhile putting his train into the loop.[761]

No sooner had the waggons been uncoupled than they began to run away down the hill. Foreman Milner grabbed hold of a moving truck but was thrown off after a few yards, Guard Marsh tried to pin down more brakes and climbed aboard a waggon but was likewise thrown off after three-quarters of a mile. Driver Simpson followed the runaways with his engine, whistling continuously.

Driver Martin Burgin and Fireman Samuel Brooks on the pilot engine of the double-headed passenger train apparently did not see the runaways before the collision but Driver Joseph Anderson on the train engine did and called for his fireman to close the regulator and apply the brake. Driver Burgin and Fireman Brooks were only slightly injured in the accident and the crew of the train engine not at all, but its results elsewhere were disastrous, especially in the carriage attached to the goods train, which mounted the pilot engine and was set on fire by coals from its fire-box. Twenty-four black passengers were killed and others doubtless injured. There were fewer casualties in the passenger train, but six died there. The newly built travelling Post Office van behind the second engine wrecked three compartments of the following Nederlandsche Zuid Afrikaansche Spoorweg Maatschappij carriage and the next carriage was also damaged, fatalities occurring in both. The 'Europeans' killed were the Rev and Mrs du Toit and their little girl; and Messrs De Villiers, Cope and Tait, the latter three described as footballers. The names of the black victims were not stated. Two Post Office employees were badly hurt while a third, who had been asleep, escaped. None of the passengers in the rear coaches of the train received more than a shaking. All the runaway waggons were wrecked. The accident occurred at about 6.50 p.m. at a point 206½ miles from Cape Town.

The accident was the subject of a departmental inquiry and an inquest. The Resident Magistrate at Worcester was instructed to proceed with a charge of culpable homicide against Guard Thomas, but the Attorney General declined to prosecute.

The Cape Government Railway Traffic Manager, Mr Price, commented that, "This serious accident would have been avoided by the exercise of ordinary care; nor is the feeling of anxiety lessened by the circumstances that most of the serious accidents to passenger trains of recent years have occurred in one section of the Cape Railway system."

The Cape Government Railway General Manager Charles Elliott wrote that, "I am

[761] Webpages – Mostertshoek Accident Parts 1 and 2

glad to say that the arrangements which I authorised some years ago for attaching the automatic brake to all carriages and rolling stock were sufficiently advanced to enable the automatic brake to be applied to all trains within a few days after the accident so that no further apprehension may be entertained with regard to the safety of the trains arising from a similar cause." This statement rather begs the question of why in that case had automatic brakes not been used before.

In his 1898 report the General Manager stated, clearly taking a very lenient view of the matter, that although the railwaymen had not fully complied with the regulations, they had believed that the steps that they had taken had secured the train from running back.[762] [763]

[762] Ibid.
[763] *The Times*, London, England, 18th August 1898

CHAPTER 8

BRIDGE FAILURES

Some of the most serious railway accidents of the 19th century resulted from the failure of bridges. Bridges may fail for a variety of reasons. The inherent design of the structure may be faulty, the materials used in its construction may be poor and inferior, the materials may deteriorate over time or its foundations may be washed by a stream or river which has become exceptionally swollen following abnormally heavy rainfall. Often a variety of factors could play a part, as was evident from the notorious Tay Bridge disaster, described in the previous volume dealing with 1811-1880 accidents.

A westbound night express train on the Rock Island division of the Chicago Milwaukee and St Paul Railroad was crossing a trestle-work bridge over the Meredosia River, a mile south of Albany, Illinois, United States, at 5 a.m. on the morning of Thursday 21st April 1881 when the structure gave way without warning. The train was travelling at its normal speed. It consisted of an engine and tender, baggage car, passenger car and sleeping car, and all but the sleeper went down, the latter ending up hanging at an angle of 45 degrees from the bridge with one end high up in the air. The baggage and passenger cars floated off down the stream, the passenger car lodging against an island 40 rods downstream and the baggage car sticking in the middle of the stream at about the same point. Both cars were submerged up to their roofs. Of the 11 passengers in the car which was carried away two men escaped by jumping onto the abutment of a waggon bridge which had been swept away and a woman and one of her two children and a boy from Pine River (also given as Blue River), Wisconsin, were rescued from the roof of the coach after it had lodged against the island. Six other passengers, including the woman's other child, the father of the boy from Pine River and Dr. L. Lunsy of Albany were drowned. Engineer West and his fireman went down with their locomotive and were also drowned, but Conductor Thomas Fuller, Baggageman Sam Hanagin, Brakeman Harry Myers and Express Messenger Dan Ellithorpe escaped, all except the conductor being unhurt. The three passengers and the coloured porter in the sleeping car also climbed out unhurt. The injured passengers were made comfortable in a hotel, the woman who had lost one of her children being prostrated by shock.

A freight train had crossed the bridge an hour or so before the passenger train, but the river was said to be in perfect torrent and had clearly washed away the foundations of the bridge and thus caused the accident. It was still pouring with rain as 20 rescue boats searched for bodies and recovered some of the baggage, holes having been made in the roof of the baggage car. The shore of the stream was lined for eight miles by people searching for souvenirs of the wreck.[764]

The narrow-gauge Morelos Railway opened from Mexico City to Cuautla, Mexico opened to the public on 18th May 1881 and to celebrate the occasion the President of Mexico and other high officials made the journey over the line, accompanied by 300 soldiers. About 100 of the soldiers returned to Mexico City on the 20th and the remainder, a proportion of whom were accompanied by their wives, returned several days later, either on the 23rd or 24th (sources vary as to the date of the accident). The train carrying the remainder of the soldiers and their families consisted of a passenger car for the officers, five wooden boxcars for the soldiers and their wives, two more wooden boxcars for the freight, which included 80 or 100 barrels of brandy, and an engine at each end of the train.[765]

There had been a severe storm along the line with heavy rain. By the time that the train reached a possibly temporary stone bridge across a ravine near Malpas it was pitch dark and the driver was anxious to stop his train and check that the bridge was safe before attempting to cross it. An officer riding on the was fearful, however, that some of the soldiers might desert were the train stopped and brandishing his pistol forced the driver to continue without stopping. The fears of the driver proved to be well founded, for the bridge, the supports of which had been undermined by the swollen waters of the San Antonio River, collapsed beneath the train. As the train fell from the bridge burning coals from the rear engine set fire to the brandy in the boxcars. The majority of those on the train were either crushed in the fall, drowned or burnt to death. In all, 13 officers and 195 privates of the 3rd Battalion of Infantry, 37 soldiers' wives and five children were stated to have lost their lives, as did the driver and fireman of the rear engine, and the total number of dead was given as 256. Forty others were injured and only 20 escaped unhurt. The driver of the leading engine and the officer who had threatened him both survived and were both arrested.[766] [767] An investigation announced on 30th June that, 'the actual and sole cause of the disaster was the very bad construction of the bridge.' [768] The rash action of the officer which had really caused the disaster had been quietly forgotten.

Abnormally heavy rain on the afternoon and evening of 6th July 1881 meant that an engine, Old No. 12, was sent out from the Chicago and North Western Railroad's Moingona depot to check the safety of the line and especially of a bridge across the Des Moines River. The engine crossed the bridge safely but further east the foundations of a

---

[764] *The Daily Ledger Standard*, New Albany, Indiana, United States, 22nd April 1881
[765] Wikipedia – Morelos Railway Accident
[766] *The Times*, London, England, 27th, 28th and 30th June 1881
[767] *The Mid-Surrey Times and General Advertiser*, London, England, 2nd July 1881
[768] Wikipedia – Morelos Railway Accident

bridge over Honey Creek had been washed away by a flash flood and when the engine reached it at 11 p.m. the bridge collapsed beneath it. Seventeen-year-old Kate Shelley lived with her widowed mother near the bridge and they heard the crash as it gave way. They also knew that at midnight an eastbound express was due which had to be stopped. With no means of communication with the outside world at their home the only thing to be done was to get to Moingana before the express left there. Kate knew that she was the only person who could do so, and she set off in the storm wearing an old skirt and a jacket and carrying a lantern. Going first to the Honey Creek trestle she found two of the four men who had been riding on the engine, Ed Wood and Adam Agar, clinging to the branches of a tree, but realised that she could not help them herself, and made for Moingana. En route she had to cross the Des Moines River bridge, not an easy task even in fine weather and daylight, especially as the railroad had removed some of its flooring to discourage trespassers. After almost falling into the river near the start she realised that she could only crawl across it on her hands and knees, added to which the continued rain soon put out her lantern, leaving her in darkness except for flashes of lightning. Eventually she reached the far side and ran on to Moingana, only to hear someone remark when she got there that, "This girl is crazy," whereupon she fainted. Fortunately the station agent realised who she was and that the express, which had 200 passengers on board, must be stopped. When she shortly came round, she insisted on accompanying the rescue party back to Honey Creek. There a rope was thrown to Wood and he was rescued, as eventually was Agar, after the floodwaters had receded. The other two men who had been on the engine, George Olmstead and Patrick Donohue, were drowned.

When news of her exploit became known Kate was awarded various sums of money, medals and a half barrel of flour, a half load of coal and a lifetime pass from the railroad company. She later became a teacher, and then the station agent at Moingana until her death in 1912. A new bridge over the Des Moines River, opened in 1900, was named after her.[769]

On 8th September 1881 a freight train fell through a fire-damaged bridge at Silver Creek, Pennsylvania, United States. The engineer and fireman were killed, and a brakeman was injured.[770]

The St Charles bridge across the Missouri River in Missouri, United States, which had collapsed in November 1879, did so again on 8th December 1881 on what had now become the Wabash St Louis and Pacific Railroad. It gave way at about 6 p.m. under eastbound freight train No. 12, consisting of 32 cars, 13 loaded with livestock and the remainder carrying general freight. It crossed the bridge just before the Warrenton accommodation train, which it was scheduled to meet at Brotherton. The whole train went down with the bridge as did three of the seven people on it.

Engineer Jack Kirksby was still missing at the time of the newspaper report, presumed drowned, but Fireman N. Chamberlain and Brakeman Charles Oblinger were rescued from the river. Conductor Andrew Durfield, Brakeman George Midcap and stockmen Anderson and Crane managed to jump off the train before it went down. Not all the

---

769 Webpage – The Kate Shelley Story

770 *Oil City Derrick*, Oil City, Pennsylvania, United States, 9th September 1881

livestock was killed outright and 'at midnight the squealing of the imprisoned swine and the hoarse bellowing of the cattle could be heard up town a mile away from the wreck.'

This time it was the easternmost span which had failed. The state railroad commissioners had reported the bridge safe only a month before. Mr C. Shaler Smith (his surname was spelt Shalersmith in the report on the 1879 accident) who had also been responsible for the second bridge, said that he could not imagine what had caused the accident and that he had been wholly taken by surprise when he had heard of it. Mr Smith's assistant, Mr Edward Bates, likewise said that he had been incredulous when he had heard of the accident.[771]

A major disaster occurred in Croatia, at the time a part of the Austro-Hungarian empire, on Friday 22nd September 1882. The location of the accident was a wooden bridge over the River Drave at Essegg. The river was in full flood when one of its piers collapsed beneath a passenger train. The engine, four baggage waggons and two passenger carriages fell into the river and had a coupling not given way the remaining passenger carriages would have followed. As it was, they remained standing on the undamaged portion of the bridge and the passengers in them escaped with no more than a fright. The carriages which fell into the river had been filled with soldiers, Hussars returning from Bosnia who were going home on furlough. A roll call taken after the rescue of the last survivor showed that 27 men were missing, drowned in the torrent. Two of the survivors were seriously and 16 slightly injured. The fate of the driver and fireman was not reported but very likely they also lost their lives. Not a vestige of the pier which had collapsed remained. As in the second St Charles accident in the United States the previous December the bridge had been examined by a technical commission which had declared it safe, in this case no more than 24 hours before. The accident was blamed on the formation of an island of alluvial wood brought down by the floodwater and which had not been removed.[772] [773]

The Macduff and Turiff section of the Great North of Scotland Railway suffered a serious accident on 27th November 1882, which although not a major disaster was the worst in the history of the small Scottish company. The 4 p.m. up mixed train from Macduff to Inveramsay consisted of an engine and tender, five loaded waggons, a brake van, two third-class, a first-class and a third-class brake composite in charge of Guard James Stewart, with Driver William Pratt and Fireman Joseph Brown. The train had left Auchterless station and was rather more than halfway between there and Fyvie station when, as it was crossing a bridge over the Inverary-Turiff turnpike road at Inverythan, Scotland, the structure collapsed without warning. Yet again the engine cleared the bridge safely, but its tender was derailed. The waggons and carriages behind went down the embankment or fell into the roadway below except the brake composite at the rear of the train which remained on the line. Four passengers were killed outright and a fifth died half an hour after he had been released from the wreckage. Fourteen passengers and the assistant guard in the front brake van were injured, five of them seriously. The five

[771] *Terre Haute Weekly Gazette*, Terre Haute, Indiana, United States, 15th December 1881
[772] *The Times*, London, England, 25th September 1882
[773] *Colonies and India*, London, England, 29th September 1882

passengers who lost their lives were James Alexander, a railway canvasser, of Aberdeen; David Bain, a joiner from Fyvie; William Gilchrist, a forester, of Cluny; Samuel Barron, farm servant, from Inveramsay and 18-year-old Jessie Wilson of Banff, a domestic servant bound for a farm at Fyvie.[774 775]

The bridge involved was a cast iron girder span 39 feet in length, with abutments and wing walls of Old Red Sandstone masonry and a clearance of 15 feet 4 inches over the road. A girder of the bridge had failed due to a vertical fracture, the result of a casting defect. The Board of Trade report described, 'a large honeycomb, or draw hole, which measured 3½ ins across at the face, and narrowed to about 1½ins inside, 2⅜ in extreme depth, and 6 ins in extreme length, extending inwards and downwards into the boss, but not within 1½ ins of the outer face of the joint flange. The total capacity of the whole, measured by filling it with water, was 15½ cubic inches, and, as the metal immediately round the hole was not sound, this amount does not measure the actual extent of the flaw.' Major F.A. Marindin, the Board of Trade inspecting officer, noted other defects in the girder and commented that the defects were 'evidence of the want of necessary precautions in the manufacture of this particular girder', and he concluded that, 'It would appear that the failure of this girder was due to a gradual extension of a flaw commencing at the honeycomb and working outwards until the strength at this point was so reduced that the girder was unable to bear the weight of the engine, and gave way suddenly; that the honeycomb, or draw-hole, which was the beginning of the evil, was due to want of proper care and attention in the casting of the girder; and that the defect in the girder was of such a nature that it could not be detected by outward inspection, however careful.' The inspecting officer did not feel therefore that any blame attached to the company engineer or his subordinates, but was concerned, 'for, if one such girder, which was of sufficient theoretical strength when cast, and which, being in constant use for 25 years, has never shown any signs of weakness or outward flaw, suddenly breaks down, owing to a defect which must have existed when it was originally made, and laid latent ever since, it is very difficult to know what steps should be taken to guard against the recurrence of a like disaster.' [776]

Eight similar bridges on the Great North were replaced as soon as possible and a warning of the danger was circulated to other British railway companies.[777]

A bridge on the Ferrocarril Central Mexicano (Mexican Central Railway) near Aquas Calicutes, Mexico, collapsed under the weight of a double-headed construction train on Sunday 14th October 1883. Five people died and the train was wrecked. [778]

The morning of Christmas Eve, 1883, saw an accident on the Louisville New Albany and Chicago Railroad, known since the previous year as the Monon route. Heavy rain which had fallen on Saturday night and throughout Sunday 23rd had much swollen the Blue River, a creek which the Monon line bridged a mile south of Salem, Indiana, United

---

774 Board of Trade report
775 *The Times*, London, England, 29th November 1883
776 Board of Trade report
777 Wikipedia – Inverythan rail accident
778 *London Evening Times*, London, England, 17th October 1883

States, and had seriously undermined the structure, which failed under southbound passenger train No 4. The train consisted of an engine, baggage express and mail car, smoker, day car and Buffet sleeper. The engine all but cleared the bridge but then seems to have fallen back into the river and three of the four coaches also went into the water. The cars piled up in a heap and overturned stoves set them ablaze.

Five passengers died in the accident; Rev Jacob Helfrich, a German who was a preacher at the local Methodist Episcopal church, Charles Sanford, Boone Thompson, a German named Ulrich and a man named possibly J.M. Whalling, whose body was washed two miles down the river. Another passenger, Charles Collins, was lying dangerously injured at the time of the report. Other passengers were less seriously hurt, and Engineer John Vaughn and Fireman James Meyers were also quite badly injured. Conductor Crit Cox, who had been in another accident near White River bridge a year or so earlier, was considerably bruised.

After the accident Brakeman Hamilton Armfield swam 40 feet to the shore of the river, procured a signal lantern from the sleeper and ran a mile back up the line to Salem, which he reached in time to stop a following southbound freight train and thus prevented a worse disaster. A newspaper commented, 'He deserves promotion for his bravery and forethought.' He had been in an accident at Horseshoe Bend, where another conductor was killed, the previous summer. Life for a 19th-century American railwayman was full of danger.[779]

Just over a month after the Salem accident the Louisville New Albany and Chicago Railroad suffered a second bridge accident. On 31st January 1884 the Air Line southbound accommodation train due at Broad Ripple, Indiana, United States, eight miles north of Indianapolis, at 10.2 a.m. went through the White River bridge a quarter mile from that place. At the moment of the accident the engineer had gone back into the baggage car, leaving the locomotive temporarily in charge of the fireman. As the engine reached the middle span of the bridge, and the train was still on the north span, the fireman heard the bridge timbers cracking and felt the bridge beginning to give way. He threw open the throttle, hoping to get the train clear of the bridge in time, but instead the sudden jerk snapped the coupling between the tender and the leading baggage car. The baggage car fell into the water, a distance of 20 feet, and the smoking car and ladies' car followed. As in the Salem accident a month earlier the wreckage then immediately caught fire. There was no escape for those in the baggage car and Engineer John Brewer, Express Messenger S.E. Smith and baggagemaster J.E. Rickets were killed. Of the 18 people on the train three others were also killed outright: Brakeman George Lowery; John Bray, a farmer of Westfield; and Thomas Parr, a bridge builder. Many others were injured, and it was feared that Linley Clark, of Kansas, who was spitting blood, was fatally hurt.

Locals from Broad Ripple and the surrounding area quickly gathered and formed a bucket brigade to try and extinguish the flames, but it was a hopeless task and the three train cars were burnt out. The bodies of the victims were so badly burnt as to be unrecognisable. Thomas Parr had not been riding on the train but was a foreman in charge of three men who were working on the bridge. With his men Foreman Parr was thrown into the river when the bridge collapsed. He drowned but his three subordinates

[779] *New Albany Ledger*, New Albany, Indiana, United States, 24th and 26th December 1883

escaped with slight bruises.

A passenger on the train, Mrs Sullivan, the wife of a roadmaster, described her experiences to a reporter. "I boarded the train at Carmel, with the intention of coming to Broad Ripple, to spend the day with my friend, with whom I am now stopping. I had my baby of ten months with me. I took a seat in the rear end of the car. I entered the front end but passed on down to the last seat. I do not know why I did this, for very few of the seats were occupied. There were four ladies besides myself on the train, and I think four or five gentlemen. The ladies were all seated well towards the rear of the coach, and the gentlemen further towards the front. Nothing unusual happened till we reached the bridge. I was talking to my baby and did not hear any crash. The first thing I knew I was thrown headlong over the ladies sitting just in front of me. I struck the back of the third or fourth seat about the time the end of the car struck on the smoker. It then stood partially on end. During this time my baby was in my arms, but I was unable to extricate myself. I did not think of myself but was afraid that my baby was hurt. A gentleman came and assisted me out from between the seats, and some lady took my baby. We then escaped by the rear end of the car to the north side of the stream. As soon as we got out, I looked back at the train and saw smoke issuing from the wreck. In fact, the smoke was so dense that I could not see the baggage car or smoker. In a few moments the whole train was in a blaze and burned until entirely consumed. Of the occupants of our car only one was injured and he but slightly cut on the chin. My neck was slightly strained in my fall over the seats. The gentlemen who escaped started to give assistance but were driven back by the flames. It was impossible to render any help, and we had to be the silent spectators of the scene without power to aid the sufferers." [780] [781]

Coincidentally another bridge over the White River near Indianapolis was the scene of an accident on 26th June 1884, this time on the Cincinnati Indianapolis St Louis and Chicago Railroad. Southbound freight train No. 24 was flagged by section men who were working on the bridge but it failed to pull up in time, two spans of the bridge collapsed, and six heavily loaded boxcars fell into the river. The engineer and fireman jumped but their engine passed over the bridge anyway. The only casualty in this American version of the 1865 Staplehurst disaster was a brakeman who was slightly bruised.[782]

Then on 28th July 1884 on the narrow-gauge Bedford and Bloomfield Railroad yet another bridge over the White River collapsed under the cars of what was probably a mixed train near Switz City, Indiana, United States and they fell 30 feet into the water. Bridges over this river were clearly fated. In this third accident passenger William Hibbets of Switz City was killed instantly and a Dr Cole was seriously injured. A James East was badly hurt, a brakeman suffered a broken leg and jaw and a number of other passengers were slightly hurt. [783] [784]

---

[780] *Madison Herald*, Madison, Indiana, United States, 1st February 1884

[781] *Columbus Herald*, Herald, Indiana, United States, 1st February 1884

[782] *Lebanon Patriot*, Lebanon, Indiana, United States, 28th June 1884

[783] *Greensburg Standard*, Greensburg, Indiana, United States, 1st August 1884

[784] *Brazil Clay County Enterprise*, Brazil, Indiana, United States, 6th August 1884

Two trains left Melbourne, in the colony of Victoria, on Saturday 25th January 1885 for Albury, in the colony of New South Wales. Monday 26th was Foundation Day, a public holiday in New South Wales in celebration the birth of the colony in 1788, and races were held at Randwick (Sydney). The mail train which left Melbourne at 2.55 or 3 p.m. on Saturday was thus followed by a special train carrying a large party of bookmakers bound for Sydney which departed at 7 p.m. The two trains were combined into one at Albury.[785]

Between 5 and 7 p.m. on the evening of Sunday 25th January 1885 extremely heavy rain fell in the vicinity of Cootamundra, New South Wales. Ganger James Cambourne, whose length included a culvert over Salt Clay Creek, set off on his regular inspection of the line at 6 p.m. Near Cootamundra station he met a man named Mullaly who told him that the water was over the line at the culvert at his mill (Hurley's Creek). Ganger Cambourne immediately got his trolley from the station yard and with three of his men proceeded to the culvert where they found that the floodwater had damaged the track. He sent his men back to the station to tell the stationmaster, Haines Giddy, to telegraph the next station at Bethungra to stop the Albury mail train, but the men soon returned to say that they could not contact Bethungra (the telegraph wires were down). The flooded creek seemed completely impassable, but the ganger tried to shout to some men on the opposite bank to stop the mail train which he knew was shortly due and then with his men proceeded upstream to try and find a point to cross. A little later an ex-sailor named George Hawke did manage to struggle across the creek by crawling across the rails and sleepers on his hands and knees. He was handed a red lantern by Stationmaster Giddy and went back the line with it to warn the train.[786] A man named Joseph Barrett on the Bethungra side of the creek had also, when Ganger Cambourne had indicated for him to do so, lit the red lamp of the Cootamundra distant signal, 200 yards on the Bethungra side of Hurley's Creek.

Stationmaster James Glynn at Bethungra told the inquest into the accident that the train from Albury had arrived at 7.9 p.m., in charge of Guard Samuel Murray with Driver Andrew Moody and Fireman Richard Wall. He had duly despatched it with a staff and it was only when he tried to notify Cootamundra of its departure that he had realised that the telegraph was out of action. The single line was worked by staff and ticket and it was not necessary to contact Cootamundra prior to letting the train go as long as he was in possession of the staff.[787] Bethungra was about 15 miles from Cootamundra. The train continued towards the latter place and at about 7.45 met with disaster at Salt Clay Creek, 3½ miles south of Cootamundra, where unknown to the men there another washout had taken place. Driver Moody had little warning of danger and was travelling fast on a falling gradient, but he made a last-minute brake application. A newspaper report read in part, 'The engine seems to have crossed the rails leading to the Cootamundra side, when the bridge gave way, precipitating the train into the creek. The engine fell in front, standing in its own proper course, and nearly upright; then a second-class carriage and the mail van dashed up alongside, and higher up the stream came a sleeping car, which was telescoped by a first-class carriage, which telescoped the sides and roof, with the exception of one side, which was left standing, while the brake van fell at the end, lying

[785] *Sydney Morning Herald*, Sydney, New South Wales, 26th January 1885
[786] *Cootamundra Herald*, Cootamundra, New South Wales, 1st July 1885
[787] Ibid, 31st January 1885

up and down the stream, and forming the letter T in shape. It was at this time that the fatalities occurred, and there were many narrow escapes. The escape of the people in the sleeping car was miraculous. The car was plunged into midstream. The water surged into it, and the occupants must have been drowned had not a large first-class carriage telescoped it, breaking off the roof and freeing the people. Several of them were badly injured.'[788]

Six people were killed outright in the accident; Conductor Alfred Wilson, Mail Guard John O'Dwyer and passengers Joseph Compy and James Simpson, both railwaymen travelling as passengers; a Mrs Hodson and four-year-old John Hode. The little boy was travelling with his uncle, William Bergin, and they were returning from a visit to Bergin's parents at Wagga Wagga. Bergin suffered two broken thighs and died of his injuries on Wednesday 28th, making the seventh fatality.[789] Over 20 other passengers were injured, and Driver Moody and Fireman Wall were also badly hurt and do not seem to have testified at the inquest.

The inquest jury returned a verdict that the deaths of the seven persons killed in the accident, "were caused by the mail train in which they travelled meeting with an accident through the washing away of an embankment at Salt Clay Creek; and that the Government have shown negligence in not having caused the culvert at Salt Clay Creek to have been built on a much larger scale; and we are also of opinion that the accident might, in all probability, have been prevented had there been more lamps than one provided for each gang of fettlers." The jury added a rider that, "Mr George Hawke is worthy of great credit for his efforts on the night of the accident, and that he merits some recognition from the Government for his actions during the whole of the night." [790]

The Cincinnati and Eastern Railroad trestle over Nine Mile Creek, nine miles from New Richmond, Ohio, United States, had reportedly been considered unsafe for some time when on 8th August 1885 it collapsed under a train in charge of Conductor E.S. Durham, with Engineer William Gallion and Fireman Ellis Smith. The short train consisted other than the engine of only a passenger car and two empty coal cars. After the accident the passenger car ended up resting at an angle of 45 degrees, its rear end remaining on the trestle. Mrs Donaldson, the wife of the master mechanic of the road, was killed instantly, William Smith, of nearby Locust Corners, died half an hour after the accident and Isaac Smith (coloured), who had suffered a fractured skull, died the next day.

Conductor Gallion, young schoolteacher Henry Sutton and the two- and four-year-old children of Mrs Donaldson were seriously if not fatally injured and the engineer and fireman were also badly hurt.[791]

On the morning of Monday 14th March 1887 an accommodation train on the West Roxbury branch of the Boston and Providence Railroad left Dedham, Massachusetts, United States at 7 a.m. for Boston. It consisted of the engine *Torrey*, a baggage car and

---

[788] *Sydney Morning Herald*, Sydney, New South Wales, 27th January 1885
[789] Ibid, 29th January 1885
[790] Ibid, 4th February 1885
[791] *East Liverpool Evening Review*, East Liverpool, Ohio, United States, 10th August 1885

nine passenger coaches, and was in charge of Conductor W.H. Alden, assisted by Conductors Webster Drake and Myron Tilden, with Engineer Walter White. It called at Spring Street, West Roxbury, Highland, Central and Roslindale stations to pick up passengers. By the time that it left Roslindale it was carrying between 300 and 400 passengers, many of whom were women and the remainder mostly clerks and labourers, only a few businessmen choosing to set out at such an early hour.[792]

Midway between Roslindale and Forest Hills stations the railway crossed South Street, Roslindale, by the Bussey Bridge, 126 feet long and 26 feet and at its centre 26 feet above the road. Known as the "Tin bridge" the structure had originally built as a Howe wooden truss covered with tin, hence its nickname. The northern truss was replaced by an iron Parker truss in 1870; in 1876 this was moved to the southern side and a new iron truss installed on the northern side. Although built for double track only a single line was laid on the northern side of the bridge. Engineer White had passed over the structure with his engine, the leading baggage car and two passenger cars when he felt a sudden jar and looking back saw to his horror that the remaining cars were falling into the roadway below. By opening the throttle fully, he probably saved the first three cars from falling back off the bridge. His doing so also separated the engine and baggage car from the two passenger cars and he hurried on to Forest Hills station to raise the alarm.

Most fortunately a police officer witnessed the accident and he immediately pulled an alarm from Box 528, which resulted in the dispatch of the fire apparatus from Roslindale and the steamer and ladder truck from Jamaica Plain. This was especially opportune as a fire had broken out in the wreckage which the firemen were able to extinguish before it had caused much harm. When news of the disaster reached Boston ambulances were sent from the General and City hospitals, Superintendent Folsom sent a corps of surgeons and Deputy Superintendent Folsom sent 100 police officers from different stations to assist the fireman.

After the accident cars Nos. 52 and 18 remained on the embankment, reasonably undamaged. Car No.28 had retained its shape, 'but was easily pulled to pieces'. The next car, number unknown, had almost reached the far abutment when the bridge collapsed, was dashed against the solid stone wall, 19 bodies were taken from it, and at the edge of the abutment was that of Conductor Tilden. Cars Nos. 87, 54 and 80 fell into the roadway, car No. 87 crashed against the stone abutment and car No. 54 fell upon it. Car No. 80 fell partly upon car No. 54. The last two cars were No. 81, which was practically demolished, and car No. 82, the smoker, which rolled down the embankment and was completely demolished. Few passengers were killed in these two cars but many injured.

By the following day the death toll had supposedly reached 32, comprising 30 passengers and Conductors Drake and Tilden (though Conductor Drake was also listed as injured). About 120 passengers and the baggagemaster were listed as injured.[793]

The Philadelphia Enquirer reported that, 'The only person known to have been an eyewitness to the disaster, who was not on the train, was J.H. Lennon, a fish dealer, whose stable was on the hill just above the bridge. Lennon was harnessing his team when the train came along and turned to see it pass. He was horror struck when he saw the train take its awful plunge through the bridge.

[792] *Boston Post*, Boston, Massachusetts, United States, 15th March 1887
[793] Ibid.

'For a moment, Lennon says, there was perfect quiet, and then the cries of the injured were heard issuing from the debris. Seizing an axe from the barn, Lennon started for the wreck, which he reached in a moment. He climbed into the window of one of the coaches that had landed against the wall and set to work to release those persons who had been pinned down by the broken timbers.

'Lennon with his axe released four men who were badly hurt and handed them out of the window to other men who had come to the rescue. He also handed out the bodies of two dead women, one of whom was nearly decapitated and had both arms severed from her body. During all this time, Lennon says, the shrieks of the wounded and the groans of the dying made a pandemonium around him that was nothing short of horrible.' [794]

The *Boston Post's* coverage of the accident included accounts from various passengers. 'Mr Weld Heinsen and Emil Heinsen, brothers, Roslindale, were in the second car from the rear. Said the former, "I was thrown from my seat but held to the arms and crouched in the aisles till the car struck. Two men were thrown over my head. One was the conductor. There was a terrible scream, and my brother and I made for a window and crawled out. When we were out, we found that we had sustained but slight bruises. We then pulled out Conductor Drake and several others whom I did not know. We next went to Roslindale for help, and secured teams and physicians."' [795]

'Mr Fred P. Smith, of Dedham, related his escape as follows: "I was seated in the middle of the train, in car No. 80, and when we got on the bridge, I heard a snapping sound, as of breaking timbers, and immediately our car pitched forward and fell. After some trouble I crept out of the car. I was dazed at the magnitude of the wreck and was called to my senses by the shrieks of the dying and wounded. I immediately turned to the work of rescue, and, after a great amount of trouble, succeeded in helping out Mr James H. Dayton, of Dedham, whose wrist was broken, and fingers smashed. He, in turn, helped me extricate Mr F. Nichols, of Dedham, who was internally injured. A Mr Burnett, of Brown, Durrell & Co, of Boston, was wedged in the wreck, and it was some time before we could get him out. We succeeded, however, and found him to be badly injured. In the meantime, fire had caught in one of the cars, but it was put out before any damage was done."' [796]

By 16th March the number of dead had been corrected to 25, comprising 24 passengers and Conductor Myron Tilden. The number of injured was stated to be 125, and it was feared that several of them might die.[797] The final toll according to the Railway Commissioners was, however, only 23.[798]

The disaster was fully investigated by the Massachusetts Railway Commissioners, who found that the northern truss, the one which had failed, had been installed in 1876 by Edward H. Hewins, whose tender had presumably been lower than that of Mr Parker, of the National Bridge Company, which had installed the 1870 truss, which was then moved to the south side and had not failed. Mr Hewins was allegedly an agent for the

---

794 *Philadelphia Inquirer*, Philadelphia, Pennsylvania, United States, 15th March 1887

795 *Boston Globe*, Boston, Massachusetts, United States, 15th March 1887

796 Ibid.

797 Ibid, 16th March

798 Bussey Bridge Investigation: Special Report by the Massachusetts Railroad Commissioners, 1887

Metropolitan Bridge Company, though in fact no such company existed. Mr A.A. Folsom, the superintendent of the line, and Mr George F. Folsom, the superintendent of construction, had apparently not made any real enquiries into Mr Hewins' competence as a bridge builder or as to the existence of the Metropolitan Bridge Company, but had taken these things on trust. Equally the Phoenix Iron Works and the Trenton Iron and Steel Co. had left the work of manufacturing the bridge components to an agent of Mr Hewins. Mr George Folsom, the superintendent of construction, was a mechanic who had no experience of bridge building and had left matters to Mr Hewins, while the company had employed no experts to either review the original design or to inspect the bridge after its construction.[799]

The commissions found that in fact the material and workmanship of the compression members whose failure had caused the bridge to collapse was good, it was the bridge design which was faulty. The commissioners were able to point out that they had written to the railway company at the end of 1881 describing the bridge as an oddity and suggesting that it would be wise and prudent to test it under a given load, but this had never been done. Mr Gorge Folsom examined the bridge every spring and fall but he did not understand its construction and had never examined the strength and condition of the hangers, a vital part. In addition, the commissioners noted that seven of the cars of the train were supplied with the Westinghouse Automatic brake, but the brakes did not come on automatically because the other two cars only had the Westinghouse Straight brake. If the brakes had come on at once automatically when the train became divided the results of the accident would have been less serious. There should also have been five brakemen on the train but there were only three; the assistant conductors could not be considered as brakemen as they had other duties to perform. Fire had in fact broken out in three separate places, one in the third car on the embankment had been extinguished by the passengers and two in the cars in the street had been put out using water available nearby, supplemented by Chemical Engine No. 4 of the Boston Fire Department, which had arrived in eight minutes.[800]

In the summary of their report the commissioners stated, 'In the erection and inspection of bridges the management of a railroad is bound to exercise the utmost care. Had such care been exercised, there is every reason to believe that the disaster would have been prevented.' [801] The direct cause of the accident was the facture of two hangers which suspended the track structure from the top chords. The hangers were poorly designed and not strong enough for the loads to be carried by them. Seven-eighths of the fracture of one of the hangers appeared to have been present for some time and the passage of the ill-fated train was merely the final straw which caused the critical failure. It also transpired that a previous bridge designed by Hewins at Kennebunk, Maine, on the Boston and Maine Railroad had also failed beneath a train, in this case supposedly after the train was derailed.

Slightly less than seven months after Bussey Bridge disaster the United States suffered a considerably worse accident near Chatsworth, Illinois, on 10th August 1887.

[799] Ibid.
[800] Ibid.
[801] Ibid.

The financially hard-up Toledo Peoria and Western Railroad, which ran over 220 miles through central Illinois without serving anywhere of any great importance, decided to generate some revenue by putting on a cheap excursion train to Niagara Falls. The round trip from central Illinois was advertised at the very low fare of $7.50 and thus attracted almost a thousand excursionists. To cater for them a double-headed train was laid on, consisting besides the engines of six sleeping cars, two chair cars, five passenger coaches, Superintendent Armstrong's special car and a baggage car.[802] The train was in charge of Conductor J.W. Stillwell with Engineer Dave Sutherland and Fireman John Rogers on the pilot engine, No. 21, and Engineer Ed McClintock and Fireman Axel Applegreen on the train engine, No. 13. It had pulled out of Peoria, Illinois at 8 p.m. had been delayed near Peoria by brakes sticking, then near Forest by a drawbar pulling out, and was about an hour and a half late as it left Chatsworth at 11.45 p.m. that Wednesday evening.

Superintendent Armstrong told the inquest into the accident that he had retired to his own car as the train left Chatsworth and had been sat there three or four minutes when he felt a fearful shock to the train. Then he was thrown out of one of the windows into a hedge fence. He got up and asked one of the firemen where the pilot engine was, to be told that Engineer Southerland had taken it forward to Gilman to get help. The engine had returned in three quarters of an hour with doctors and the injured were taken to Piper City and Chatsworth and the bodies of the dead to Chatsworth.[803]

The train being late, it was running perhaps a little faster than it might otherwise have been, though Superintendent Armstrong did not consider that the engineers had been trying to make up time. He estimated the speed of the train at 30 or 35 miles an hour. As it neared the point of the accident Engineer Sutherland had glimpsed what he thought was a brush fire up ahead, then realised that it was small wooden bridge itself which was on fire. His fireman told him to jump and did so himself; Sutherland considered it but, in the end, did not do so. When he first saw the bridge it seemed, the engineer said, like embers but seemed to burst into flame as the train ran onto it. He told the inquest jury that his engine had no air brakes and there was no chance of stopping when he saw the fire, nor had he had time to whistle.[804] The train engine crossed the burning bridge, but the trestle collapsed beneath the tender, breaking the couplings so that the engine continued its course down the line. The train engine fell with the bridge and ended up on its side. Engineer McClintock was killed but Fireman Applegreen somehow jumped or fell clear. Behind the two engines the first nine cars piled up in a confused mass, with the second car completely telescoping the first. The tenth car, the first sleeper, came to rest with its leading end on the edge of the burning trestle. McClintock's watch had stopped at 11.49. Engine No 21 ran 1,000 feet before Engineer Sutherland stopped it, then seeing what had happened behind him he went on for assistance.

The summer had been a very dry one, and the Peoria line had been much troubled by brush fires along its length. A policy had therefore been adopted of burning off grass and weeds near the track in a controlled manner to pre-empt an uncontrolled fire outbreak. On the afternoon of the 10th a work gang in charge of Timothy Coughlan had been doing this between Chatsworth and Piper City, the next station to the east. When the work was finished for the day Coughlan told one of his men to check carefully that any fires in the

---

802 *Philadelphia Inquirer*, Philadelphia, Pennsylvania, United States, 13th August 1887
803 Ibid.
804 Centralia Democrat, Centralia, Illinois, United States,19th August 1887

vicinity of the culvert were completely extinguished and the man had had a look and assured him that they were completely out. They had then gone home. That was at 5.30. Traffic on the line was obviously very light as the last train that had passed over the bridge before the arrival of the excursion train was a freight train at about 3.30 p.m., which had passed the spot safely. Roadmaster Ennis had ridden on the freight to inspect the line and with regard to the fire risk Superintendent Armstrong told the Illinois Railroad Commissioners inquiry into the accident that he had sent out an order by telegraph the previous day:

August 9. 1887

T.E. ENNIS: Have your men run over their fire sections tomorrow p.m. the last thing quitting, and see that bridges are in no danger of fire.

(Signed). E.N. ARMSTRONG [805]

Bad though it was, the accident could have been even worse, for a large number of passengers were trapped in the wreckage of the wooden cars, and the burning bridge was threatening to set fire to the train and roast them all to death. No water was available to put the flames out, but uninjured passengers from the cars at the rear of the train, assisted by Foreman Coughlin and his four section men and rescuers who had come from Piper City or Gilman on the returning engine managed to extinguish the fire by tearing up the dry soil with their hands and throwing it on the wreckage. It took four hours to finally put the fire out. Even though the horrors of an Angola or an Ashtabula had thus been averted the accident was still a very bad one, the final death toll being 82, making it almost as great a catastrophe as that at Ashtabula.

Foreman Coghlan had 8½ miles of track to maintain with his four men. According to Superintendent Armstrong's order he should have checked the bridges and track over the whole of the section before the passage of the excursion train. He had gone over the eastern half of his section and burned the grass along the track for half a mile and half a mile from the bridge. He had checked the bridge about 5 o'clock. About three weeks earlier the grass under the bridge had been cut away for about 10 feet from the bridge timbers and he had no idea how the bridge could have caught fire.

Roadmaster Christopher Ennis was of the opinion that the bridge had been deliberately set alight. He had seen no signs of fire or smoke when he passed over it on the freight train rather before 4 o'clock. He recalled that about three years earlier two attempts had been made to ditch the 10 o'clock passenger train at the same bridge by placing obstructions on the line, after which a watchman had been stationed there for six weeks.[806] The arson theory was never substantiated but there were certainly attempts to rob the dead and wounded after the accident. Superintendent Armstrong himself said that he had witnessed 'many thieves at work and had stopped them while {they were} despoiling the wreck of property and money.' One apparent victim was an Edwin A. Adams, who was known to have started the trip with a large amount of money, but after the accident his friends could find neither his pocket book nor his gold watch on his

805 *Bloomington Weekly Pantagraph*, Bloomington, Indiana, United States, 19th August 1887
806 *Philadelphia Times*, Philadelphia, Pennsylvania, United States, 13th August 1887

body, and only $30 was found on his body, hidden in the watch pocket of his trousers.[807]

The inquest jury heard evidence from Chatsworth residents who stated that they had seen a light on the railway track on or near the bridge two or three hours before the accident. It also appeared that testimony by the foreman that he had not set fire to any grass or rubbish within 20 rods (330 feet) of the bridge was untrue as James Taggart swore that he had been ordered to set fire to grass 100 feet from the bridge and James Heald testified that he had been told to set fire to grass and weeds only a rod (16½ feet) from the bridge. Everything was very dry, and the wind was blowing from the southwest, from the fires directly towards the bridge. The conduct of Agent S.A. Mason at Piper City was also severely criticised. The excursion was due there at 10 but he had not left a prayer meeting until 9.30. When he arrived back at the depot several residents had allegedly expressed their fears to him as to the nature of a peculiar light down the line towards Chatsworth. He had ascertained by the telegraph that the excursion train was then at Weston, 24 miles away and 40 minutes late. There were two hand cars and a flat car at Piper City which he could have used to go down the line to investigate but he had then, it was said, either through indifference or stupidity, done nothing, and the bridge had been left to burn.[808] Called to the witness box Agent Mason denied point blank the statements of Messrs Jones and Clark that they had called his attention to any light, said that he had ascertained from the telegraph that the train was at Gridley (three stations further away than Weston) and had then gone out and looked down the track but seen no light.

At the end of the inquest the jury, after some considerable debate, found that the bridge had burnt out before the train struck it, that Coughlan had disobeyed positive orders to inspect the track and bridges, that he was guilty of gross criminal negligence in leaving fires burning along the track, and recommending that he be held for examination before the grand jury. In a rider the jury expressed the opinion that leaving the track unsupervised for six hours before the arrival of the excursion and the starting of fires by section men were guilty of gross censure. Coughlan was arrested and being unable to find bail as carted off to jail, claiming that the verdict was unjust.[809]

In October the case against Coughlan was dismissed by the grand jury.[810]

Browning or "Fat Nancy" trestle, two miles south of Orange Courthouse, Virginia, United States, was known to be weak and was in the process of being filled in, presumably being replaced by an embankment and culvert. It was about 350 feet long and 47 feet high at its highest point. On the morning of 12th July 1888 southbound Virginia Midland Railroad passenger train No. 52, which had left Alexandria at 11.20 p.m. the night before, ran onto the structure at 2.15 a.m., when the trestle gave way beneath it. The train consisted of an engine, mail car, baggage car, smoker, ladies' car and three sleepers. The engine and the whole of the train except the last two sleepers went through the trestle, piled up on top of one another and were broken up. 'The rails were bent in every conceivable shape. The hillsides this evening were strewn for hundreds of yards around with all kinds of timber, iron, baggage, boxes, paper, etc. Hundreds of people from

---

807 Ibid.

808 *Defiance Democrat*, Defiance, Ohio, United States, 25th August 1887

809 Ibid.

810 *San Antonio Light*, San Antonio, Texas, United States, 17th October 1887

Orange and the neighbourhood visited the place during the day.' [811] Eight passengers were killed outright, and Mail Agent H.C. Brightwell was fatally injured and died in the College hospital at Charlottesville the next day, five minutes before his wife and brother arrived from Prospect, Virginia.

The accident caused much indignation when it became known that the wooden trestle was 'rotten beyond repair'. The coroner's jury quickly returned a verdict that the accident was due to the decayed timbers in the bridge, a local newspaper dismissing any suggestion of a derailment and stated that the train had only been moving at two or three miles an hour. The paper added that 'The fact that the bridge is rotten is sufficient to account for the horrible disaster,' and, 'It is said there was not a sound timber in the bridge.' [812]

A bridge over a small stream which was swollen by abnormally heavy rain collapsed under a southbound mail train a mile north of Tensas, Alabama, United States at 2.45 a.m. on 27th August 1888. The engine and the baggage, postal, passenger (two) and sleeper cars were wrecked. Engineer John Morgan, negro Fireman Sam Williams and two tramps were killed outright and Mail Messenger Alonzo Davis (coloured) was fatally injured and died the same day. baggagemaster Taylor and four passengers were slightly hurt but the other mail messenger, Thos. Jackey, escaped injury. The train had left Montgomery, Alabama at 10 p.m. the night before in charge of Conductor Captain Coleman. [813]

An unusual tragedy occurred on the Lebanon branch of the Nashville Chattanooga and St Louis Railroad at Stone River, about nine miles from Nashville, Tennessee, United States, at about 4 p.m. on the afternoon of Saturday 18th May 1889. The bridge across the river had been set on fire by a passing train. Seventeen men, 15 negroes and two whites, who had been working on the track nearby, went to the bridge to put out the blaze with water from the tender of an engine which they took with them. Unfortunately, the fire had already damaged the bridge more badly than they realised and the extra weight of the engine moving onto it was enough to cause it to give way. The engine and men fell 50 feet to the river below. Three of the negro labourers were killed and six injured, four dangerously. Engineer William Morgan, white, suffered serious internal injuries and his white fireman, who was not named, suffered a broken leg and other serious injuries. [814]

The Ferrocarril Central Mexicano (Mexican Central Railway) passenger train which left Paso del Norte at 6 p.m. on the evening of Monday 8th July 1888 in charge of Conductor Jerry Sullivan, was wrecked by a bridge giving way five miles on the El Paso side of Chihuahua, Mexico. The train had been running cautiously all night because of the danger of washouts. Having apparently passed the dangerous part of the line it had

[811] *New Market Our Church Paper*, New Market, Virginia, United States, 25th July 1888
[812] *Harrisonburg-Rockingham Register*, Harrisonburg, Virginia, United States, 19th July 1888
[813] *Columbus Enquirer Sun*, Columbus, Georgia, United States, 28th June 1888
[814] *Cambridge Herald*, Cambridge, Ohio, United States, 23rd May 1889

been running fast to try and regain some of the lost time. There was no rain at the bridge itself but a cloudburst up in the mountains had resulted in a torrent which had undermined the structure. The baggage car and a third-class carriage were telescoped in the accident and a third-class passenger and a Wells Fargo Co. guard were killed outright. Twenty-five other passengers were injured, the conductor sustaining four broken ribs and a newsboy having his leg broken in two places. Nineteen of the injured were taken to hospital in Chihuahua, two of whom had reportedly died. The newspaper reporter added that, 'The railroad people are very reticent about the accident.' [815]

On the evening of Thursday 2nd January 1890 westbound local freight train on the San Antonio and Aransas Pass Railway, consisting of 26 boxcars, all but two of them loaded, engine and caboose, in charge of Conductor Ferrier, stopped at Hallettsville, Texas, United States. The engine was detached and taken forward for water, crossing and re-crossing a trestle bridge over the Lavaca River in doing so. The train then restarted, but in the intervening few minutes the river, which must have been greatly swollen by a cloudburst further up, had riven seven feet and had damaged the foundations of the trestle. The engine had just crossed the bridge when its western portion gave way, jerking the engine back into the water, and the whole of the remainder of the train went into the river. Conductor Ferrier shouted, "Look out, boys!" to two men, Pat Dunn and James Lynch, riding with him in the caboose. The conductor and Lynch went into the water, but Pat Dunn jumped from the rear of the caboose and landed at the river's edge. He then saw a brakeman struggling in the water and after an effort was able to save him.

Pat Dunn was the only person on the train who escaped being carried down the river. A stockman from McGregor, Texas, and two negroes succeeded in grasping the top of a willow tree about a quarter mile down river. The entire contents of the stockman's household, his buggy and two fine horses had been on the train. Fifty or sixty rescuers quickly went to the riverbank but could do nothing to help the three men in the seething waters until Friday morning, when they were finally rescued using rafts and ropes. Also found on Friday, seven miles downriver, where his body had lodged in a tree, was Extra Fireman George Burke, one of whose legs was badly injured, so that his foot had to be amputated, and who died on Saturday morning. Engineer Ed Palmer and Fireman B.B. Todd were rescued about a mile downriver. The engineer had serious head injuries, which subsequently proved fatal, and the fireman suffered a broken arm and jaw. On Saturday the bodies of a railwayman named Fitzpatrick and a 13-year-old boy who had been deadheading were recovered from the river 13 miles from Hallettsville, while the body of Lynch had still not been found and was thought to be somewhere under the debris of the bridge. Conductor Ferrier, on the other hand, after being carried about 12 miles down the river, was pulled out unhurt. The engine ended up buried in the sand with only a small portion of its boiler visible while goods were scattered along the banks of the river for miles.[816] [817]

Five railwaymen from Toronto lost their lives in an accident near Claremont, Ontario,

---

815 *New Philadelphia Times*, New Philadelphia, Ohio, United States, 18th July 1889

816 *Galveston Daily News*, Galveston, Texas, United States, 4th January 1890

817 *San Antonio Daily Express*, San Antonio, Texas, United States, 8th January 1890

Canada, on the Canadian Pacific Railway, on the night of 17th June 1890. A washout had been reported and a light engine was sent out from Toronto to inspect the line and repair any damage. Riding on the engine in addition to Engineer James Atton and Fireman Tom Oliver were Bridge Inspector John Wanless Jr. and three section hands named Lott or Latton, Myrarity and Nighswander. The engine was running cautiously but, it would seem, not cautiously enough, for it went through a bridge. Five of the men on the engine were drowned but D. Nighswander miraculously survived with a few bruises. At the time of the report on the accident the next day the body of Wanless had been recovered but those of the other four victims remained buried in the wreckage. The bridge inspector left a widow and three young children.[818]

A student from the Chemawa Indian school, near the scene of the occurrence, was the first person to bring news of a train accident to the town of Salem, Oregon, United States on Wednesday 12th November 1890. According to his report one hundred people had been killed and every physician in the city hurried to the scene at Lake Labish, together with a large number of helpers and curious onlookers. A relief train was ordered from Albany and a wreck train from Portland.

The train involved in the accident was a southbound Southern Pacific Railroad passenger train. It was crossing the low trestle at Lake Labish, just north of Sale, which was about 600 feet long and from 16 to 20 feet high. It was supposed that the engineer felt the trestle giving way, for he sounded the whistle and applied the brakes. The train ran about 50 yards further before the engine overturned and ended up buried in the mud. The tender and the mail, baggage and express cars followed the engine, but the first-class coach was saved from going over by coming up against an old tree. The accident was nowhere near as bad as the Indian student had thought but three passengers were killed outright as was a tramp riding on the trucks of the express car. All but three of the 30 passengers in the tourist sleeper, all but three of the 17 passengers in the Pullman car *Alatia*, a large proportion of those in the fully occupied sleeper and fifteen in the Pullman car *Roseberg* were injured. Twenty-five of the injured were taken to the Willamette hotel in Salem, which was converted into a temporary hospital and others were taken to Portland. As the relief train arrived at Portland one of the injured, Mr C Griebel of Milwaukee, Wisconsin, a travelling agent for the Val Blatz Brewing Company, died, making a fifth fatality. It was feared that the final death toll might reach 10.[819]

A Maryland Central Railroad train consisting of three engines and 14 cars fell through the Overshot trestle at Fallston, Maryland, United States on Thursday 16th April 1891. At the coroner's inquest into the death of John Martin, one of the four men killed in the accident, witnesses testified that the trestle's supports had been rotten and the jury returned that the death had been due to the giving way of the trestle due to decaying timbers and neglect on the part of the railroad company to keep it in proper repair, also that there had been unnecessary delay in the rescue of the victims.[820] John Martin had been a brakeman. The other men killed outright were Brakeman Benjamin Sprucebanks

---

[818] *Winnipeg Free Press*, Winnipeg, Manitoba, Canada, 19th June 1890
[819] *Fresno Republican Weekly*, Fresno, California, United States, 14th November 1890
[820] *Philadelphia Inquirer*, Philadelphia, Pennsylvania, United States, 18th April 1891

and Conductor James Dodson. Two other men, Charles Watts and Walter Hawkins were seriously injured, and Watts later died. The remaining four men riding on the train were able to jump clear in time. The trestle was about 350 feet long with a maximum height of 50 feet above a stream. The bodies of Sprucebanks and Dodson could not be found, and it was supposed that they had been consumed in the fire which had broken out immediately after the accident.[821]

On 1st May 1891 as the 8.45 a.m. London Brighton and South Coast Railway express train was passing over the Portland Road bridge near Norwood Junction station, England, the whole train was derailed. The train consisted of Engine No. 175 *Hayling* and tender, a brake van, six first-class carriages, a Pullman car, three first-class carriages and a brake van and was in charge of Guard Mark Sayers, with Driver Harry Hargreaves and Fireman Joseph Grover. The three rear vehicles making up the train leaving Brighton had been slipped at East Croydon for London Victoria. The results of the accident were not very serious, the worst injury was a dislocated ankle, while a good number of passengers were shaken. The accident was, however, the subject of a comprehensive Board of Trade report, for it was caused by the failure of the cast-iron Portland Road underbridge.

Major-General C.S. Hutchinson, the Board of Trade inspecting officer, found that there was no doubt that the accident had been caused by the failure of a cast-iron girder, which had possibly initially given way under the preceding South Eastern Railway train. The girder which had failed had contained a very serious flaw, abstracting at least one quarter from its strength. Independent of this flaw the girder had not possessed a sufficient theoretical margin of safety for the weight of engines then passing over the line. He noted that, 'The attention of the Brighton Company was drawn by the Board of Trade to this deficiency of strength after the occurrence of the accident on this bridge in December 1876, when two identical girders at a different part of the same bridge were broken by an engine getting off the rails, and they were then recommended to substitute stronger girders in their place, a recommendation to which unfortunately no attention was paid, or the present serious accident would have been prevented; the Brighton Company is therefore, in my opinion, deserving of much blame for having omitted to substitute stronger girders for the existing ones after attention had been thus specially directed to the weakness of the latter.' [822]

The inspecting officer recommended the replacement of the other cast-iron girder bridges on the Brighton Company's main line by wrought-iron underbridges without any unnecessary delay, as well as the replacement of all similar bridges elsewhere on the company system. The Board of Trade also sent out a circular requesting details of all cast-iron underbridges on the national network and found that there were many thousands of them. Most were eventually replaced.[823]

On 14th June 1891 an annual musical festival was being held in the village of Mönchenstein (contemporary spelling – later Münchenstein) near Basle, Switzerland. It

[821] *Boston Daily Globe*, Boston, Massachusetts, United States, 17th April 1891
[822] Board of Trade report
[823] Wikipedia – Norwood Junction rail accident

was a very popular event, attended by hundreds of people, including, not surprisingly, a good many from Basle. To cater for the traffic from Basle local train No. 174 of the Jura-Simplon Railway, due to depart from Basle at 2.15 p.m. for Delémont, was strengthened by the addition of two extra carriages, and a second locomotive was provided to pull it. On leaving Basle the train consisted, according to the report in *The Times*, of engine Nos. 203 (pilot) and 209 (train), a first-class carriage, a second-class carriage, a mail van, a luggage van and seven third-class carriages, but there appears to have been another four-wheeled carriage at the front of the train and also another carriage at the rear of the train, making 13 vehicles in all.[824] [825] The train was in charge of Head Conductor Wenger with Driver Bodmer and Fireman Frey on the pilot engine and Driver Steinmann and Fireman Eichenberger on the train engine, and was carrying between 530 and 550 passengers.

Approaching Mönchenstein station the railway crossed the River Birs, a tributary of the Rhine, by a wrought iron girder bridge, and it was here that disaster struck as the train was braking for the station. The two engine crews could feel the bridge giving way beneath them and the next moment the two engines and the seven leading carriages were in the river, with the eighth left hanging from the bridge abutment. The severing of the brake pipe had applied the brakes fully and prevented the whole of the train leaving the bridge and the passengers in the rear five carriages escaped unhurt. Even so, it was a major catastrophe as many of the passengers in the carriages which went into the water were drowned, along with Driver Bodmer and Head Conductor Wenger.

Upon news of the accident reaching the village the festival was immediately abandoned and villagers went to assist in the rescue work while help was summoned from Basle and other towns.[826] The river was swollen, which made the work of recovering the bodies of those who had been drowned very difficult, while, 'The corpses, on being brought up, were laid out in an orchard on the bank of the river, where most heartrending scenes ensued, wives looking for their husbands and parents for their children.' Another newspaper correspondent wrote, 'Among the victims of the railway accident whose bodies have been recovered are Herr Bubeck, director of the Basle Polytechnic; the wife and two sons of a master tailor named Deilmann, another son still being missing; and Dr Voegtli with his three children. Herr Jacob Sutter, proprietor of the Wilden Mann Hotel, had both his legs broken, but it is hoped that he may recover. An engineer named Perdriset is also among the victims, and his wife and daughter have both been injured. The driver of the first locomotive still lies under his overturned engine. The stoker had a marvellous escape. He was precipitated into the river clear of all the fallen fragments and saved himself by swimming ashore.' [827]

By the end of 15th June 60 bodies had been recovered and by the end of the following day 65. Great difficulty was being experienced in raising the front carriages and special care was necessary in extricating the bodies in order to preserve the identity of the victims. 'A crane has been erected at the head of the bridge for raising the wreckage, but as the waters of the Birs continue to rise, this task is rendered very difficult. A second-class coupé, probably occupied by foreign visitors, lies undermost, and it is feared that it

---

[824] *The Times*, London, England, 16th June 1891
[825] *Railway Accidents of Great Britain and Europe* by Ascanio Schneider and Armin Masé, pg. 245
[826] *The Times*, London, England, 15th June 1891
[827] Ibid, 16th June 1891

contains many victims. An unofficial subscription has been opened for the relief of the sufferers by the disaster.' [828] By 18th June 71 bodies had been recovered and a dam had been constructed to prevent the strong river current from carrying the corpses away.[829] The final death toll has been variously quoted as 71, 73 and 74. About 170 other passengers were injured to various degrees.

The single-track wrought iron bridge was 42 metres long and crossed the river on the skew about five metres above water level. It was described as having main bearers 'formed as statically determinate girders in triangular framework with parallel chords 6.2m apart and upper connecting members.' The design was markedly different from the more usual ones with close-meshed lattice girders and appeared extremely light.[830] Like the others on the line the bridge had been built by the firm of Gustave Eiffel, the designer of the famous Eiffel Tower in Paris, but after the accident M. Eiffel disclaimed all responsibility for it, saying that it had been built from plans supplied by the railway company and to a design conceived by Swiss Government engineers.[831] The Basle newspaper *Der Bund* pointed out given that the plans bore M. Eiffel's signature and the materials for its construction were obtained under his firm's consent it was purely and simply an Eiffel bridge.[832]

The bridge had opened for traffic in 1875. Six years later, in the autumn of 1881, the abutment on the left side had been undermined and the masonry seriously damaged when the river was in flood, leaving the ironwork resting on only three points, the fourth supporting point sinking by no less than 75 cm, putting the iron parts under great stress. Numerous cracks had appeared, and the affected part had had to be replaced. No further problems had appeared and in 1890, following the introduction of heavier locomotives, the cross-members, but not the main bearers, had been strengthened.[833]

The government inquiry into the disaster published its results in just over two months. It found that:

The individual components of the bridge had been too weak and defectively constructed to begin with.

The iron used did not, for the most part, meet the necessary requirements regarding tensile strength and toughness.

The 1881 flood damage had caused a permanent weakening of the load-bearing capacity of the structure.

The strengthening carried out in 1890 had only extended to individual parts of the bridge; other and important weaknesses had remained.

No derailment had occurred prior to the failure of the bridge.

---

[828] Ibid, 17th June 1891
[829] Ibid, 18th June 1891
[830] *Railway Accidents of Great Britain and Europe* by Ascanio Schneider and Armin Masé, pg. 246
[831] Webpage – An Engineer's Aspect: The Moenchenstein Railroad Bridge Collapse – June 14, 1891
[832] *The Times*, London, England, 29th June 1891
[833] *Railway Accidents of Great Britain and Europe* by Ascanio Schneider and Armin Masé, pg. 246.

The principal cause of the collapse was the insufficiently strong central struts, subsequently the eccentric fixing of the struts and the poor quality of the iron had been enough to precipitate the collapse.[834]

At 1.45 a.m. on the morning of Thursday 5th May 1892 an eastbound Atchison Topeka and Santa Fe express was wrecked at a trestle near Revere, Missouri, United States, not many miles from Medill, Missouri. Some reports identified the train as No. 4, the California vestibule express, but it was actually train No. 6, the Colorado vestibule express, running ahead of the former. The California train was due an hour earlier than the Colorado one but had been delayed at Kansas City awaiting connections. This turned out to be fortunate, for the California train carried a heavier load of passengers, and had it been involved in the accident the number of casualties would probably have been greater.[835]

Engineer Charles Crane told a newspaper reporter what happened when his train reached the fateful trestle, "We had had no rain from Kansas City to Revere. At Revere there was a light rain and it was not raining when the train struck the bridge. A perfect torrent fell immediately after the accident. I was running at a twenty mile per hour speed. It was pitch dark. My fireman, Noah Price, was in the cab. The trestle is about 150 feet long. The main flow of the creek is under the west end. The light from the headlight showed that the bridge was swayed out of line. It was, perhaps, two feet out of line. This could not be seen until the engine was on the bridge. I saw an accident was a certainty and expected to be plunged into the creek. Thinking of those behind I instantly applied the air. The engine passed over safely, but the tender wrenched off and slid away several hundred feet into the creek bed. The engine stopped one hundred and fifty feet east of the bridge. The crash followed instantly. I can't well describe the scene as all was chaos. Price and I ran back. We – all of us – worked at great disadvantage as there was no light. Six cars, the postal, baggage, smoker, tourist sleeper, chair car, and one sleeper were piled in one mass. As much haste was used as possible. All helped, even the injured ones. Most feared fire. Fire once well started would have meant much worse than had already happened. We had no buckets but fought out three incipient fires with mud. My dinner buck was finally used. Many were under the wreck and they would have been burned to death but for our efforts. It was not till day break that we could realize the real situation. The last sleeper remained on the western approach. 'For God's sake, help me,' came a cry. It was from the baggageman pinned under his car. He was extricated and was not badly hurt but was being tightly squeezed. It is presumed that earlier rains of the evening had caused driftwood to press the trestle out of line. I've never experienced such a wreck."[836]

A passenger, Samuel Watson, of Kansas City, also described his experiences. "There were seven cars in the train-mail, baggage, smoker, chair, tourist, and two Pullmans. I was in the last sleeper. We were running on No. 4's time, our train being regular No. 6, the fast train from Kansas City. Both trains were late, and we were running very fast. We passed a meat train at Revere, and it followed us out. Our engineer says that when he

[834] Ibid, pg. 253
[835] *Burlington Gazette*, Burlington, Iowa, United States, 6th May 1892
[836] *Burlington Hawkeye*, Burlington, Iowa, United States, 7th May 1892

approached the trestle, he saw it was out of plumb, and it flashed through him that we were going down. He said to the fireman, 'Jim, we're in for it this time,' and threw on all the steam he had, thinking he might be able to pass over. As he struck the bridge he instinctively shrank from his original purpose and threw on the air brakes, the conviction coming to him like a flash of lightning that the train could not get through. The engine passed over by so close a chance that it lost its tender. The train went down, the baggage, mail, smoker and chair car on one side, the tourist car on the other, and the sleeper standing upright on its front end. The last sleeper, in which I was, stopped on the remaining portion of the trestle.

"When the train took the plunge most of the passengers were asleep. I woke in an instant and was the first man to get out of our car. When I reached the rear door and looked out, I saw a bright light approaching around the curve, and my heart almost jumped into my throat at the thought that the meat train which followed us out of Revere was about to dash upon us. It was so black outside that one could not see anything except what came within reach. I got out upon the trestle and crept along as fast as I could towards the nearest bank. When I reached the ground and rushed back to stop the coming meat train, I found that the faithful flagman of our train had also escaped and had preceded me. The meat train was stopped in ample time.

"The feeling of exultation mingled with the deep sadness that I felt for the others whom I knew must be dead or dying when I trod on *terra firma*, can hardly be described. A friend of mine, a railroad operator from Kansas City, came out next. The end of the broken trestle upon which our car stood was weaving to and fro, and we immediately proceeded to aid the passengers in reaching the ground, several of the occupants of our car being ladies. It was raining as if whole heavens of water were falling, and we were drenched in a moment. The terrible distress of the ensuing few hours I shall never forget. The baggage, mail, tourist, chair and smoking cars were lying in shapeless masses about in the water or on the ground thirty feet below, and the uninjured in them were trying to get the wounded out. The swollen stream was probably 60 feet wide and not very deep but running like a racecourse and rising steadily. We on the saved sleeper were unable to lend a hand, it being wholly impossible to reach the others, or even to see them with any intelligible view except when the lightning flashed. The tourist car was the only one that lay entirely in the water, and in it a child was drowned. Most of the others escaped. The smoking car was entirely crushed, only one man escaping from it alive. One of the killed in this car was a big man from Oklahoma City, travelling with whom was a man named Burns, who was the hero of the night's work. He escaped uninjured and did Herculean work in rescuing others and caring for the wounded. The tourist's car caught fire three times, and as nearly as I could make out it was also Burns who did the directing or the extinguishing of the flames. A leather mail pouch was torn open and used to carry water to the fire. Other cars also caught fire but were extinguished by the alert Burns and his co-workers.

"Burns' friend from Oklahoma City was killed through his accidental return to the smoker for a bag that he had left there. He and Burns had tried to get a berth in the rear sleeper, but there was none left except an upper berth, and the large man requiring more room than could be divided between them, neither would accept the berth and they returned to the chair car, when Burns' friend remembered that he had left his valise in the smoker. He went after it, and a moment later went down to his death.

"The baggage car seemed to collapse sideways. It flattened out just like a paper box folds together when the ends are removed. There was an extra train crew asleep in this car on their way to Chicago. One of the brakemen and a baggageman were killed. One of the men killed in the chair car was found sitting erect in his chair, as naturally as if only asleep, and without a scratch on his body. "Why were not more of the passengers drowned? Well, it seems the cars containing most of the people jumped clear over the stream, and with the exception of the tourist's car none were very much in the water. We remained at the wreck most of the day. The water subsided during the forenoon and left the wrecked cars and the broken trestle exposed. The trestle ordinarily spans only a dry ravine, but the storm substituted a raging torrent that carried out the piling. There was no bridge. It was only a trestle work, about 150 feet long." [837]

In total seven passengers died in the accident. They were Lon Marker, of Kansas City, Missouri; S.E. Verkeler, of Westpoint, Missouri; Brakeman Lester Cornelius, of Knoxville, Missouri; J.C. Graves, of Macon City, Missouri, William Hines, of Oklahoma City, Oklahoma Territory; and Mrs G. Bausch and her child of San Bernadino, California, who were the last two to be identified. About 30 passengers, two postal clerks and Conductor Tucker of Chicago were badly hurt, the latter possibly fatally. Another 50 passengers were slightly hurt. W.C. Smith of Hartford, Missouri, claimed to have lost $325 in the wreck.[838]

The next significant bridge accident occurred on the Atlantic and Danville Railroad at 2.50 a.m. on the morning of 16th August 1893 at a trestle just east of Milton, Virginia, United States. A regular passenger train had left Danville, Virginia, at 1.35 a.m. for Portsmouth in charge of Conductor H. Morris, who was one of those who lost their lives. It had passed through Milton station and was running over the trestle when Engineer Peyton Tunstall felt the structure giving way. Engineer Tunstall said that the train had only been travelling at 10 miles an hour at the time. He had opened the throttle in an attempt to get the train clear of the trestle and the engine, tender and a boxcar got safely onto the embankment, but the following two passenger cars and the sleeper went down 60 feet to the creek below. 'The cars were shivered into kindling wood and the escape of any of the passengers was a miracle.' [839]

Apart from Conductor Morris the dead were all passengers. They were W.R. Ellison, of Alton, Virginia; J.G. Davies, a farmer, of Baskerville, Virginia; W.H. Elam, of Durham, North Carolina; James S. Lowe, a railroad conductor from the West, travelling as a passenger; Thomas Lee, (coloured) of Winston, Virginia and Francis Jenkins (coloured) of Salem, North Carolina. These were all the persons given in the list of killed but the newspaper reports then go on to say that a nurse of Mrs Harvey Giersch, travelling in the sleeper with her two children from Winston, North Carolina, was found under the water at the rear of the car, drowned, indicating an eighth fatality.

The noise of the crash was heard by the depot agent at Milton, who immediately gave the alarm by ringing the station bell. Locals immediately turned out in full force to rescue the victims from the wreckage.

---

837 *Burlington Daily Gazette*, Burlington, Iowa, United States, 6th May 1892
838 Ibid.
839 *Philadelphia Times*, Philadelphia, Pennsylvania, United States, 17th August 1893

The Pullman Conductor, J.L. Sizer, was interviewed by a reporter and related, "I had just gotten through with my collections and had reported to Captain Morris, who had passed forward on the train, and was preparing to retire when the crash came. Fortunately, it was in the forward end of my car. Mrs Giersch's nurse was in the rear end of the car, where she was found with her head and chest under the water and was most probably drowned, as there was no heavy timber or anything else on her body that might have caused her death."

The newspaper reporter also found a young heroine. 'While the rescuers were at work, they came to the little 13-year-old daughter of Mrs Giersch and tried to extricate her before turning their attention to her mother, but she cried out: "Don't mind me; I am not hurt, but please take mamma out."' Neither Mrs Gleisch, on her way with her children to Lawrenceville, Virginia, to visit friends, nor any of the other injured passengers was badly hurt.

The water in the creek had risen to a height of 12 or more feet and the general belief was that the rising water had undermined the foundation of the iron piers, causing the trestle to settle. The loss to the railroad company was expected to be very heavy. [840]

On 31st August 1893 Boston and Albany Railroad train No. 16 from Chicago to Boston, known as the Chicago special, was wrecked near Chester, Massachusetts, United States. The train was in charge of Conductor Guy Gray, with Engineer W.A. Horton and Fireman Louis Mitchell, and consisted of an engine and tender, a combination car, two Wagner sleeping cars, a dining car, two coaches and a smoking car. It had left Chester station, running a few minutes late, and had reached an iron truss bridge over the Westfield River a mile and a half from the station. The river ran over a rocky shallow bed 15 or 20 feet below the bridge. It was 12.30 p.m. and sat on the bank by the bridge were a gang of workmen from the R.F. Hawkins iron company of Springfield who had been a work strengthening the structure with iron plates to cope with the weight of new heavier engines. The bolts had been removed from one of the lap-plates, preparatory to putting a new plate in, when the men stopped for dinner.

The train ran onto the bridge and as its engine reached the second span it suddenly rose into the air, bounded forward ran about 15 feet and buried itself in the bank on one side of the line, while its tender broke loose and rolled down the embankment towards the river. The east span of the bridge then collapsed, taking with it the cars at the front of the train down to the bottom of the gorge with a tremendous crash. The two coaches and the sleeping car remained on the track. The buffet car was smashed on the rocks and the *St Elmo* sleeping car was turned upside down; it was in these two vehicles that most of the casualties occurred.

A rescue party formed at once comprised of workmen, uninjured passengers and local residents. As the bodies of those who had lost their lives were removed from the wreckage they were laid out in a nearby orchard and covered with blankets. There were soon 14 of them. Killed in the accident were 10 passengers, Wagner Car Conductor George Morse, baggagemaster J.H. Murray, Brakeman James McMasters and Express Messenger T. Everett Sedgwick. The injured were cared for in nearby farmhouses and at

[840] Ibid.

White's hotel in Chester. About 30 passengers were injured. The engineer and fireman had a lucky escape; Engineer Horton found himself pinned down by an iron rod, but his fireman was able to release him.[841] [842]

The Massachusetts Board of Railroad Commissioners investigated the disaster. Its report stated that the bridge involved, No. 131, known locally as Willcuts Bridge, had been constructed in 1874 by the Niagara Bridge Company. It had been regularly inspected but was reported to need strengthening to bear the weight of the heavier engines to be used on the line and this work was in progress at the time of the accident. The bridge had failed beneath a train which although it was being pulled by one of the heavier engines, was far from the heaviest train run over it. It had failed because in the course of their work to strengthen it the workmen had left it fatally weakened by disconnecting the old cover plates and permitting a train to cross it before the new plates had been put in place. An expert stated, 'To sum up the matter, a bridge admittedly too weak and badly designed to safely carry its loads was being reinforced (?) by a method which reduced the strength of one of its main members by a most serious, but unknown and unknowable amount through a violation of well-known engineering principles, probably aggravated by the carelessness of workmen, and yet was required to carry one of the heaviest passenger locomotives and trains at a high rate of speed; hence it fell.' [843]

At the beginning of November 1893 heavy rain had caused flooding in parts of Italy with floodwater four feet deep at Cassino and in the surrounding countryside. A mixed train from Ceprano consisting of two engines, four cattle waggons and some passenger carriages was crossing a bridge between Cassino and Rocca d'Evandro when the structure, whose foundations had been undermined by the floodwaters, collapsed. The engines and cattle waggons fell into the river, but the passenger coaches fortunately remained on the track. Even so the two engine drivers, the two firemen and four cattle drovers lost their lives.[844] [845]

On the evening of Friday 15th December 1893 the westbound passenger train on the Western New York and Pennsylvania Railroad which had left Buffalo, New York, at 5.50 p.m. went through a trestle near Sheridan, New York, United States, just east of Dunkirk. Heavy rains had swollen a creek which had weakened the trestle and the structure gave way beneath the weight of the train. Initial reports spoke of the deaths of only three people: Conductor Harry Hodge, of Oil City, Pennsylvania; baggagemaster William McKane of Buffalo and passenger Oscar Porter of Brocton, New York, but a later report indicated that the bodies of six passengers had been recovered from the wreckage. Three of the other dead were identified as Mrs Porter, of Brocton, Oscar Porter's mother; George Ryman, of Fredonia, New York and a Mrs Stockholm of New York. [846] [847] [848] The

---

[841] *Boston Daily Globe*, Boston, Massachusetts, United States, 1st September 1893
[842] Ibid.
[843] Ibid, 9th September 1893
[844] *Evening News and Post*, London, England, 9th November 1893
[845] *St James Gazette*, London, England, 9th November 1893
[846] *Bradford Era*, Bradford, Pennsylvania, United States, 16th December 1893
[847] *Logansport Pharos Tribune*, Logansport, Indiana, United States, 16th December 1893

engine passed over the trestle, but the tender, baggage and smoking cars went down to the creek and were crushed by the cars behind them. Several other passengers and the engineer and fireman were injured.

An accident similar to the Canadian one near Claremont in June 1890 occurred on about 14th January 1894 on the narrow gauge North Pacific Coast Railroad between Duncan's Mill and Cazadero, California, United States. Because of fears that the track might have been damaged by washouts following heavy rain engine No. 61 was detached from its train and sent forward to examine the line. It appeared that the engine had crossed a bridge over Austin Creek safely and proceeded further down the line before returning towards its train. The piling of the bridge having meanwhile been undermined by the floodwaters it collapsed beneath the engine as the locomotive attempted to re-cross the structure. In addition to Engineer Briggs and Fireman Collisten, riding on the engine were Conductor Baker; Station Agent Sabine; Cazadero postmaster Tom Gould; Frank Hart, proprietor of the Cazadero hotel; William Bramer, a clerk at the hotel and a man named John Rich. The engine fell about 20 feet into the water. Conductor Baker jumped and survived but the other seven men all drowned. What the four non-railwaymen were doing riding on the engine was not explained.[849] [850]

At 9 a.m. on the morning of 12th April 1896 a serious accident occurred near Louisville, Kentucky, United States, half a mile from a location known as Bedford Junction, on a belt railway on which stone from Bedford Junction quarry was transferred from one line to another. A train consisting of an engine and just two gravel cars was running over this line at high speed over what was known as the Standard Trestle when the last span from the Bedford side collapsed. The engine had already cleared the trestle but was pulled off the line by the weight of the cars behind. The two cars fell 75 feet with the trestle and the derailed engine fell down the embankment. Riding on the train were a conductor and workers engaged in ballasting the track. Fireman Sherman Carpenter, Conductor Charles Ogden, Warren Leonard and men whose surnames were Masterson and Bevins were all killed instantly. A man named Charles Davis died the same afternoon and Engineer Mensener was so seriously injured that he was not expected to survive long.[851]

Two days after Christmas 1896, on Sunday 27th December, the Cahaba River Bridge, 27 miles south of Birmingham, Alabama, United States, was the scene of a major disaster. The disaster occurred at 7.50 a.m. and involved local passenger train No. 40 of the Birmingham Mineral Railroad. The train had left Birmingham at 6.30 a.m. and was scheduled to make a circuit of the Birmingham Mineral branch, calling at all the important mining towns along the route. It consisted of an engine, baggage car and two coaches and was in charge of Conductor A.P. Connell, one of those who lost his life in the accident. The passengers on the train were mostly miners and local residents and their

---

[848] *Castle Rock Journal*, Castle Rock, Colorado, United States, 20th December 1893
[849] *Columbus Daily Times*, Columbus, Indiana, United States, 17th January 1894
[850] *Bloomington Weekly Leader*, Bloomington, Illinois, United States, 18th January 1894
[851] *Salt Lake City Herald*, Salt Lake City, Utah, United States, 14th April 1896

families with round trip holiday tickets returning home after visiting friends and relatives in Birmingham.[852]

Leaving the main line of the Louisville and Nashville Railroad at Taccoa the Birmingham Midland branch passed through Gurnee, and six miles beyond there crossed the Cahaba River, a stream only about three or four feet deep at the time of the accident. Details of the bridge are hard to ascertain but it was either 800 or 1,200 feet in length and had at least two main iron spans across the river approached by wooden trestles. The main spans were reportedly 110 feet long and the maximum height of the bridge was supposedly also 110 feet.[853] [854] The Keystone Bridge Company of Pittsburgh, Pennsylvania, owned by steel magnate Andrew Carnegie, had built the bridge.

Engineer Frank White apparently realised that something was seriously amiss as his train got onto the bridge and reversed his engine, but it was too late. Like Conductor Connell he was killed, but Sam Spencer, his coloured fireman, jumped from the engine while it was in mid-air and escaped with a broken arm. Doubtless much traumatised by his experience the fireman ran from the scene soon after the accident. The whole train fell from the bridge to the creek below, and fire broke out from damaged stoves in the coaches. consuming the wreckage and killing a number of those who had survived the fall. The number of people on the train was unclear but it was estimated that five railwaymen and 17 passengers lost their lives,[855] with only Fireman Spencer and eight passengers surviving the crash.

'Of the eight other survivors three were children, all of whom had their feet burned and smashed, and a lady who had both feet crushed. The other three were Henry Handberry, a conductor from Birmingham., who was taking his wife and two children out for a ride around the circuit, and Will Gardner and Andrew Bryson, miners from Blocton. Handberry was on the same seat with his wife and children when the crash came. The latter three were killed and he was pinned down and would have been burned alive had he not been rescued by Bryson and Gardner. These two men also saved the unknown woman and the three children, but by the time they had rescued these five persons the heat had become so intense that they were compelled to desist in their work of rescue.

'It is thought that fully three-fourths of the dead were killed outright in the crash, while the other five were pinioned down and cremated. Gardner and Bryson were both badly hurt, while Handberry will probably die.'[856]

A passing farmer heard the crash and going nearer saw the two fallen spans of the bridge, then the burning wreckage in the shallow river below. Then he heard the groans of the injured and dying victims and without waiting further rode his horse as hard as he could to Hargrove, a telegraph station four miles away. From Hargrove news of the disaster was telegraphed to Blocton and to Birmingham. Meanwhile a few country people had gathered at the scene but found that there was little that they could do.

'When the relief train from Birmingham arrived, nothing was left of the wreckage but

---

852 *Tyrone Daily Herald*, Tyrone, Pennsylvania, United States, 28th December 1896
853 Ibid.
854 Webpage – The Cahaba Bridge Train Wreck
855 Ibid.
856 *Tyrone Daily Herald*, Tyrone, Pennsylvania, United States, 28th December 1896

the smouldering ruins. Charred corpses were packed in between the iron framework where the seats had been, most of the bodies having been burned beyond recognition. Some had their heads burned off and of others nothing was left but the mere skeleton. As rapidly as possible the remains were taken out and laid in a row on the river bank. At 4 o'clock twenty bodies had been found and no more were in sight. It is possible, however, that there were others under the wreckage, and also that some may be hid under the water.' [857]

At 1 p.m. the second accident followed as an indirect result of the first. A wrecking train which had been sent to the scene was standing on the line close to the bridge when it was run into from behind by a construction train from Birmingham carrying men and timber to repair the structure. James Estis, of Boyles, was fatally crushed in this second accident and several workmen were more or less hurt.

Initial reports stated that the disaster must have resulted from an act of sabotage, and one of the survivors reportedly caught a glimpse of those responsible. Thus: 'Dr Ray, a Blocton physician, who attended Gardner, one of the injured survivors, telegraphs the statement made by Gardner as to the cause of the wreck. Gardner says that when he felt the cars leave the track he looked out and saw three savage looking men rushing from a hiding place down towards the water's edge, and that after the wreck they went through the wreckage, robbing the dead and wounded, and then fled into the woods. They did their work quickly and offered no assistance whatever in rescuing the imprisoned and injured passengers. This story, however, has not yet been corroborated, but several other facts tend to show that it was the work of train wreckers.

'An examination of the engine shows that Engineer White had shut off steam and reversed before it went down, indicating that he saw danger ahead as he approached the bridge, and tried to stop his train. His charred body was found with his hands on the throttle. The survivors all agree that the train left the track and bumped along the ties, and then with a crash bumped through the bridge. Expert engineers say that there is every indication that a rail was removed, which derailed the train and caused it to pull down the bridge. The crossties show marks on the wheels.' [858]

That the disaster arose from a deliberate act thus seemed clear and the newspapers were writing of 'fiends in human form', while on 31st December it was reported that a reward of $10,000 was being offered for information leading to the arrest of the perpetrators.[859] No one was ever caught but the theory that sabotage had occurred persisted. Eighty-two years later Shaw in his book included the accident in his 'Sabotage and Vandalism' chapter, though he added that 'There seems to have been nothing on the excursion {sic} train worth robbing by such extreme methods.' [860] More recent research, however, has revealed that the final Louisville and Nashville report on the disaster, prepared by their chief industrial engineer, had concluded that it had resulted not from a wicked deliberate act but from the faulty construction of the bridge by the Keystone Bridge Company. Lawyers had pointed out to the railroad company that should this become public there would be lawsuits and litigation directed at both it and the bridge

---

[857] Ibid.

[858] Ibid.

[859] *New Brunswick Daily Times*, New Brunswick, New Jersey, United States, 31st December 1896.

[860] *A History of Railroad Accidents*, Safety Precautions and Operating Practices by Robert B. Shaw, pgs. 274-5

company, so the decision was made to conceal the truth and perpetuate the story of the saboteurs, who, not surprisingly, were never apprehended.[861]

Wabash Railroad New York fast mail and express train left the Union depot at Kansas City, Missouri, United States at 6.20 p.m. on the evening of Saturday 26th June 1897, in charge of Conductor C.G. Copeland with Engineer W.P. Carlisle and Fireman Rufus Bramer. The train was well filled with passengers and had one more car than usual. As the express crossed the Hannibal Bridge over the Missouri River just after setting off it began to rain hard and there was incessant lightning. At Randolph, about six miles out, the rain was pouring down in torrents and at Liberty Landing, about the same distance further on, it was raining even heavier. About 20 miles from Kansas City, near Missouri City, the line crossed a creek by a trestle about 80 feet long and 15 feet high. Except in wet weather there was no water in the creek. The exceptional rainfall this evening had transformed the dry creek into a raging torrent which was powerful enough to sweep away a waggon bridge a short distance upstream of the trestle and the debris of the road bridge struck the wooden piers of the railroad trestle and they were 'snapped off like pipe stems.' [862]

Wary of possible washouts due to the heavy rain Engineer Carlisle had reduced speed from 40 to 20 miles an hour but this was not enough to prevent the accident. The speed of the engine, although not that high, carried it across the trestle but the tender fell into the stream, followed by the baggage, mail, smoking, chair and sleeper cars, one after the other. The cars piled up on one another and the baggage and mail cars were smashed. Two Excelsior Springs coaches at the rear of the train remained on the track.[863] No one in the sleeper car was much injured and no one in the chair car was killed. The mail car next to the tender suffered most and was crushed by the heavier cars behind it and the five postal clerks in it were all killed. They were: G.M. Smith, Frank Brink, W.S. Mills, Gus Smith and Saul Winters, all of St Louis. Two railwaymen, Brakeman Edward Grindrod and Brakeman Charles Gressley, both of St Louis, also lost their lives. Conductor Copeland, who had been in the smoking car, was rendered unconscious in the accident and a little later it was thought that he had died, and his body was tenderly laid out. Fortunately, upon the arrival of medical men from Moberly his body was examined, and it was found that he was still alive. He regained consciousness and was taken to the Wabash hospital at Moberly where it was thought there were goods chances of his recovery. About 30 passengers were injured and were taken to Kansas City, Missouri City and Excelsior Springs.

Seeing what had happened behind him Engineer Carlisle had sped on to Missouri City for assistance.

A boy or a farmer (newspaper reports vary) living near the creek had seen the waggon bridge carried away and the trestle piers undermined. Knowing that the express was due he had run back along the line but had only managed to cover a short distance when the train came along, and his frantic cries and arm waving had come too late to

---

861 Webpage – The Cahaba Bridge Train Wreck

862 *Emporia Daily Gazette*, Emporia, Kansas, United States, 28th June 1897

863 *Salt Lake Tribune*, Salt Lake City, Utah, United States, 28th June 1897

save it.[864]

Almost all of the valuable Kansas City mail in the car was lost or destroyed. The clerks would have had all the pouches open sorting the mail at the time of the accident and the mail would have floated away in the current.[865]

Coincidentally, the day after the Missouri City accident a very similar accident occurred in Galicia, Austria-Hungary, near the town of Kolomea (now Kolomyia). On 27th June 1897 a very severe thunderstorm, with torrential rain and hail, accompanied by much thunder and lightning, was raging as the 1 a.m. overnight train from Kolomea to Lemburg was crossing a bridge midway between Kolomea and the next station at Turka. The bridge crossed a tributary of the River Pruth which was much swollen by the excessive rain and the floodwater had damaged two piers of the bridge, which failed beneath the train. The engine, mail van and five carriages fell into the water but a report from a Vienna correspondent stated that two conductors succeeded in detaching five other carriages from the train and saving the lives of many of the passengers.[866] Even so six or seven postal workers, five passengers and presumably also the driver and fireman were killed, and many other persons were reportedly missing, presumed drowned. One of those who lost their lives was an army doctor going on honeymoon with his wife the day after the marriage. She survived after struggling in the water for four hours. This led the Vienna correspondent to criticise the rescue arrangements. He commented that, 'The accident was not known of at Kolomea, which is one and a half miles off, for several hours afterwards, and as the town was flooded in parts, everybody thought of himself, and left the unhappy passengers to their fate.' [867] However, another report stated that a detachment of soldiers with doctors and the local authorities had left Kolomea for the scene of the accident as soon as news had been received there of it and had added that the flood in Kolomea had swept away hundreds of houses there, leaving countless families homeless and devoid of possessions, and that the whole population had spent a night of extreme terror.[868] Clearly the passengers on the train had not just been selfishly neglected.

On about 24th September 1897 a major catastrophe occurred on the Mysore State Railway in southern India.[869] It was the end of the monsoon season and the River Shimsha at Maddur, near Bangalore, had been partly washed away by the swollen river. A night train ran into the gap and the engine, a second class and four third class carriages were submerged and most of their occupants perished.[870] One report stated mysteriously that 'when the train plunged into the gap where the bridge had stood the driver and stoker of the engine were blown into fragments.' [871] Possibly the locomotive boiler had exploded. The death of 150 passengers was reportedly confirmed, 14 others were injured

[864] *Emporia Daily Gazette*, Emporia, Kansas, United States, 28th June 1897
[865] *Salt Lake Tribune*, Salt Lake City, Utah, United States, 28th June 1897
[866] *London Standard*, London, England, 29th June 1897
[867] Ibid.
[868] *St James Gazette*, London, England, 28th June 1897
[869] *Wagga Wagga Advertiser*, New South Wales, 28th September 1897
[870] *The Leader*, Melbourne, Victoria, 20th November 1897
[871] *Liverpool Courier*, Liverpool, England, 26th September 1897

and 70 escaped. 'Most of those killed and injured were 'natives.' [872] The train was apparently running at speed at the time of the occurrence. It was expected that many of the bodies would not be recovered as the current, flowing with tremendous strength, had swept away both bodies and wreckage.[873]

A colliery line in British Columbia, Canada, was the perhaps rather unlikely location of a serious bridge accident on 17th August 1898. The Union Colliery Company on Vancouver Island owned a branch of about 10 miles in length which ran between its mines at Cumberland and the sea at Union Wharf. About seven miles from Cumberland the railway crossed the valley of the River Trent by a wooden trestle bridge 95 feet high and 130 feet long. The line was mainly used for coal trains, but passenger trains ran as well. The coal trains ran without a conductor and were not meant to carry passengers, but the habit had grown up of passengers, both railway staff and others, riding on them anyway.

At 7.20 a.m. on the morning of 17th August 1898 the first coal train of the day left Cumberland for the Union Wharf with new, and heavy, engine No. 4, 20 coal cars and one car loaded with lumber. On the engine were Driver Alf Walker and Fireman Hugh Grant and the other railwayman on the train was Brakeman Alex Mellado. There were also seven passengers: Richard Nightingale, a contractor who was being employed to add two additional piers to the bridge which failed; Walter Work, the son of the works superintendent at Union Wharf; a man named Matt Piercy; two Japanese workmen named Nanko and Osana; and two women named Miss Frances Horne and Miss Lavilla Grieves. [874] [875] As the train entered onto the bridge it gave way and the engine and cars fell into the valley below, narrowly missing two men named Walker and Bell who were working on the construction of the new piers. Matt Piercy escaped by jumping off the rear of the train as it went down, but the other people went down with it and six of them were killed outright. Fireman Grant and the two women were seriously injured, and Miss Frances Horne succumbed to her injuries the next day.[876] The accident was the subject of a lengthy inquest, at which evidence was given that Douglas fir timbers of the bridge were rotten, and it seems likely that the deterioration of the structure combined with the weight of the train caused it to fail.[877]

Various Australian and British newspapers contained brief reports of what was apparently a very serious bridge accident in Chile in August 1899, with 50 or 60 deaths, [878] [879] but further details of the accident have been tantalisingly hard to come by, even from Spanish language sources. It appears that the accident occurred on Sunday 13th August 1899 to a train crossing the Mapocho River by the Talagante bridge near Santiago. The area had been affected by severe storms and the flooded river had

[872] *Boston Sunday Post*, Boston, Massachusetts, United States, 26th September 1897
[873] *Liverpool Courier*, Liverpool, England, 26th September 1897
[874] Webpage – Trent River Train Disaster
[875] Cumberland News, Cumberland, British Columbia, Canada, 20th August 1898
[876] Ibid, 20th and 27th August 1898
[877] Ibid, 13th September 1898
[878] *Adelaide Observer*, Adelaide South Australia, 26th August 1899
[879] *The Times*, London, England, 17th and 26th August

undermined one of the piers of the bridge, which collapsed beneath the train.

Most of the information regarding the events leading up to the accident comes from a civil court case of 1904 brought by the widow of the second Driver (Fireman?), Miguel Angel Molina, one of those who lost his life in the accident. It was alleged in her case for compensation that there had been rumours that the bridge was unsafe for six months before the accident and that several days earlier the driver had spoken to an inspector of his fears of a derailment because heavy rain had undermined the track, leaving sleepers unsupported, while there were also signs of damage to the third pier of the bridge. At 12.40 p.m. on Sunday 13th Señor Vivanco, the Talagante stationmaster, had reported to his superiors that the driver of the last train from Melipilla to Santiago had told him that the bridge had subsided several inches as his train passed over it. Traffic over the bridge was then suspended. Passengers and crew for the 5 p.m. mixed train to Melipilla found a notice to this effect displayed at Santiago station and a number of them then returned to their homes and hotels. The train crew were told to remain at the station but at 7 p.m. were ordered to leave, which they did, despite their misgivings. The train reached Talangante safely, and here the driver, his fireman (? described as an assistant in the Spanish account) and a brakeman went forward on foot with lanterns to examine the condition of the bridge. Unhappy that it was safe to proceed they then hesitated for over half an hour before a telegraph message from headquarters demanded that they proceed. They went on, and the anticipated accident happened. The bridge collapsed, the engine and carriages fell into the river, and those who were not killed outright clung to the roofs of the carriages all night but were eventually swept away by the floodwaters, it being impossible to rescue them. This alleged sequence of events leading to the accident was however disputed by the counsel acting on behalf of the State Railway Company.[880]

It has not been possible to ascertain details of the dimensions of the bridge but from a photograph of it taken after the accident it was clearly a fairly tall and lengthy structure with several metal lattice girder spans supported by stone piers. It had been built not long before the accident by the French Schneider et Cie company. After the accident other recently built "French" bridges were examined and discovered to have cracks and defects, whereupon the Chilean government required the French company to enter into a partnership with the Chilean Lever Murphy company in an effort to guarantee the safety of bridges in the future.[881] The correct number of dead was apparently 22. It is not clear if anyone on the train survived the accident, or if it was another Tay Bridge disaster with all hands lost.

On 4th August 1900 a bridge gave way under a passenger train between Guanjato and Silac, Mexico, on the Ferrocarril Central Mexicano (Mexican Central Railway). The engineer, two passengers and four soldiers were killed, and other passengers injured. Four days later a northbound passenger train was somewhat luckier. This time a washout was encountered just short of Queretaro. The engine and remainder of the train were

[880] Webpage - vlex.cl – apelacion de 25 de Agosto de 1904 Higinia Moyano viuda de Molina contra Ferrocarriles del Estado (in Spanish)
[881] Wikiwand – Lever Murphy & Co (in Spanish)

derailed, and many passengers were more or less injured, but no lives were lost.[882]

A very brief report via the Reuters news agency mentioned an accident to a goods train on about 27th August 1900. The train fell through a bridge at a location described as 13 hours' journey (by passenger train?) from Salisbury, Rhodesia. Five persons were killed.[883]

[882] *Delphos Daily Herald*, Delphos, Ohio, United States, 16th August 1900
[883] *The Times*, London, England, 29th August 1900

CHAPTER 9

OPENING BRIDGE ACCIDENTS

Accidents involving opening bridges appear to have occurred in only a handful of countries, with the majority occurring in the United States. It seems that most countries managed to a large degree to avoid building such bridges, which from a railway point of view are a constant source of expense and worry. Shaw states that the early railroads on the eastern seaboard of the United States were forced by opponents of the new form of transport to build such bridges over even insignificant waterways where they were never really needed.[884]

A minor spate of opening bridge accidents in North America in the late 1870s ceased after one at Woodbridge, New Jersey, in January 1880, and the next one does not seem to have happened until 1885, when there was one in Canada on the Grand Trunk Railway at Merriton, Ontario. Tragedy struck on 10th August 1885 when in a double-headed Pacific express running into the station the air brakes failed to work, and the hand brakes being insufficient to stop the train it overshot the station where it should have stopped. The bridge just beyond was open and both engines and the leading vehicles of the train fell into the Welland Canal and an adjoining pond. At least one of the engineers was killed and several people were injured.[885]

Bridgewater, Tasmania, was the scene of an opening bridge accident on 22nd July 1886. The bridge over the Derwent River, 12 miles from Hobart, had last been opened for a boat at 4.55 p.m. on the day of the accident. Edward Roach, the North Bridgewater stationmaster and bridge tender, told an inquest the day after the accident that he was sure that he had closed the bridge properly after the passing of the vessel, that the pawls securing the track at each end of the span had been in place and he was quite sure that he had secured two screws on the southern side. After he had closed the bridge the midland

[884] *A History of Railroad Accidents, Safety Precautions and Operating Practices* by Robert B. Shaw, pg. 187
[885] *Oshkosh Daily Northwestern*, Oshkosh, Wisconsin, United States, 10th August 1885

train from Hobart and the ordinary train from Launceston had passed over it safely before the arrival of the Launceston to Hobart express, the train involved in the accident.

There was a permanent speed restriction of six miles an hour over the bridge and driver John Robertson told the inquest that he was certain that he had not exceeded a speed of five miles an hour over the structure. He had received a green light from the stationmaster and as far as he was aware his engine had not become derailed until it reached the southern (Hobart) end of the bridge. The engine had overturned as it became derailed and Driver Robertson's fireman, William Shaw, fell from it into the river and was drowned. The other fatality in the accident was a passenger, Daniel Turner, who had opened the door of his carriage after the train had stopped on the bridge, and not realising where it was, had fallen from it into the river, and was also drowned. Another passenger, an old man named William Gunn, made the same mistake of trying to leave the train and was very lucky to be rescued from the water by four men in a boat after he had attempted to swim to a bridge pile. He was only resuscitated after 15 minutes' effort by a Dr Silvester. Mr Turner, it appeared, had fallen through a gap between the sleepers on the bridge, had struck his head against a bridge girder and was probably unconscious when he hit the water.[886]

The cause of the train's derailment remained a mystery. There was no evidence of any defect with it or of a broken rail and Charles Cameron Nairn, resident engineer of the Tasmanian Main Line Railway, said that he was unable to account for it, while the general manager, Mr C.H. Grant, stated that the speed of the train had nothing to do with the accident and that it could only have been caused by the bridge having been tampered with, which seems a rather unlikely possibility. The inquest jury returned a verdict of "Accidental death" with a rider that the crossties (sleepers) on the bridge should be boarded over.[887 888]

On 16th September 1886 the engine and two freight cars of an Illinois Central Railroad train fell from an open drawbridge at East Dubuque, Illinois, United States, but no one was hurt.[889]

Baltimore and Ohio Railroad engine No. 654 went through an open drawbridge, the Schuylkill Bridge, near Gray's Ferry, Pennsylvania, United States, on 8th August 1888. Engineer James Bell, Brakeman Harry Sawyer and a man named McCord went down with the engine, Sawyer being drowned, and Bell seriously injured. Fireman McMichael jumped and escaped unhurt.[890 891]

A blizzard was raging as a Boston and Maine Railroad freight train approached a drawbridge across the Merrimac River at Newburyport, Massachusetts, United States, on

---

886 *The Launceston Examiner*, Launceston, Tasmania, 26th July 1886
887 *Tasmanian News*, Hobart, Tasmania, 24th July 1886
888 *Australian Railway Disasters* by Ken Pearce, pgs. 22-4
889 *Mitchell Daily Republican*, Mitchell, Dakota Territory, United States, 17th September 1886
890 *Tyrone Daily Herald*, Tyrone, Pennsylvania, United States, 10th August 1888
891 *Mitchell Daily Republican*, Mitchell, Dakota Territory, United States, 10th August 1888

19th March 1890. As a result, the engineer failed to see that the span was open until he was almost on it and then with slippery rails, he was unable to pull up in time. The draw tender got the bridge to within three feet of being closed before the train passed on to it. The train came to a stand with the engine and one truck derailed, but with no further damage done. [892]

The first opening bridge accident to happen on the west coast of the United States was a major disaster which occurred at Oakland, California, on 30th May 1890. Here the narrow-gauge South Pacific Coast Railroad was carried by a swing bridge over San Antonio Creek. The only signal protection provided was a red flag flown by the bridge tender before he opened the span, a measure which was considered sufficient given that all trains were ordered to approach the bridge prepared to stop if necessary, before crossing. On the afternoon of 30th May Bridge Tender James Dunlop had opened the bridge for the yacht *Juanita* after hoisting the red flag. The yacht passed through and Dunlop was just about to close the span when he saw and heard a northbound passenger train bound for Alameda approaching at too high a speed to pull up. He desperately tried to close the bridge but the time remaining was just too short and engine, tender and packed leading passenger car fell into the water. Engineer Sam Dunn and Fireman O'Brien went down with their locomotive. The couplings snapped behind the first cars and the remaining two coaches remained on the bridge, but the sudden jar broke open the front of what was then the leading one, spilling a number of its passengers into the water as well. The car which had followed the engine into the river soon rose to the surface and those passengers who were able to escape from it were rescued by yachts and small boats. When a breakdown train arrived from Oakland the car was dragged to shallow water and its roof cut off to retrieve the bodies of victims inside, while small boats dragged the creek for more bodies. The number of dead totalled 13.[893] [894]

Engineer Dunn and Fireman O'Brien both survived the accident and disappeared afterwards, probably because they feared being lynched.

F.F. Finley of San Francisco, a survivor who was in the coach which went down with the engine told his story to a newspaper reporter, 'I was on the front seat facing the engine. Just as we approached the drawbridge it seemed to me that the draw was open, and a fearful accident was inevitable. Just then a man jumped from the engine into the water and then came a crash. A horrible crushing of timber and the snapping of heavy ironwork followed and at once consternation prevailed in the car. The next thing I knew the car was in the water and I found myself grasping for the door which I fortunately reached when the end of the car rose out of the water and with quite a number of people I escaped in this manner. Those who escaped were principally women and children. The car was about two thirds full when we left the wharf, and I should judge there were at least twenty-five to thirty people in it. There was a fearful outcry when the car began to fill, but this was almost immediately hushed to one final wail of despair.' [895]

---

[892] *Salem Daily News*, Salem, Ohio, United States, 20th March 1890

[893] *Colorado Spring Gazette*, Colorado Springs, Colorado, United States, 31st May 1890

[894] *A History of Railroad Accidents Safety Precautions and Operating Practices* by Robert B. Shaw, pg. 193

[895] *Colorado Spring Gazette*, Colorado Springs, Colorado, United States, 31st May 1890

The accident was clearly due to Engineer Dunn failing to approach the bridge at a speed low enough to be able to pull up short of it if required. Shaw points out that while a single flag or lamp might have been adequate in 1840 it was hardly satisfactory 50 years later, especially as although the flag should have been visible for some distance the bridge itself was out of sight around a curve. Contributing to the accident was the fact that because of the heavy Memorial Day traffic extra train crews had been pressed into service and the engineer and fireman did not know the route very well.[896]

An early afternoon freight train from New London to New Haven on the Shore Line Railroad was in a serious accident at 3 p.m. on 12th June 1890 near East River, Connecticut, United States, when it entered a partially open drawbridge across the said East River. The engine and several cars left the track, and had it not been for their being held by the narrow framework of the bridge they would have gone over into the water. Considerable damage was done to both the train and the bridge structure, and all traffic on the line was seriously delayed while the mess was being sorted out. No one was hurt. A pontoon bridge was constructed as a temporary measure for the transfer of the passengers from the local trains and the early express trains but 'all of the later express trains were diverted between Saybrook and New Haven via the Connecticut Valley and Air Line Roads.' [897]

A section of a freight train consisting of six cars fell into creek four miles from San Rafael, California, United States, on 9th May 1891 because the drawbridge was open. No casualties were reported.[898]

A brief and rather vague newspaper report in mid-July 1891 spoke of a drawbridge accident averted 'the other day' at Cos Cob, Connecticut, United States on the New York, New Haven and Hartford Railroad. The engineer of an eastbound freight train failed to see the danger signals and his train was derailed on the derail protecting the drawbridge.[899]

Another drawbridge accident occurred in Ontario, Canada at the end of August 1891. A Grand Trunk Railway train of 23 coal cars left Port Erie for Toronto on the afternoon of 30th August 1891, in charge of Conductor Campbell with Engineer William Hall, Fireman W.J. Crouch, Head Brakeman Hugh McKeown and Tail-end Brakeman Robert Shaw. Before the train had set off there was talk of asking for a pilot as the crew stated they were unfamiliar with the line, but none was actually requested. The train proceeded via Welland Junction, Merriton and Stoney Creek, where the front and rear brakemen swopped over as McKeown had never before travelled over the next piece of line. The train then took the Burlington Beach branch, by which time the night had got very dark,

---

896 *A History of Railroad Accidents Safety Precautions and Operating Practices* by Robert B. Shaw, pg. 193

897 *New York Times*, New York, New York, United States, 13th June 1890

898 *The Mountain Democrat*, Placerville, California, United States, 16th May 1891

899 *Hornellsville Weekly Tribune*, Hornellsville New York, United States, 17th July 1891 – reproduced from the *Hartford Courant*

with a north easterly storm and heavy rain. By this date most swing bridges were protected by distant as well as stop signals but that over the Burlington Canal on this branch was an exception, having no distant signal (cf Oakland accident of 1890). Engineer Hall apparently forgot this and passed the stop signal at 15 mph without braking until he suddenly saw the points near Ocean House and realised that he must be close to the bridge, and he had not stopped his train as per the rule book. Realising that the swing bridge was open (it was in fact just in the process of being closed for his train) he sounded several blasts on his whistle, shouted to his fireman to jump and took off out of the window, landing safely in the sand below the embankment. Fireman Crouch did not jump and lost his life along with Brakeman Shaw when the engine and nine cars went into the canal.

At the inquest into the accident the coroner was critical of Conductor Campbell, who had sworn that he had only ever been over the line once before, whereas it was shown that he had in fact been over it four times. The coroner's jury returned a verdict against the Grand Trunk Railroad by a majority of ten to six. The wreckage did not obstruct the canal, and the cars, though not the engine, had been retrieved by 3rd September.[900]

The year 1892 began with a swing bridge accident in England, on 1st January. In the early morning of New Year's Day, a London and North Western Railway train of 15 empty waggons left the Egerton Dock siding for Birkenhead, the route taking it over an opening bridge spanning the docks. The newspaper report does not indicate how the bridge was protected, presumably there must have been some sort of signal protection, but at any rate in the darkness the driver entered the bridge, which had been opened for some flats, and the whole train fell into and was swallowed up by the water in the dock, which was 25 or 26ft deep. The driver and a shunter had time to save themselves by jumping off, but the fireman went down with the engine. Fortunately, he was a good swimmer and was able to stay afloat until rescued.[901]

Special mail train No. 21 on the New York Central and Hudson River Railroad, which left New York Grand Central terminus at 4.25 a.m. met with an accident at the New Hamburg drawbridge, New York, United States, at 6.10 a.m. on 1st September 1892. The bridge had been opened for the little steamer *Young America* and was in the process of being closed again when the train appeared, travelling, it was stated, at almost 60 miles an hour. The engine cleared the gap, but its tender did not, and the leading mail car crashing into the tender pushed it against the engine, killing Engineer Joseph Owens and Fireman Edward Best instantly. Mail Clerk J.K. Kane also lost his life, but the other mail clerk in the car with him was unhurt. Following the mail car were two baggage cars, in the first of which was Conductor William Todd and Trainman John Hitchcock, with Trainman J.H. O'Neill in the second. Todd and Hitchcock were thrown to the end of the car and somewhat hurt but Todd was able to walk to New Hamburg station and send off despatches about the accident. Half a dozen passengers were injured.

Ten minutes after the accident the southbound train which left Poughkeepsie at 6.20 came along. It was carrying a large number of labourers on their way to work who at

[900] Webpage – Hamilton Harbour 1826-1901 – The Electric Era 1891
[901] *The Times*, London, England, 2nd January 1892

once got to work to get the drawbridge closed and succeeded in doing so in half an hour. A down freight train which had been following the mail was flagged down 300 ft from the wreck. The engine, tender and mail car were wrecked and much of the mail was badly damaged. The wreckage initially blocked both lines, but the up line was soon cleared, and trains worked past the wreck over it in both directions.

The bridge tender, Richard McGoldrick, stated that the warning signals had been properly displayed. 'The dead engineer was one of the best employees on the road.' Kane had only been in the mail service eight months.

The same drawbridge had previously been the scene of a major disaster in February 1871.[902]

A freight train on the Jacksonville Tampa and Key West Railroad became derailed and fell into the river from the Buffalo Bluff, Florida, United States drawbridge on about 5th April 1893, the engine and six cars taking the plunge. Engineer A.C. Knox, Fireman J. Holstead and stockman J.A. Lewis, in charge of horses and mules bound for Barbersville, Florida, were all killed.[903]

Another drawbridge accident occurred on about 29th April 1893, this time on the Illinois Central Railroad at the 92nd Street drawbridge in Chicago, Illinois, United States. A freight train dashed onto the bridge just as Bridge Tender James Jenson was swinging it. The bridge was open six inches, which was sufficient to derail the engine of the train, which ran into Jenson's house and then fell into the river, carrying the bridge tender with it. Jenson was drowned, but the engineer and fireman, who had jumped from the engine as it left the bridge, were able to swim ashore.[904]

On 1st November 1894 an engine and three cars on the Duluth Transfer Railway went through a drawbridge on the Wisconsin side of the St Louis River at West Superior, United States, 'and buried themselves in the eighteen-foot channel'. The brief newspaper reports did not mention any casualties.[905]

In an accident with similarities to that at Merriton, Ontario in August 1885, the Rockport express of the Boston and Maine Railroad, which had left Boston at 4.10 p.m. had a narrow escape from a possible disaster on 30th October 1896 near West Lynn, Massachusetts, United States. The draw bridge over the Saugus river had been opened for the passage of a coal barge. It was protected by signals, which according to the bridge tender were at danger. The barge had passed through, and the bridge was being closed again, when the Rockport express appeared in sight. The train slowed down, but did not stop, owing it was said to a failure of its air brakes to work. Engineer Doherty struggled to get the brakes to work and whistled for hand brakes. The train was finally pulled up,

---

902 *Sandusky Daily Register*, Sandusky, Ohio, United States, 2nd September 1892
903 *The News*, Frederick, Maryland, United States, 6th April 1893
904 *The Evening News*, Lincoln, Nebraska, United States, 29th April 1893
905 *Stevens Point Gazette*, Stevens Point, Wisconsin, United States, 7th November 1894

but not before the engine had reached the bridge and fallen into the river, while the tender and four passenger cars were derailed, the tender ending up hanging over the edge of the bridge. When the engineer and Fireman Jameson saw that their engine was going into the river they jumped and were rescued from the river somewhat injured. There were about 300 passengers on the train.[906]

In the first accident of this type involving an interurban train, one running between Bay City and Saginaw came to grief two miles south of Bay City, Michigan, United States, when it fell through an open draw at about 11 a.m. on 7th July 1897. The seven passengers on the service were all carried down the river and a woman, Mrs William McClelland, and three children were drowned. The three male passengers were seriously injured and one of them, a T.P. Klump, described as a travelling man from Geneva, Ohio, who was taken to St Mary's Hospital in Bay City, was stated to be so badly hurt that he could not recover. Motorman Herbert Riley and Conductor James O'Brien jumped from the falling car and were only slightly hurt. The accident was allegedly caused by the fact that Motorman Riley racing a Flint and Pere Marquette Railroad train, the latter company being a competitor for the suburban traffic and being so preoccupied with doing so that he forgot about the drawbridge.[907] [908]

A train had a narrow escape at Green Bay, Wisconsin, United States, on 26th October 1897. The Chicago Milwaukee and St Paul Railroad log train was being backed over a drawbridge across the East River which was not quite closed, and several cars were derailed. No one was hurt.[909] The engineer apparently mistook a swinging lantern on a passing milk cart for a proceed signal from the bridge tender.[910]

[906] *The Marion Daily Star*, Marion, Ohio, United States, 31st October 1894
[907] *The Steubenville Herald*, Steubenville, Ohio, United States, 8th July 1897
[908] *The Nebraska State Journal*, Lincoln, Nebraska, United States, 8th July 1897
[909] *Centralia Enterprise and Tribune*, Centralia, Wisconsin, United States, 30th October 1897
[910] *The Daily Northwestern*, Oshkosh, Wisconsin, United States, 27th October 1897

CHAPTER 10

TRAIN ACCIDENTS INVOLVING BRIDGES

Bridges could feature in train accidents in one of two ways. Either the bridge itself could collapse for a number of reasons such as faulty design, faulty construction or because its foundations had been damaged; or a train could fall from a bridge which remained intact. Bridge failures have been dealt with in a previous chapter and this one will concern itself solely with accidents where the bridge itself was not at fault. It may seem perverse to list such accidents separately instead of in those chapters dealing with primary causes of the accidents, the method adopted elsewhere in this book, but the fortuitous presence of a bridge has such a potential for transforming what might have been only a minor accident into a major disaster that it has been decided to treat such accidents separately, in the same way as accidents occurring in tunnels have been given a chapter of their own.

At 2.30 a.m. on Tuesday 1st March 1881 passenger train No. 3 on the Hannibal and St Joseph Railroad was derailed by a broken rail between Summit and Bevier, Missouri, United States, and the engine and tender, baggage car, three coaches, chair car and a Pullman car were thrown down an embankment. The baggage car was smashed but the passenger cars were not much damaged and none of the hundred or so passengers was killed, though about 10 of them and Fireman Antony Sallg were injured. A fire which broke out in the wreckage was extinguished before it could do too much damage.

Unfortunately this lucky escape from a worse accident was not the end of the affair. News of the accident was telegraphed as soon as possible, and a wrecking train was despatched to the scene from Brookfield consisting of an engine and tender, two flatcars, a derrick car, a tool car, passenger coach and a caboose and carrying 35 or 40 people; railwaymen, doctors and citizens of Brookfield. The front part of this train crossed Brush Creek bridge, two miles west of New Cambria, safely, but a truck of the derrick car broke, and the derrick swung over against the bridge, striking its trusses and timbers and causing it to collapse. The tool car, caboose and passenger coach fell 20 feet to the creek below and the caboose, in which most of the men were riding, was smashed to pieces. As in the earlier derailment fire broke out in the wreckage but it was put out by Engineer Richard Ash with buckets of water. Bridgemen W. Hallet, George Zwick and M. Jury,

Brakeman Frank Locke and Foreman J. Connors were killed outright, Dr. C.R. Wood of Brookfield was fatally injured, and the recovery of a D. Fitzgerald was considered doubtful. Almost everyone else on the train was more or less injured. The dead and injured were removed to New Cambria. Subsequent to the accident trains terminated on either side of the bridge, but it was hoped that it would be repaired in a few days.[911]

At about 10 a.m. on the morning of 19th July 1881 a Chicago St Paul Minneapolis and Omaha Railroad wild freight train of 43 cars hauled by two engines left the depot at North Hudson, Wisconsin, United States. It was in charge of Conductor Frank Fowler, with Engineers Mike Sullivan and John Griffith. About a mile and a half past North Wisconsin Junction, or about 4½ miles from Hudson, the train was stopped to wood up, and when it restarted a coupling link broke and the 14 rear cars began to run back down the gradient of about 1 in 70 towards North Hudson. To make matters worse another double-headed wild freight of 45 cars was at the depot just about to follow the first one eastwards. Master Mechanic Ellis and his brother caught sight of the runaway waggons and ran across the high trestle over the Willow River outside of the station to warn the engine crews, and their shouts of alarm were taken up by a crowd standing on the platform. All the crew thus escaped except for Engineer George Tridor who apparently did not realise at first what all the shouting was about and then attempted to reverse his engine out of the way. When he did finally try and jump it was too late, for he was struck on the forehead, chest and arms by pieces of the runaway caboose and killed, while some of the wreckage fell from the trestle. The loss was estimated at $70,000.[912]

Animals on the line were not often a serious hazard to a train but occasionally a collision with a large one could result in a bad accident, such as the one which happened near Anchorage, Kentucky, United States, on 9th September 1881 on the Louisville and Nashville Railroad. Although described in the newspaper reports as having happened near Anchorage the accident appears more precisely to have occurred some distance on the far side of the next station at Avoca, on the Shelbyville branch, at a bridge over Floyd's Fork. The train involved was the 5 p.m. from Louisville to Shelbyville, consisting of an engine, baggage car and two coaches, in charge of Conductor Petry, Engineer Honnaker and Fireman Gibson. About 70 (different reports stated 50 and 90) feet from the end of a bridge over Floyd's Fork on the Shelbyville branch, about four miles from Anchorage, Kentucky, its engine collided with a cow and was thrown from the track. Striking the bridge, the engine brought the structure down and the train fell 30 feet to the bed of a creek and was wrecked.[913] Known to have lost their lives in the accident were Mr Perry, of Louisville; Colonel Fielding Neel, of Shelbyville; Captain William Maddox, of Shelbyville; Walter Scarce, of Shelbyville; Mr Jones, of Shelbyville; Thomas Buchanan, of Shelbyville; and 'a stranger from California'. There were supposed to be two or three more bodies in the wreckage. A number of other passengers and Fireman Gibson were seriously hurt while the conductor, engineer and baggagemaster

---

[911] *Macon Republican*, Macon, Missouri, United States, 3rd March 1881
[912] *Bismark Tribune*, Bismark, Dakota Territory, United States, 22nd July 1881
[913] *Galveston Daily News*, Galveston, Texas, United States, 10th September 1881

Humberton were slightly hurt. [914]

Communications after the accident appear to have been delayed because there was no night operator at Anchorage, and it was not until about 10.30 p.m. that a message was received at Louisville giving notice of the accident and asking that assistance be sent. A relief train was then despatched with Dr Leber, the company surgeon. A Courier-Journal reporter attempted to accompany the train but the master of transportation in charge refused to allow a reporter to travel by it under any circumstances. Undeterred, two reporters set off on fast horses for the scene.[915] [916]

Engineer Honnaker told a reporter that he had only been travelling at 18 or 20 miles an hour when he saw a cow crossing the line from the left side and reduced speed. The animal crossed safely in front of his engine but was followed by a second one which he struck about 26 feet from the bridge and his engine left the rails, pushing the cow in front of it to the bridge. His headlight had been burning.

The owner of the cattle, Mr Richard Isaacs, saw the whole thing. The injured were treated by doctors in houses nearby and those living in Louisville were taken home by train at 4 a.m. the next morning.[917]

At about 7.30 p.m. on 21st March 1882 a snow plough left Sweet Briar station, Dakota Territory, United States, travelling eastwards on the Northern Pacific Railroad. It was followed by a work train and then a passenger train. The work train consisted of an engine, a flat car, two sleeping cars with accommodation for 30 men each, a dining car, a kitchen car and a store car in that order. Two miles beyond Sweet Briar station the line crossed Sweet Briar Creek by a ten-bent bridge (bent is an American English term meaning a transverse rigid frame). The snow plough passed over the bridge safely, as did the engine of the work train, travelling at 10 or 15 miles an hour, but its engineer heard a crack and felt a jar behind him, and before he had time to react disaster had struck. A wheel on the rear bogie of the flat car had broken about a mile before the bridge, and on the sharp curve approaching the structure the strain on the bogie was such that it broke to pieces and some of the pieces striking the sleepers knocked them out of position, derailing the leading sleeping car, which fell 30 feet from the bridge to the river below. The second sleeping car, dining and kitchen cars followed, the office and store car was partly derailed. Four bridge bents were broken to splinters and the two sleeping cars were completely demolished. There were about 50 men on the train at the time, 24 of them in the leading sleeping car, some of them sleeping in their bunks, others smoking, reading and playing cards. As soon as the car fell from the bridge and overturned the bedding caught fire, and all the eight dead were in this car. 'Those who escaped immediately grabbed such buckets as they could and worked like heroes to extinguish the flames, but the straw bedding, blankets and other combustible material made it an impossibility. It is supposed that seven of the eight men killed were killed instantly before being burned, as the voice of but one man crying for help was heard. He was lying with his feet to the flames and held in the angry toils of broken pieces of timber. His comrades could do

914 *Indianapolis News*, Indianapolis, Indiana, United States, 9th September 1881
915 *Indianapolis Journal*, Indianapolis, Indiana, United States, 10th September 1881
916 *Fort Wayne Daily Gazette*, Fort Wayne, Indiana, United States, 10th September 1881
917 *Indianapolis Journal*, Indianapolis, Indiana, United States, 10th September 1881

nothing and were obliged to see him gradually roast to death. The scene beggars description. Scattered here and there promiscuously among half-burned blankets, valises, clothing, broken car wheels and iron rods, were the charred remains of the unfortunate victims. One peculiar feature of the accident was the fact that but two of the twenty-three injured and not killed, were seriously hurt. They had their shoulders dislocated, and one of them had a broken arm.' [918]

The newspaper then went on to list the victims, with rather lurid descriptions of their injuries, after which it added a few more details of the accident. 'Messrs Simmons and Duncan who run the boarding train for the North Pacific, had over $700 in the rear car which was saved.

'Thos. O'Connor, engineer of the fated train, says the wreck caught fire almost immediately.

'The coroner's jury found the wreck to have been accidental and both the train men and North Pacific officials are held guiltless.

'A partially burned book with the following rather suggestive title was found in the ruins, "Knowledge is Safety."

'The wreck is being rapidly cleared away and Superintendent Taylor says trains will be running on time today.

'The ladies of Mandan are taking an active part in attending the wants of the injured, who are mostly foreigners, having no friends in this country.' [919]

On 16th June 1882 train No. 2 of the Clarksburg Weston and Glenville Railroad had left Clarksburg for Weston and was entering the trestle at Walden's, West Virginia, United States, at about 8.30 a.m. when the passenger car was pulled off the track. Before the train could be stopped the passenger car, the gondola car and one house car had fallen from the trestle, the passenger car overturning in its fall and landing on its roof.

'Word was immediately communicated to Clarksburg, and all the physicians in town were soon on the ground, {and} ready and willing people repaired to the scene in carriages and buggies with medicines and nurses.' [920] Of the 20-odd passengers in the car more than half were hurt, some seriously, and a Mr Carey, supposedly of New York, died at 11.30 a.m. and Mr Goldsborough, of Baltimore, at 1 p.m. Conductor J.W. Smith was also severely injured. The accident was supposedly caused by a rail turning over.[921] Engineer Samuel Steel, Fireman Jack Crogan and baggagemaster Ben Jackson all escaped injury, Jackson having jumped from the train.

On 29th June 1882 a similar accident occurred on the New York and Long Branch Railroad at Parkers Creek, New Jersey, United States. Lightning express No. 7 of the Central of New Jersey had started from Point Pleasant, New Jersey, at 7.35 a.m. that morning, and consisted of Engine No. 163, a baggage car, drawing room car, smoking car

---

918 *The Bismark Tribune*, Bismark, Dakota Territory, 24th March 1882
919 Ibid.
920 *Janesville Daily Gazette*, Janesville, Wisconsin, United States, 17th June 1882
921 *Philadelphia Inquirer*, Philadelphia, Pennsylvania, United States, 17th June 1882

and four passenger cars in charge of Conductor C.F. Tuck with Engineer George Chickener and Fireman Charles Foster or Forster. As far as Long Branch it ran as an accommodation train, stopping at Manasquan, Sea Girt, Ocean Beach and the other little stations to pick up businessmen from their summer houses and take them into New York for the day's work. Beyond Long Branch it ran as an express to Jersey City, stopping only at Red Bank.

Parkers Creek trestle was situated slightly less than four miles beyond Long Beach, between Branchport and Little Silver stations. It was all but 1,000 feet long but only about seven feet above the creek, which was a branch of the Shrewsbury River. The creek was about two feet deep at low tide, three or four feet at high tide, with black sticky mud at the bottom. To the south of the trestle the two main lines were crossed by a spur from the racecourse at Monmouth park, which continued across the west side of the trestle, and for some days, workmen had been occupied installing connections between the racecourse spur and the two main lines. It was at the connection between the spur line and the eastern northbound track that the express was derailed.

The train was travelling at about 40 miles an hour as it approached the trestle, fast enough according to a newspaper report for the workmen to have to quickly scramble out of the way to save their lives, though this may be an exaggeration. The engine derailed on the connection, but it and the leading six cars continued over the trestle until the rear and seventh car jumped off the structure and pulled the cars ahead of it off in turn. The couplings of the three rear cars held and they fell into the creek on their sides together; the two cars ahead fell into the creek separately, also on their sides. The driving wheel of the engine was off but close to the track, and the tender, baggage and drawing room cars were also derailed, the last-named coming to a stand overhanging the end of the trestle.

As soon as the train stopped Engineer Chickener and baggagemaster Leake ran back to aid the passengers who were screaming in the water. One of the first passengers to climb out of the smoking car was ex-Union commander and former U.S. President Ulysses S. Grant, now a businessman, who had boarded the train at Long Branch. He was only slightly hurt, with a cut to the leg, and immediately began directing the rescue operations. Planks were lowered from the trestle to the cars and passengers scrambled up them to safety, some of them wet through and others cut by broken glass from coach windows. The *New York Times* reported that there had been three fatalities; Salisbury Bradley, a New York contractor; George Demarest, a wealthy produce merchant; and Cephas Woodruff, President of the Newark Fire Insurances Company, who died from his injuries three hours after the accident. A large number of passengers were injured, a number seriously, two of whom seem to have died later.

News of the accident was telegraphed from Little Silver to Long Branch, and Superintendent Randolph, and an engine and two cars, carrying the doctors who were closest at hand, was despatched to the scene, while the wrecking (breakdown) train was telegraphed for from Elizabethport. The special train took the passengers back to Long Beach or to whatever other station they had set out from earlier, the injured all being asked to give their names and addresses. Within three hours the wrecking train had arrived with between 300 and 400 men, a temporary track was laid down the embankment and the afternoon and evening were occupied dragging the cars out of the water. The paper also noted that, 'it was noticeable that all vestiges of rails that might have spread apart or of faulty spiking along the track had disappeared soon after the accident.' [922]

The day after the accident the coroner's jury were at the scene and found a number of places where the spikes were missing. They were watched by Foreman Alexander Kier, and one of the jurymen said to him, "Well, I think you are responsible for the death of these men," to which Kier allegedly replied, "I wish I had been on the train and been killed. I know this will all be blamed on me, though it is not my fault at all." Another juryman was told by a Kier that was customary to unspike every other tie (sleeper) prior to putting in new rails. The sleepers themselves also appeared to be rotten.

A passenger, Mr John Scott, City Clerk of Jersey City, described his experiences to a New York Times reporter. "I was sitting in the smoking car. I had observed that Gen. Grant was sitting in front of me contentedly smoking. Glancing around the car, which was about one third full of passengers, I saw no one with whom I was acquainted or whom I recognized, so I opened my morning paper and began to read. The train was fairly under way, and we were nearing the bridge, when we felt a slight quivering of our car, which soon became a disagreeable jolting. I knew the car was off the track and had not made up my mind what to do about it when Gen. Grant got up, and stooping down in the aisle, grasped the iron work of the seat on each side. I said to myself, 'Old man, your head is level, if you are a General,' and I got up and followed his example. Hardly had I changed my position when the car gave a lurch to the right, and then went down what seemed to be an interminable distance.

"It finally settled in the water, lying on its left side. The right side was above the water, and Gen. Grant, poking out the window, drew himself up and climbed out of the car. I was submerged up to the neck, but as soon as I could disengage my right hand I struggled to my feet and found that the water was up to my waist on a level. I followed Gen Grant out of the window and discovered where we were. Gen. Grant lost his hat, but he was not injured. The story about a heroic brakeman pulling the General out of the car window is not true. He helped himself as rapidly as possible. Some of the passengers in the smoking car received bruises, but none of them were so badly injured as to be unable to help themselves out of the wreck. The first thing I saw on emerging out of the car was persons pulling the dead body of a man out of the car behind me. He had evidently been fastened down and had been drowned." [923]

Engineer Clickener and Conductor Tuck's statements were also reported. The engineer said, "I am engineer of the Long Branch train No. 7. This morning I left Long Branch at 8.15 o'clock on time. When we were near the Oceanport draw, I shut off steam as usual, and crossed the draw without incident, and was about putting on steam again when we struck the south end of the Parker's Creek bridge where they were putting in a new switch. My left back driver dropped off the rail. I shut off steam right away, applied the brakes, and reversed the engine. The baggage, parlor car, and smoker, with four cars, dropped off the rail and followed along on the ties until the east rail (outside rail) gave way, then the last coach dropped off first into the water. The others following one after the other altogether, until five were in the water. I tried hard to save the parlor car and threw the engine and thus saved it. The tank and engine lay close to the rail."

---

[922] *New York Times*, New York, New York, United States, 30th June 1882
[923] Ibid.

Conductor Tuck said, "The train left Point Pleasant and reached the south end of Parker Creek at 8.19.30. The first I knew of the accident was that the engine was shutting off steam. I stepped down the steps of the forward end of the last car and looked down to see if we were off the track. I had every reason to believe that we should stop before we went over the bridge. In a moment we were off the track. The three rear cars were still coupled and were separated from the next two, which were also coupled together. They were uncoupled from the drawing -room car. The latter remained on the bridge with the exception of half of the rear truck, which laid over the bridge. The baggage car and engine were off the track, but on the trestle. The next thing, of course, was to protect my train. As soon as I got off, I sent men to the rear and front of the train and then telegraphed to Long Branch for all the assistance necessary. We took all the injured but two back west. We also took the names of all persons injured and the extent of their injuries, except those who refused to give their names." [924]

Chicago Milwaukee and St Paul Railroad northbound freight train No. 7, in charge of Conductor Frank Sheridan, Engineer Ed Clough and Fireman Myron Hibbard, was due to meet southbound freight train No. 10, in charge of Conductor James Henry, with Engineer Matthew Trumbull and Fireman Dennis Shaffner, at Sni Magill siding, Iowa, United States, at 8.45 p.m. on Wednesday 26th July 1882. The northbound train was apparently superior and train No 10 should therefore have waited at Sni Magill until 8.50 if No. 7 was not there. No. 10 did not, however, wait until 8.50 and the two trains collided head-on on a bridge a mile south of Sni Magill. The crew of No. 7 stuck to their engine and in the collision Fireman Hibbard was killed instantly and Engineer Clough seriously injured. Engineer Trumbull and his fireman escaped but No. 7 was also carrying 10 passengers, members of a wood-sawyer's gang on their way to Dubuque along with their boss, Mr Grover, and his wife, several of whom were hurt. Also injured were Brakeman Frank Carney and (slightly) Head Brakeman William Miller of train No. 7, and a tramp named John Cayton, of Monroe, Wisconsin, who suffered a broken thigh. Nine cars of train No. 7 were thrown from the bridge into the creek below, making 'a massive wreck'. Fire broke out after the collision but was extinguished before much damage was done.

Two days after the accident, after the operator at McGregor had testified No. 10 left there at 8.17 p.m., and his clock was four minutes fast (the crew of No. 10 alleged it was 8.25) and a watchman at Sni Magill stated that he had been surprised to see No. 10 pass there at 8.40, 10 minutes early; the coroner's jury found that the crew of train No. 10 had caused the wreck, and an order was issued for the arrest of Engineer Trumbull and Conductor Henry.[925]

The American accidents involving bridges continued with one on Saturday 11th November 1882 on the 3ft gauge Bradford Eldred and Cuba Railroad. A train was derailed on Courtney Trestle, Pennsylvania, the locomotive and combination baggage and passenger car falling from the structure. Engineer James France was crushed under his engine and killed, and several passengers were hurt. One source attributed the accident to

[924] Ibid.

[925] *Dubuque Times*, Dubuque, Iowa, United States, 28th, 29th and 30th July 1882

excessive speed and track defects but at the inquest on Monday a verdict was passed attributing the derailment to an unknown cause. It was stated that the only witnesses to the accident were Conductor Mulcahy and a farmer Smith of Cuba.[926] [927] A few days later it was stated that the train had been thrown from the track by a broken wheel on the combination baggage and passenger car.[928]

Then on 22nd December 1882 a passenger train on the New York Lake Erie and Western Railroad was thrown from a trestle just west of the bridge over the Allegheny River on the Bradford branch near Carrolton, New York, United States, by the rails spreading and the baggage car, two coaches and a caboose fell a distance of 15 feet. Conductor Barnes was killed, and a number of passengers were injured. [929] [930]

Moving on into 1883 there was a fairly minor accident on 16th January at Petersburg Junction, New York, United States, on the Troy and Boston Railroad. The scene of the accident was a 200-foot-long two-span bridge over the Hoosick River. An evening freight train in charge of Conductor Coneally and Engineer Morty was crossing the bridge when it was derailed by a broken journal and eight cars fell from the bridge and piled up on top of one another in the river. A brakeman escaped by jumping. The bridge was expected to be out of use for a week or more.[931]

A Union Pacific Railroad iron bridge across Black's Fork, near Granger, Wyoming Territory, United States, was considered one of the best on the road, capable of sustaining any load that could be placed upon it, but an accident at 11 p.m. on the night of Saturday 8th September 1883, proved too much for it. There were long, steep descents to the bridge on either side. An eastbound freight train was crossing it when it became divided and the division was not noticed in the dark by the train crew. The front portion of the train had just cleared the bridge and was slowly climbing away from it when the rear part crashed into it. The cars were thrown against the sides of the truss irons, breaking them and making a complete wreck of the 145 ft span, which collapsed, taking 11 loaded cars with it. Damage to the bridge was estimated at between $16,000 and $20,000, in addition to which the cars were wrecked. Their cargo was not especially valuable, consisting of ore, wool and hides. It was rumoured that a tramp riding the train had been killed but in fact, although he had been injured, it was only slightly.[932]

Less fortunate than the tramp at Granger were two tramps on a Southern Pacific Railroad train on 1st November 1883 who were killed in an accident on the Santa Anna River Bridge in California, United States. The wheel of a car of a Southern Pacific train

---

[926] *Granville Times*, Granville, Ohio, 17th November 1882
[927] *Bradford Daily Era*, Bradford, Pennsylvania, United States, 14th November 1882
[928] *Madison Dollar Weekly Star*, Madison, Indiana, United States, 17th November1882
[929] *Logansport Chronicle*, Logansport, Indiana, United States, 23rd December 1882
[930] *Philadelphia Times*, Philadelphia, Pennsylvania, United States, 23rd December 1882
[931] *New York Times*, New York, New York, United States, 18th January 1883
[932] *Salt Lake Daily Tribune*, Salt Lake City, Utah, United States, 11th September 1883

broke as it was crossing the bridge and 18 cars were precipitated into the river. Two brakemen were badly hurt as well.[933 934]

Chicago Burlington and Quincy Railroad train No. 3 which left Burlington for the west at 6.55 a.m. on Sunday 9th February 1885 was in an unlucky accident about 4 p.m. that day when it was derailed by a broken rail just short of Bridge No. 197, between Cromwell and Prescott, Iowa, United States. The front of the train passed over the bridge unscathed, but two coaches and a Pullman car were thrown from the bridge, which had a maximum height of about 30 ft, and were wrecked. Six passengers were killed outright and about 20 were injured, some seriously. Conductor J.R. Hawkins, Divisional Passenger and Freight Agent W.J. Davenport and Trainmaster Henry Stoores were also hurt, but according to one report the latter two nevertheless walked to Prescott to raise the alarm, though the engine of the train had already proceeded there. News of the accident was telegraphed from Prescott to Superintendent Duggan at Creston, from where a relief train was sent, carrying railwaymen, all the local physicians and apparently a considerable number of sightseers, whom after an effort the railway officials had given up trying to stop boarding the train. A second relief train later followed.[935 936]

Commenting on the accident the *Fairfield Tribune* newspaper wrote, 'Such accidents as this one are liable to happen at any time, and the closest and most careful watching cannot prevent them. The chrystalization {sic} of the steel rails, which may crumble into many pieces at any time during the cold weather, is one of the most serious things the railroads must contend with. Though every piece of track is carefully inspected several times each day, says the *Burlington Gazette*, it is impossible to detect a rail which an hour later may break into foot lengths and hurl a train to destruction. It is one of the risks which must be met when travelling by rail, and as yet there is no way of preventing it. Chief Clerk Harrington of the Roadmasters office, to whom the section foremen report the broken rails found in Iowa, says, more than seventy-five on the main line alone, were reported during January. A record of these broken rails, describing the nature of the break, size of the material, where made, when laid, and the exact location is kept for reference and comparison. In this office are fragments of a rail which broke in about fifty pieces by the weight of a train, which was ditched by this defective rail. The main track of the C.B.& Q. is laid with the best steel rail, sixty pounds to the foot, which is replaced whenever worn or perceptively defective. These changes in the temperature are full of dangers and destruction for the railroads, and they must be ready at any time to contend with the fierce elements.' [937]

What a newspaper described, rather melodramatically, as 'a fearful accident' occurred on 9th February 1886 on a bridge over the Flint River near Wilhites, Alabama, United States, on the Louisville and Nashville Railroad. A northbound freight train was being run in two sections. The first section became divided, part of it remaining on the bridge,

---

[933] *Atchison Globe*, Atkinson, Kansas, United States, 2nd November 1883
[934] *Cambridge Herald*, Cambridge, Ohio, United States, 8th November 1883
[935] *Humeston New Era*, Humeston, Iowa, United States, 12th February 1885
[936] *Chariton Democrat Leader*, Chariton, Iowa, United States, 11th February 1885
[937] *Fairfield Tribune*, Fairfield, Iowa, United States, 12th February 1885

and this the second section ran into. The shock of the collision was too much for the bridge, one hundred feet of which was under repair, and it collapsed, taking with it five cars of the first section and the whole of the second; the engine and 17 cars. The wreck then caught fire and 18 cars were burned.

Fireman John Johnson fell under the engine and was drowned, and Brakeman Henry Boteler was trapped under a car and burnt to death. W.D. Johnson, brother of the fireman, was burned to death and Conductor George Young and Brakeman Thomas McCreay were badly burned. The freight consisted mostly of pig iron 'which will be saved'. Conductor L.O. Harris of the first section showed that he was a true railwayman by swimming across the almost frozen river in time to flag a following northbound passenger train which was due in a few minutes.[938]

A Northeastern Railroad passenger train which left Charleston, South Carolina, at 12 noon on Monday 7th June 1886 plunged through the Santee River Trestle, between St Stephens station and Santee River Bridge, smashing up the coaches and killing six passengers. Conductor B.G. Masock and Mail Agent F.W. Renner were badly hurt. It was supposed that the derailment had 'resulted from insecure fastenings of the rail on changing the gauge'. On news of the accident reaching Charleston a special train was despatched to the scene carrying Superintendent Royall and Dr Ancrum. The line was blocked for the remainder of the day.[939]

There was a lucky escape for the men involved in an accident at Parkersburg, West Virginia, United States, on the Cincinnati Washington and Baltimore Railroad on 29th October 1886. A heavy freight train collided with a cow on a bridge and the engine and several cars were derailed. The bridge then caught fire and collapsed, together with the engine and 17 cars, nine of them loaded with oil. Both train and bridge were totally destroyed, except the part of the train which had not reached the bridge. The trainmen had what was described as a miraculous escape. Damage was estimated at not less than $100,000 and repairs were expected to take some time. Meanwhile trains were diverted over the Baltimore and Ohio Railroad via Newark and Columbus, Ohio.[940]

The series of minor or relatively minor accidents involving trains and bridges or trestles in the United States continued on 31st January 1887 with one on the Pittsburgh and Western Railroad when a morning passenger train was derailed and it engine fell from Gerty's or Gorty's Creek trestle, near Evergreen Hamlet, Pennsylvania. Fortunately, the baggage car and coaches remained on the track, but Engineer Scott Marshall was fatally hurt, and the fireman was badly scalded and bruised.[941] Much worse was to follow at the beginning of February when a major disaster occurred on the Central Vermont Railroad.

---

938 *Norwalk Daily Reflector*, Norwalk, Ohio, United States, 11th November 1885
939 *Columbus Enquirer Sun*, Columbus, Georgia, United States, 8th June 1886
940 *Hawarden Independent.*, Hawarden, Iowa, United States, 11th November 1886
941 *Logansport Journal*, Logansport, Indiana, United States, 1st February 1887

Early on the morning of 5th February 1887 a six-car northbound Central Vermont Railroad passenger with about 80 passengers was approaching a 650-foot-long wooden bridge across the White River near West Hartford, Vermont. As the train was entering the bridge Engineer Charles Pearce felt a snatch and his engine shot forward. Looking back, he was in time to see the rear Pullman car, the *Pilgrim*, had left the rails and was tilting over the right side of the bridge. Before the train's speed of 40 miles an hour could be much reduced the Pullman car had gone off the edge of the bridge, taking the preceding three cars with it. The breaking of a coupling behind the second car saved the engine and leading two coaches from sharing the same fate.

Falling 60 feet to the river below the cars were smashed and at once caught fire. Engineer Pearce and his fireman scrambled down the embankment to try and help the passengers in the wrecked cars, and were assisted by passengers from the smoking car, the second vehicle of the train, but the intensity of the fire soon made further attempts at rescue impossible. It would seem that 36 of the passengers were killed in the disaster, most of them in the fire, though a few drowned in the icy water. Because of the severity of the fire identification of the dead proved very difficult but on 9th February it was stated that of the 84 people, including trainmen, known to have been on the train there were 34 dead, 36 injured and 12 known to have gone home, which left only two missing, who were doubtless among the charred bodies, as only 15 of the dead had been identified.[942]

The house of Mr and Mrs Oscar Paine was nearest to the scene of the disaster and at about 2.30 a.m. he and his pregnant wife were awakened by a violent rapping at the door. Mr Paine got up and went down and opened it, 'where he was met by a man half-dressed and in his stockinged feet who said, "For God's sake come quick; there's a man in the river and the train's off the track." He was terribly excited, and it was some minutes before they could realize what was the matter. Mrs Paine dressed quickly, lit the lamps, and then looked out of the door, while her husband ran rapidly for assistance to the nearest neighbour. She again looked out and saw that the fire had extended from the burning car to the bridge, and the flames lit up the whole valley.

'About this time those of the passengers able to walk had begun to come in. They presented a pitiful sight; some were bare-footed, some with nothing on but flannels, and some in night clothes, and the thermometer was eighteen degrees below zero. Some were completely exhausted from the effects of the shock and the tremendous efforts made to extricate their fellow beings out of the wreck, and they fell onto the floor and then crawled around to make room for others as they arrived. In the meantime her husband had called Mrs Paine's mother, Mrs A. P. Howe, who had hastened to her daughter's, bringing with her a small bottle of Jamaica ginger, the only stimulant she had in her possession.

'The scene at Paine's house was trying even to men of the stoutest nerve, but Mrs Paine and her mother, actuated by womanly sympathy, did everything the limited means of the house would permit to relieve the sufferings of humanity. Nothing but moans and groans could be heard, mingled with pitiful appeals to God to restore them to their friends of whom they could learn nothing, and who even then might be slowly burning up in the flames that were consuming the wrecked cars and bridge. Nearly every one was chilled through, and were begging, "For God's sake give me something warm; I'm freezing."

[942] *Boston Daily Globe*, Boston, Massachusetts, United States, 9th February 1887

Faithfully these two brave women worked, giving hot ginger tea, camphor, hot tea and coffee until their supply was exhausted. In the meantime, two or three died, but no one was present to remove them until the train came from White River Junction with the surgeons on board.

'After the floor and beds downstairs had been covered by people, five or six went upstairs and lay down. The people, as they lay thickly on the floor, were covered with blood, and they were cut and bruised and burned, deathly faint, partially delirious, sick at the stomach and in other ways suffering from the accident and the extreme cold, and the quiet house was transformed into a hospital, in which the sufferers could get nothing but temporary relief until the doctors arrived.

'Oscar Paine, after he left home, visited the house of every neighbour in Centreville, the name of that part of Hartford where the disaster took place. He rapped at every door, and conveyed to them the awful tidings, and aroused everyone, who at once repaired to the scene of the accident and did all they could to give comfort and assistance to the injured passengers.

'At the house of Mr Paine the freezing and suffering people would whisper to each other, asking, "Who are you?" "What is your name?" "Have you any friends on board the train?" And the replies would be greeted with pitiful sobs and fervent supplications to the Almighty for relief. In the kitchen sat a man clad in nothing but his flannels, who was apparently in a dazed condition. He said nothing but looked into vacancy. An examination revealed the fact that both his feet were frozen, and it was ascertained that he had worked so long in assisting others that he had become chilled and partially delirious. Among others in the Paine house was Ann Murphy of Boston, half-clothed, cut, bruised and burned, who, when taken to Peare's Hotel, rent the air with shrieks and moans. One of the occupants of the Paine house was O. Bolsvert, of Angeliana, with his wife, the latter not much injured. The husband lay on a mattress with a chair under his head under which was a pillow. He said he had been to New York, thence came to Boston, and was on his way home. He was asleep in the Pullman when the crash came, and as soon as he heard the sound knew the car was off the track, as he felt it bob up and down. The next thing he knew, he felt the knees of his wife grinding into his stomach. The car had turned bottom up, and he thought it turned over twice in going down. The pressure of his wife's knees hurt him terribly, but he couldn't move, and she couldn't get away from the position. The next thing he knew he was out on the ice, with his wife beside him. She had on a white nightdress, and it was covered with blood, which came from a gash in her head, but aside from that she was apparently unhurt. She seemed very anxious about the condition of her husband, who complained considerably about his stomach and bowels.

'At the Paine house lay Conductor Smith Sturtevant, who ran the train from Windsor, the most popular man on the road, well known from his twenty-five years of service. He was partially unconscious, and in his moments of consciousness would inquire anxiously about his passengers. On the night of the accident he started through his train after leaving White River Junction, and began to collect and punch tickets, and had arrived in the second passenger car, nearly to the rear end, and was in the act of punching the ticket of a young man when the accident took place. He went down with the car and was pulled out by Engineer Pearce when the car was burning. He had his left thigh bone broken, a fracture of three right ribs, a circular gash under the right eye, reaching nearly to the ear, a severe scalp wound on the back of his head, and about one-sixth of his body was burnt over, besides the end of his right toe burned off. The severe burns are on his back and right leg. The chances are against his recovery, but at 9 o'clock Saturday evening he was better than a man in his condition could be expected to be.' [943]

Examination of the site after the accident disclosed a broken rail and also a broken journal. According to Shaw in his book the evidence pointed without doubt to a broken rail having caused the derailment [944] but from a contemporary newspaper report of the inquiry conducted by the Railroad Commissioners there was some uncertainty as to whether it was due to the broken rail or the broken journal. As is often the case with derailments the problem was deciding what was cause and what was effect.

Section Hand Charles Sturtevant testified to helping the repairs to the track after the accident and of seeing the broken rail, which was about two feet long.

Roadmaster A.C. Dean told of finding the safety strap that went under the journal to hold it if the journal broke six or eight feet in the rear of the broken rail. A little further on he found the broken jaw that held the journal. After then describing how he had looked after some of the passengers he continued, "At the derailment the rails were all torn out of their positions for some distance, and some were twisted and bent. Six feet of the second rail was broken square off. The first fracture in the first rail was nineteen feet from one end. The ten feet remaining was broken into small pieces. The rails were rolled in St Albans. The blooms came from Germany, of Bessemer steel. The forward trucks of the *Pilgrim* were those that left the track. There was only one set of trucks that left the rail. The marks of the trucks were five or six inches from the rail at the bridge. I think the cars were held together when they went on to the bridge. I think the second car, the St Albans, stayed on the rail until the weight of the *Pilgrim* pulled it over. I have no doubt but that the *Pilgrim* turned over when it left the abutment. The trucks with the broken journal fell on each side of the bridge. I think that the broken journal broke the rail. There are indications that a hard substance scratched the left rail 100 feet south of the bridge. That was done by the broken journal from the safety strap. I think the journal broke ten or twelve feet from the broken rail; the curve is 3^0 45′. I have never known a journal broken from a truck being derailed, while I have known rails break from broken journals. The journals is a four-inch standard Pullman journal. I could see nothing to indicate any flaw in it."

John Mulligan, Superintendent of the Connecticut River Railroad, and George Todd, Superintendent of the New London Northern Railroad, agreed with Roadmaster Dean as to the cause of the derailment having been a broken journal.

The inquiry then considered the fire which had caused so many deaths and injuries in the accident and was caused by either an overturned stove of a broken oil lamp. It was stated that in New England the Connecticut River and Boston and Albany Railroads used steam heating, as did a few companies in the West, while 'The New York New Haven and Hartford are lighting successfully with compressed gas. Steam heating is not yet perfected enough for universal adoption, but the patentees ought to be encouraged, and in a year or two steam heat will be universal and kerosene lamps will be abolished.' [945] The

---

[943] Ibid, 7th February 1887

[944] *A History of Railroad Accidents, Operating Practices and Safety Precautions* by Robert B. Shaw, pg. 59

[945] *Boston Daily Globe*, Boston, Massachusetts, United States, 9th February 1887

dangers of using gas for lighting were clearly unknown or at least not fully appreciated at this time. Lighting by electricity was still in its infancy and the technology was not yet sufficiently advanced for it to be used on trains.

The question as to whether a broken journal or a broken rail caused the derailment was more than academic, even though given the technology of the time neither was really preventable. If the derailment was caused by a broken rail the Central Vermont would have to pay compensation to all the victims of the disaster, but not if it was a failure of a Pullman car journal. The following day the inquiry heard evidence from a Robert Thayer, a Professor of Civil Engineering, who stated that he could find no fault with the journal but said that the rails had been very defective and cleared considered a broken rail to have caused the accident.[946]

Another factor which led to the disaster, and which was brought up on the final day of the inquiry was the lack of guard rails or their equivalent on the approaches to the bridge. There were no guard rails as such on the bridge itself but eight-inch square pieces of timber fastened 18 inches outside of the running rails served the same purpose on the bridge itself but did not extend beyond the ends of it. [947] [948]

After the seemingly endless succession of accidents involving bridges in the United States, on 9th September 1887 one occurred in New South Wales. The 4.40 a.m. morning newspaper train from Sydney, consisting of 15 waggons and two brake vans, was crossing a viaduct between Lewisham and Petersham, and nearer the latter place, when it became derailed and all but four of the waggons were derailed and fell between 60 and 70 feet from the viaduct to the stream below. One or more explosions were heard within the wreckage and fire broke out almost immediately, consuming the waggons and damaging the wooden viaduct. The Metropolitan, Burwood, Leichhardt. Marrickville and Newtown Fire Brigades all went to the scene while 'All this time willing hands were busy emptying the trucks, many of which contained combustibles such as petroleum, kerosene, gasoline and dynamite.' Unloading the waggons of such dangerous substances while a fire was raging close by seems a highly hazardous occupation but the motives of some of the volunteers were perhaps not entirely free of self-interest, for the report continues, 'The freight otherwise contained a large quantity of fruit and general merchandise, but although every endeavour was made to remove these out of the danger a good portion of the goods was destroyed. As usual on such occasions, a great crowd of larrikins had formed a cordon around the burning train, ready to snap up any unconsidered trifles in the shape of boots, fruit, cases of spirits, articles of clothing, and miscellaneous property for the lawful possession of which they could not have accounted. One of their number was taken in the act by a constable of police and received a sentence of imprisonment at the Newton Police Court yesterday.'

Given the early hour of the train's departure it is not surprising that there was only one passenger on it, Charles Mann, a Sydney coachman. The carriage he was in was thrown into the creek, but he was not hurt and made light of the accident. The only casualty was

---

[946] Ibid, 10th February 1887

[947] Ibid, 11th February 1887

[948] *A History of Railroad Accidents, Operating Practices and Safety Precautions* by Robert B. Shaw, pg. 59

Guard Dunn, who knew nothing of the accident until he was thrown out of his van. He was considerably shaken up but had recovered sufficiently to help in the clearing up work later in the day.

The up line was cleared at 6.30 a.m., an hour and a half after the accident, a single-line working was in operation until the down line could be cleared at 10.30. Considerable disruption was meanwhile caused to the traffic. The daily newspapers for the country districts were also on the train and all were consumed, and the telegraph wires were brought down. The total damage done was estimated at about £3,000 in terms of damaged rolling stock and a similar amount in terms of lost goods.[949]

The inquiry into the derailment determined that it had been caused by a broken axle of a van, the axle being of poor design and manufactured from inferior material. It was recommended that a record should be kept of the mileage run by each axle and of what repairs it had received. The inquiry also found that the brake power of the train had been inadequate and that had it been fitted with continuous automatic brakes the result of the accident would have been much less serious and probably confined to a broken axle. The inquiry was unable to determine the origin of the fire.[950]

A serious collision occurred on the Canadian Pacific Railway in Ontario, Canada, on Wednesday 4th January 1888. Eastbound and westbound freight trains collided head-on as they were crossing the 80-foot-high Red Succor Trestle, near Stewart, and the whole of the eastbound train and the engine and tender, together with 30 feet of the trestle itself, fell into the valley below. According to one newspaper report Engineers Hollwood and Charles Atkinson, Fireman Mitchell and Brakeman Oates were all killed and Fireman Hayes, who jumped, was seriously injured and not thought likely to survive. Other reports stated that seven railwaymen lost their lives, both engineers, both firemen and three brakemen. Another brakeman who was riding on top of the cars of the eastbound train at the moment of the accident ran back the train, jumping from car to car as it fell from the trestle, and finally jumped from the rear car to the ground, badly shaken up and badly scared. The two conductors in their cabooses at the rear of the trains both survived. Eight of the cars which fell from the trestle were loaded with wheat and the next day passengers transferring between trains found themselves walking knee-deep through wheat.[951]

As to how the collision occurred a newspaper report stated, 'It is said that the accident was caused by the mistake of a conductor on the train going west. He is said to have comparatively new to the route and neglected to get his orders made O.K. at Schrieber. He consequently did not receive the order of the train dispatcher to side-track his train at Stewart station and allow the eastbound train to pass him there. The place where the accident occurred, Horse-Shoe Curve, is one of the wildest and most dangerous parts of the new line between Sudbury and Winnipeg. The Horse-Shoe is an immense series of trestle work over a heavy ravine some three-quarters of a mile long and about one hundred and twenty feet high in the centre.

949 *The Sydney Morning Herald*, Sydney, New South Wales, 10th September 1887

950 Ibid, 28th October 1887

951 *New Westminster Mainland Guardian*, New Westminster, British Columbia, Canada, 11th January 1888

'Another account throws the responsibility on the train dispatcher at one of the stations, who, it is said, failed to cancel one of the freight trains as ordered. The officials of the road here cannot or will not give any information about the disaster. It is a standing order of the company that no information of what happens on the road shall be divulged by their employees, and such information as has been obtained came through mail conductors and clerks running on the northern division of the road.' [952]

Just over two months after the above accident in Canada the United States suffered another major disaster near Blackshear, Georgia, on the Savannah Florida and Western Railroad. A newspaper reported rather patronisingly that, 'The news of the terrible wreck near Blackshear caused the wildest excitement among the simple folk of the Pearce county pines and white and black and old and young journeyed to the awful scene of death and destruction,' adding that, 'Twenty-seven were found to be dead at daybreak this morning and forty-three to be in a wounded condition, some of them serious.'

The train involved was New York and Southern express No. 23 (also known as the *West India Mail*), which had left Jersey City, New Jersey, at 9.45 p.m. on Thursday 15th March 1888, 45 minutes late, and Philadelphia, Pennsylvania at 12.45 a.m. on Friday 16th. On departure from Washington D.C. the train had consisted of a mail and baggage car, Pullman car, first and second-class passenger cars and the private car of President Wilbur of the Lehigh Valley Railroad and was in charge of Conductor W.L. Griffin. A large number of passengers boarded at Savannah, Georgia and the train was 'running at a brisk pace' when it reached the Allapaha Creek trestle (also confusingly known as the Hurricane River trestle) near Blackshear early on 17th March. The engine and tender had almost crossed the 800-foot-long, 50-foot-high structure and were nearing the embankment at its eastern end when the leading baggage car jumped the track and the trestle collapsed. The baggage car dragged the tender off the track in turn and both fell from the trestle along with the other four coaches of the train, but Engineer Dick Welch just managed to get his engine to safety. He then despatched Fireman Campbell to Blackshear on the engine for assistance and climbed down the slope to the wrecked train. His first task was to drag Porter Williard from the wreckage of the Pullman car *Minerva* and the two of them then began to fight the fire which had broken out and already consumed part of the baggage car.[953] Some of the injured male passengers helped the driver and porter quell the fire.

A newspaper account of the accident continued, 'The private car of President Wilbur was a smashed heap of wood piled on the apex of the wreck, and it was a herculean feat to remove the Lehigh Valley president, who was pinned beneath the roof of the car. His little son's mangled body was near him and two other sons were crushed, and bleeding lay a few yards distant. As if saved by a miracle George Gould and his wife were only held down by light timbers and were easily brought out of the wreck…There were many touching scenes. The doctors that came from Savannah and Jacksonville had more victims than they were able to attend to, and many suffered intensely while waiting to have their wounds dressed.

[952] *New York World*, New York, United States, 7th January 1888

[953] *Philadelphia Times*, Philadelphia, Pennsylvania, United States, 19th March 1888

'An undertaker from Waycross nearby, with assistance from Blackshear, began to embalm the bodies, and most of the injured were removed to Waycross and Blackshear and placed in the cabins of the primitive towns. When a special car reached Waycross the Wilbur family and George Gould and wife were put aboard and taken to Jacksonville. The accident is the first of its kind that has ever occurred on the road since it was built, thirty-five years ago.'

The derailment was caused by the fracture of the front axle of the baggage car.

A disaster in Mexico on 5th June 1888 had several notable features. It occurred 40 kilometres from Tampico on the Ferrocarril Central Mexicano (Mexican Central Railway). A construction train consisting of a car carrying 60 labourers and seven flat waggons loaded with a large quantity of wood was being propelled along the line when it collided with a cow only five yards from a bridge. The collision derailed the leading car, which broke through the 45-foot-long and 12-foot-high bridge. The remaining cars fell from the bridge in turn and piled up in a mass of wreckage surmounted by the engine. Thirteen of the labourers were killed, 20 seriously and 17 slightly injured and an unknown number buried in the wreckage. The American engineer and conductor jumped and were seriously injured. The labourers alleged that the train crew had been drunk and had ignored the labourer's frantic warnings.[954] Other reports spoke of the train having struck both a cow and a donkey and put the number of dead at 18 with 41 injured.[955]

A rear collision near Newton, Illinois, United States, on the Peoria Decatur and Evansville Railroad on 16th February 1889 did not result in more than slight injury but was expensive in terms of damage to property. Two freight trains collided on a high trestle over a piece of swampy ground and engine No. 8 of the colliding train went through the trestle and fell into the swamp, as did the caboose and a freight car of the first train. The engineer and fireman of the second train jumped and escaped with a few bruises and scratches. The freight car had been carrying a load of whisky, most of the barrels of which broke, and the whisky was lost in the swamp. 'One barrel remained intact and was seized by the natives and the whisky poured down their throats with avidity.'[956]

Canada suffered another bad accident on 27th February 1889 at St George, Ontario. It involved train No. 54, a Grand Trunk Railway working known as the St Louis express and running from Windsor to the Queenston Suspension Bridge, in charge of Conductor Revell with Engineer William Blackwell and Fireman Henry Angeles, the latter possibly the same fireman as one of those on one of the trains involved in the St Thomas disaster a year and a half earlier.

St George was only a small roadside station with a siding but as the express was passing the siding at the east end of the station a steel tyre on one of the main right-hand driving wheels of the engine snapped, the driving rods broke, and the engine was

954 *Laredo Times*, Laredo, Texas, 9th June 1888
955 *Kokomo Evening Gazette Tribune*, Kokomo, Indiana, United States, 6th June 1888
956 *Decatur Saturday Herald*, Decatur, Illinois, 23rd February 1889

derailed and ripped up the track as it continued along the line. To make matters worse, immediately beyond the siding was a wooden trestle bridge across a deep, wide creek with a maximum height of about 80 feet. The Pullman car and dining car went off the trestle, the Pullman car turning a complete somersault and finishing right side up while the dining car ended up almost vertical with the top of it leaning against the pier of the bridge. The following passenger car was left hanging off the edge of the trestle but did not fall. Eight passengers were killed more or less outright and over 30 injured, two of whom died later. The accident would have been considerably worse had it happened a little later as apart from the waiters there were only seven people in the dining car, though supper had just been announced and a few minutes later it would have been full.[957]

Some of the other passengers had lucky escapes because they had gone to the smoking car just before the accident. Woodstock town clerk G.C. Eden told a reporter for the Toronto Mail, 'I was in the smoking car, and feeling an unusual jolting in the car I turned to the window and saw we were just passing over a high bridge, when I immediately became aware that the hind carriage had left the track. I rushed out of the car and down the embankment, and there stood the car which had run immediately behind the smoker, standing squarely on its bottom, the inside presenting a terrible scene. Stoves, seats and people were all huddled together, and but two escaped unhurt – one a woman, of about twenty, the other a man of about the same age.'

Also in the smoker was C.A. Deslisle of Toronto. When it began to shake violently and was clearly derailed, he made for the door preparing to jump but found the train was over a deep ravine, "Shall we jump?" he shouted to a brakeman. "Not much," the latter replied, "as long as the old thing hangs to the rails." [958]

The accident was witnessed by a number of the local townspeople and so the alarm was quickly given, and help arrived at once. The dead and injured were removed to nearby hotels and private houses and attended by doctors and womenfolk.

Grand Trunk Superintendent Charles Stiff subsequently found the broken tyre in a field and concluded that the locomotive, No. 753 (which was a new one), had lost the tyre when it was 50 yards from the bridge, while examination of marks on the track indicated that the initial fracture of the tyre had taken place about 150 yards from the bridge.

Assistant Mechanical Superintendent Arthur Smith examined the broken tyre and was hard put to identify any flaws or cracks which might have been present before the accident, from damage caused by the accident itself, while the fractures were not rusty, suggesting that they had occurred at the time of the accident. Basically, given the technology of the day, the accident was unavoidable.

The inquest jury returned a verdict of "Accidental death" with a recommendation that the speed of trains over the bridge should not excess 12 miles an hour.[959] The speed of the train had nothing to do with the accident of course, it seems likely that the jury decided that they must recommend something and that was the best that they could think of.

[957] *Ironwood Times*, Ironwood, Michigan, United States, 2nd March 1889
[958] *Wreck!* by Hugh A. Halliday, pg. 73
[959] Ibid, pgs. 75-6

On 21st April 1889 an eastbound Iowa Central Railroad train was wrecked four miles west of Abingdon, Illinois, United States. The train consisted of two freight cars and two passenger cars, one of which was partly a mail car. One hundred feet or so short of a trestle over Cedar Creek the flange of a wheel of the second freight car broke and it left the rails and was dragged along the line, ripping up the track as far as the trestle, from which the two coaches fell 30 feet to the creek below. The passenger and mail car caught fire almost at once and the blaze spread to the second coach as well as to four spans of the wooden trestle. One passenger, Will Savage, of Berwick, was killed outright and four other passengers were seriously injured. A woman whom it was thought was on the train could not be traced afterwards and possibly was burned to death. Conductor Robert Colville was stated to be dying, Express Messenger Rogers suffered broken limbs and the unnamed mail agent a broken shoulder blade. The accident occurred at about 9.15 p.m.[960]

At 2.30 a.m. on the morning of 26th June 1889 shunting operations were in progress near Latrobe, Pennsylvania, United States, and the Derry shifter (shunting engine) had left 18 cars standing on the Loyal Hanna bridge while more loaded cars were taken out of the Latrobe Coal Works to add to them. Flagman David Heacock had gone to the signal tower, but it was not clear if he had gone out far enough to protect the obstruction. At any rate it was raining hard at the time and Engineer Elmer Caldwell in charge of the engine of a heavy 42 car train and coming down a falling gradient at 25 miles an hour apparently did not see the flagman's signals, or could not pull up in time, and collided with the stationary wagons. The collision tore up the rails and sleepers and the engine and a number of cars went over the side of the stone bridge and plunged 40 feet to the bed of the Loyal Hanna creek. At the same moment an eastbound freight train was passing and one of the derailed cars caught its caboose and pushing engine, badly damaging them. The wreckage caught fire and 18 loaded cars, 13 empty cars and engine No. 1318 were described as being a total wreck.

Engineer Caldwell and Fireman George Frolich went down with their engine and were buried in the wreckage, but they were by no means the only fatalities as there had been between 40 and 50 men on the westbound train who had been working at Johnstown, had just been paid off and were trying to make their way home. Initially only four of their bodies had been recovered but it was believed that there were between 15 and 20 more bodies buried in the wreckage.[961]

Another report the day after the accident stated that the bodies of 10 tramps had been recovered but that the water in the creek was about 12 feet deep and it was expected that there were more bodies to be found. The bodies of the engineer and fireman had also yet to be recovered. About 40 workmen had boarded the train at Bolivar; Conductor Barnhardt had put them off at Derry but many of them had then re-boarded the train. The statements of surviving tramps also indicated that there must be a number of their number unaccounted for. One or two of the bodies were badly burned by the lime. The loss to the Pennsylvania Railroad Company was estimated at $100,000. According to this report a freight train flagman named John H. Miller, not David Heacock, stated that his signals

[960] *Owensville Gleaner*, Owensville, Indiana, United States, 27th April 1889
[961] *Salem Daily News*, Salem, Ohio, United States, 27th June 1889

had been answered by Engineer Caldwell but that the latter had been travelling to fast to be able to pull up.[962]

Late in the evening of the same day as the Latrobe disaster, 26th June 1889, a northbound passenger train on the St Louis Arkansas and Texas Railroad collided with a cow on the trestle bridge across the Crooked Bayou in Arkansas, United States. The engine was thrown from the track and running across the ties (sleepers) shattered the bridge to pieces. The express car, baggage car, mail car and two passenger coaches fell 28 feet into the bayou. S.O. Stofford, route agent of the Southern Express Company, was killed outright and Postal Clerks Sanguin and Jackson were fatally injured. An unknown woman suffered severe head contusions and a large number of other passengers were bruised. A relief train was sent from Pine Bluff and returned there with the passengers and the body of Stofford.[963]

The inaugural train over the new Knoxville Cumberland Gap and Louisville Railroad met with disaster at the Flat Gap Creek Trestle, Tennessee, United States, on 22nd August 1889. The train was carrying a select party of excursionists consisting of members of the Knoxville City Council and Board of Public Works, representatives of the Chamber of Commerce and 'the very flower' of Knoxville business and professional men and consisted only of an engine and two cars. It left the track at a crossing on the approach to the trestle and the rear car fell from the structure. Every passenger on the train was injured but one. It took a long time to obtain medical attention and little could be done for the victims until the train reached Knoxville. 'Many of the injured had to be brought back on flat cars and the last part of the journey was made in driving rain.'

By the time of the report on the accident three of the injured men had died and it was thought that others could not live. Those killed were Judge Andrews, the most prominent lawyer in East Tennessee; S.F. Powers, a leading East Tennessee merchant; and Alexander Reeder, a leading politician who had held many positions of trust. The injured included the President of the Chamber of Commerce, two judges an attorney and a professor. It was thought that Alderman Barry in particular, who was thrown against a window when the accident occurred, had little chance of recovery.

The newspaper report added that, 'The belief is general that the wreck can be attributed to the bad condition of the road. The city subscribed $225,000 towards the construction of the road two years ago, and the contract period expires today. It was to show the city authorities that the road was completed that the excursion was given. George Eager of Boston, the president of the company who built the road, and Lenox Smith, vice-president of the Knoxville Cumberland Gap and Louisville, escaped injury. They were in the forward car, and to this fact is attributed their escape.' [964]

The bad start to the Knoxville Cumberland Gap and Louisville Railroad was an augury of things to come, for it lasted only until 1896 before it was taken over by the

[962] *Van Wert Weekly Bulletin*, Van Wert, Ohio, United States, 5th July 1889
[963] *Kokomo Daily Gazette Tribune*, Kokomo, Indiana, United States, 27th June 1889
[964] *Syracuse Evening Herald*, Syracuse, New York, United States, 23rd August 1889

Louisville and Nashville Railroad and much of it abandoned.[965]

At the beginning of 1890 it was the turn of the Monon (Louisville New Albany and Chicago) Railroad to suffer a bridge accident of the type being described in this chapter. The accident happened near Carmel, Indiana, United States, on 27th January 1890. The locomotive of Monon through passenger train from Chicago to Indianapolis, which started at 11.55 p.m. Sunday night and was due in Indianapolis at 8.30 a.m. on Monday, was in charge of Engineer "Dad" Strope. As it left Monon he noticed that he was running a little late and after continuing at the usual pace to Westfield he opened the throttle to try and regain time on the last stretch into Indianapolis. Just north of Carmel was a long curve and in its centre a bridge over Wilkinson Creek. The train was travelling at considerable speed as it rounded the curve and when it was about 40 yards from the bridge the bogies of the tender left the rails. When the bridge was reached, the day coach and sleeper were thrown from it to the creek below. The baggage and smoking cars made it across the bridge but on the slight reverse curve beyond they were thrown down opposite sides of the embankment. The tender itself was split in two and likewise went down opposite sides of the embankment, leaving the engine alone on the rails.

Engineer Strope stopped his machine and with his fireman ran back to the wreck. Smoke began to issue from the day coach, followed by flames from the windows, and a cry of, "Fire!" was heard from those trapped in it. The blaze quickly spread and the railwaymen and the passengers who were able to extricate themselves were powerless to put it out. Conductor Angle was extricated from the wreckage of his coach, his head badly cut and severely bruised and despite his injuries set off at once to Carmel to get assistance. On his arrival there a message was sent to Monon for the wrecking train and Drs McShane and Mendenhall of Carmel set off to the scene of the accident with their surgical cases. News of the fire brought farmers with buckets and within an hour a large number of villagers were wading about in the muddy creek engaged in the rescue work.[966]

Once again, as so many times before, stoves used to keep the passengers and railwaymen warm had caused the fire. The newspaper wrote, 'In the day coach there were two stoves. Both of them dropped their burning coals into the car when it was upset. The position of the car kept the front stove above the water's edge, and this part of the car was entirely burned, leaving only a few timbers. The coals from the rear stove dropped upon a submerged portion of the car and did not do any further damage.

'The sleeper was lying directly in the centre of the stream, and the coals from the stoves fell in the water that had entered the coach.

'There was no one badly injured in the sleeper. The burning sides of the day coach gave way and the trucks fell in on the debris. When a part of the debris was removed, the charred bodies of the dead could be seen under the timbers. The bodies of the women lay close together in the front end of the car. Although badly charred, they were recognised as Mrs Nellie Eubanks of Broad Ripple and Miss Hattie Hensley, of Cyclone, Ind. A few feet from them was the body of Mary Hoover, of Horton, Ind. Her head was horribly smashed, showing that her death was almost instantaneous. The fire did not reach her

[965] Wikipedia – Knoxville Cumberland Gap and Louisville Railroad
[966] *Lebanon Pioneer*, Lebanon, Indiana, United States, 30th January 1890

body. Close to the upturned stoves were the bodies of Mrs Octavia Oldham's two children. They were undoubtedly burned to death. These children were sitting on the seat next to the stove, and their mother was in the seat directly behind them. She was badly cut about the head but was carried out of the car before it was afire by J.T. Altizer and Louis Hammen, two of the passengers. Another child of hers about weight years' old was saved. The sleeping car porters were only scratched, and they busied themselves laying blankets and cushions on the banks for the repose of the injured. When the northbound train at 7.30 this morning reached Carmel the news of the wreck was conveyed to the passengers, and the train men called for able-bodied volunteers to assist in putting out the flames, for the cars were still burning. The injured were put on the train, and it returned to the city.'

The sixth passenger who died was C.O. Deming, of Frankfort, Indiana, who was taken out of the smoker alive and conveyed to Carmel but succumbed to severe head and chest injuries about an hour later. About 20 other passengers were injured, as were Conductor Harry Angle, baggagemaster D.S. Justus and Express Messenger George Munger.[967]

The coroner's inquest on the victims of the disaster found that it was due to the carelessness of the railroad officials in using "shims" on the curve leading to the trestle, causing a rail to overturn. The person directly responsible was Section Foreman McAvoy, 'but he was probably acting under the orders of a superior officer.' [968]

On 12th May 1890 the engine of an eastbound freight train became derailed as it neared a trestle a mile east of Clayton, Missouri, United States, on the St Louis Kansas City and Colorado Railroad. The engine bounced along the sleepers and plunged off the trestle, having travelled far enough to strike and demolish the eastern masonry abutment of the latter as it fell. In its descent the engine turned completely over and landed on its wheels, but considerably damaged. Engineer R.I. Jones was killed outright, and Conductor R.W. Shellcross died of his injuries four hours later. Fireman A. Greifeld suffered serious internal injuries but it was hoped that he might recover. Several of the cars of the train were wrecked.[969]

Train No. 33, the Halifax express, bound for Montreal on the Intercolonial Railway, was derailed as it neared St Joseph de Levis, Quebec, Canada, at about 11.45 a.m. on Thursday 18th December 1890. The derailment occurred just past St Joseph de Levis station close to the bridge over the main street of the town. The train consisted of an engine and tender, baggage car, mail car, second-class coach, smoking car, first-class coach and sleeper. The second-class car jumped the track at the end of the station siding on a curve, dragging the first-class and sleeper cars with it and pulling the baggage and mail cars in front of it from the rails. The passenger cars ran diagonally along the embankment for two hundred yards until the second-class car struck the stone abutment of the bridge carrying the railway across the main street of the town, and the front of this car was 'smashed to atoms' against the solid stone abutments. The smoking car fell from the middle of the bridge to the roadway below, grazing the corner of a dwelling house as

---

[967] Ibid.
[968] *Columbus Daily Herald*, Columbus, Indiana, United States, 3rd February 1890
[969] *Centralia Daily Sentinel*, Centralia, Illinois, United States, 13th May 1890

it did so, and the first-class car was also badly damaged. Fortunately, the curve around which they were running kept the cars away from each other so that the heavy sleeper car ran alongside the first-class car instead of smashing into it. The sleeping car was less badly damaged than the other vehicles. The engine, tender and derailed baggage car continued over a second road bridge before coming to a stand, the mail car crossed the first bridge but then went down the embankment, completely overturning as it did so.[970] [971] The train was behind time and was running fast on a falling gradient when the accident happened.

The bodies of five men were removed from the wreckage of the train. They were Michael Lebel, Blair and Gauvin, all of Kamouraska, Quebec; Leclair, of Levis, and a news agent named Winner. Over 40 other passengers were injured. [972] Two of the inured probably died later.

The derailment was attributed to the breaking of one of the axles of the baggage car.[973]

Just before noon on 4th February 1891 the axle of the front bogie of a sleeping car on a westbound Canadian Pacific Railway express broke about 100 feet east of a trestle bridge near Schreiber, Ontario, Canada, exactly three years and a month after the earlier trestle accident in the same area. The derailed car ran halfway across the trestle, tearing up the sleepers, and then fell 50 or 60 feet from the structure to the ground. Almost all of the 11 passengers in the sleeping car were hurt but, almost miraculously, none of them fatally so despite the distance that the car fell. Some of the passengers who had been in the sleeping car had gone to the dining car and thus escaped injury altogether. The sleeping car porter was not hurt. A surgeon was on the spot soon afterwards and another was sent for from Port Arthur.[974] In all, the accident seems to have been a very lucky escape from what could easily have been a major disaster.

Seven miles north of Luray, Virginia, on the Norfolk and Western Railroad, at Vaughan, was the Possum Run trestle, and three hundred yards north of that structure was Jerrymy's trestle, which at the end of May 1891 was being replaced by an iron bridge. While the reconstruction work was in progress there was a speed restriction of four miles an hour over the structure. At about 4 a.m. on the morning of Sunday 31st May 1891 two northbound freight trains were approaching the trestles, the first loaded with livestock and the second carrying lumber and coke. The first train reduced speed to comply with the restriction but the second did not. It ran into the back of the first at 25 miles an hour while the trains were on the structure, and the violence of the collision caused the trestle to collapse. The caboose and two cars of the front train and the engine and all the cars of the second fell 70 feet into the chasm, and the wreckage of 21 cars was piled up over the engine. Riding in the caboose of the first train was Mr James F. Strickler, a prominent farmer, of Madison County, Virginia, who had several carloads of sheep on the train. He was killed outright while Brakeman John Reese of the second train was dangerously

---

970 *Monmouth Warren County Democrat*, Monmouth, Illinois, 25th December 1890

971 *Winnipeg Free Press*, Winnipeg, Manitoba, Canada, 19th December 1890

972 *The Times*, Philadelphia, Pennsylvania, United States, 19th December 1890

973 *The Rochester Tribune*, Rochester, Indiana, United States, 19th December 1890

974 *Winnipeg Free Press*, Winnipeg, Manitoba, Canada, 5th February 1891

injured. The other five railwaymen escaped with slight injuries, but six tramps had been seen to board the train at Luray, all of whom it was supposed had been killed. Over two hundred men were employed in clearing the wreck and presumably in rebuilding the trestle, but the line could not be re-opened until the evening of Tuesday 2nd June.

Engineer William Hoover stated that he had whistled for brakes several times as his train descended the steep bank to the trestles, but the brakes had not been applied. He supposed that either the brakes had failed, that they had failed to hold on the slippery rails or the brakesmen had been asleep.[975]

Unbeknown to anyone the sleepers of a Kanawha and Michigan Railroad trestle at Farm, West Virginia, United States, had caught fire over the night of the 3rd-4th July 1891. The heat had been sufficient to distort the rails and when a passenger train from Charleston to Columbus reached the spot at about 8 a.m. it was derailed. The engine and baggage and mail cars passed the point safely, but the track then spread beneath the two passenger cars, which were filled with excursionists, including a party from the Charleston Lodge of the Order of United American Mechanics who were going to Pocas. After running about 40 feet on the sleepers the forward car toppled over to the left and the rear car to the right and both fell from the trestle. The forward car turned over completely and landed right side up, but the rear car fell 20 feet and landed upside down, and one of its trucks fell on top of it, crushing it to splinters and causing most of the deaths. The usual heartrending scenes were reported, for example one baby lost both of its parents. Some of the victims were much disfigured, at least one man being beheaded. At the time of the initial report 13 passengers had been killed outright, four were still alive but were thought to be fatally injured, and about 60 were more or less injured. Only one passenger, John Norvell, of Charleston, who had been in the mail car, escaped unhurt.[976] There was no mention in the newspaper report of the trestle having been deliberately set alight; the fire was presumably started by a spark from a passing engine.

The railways of Tsarist Russia suffered a major disaster on 23rd November 1891 when a train was derailed near Domnino station on the Orel-Griasi section of the Kosloff, Rostoff, Orel and Griasi line. The train service on this line was very sparse, consisting of only two daily mixed trains in each direction. The accident occurred to the morning train from Orel.[977] A second report on the accident gave more details of what was stated to have been the most serious in the country for many years. The long mixed train was derailed due to the breaking of the tyre of one of its carriages. The train was travelling at high speed at the time and the derailed carriage was dragged along, swaying from side to side, ploughing up the permanent way, until the train reached a bridge over the River Optoukha. As the train was passing over the bridge the derailed carriage overturned and struck the bridge parapet, which gave way from the impact, and the carriage was precipitated into the frozen river. Four more carriages were dragged after it until the breaking of a coupling prevented further disaster. The weight of the carriages was sufficient to break the thick ice covering the river and four of them disappeared beneath

---

975 *The Herald and Torch Light*, Hagerstown, Maryland, United States, 4th June 1891
976 *Philadelphia Inquirer*, Philadelphia, Pennsylvania, United States, 5th July 1891
977 The Times, London, England, 25th November 1891

the surface of the water. A fifth carriage fell onto the other four and overturned but supported by the wreckage beneath was only partly submerged. It was only from this fifth carriage that any survivors could be rescued.

The accident occurred in a desolate part of the country near a small roadside station and was of such a magnitude that the few officials and others who were on the spot and able to give assistance scarcely knew what to do first, but telegrams were sent to Orel for assistance. Strenuous efforts were made to rescue the passengers from the partly submerged fifth coach, and a few were gradually got out and brought to the riverbank, but the main rescue effort had to await the arrival of more men and doctors from Orel. A few lucky passengers escaped with slight injuries but 15 were seriously hurt and were taken to the hospital at Orel. The work of removing the sunken carriages and the recovery of the bodies was then begun. On the 26th November it was still in progress, but 26 bodies had been recovered, the line had been repaired and the traffic resumed, the bridge having sustained little damage.[978]

A third report on the disaster appeared in *The Times* a few days later. 'The causes of the recent terrible railway accident on the Orel-Griasi line, in which four carriages of a passenger train fell over a bridge into a frozen river, are being discussed here {St Petersburg} with the greatest interest, as they seem to point to a gross negligence on the part of the Railway Administration. Several of the newspapers publish accounts of eye-witnesses in which the disaster is variously attributed to the miserable manner in which the Poliakow lines are kept up, the oldness of the material used, and the criminal negligence of the railway staff. In the case under notice, it is asserted that the driver had not any means of regulating the speed of the train, and that the guards chatted with one another and with the passengers instead of attending to their duties. It is also complained that the injured were detained for some time at a wretched country station. They were not sent to Orel, where there is a commodious hospital, until nightfall. This delay is attributed to the anxiety of the railway authorities to conceal the extent of the catastrophe from the inhabitants of that city. Although the scene of the accident was only a few miles from Orel., more than three hours elapsed before the arrival of the relief train. The police were, it is declared, more bent upon the fulfilment of judicial formalities than upon caring for the safety of the passengers, and thus greatly impeded the work of rescue. The list of casualties comprised 31 killed and 15 injured.' [979]

Initial reports that there had been 11 deaths in an accident on the Louisville New Orleans and Texas Railroad near Evansville, Mississippi, United States, on Saturday 5th December 1891, proved to be incorrect. No one was in fact killed, though about 15 passengers were injured, some seriously, and Conductor L.J. Tomas and Pullman Porter L.J. Kinchen were also hurt, the conductor having his arm 'mashed'. It was considered a miracle that no one had lost their life as having become derailed and run along the sleepers for about 150 feet the train ran onto a trestle over the White Oak bayou. The trestle gave way under the strain and the train, which consisted of two baggage cars, two day coaches and a Pullman sleeper fell with the trestle 30 feet into the bayou, which was partially filled with water. All the train except the engine fell from the bridge. A relief

---

[978] Ibid, 27th November 1891
[979] Ibid, 5th December 1891

train with four surgeons was sent from Memphis, 41 miles to the north. The derailment was blamed on a broken tender truck (bogie).[980 981]

On Sunday 21st August 1892 an accident occurred 800 metres from Lignan station, France, on the Béziers – Saint Chinian local line of the Hérault Company. The wooden sleepers on a bridge over the Orb River were reportedly old and almost rotten and the rails spread under the weight of a passenger train, throwing it off the line. The bridge was an open-work structure and once the train had left the track it could run no further. The engine came to a dead halt and fell over on its side and the following carriages piled up against it and upon one another. 'In a moment nearly the whole train was a heap of wreckage, from which the passengers were extricated with the greatest difficulty.' Five passengers died in the accident and 39 were injured. [982 983]

At 1.5 a.m. on Monday 6th March 1893 the third section of Chicago and Erie Railroad westbound freight train No. 83 registered out of Lima, Ohio, United States, Conductor Pat Herr having orders to run his train as far as Spencerville, about 12 miles west of Lima, ahead of passenger train No. 3, due in Lima at 12.50 but running a few minutes late. About a mile west of Lima was Hog Creek bridge, beyond which there was a sharp climb, so Engineer Helm set off at a good speed down the hill to the bridge to get a good run up the other side. The engine, No. 319, and seven cars, had crossed the bridge and were beginning the climb when there was a violent jerk which almost pulled the engine off the line, followed by a crash which could be heard for miles around. The engineer pulled up as soon as possible but by that time 19 cars had piled up in the creek 50 feet below.

The eighth car behind the engine apparently left the rails at a set of points a hundred feet east of the bridge and when it reached the bridge it damaged the timbers sufficiently to cause the first span to give way and both span and cars fell into the creek. About 10 of the cars which went into the creek were smashed to pieces. All but two of those which fell with the bridge were loaded, and apart from two New York Pennsylvania and Ohio (NYP&O) coal cars carrying pipes were boxcars loaded with barrels of oil, coke, salt, tiles, washing machines and so on. A cattle car loaded with coal rolled down the embankment and overturned and another NYP&O flat car loaded with water pipes was left overhanging the edge of the bridge. The pile of wreckage in the creek dammed up the water until it made a new channel for itself under the far end of the bridge.

Fortunately for the front and rear brakemen they had left the top of the train, one going to the engine and the other to the caboose, otherwise they would probably have been crushed to death in the wreckage. The trainmen believed that this was exactly the fate which had befallen several tramps who had been on the train when it left Kenton and had not been seen since. The bridge was a practically new one.[984]

---

980 *Kokomo Daily Gazette Tribune*, Kokomo, Indiana, United States, 7th December 1891

981 *Hillsdale Standard*, Hillsdale, Michigan, 8th December 1891

982 *The Times*, London, England, 23rd August 1892

983 Wikipedia – Liste des accidents ferroviaires en France au XIX siècle

984 *Lima Daily Times*, Lima, Ohio, United States, 6th March 1893

An animal on the line, more specifically a cow, caused a fairly minor accident three miles from Nebraska City, Nebraska, United States, on about 30th April 1893. A Missouri Pacific Railroad construction train was backing towards a bridge over a pond when the beast tried to cross the line in front of it and was hit by the tender. The animal was pushed about 180 feet along the line when its carcass got under the wheels of the tender and derailed it. The engineer and fireman then jumped from their engine as it was almost on the bridge. The engine remained on the track for about 100 feet as it crossed the trestle but then fell from it, overturning as they fell and ending up in the mud 20 feet below. Little damage was done to the track, but it took much effort to retrieve the engine and tender.[985]

An accident near Bessemer, Pennsylvania, United States, on 19th December 1893 involved a collision between the Baltimore and Ohio Railroad's West Newton accommodation and a narrow-gauge train of Andrew Carnegie's Edgar Thompson steelworks. As the passenger train was passing the spot the narrow-gauge engine, which was hauling a large train of empty ladle cars, came up at some speed and hit the middle coach of accommodation train, knocking it off the rails and off a bridge and into the creek 20 feet below, while the rear coach was overturned on the track, and its 10 or 12 passengers had to climb out via the windows above them. Fire broke out in the wreckage, but everyone had been rescued before the flames reached them. Ten people were injured, mostly passengers but also at least one steel works employee. One of the injured passengers died later. Dodds, the watchman, claimed to have shown a red light to the dinkey engine, but this was disputed by its engineer, named Dougherty.

As was all too often the case it was tramps who were the victims of an accident to a freight train on the Baltimore and Ohio Southwestern Railroad on the afternoon of Sunday 26th June 1894. The accident occurred on the Walnut Creek Trestle, near Chillicothe, Ohio, United States. An extra westbound freight train of 25 cars broke in two and the two parts came together as it was crossing the trestle. Two of the tramps in one car saw the danger, jumped from the 50-foot-high trestle and remarkably escaped with slight injuries. The two cars at the point of the collision reared up on end and fell from the trestle, whirling around and around, and finally striking the rocky bottom of the creek. One of these cars contained another six tramps who were buried in the wreckage. The train was stopped and the trainmen and three tramps who had been in another part of the train went down to the creek and managed to extricate four of the tramps, who were badly injured, the other two being dead. The four injured were taken to Chillicothe, where one of them died within a few minutes. Papers on him identified him as a John Snell of Toledo though he told the doctors before he died that his name was George Wilson. The other injured tramps were William McIntyre of Pittsburgh; Walter Jackson, of Baltimore, W.Va.; and John Lee, coloured, of Savannah, Ga. All were badly hurt, and it was said that McIntyre could not live. The identities of the two dead tramps were not known. None of the trainmen was hurt. Six cars were wrecked, and the loss was said to be heavy.[986]

---

985 *Nebraska State Journal*, Lincoln, Nebraska, United States, 2nd May 1893
986 *Norwalk Daily Reflector*, Norwalk, Ohio, United States, 25th June 1894

Three coal miners were the victims of an accident on a half-mile branch from the Pennsylvania Railroad main line to the New Larimer colliery of the Westmoreland Coal Company near Larimer, Pennsylvania, United States, on Monday 19th November 1894. The branch included an iron trestle across a gully which had been constructed by the railroad company to divert the course of a small stream. About 4 p.m. a train of 20 loaded coal cars left the mine on a transfer trip to the main line, carrying in addition to the train crew a number of miners making their way home at the end of the day's work. The engine and some of the cars had passed over the trestle when the axle of a car broke just as it was entering the structure, the derailed car caught the supports of the trestle and it collapsed, taking 16 of the cars down with it. The trainmen and some of the mines saved themselves by jumping to the sides of the gully when they felt the trestle giving way but miners John Rice and August and Frederick Thorm (his surname was also given as Lyon and Loyn) were killed, the latter two being father and son. It was thought possible that one or two other miners might also have lost their lives.[987] [988] [989]

The Bellaire Zanesville and Cincinnati Railroad had a bad accident at Whigville, Ohio, United States, on the morning of Friday 4th April 1895. The single combination coach of a westbound passenger train jumped the track just as the train was approaching a 30-foot-high trestle over a small stream. The trestle was situated on a curve and the derailed coach managed to make it almost to its far side, running on the ties, but it then dragged the engine from the rails and the trestle failed, taking the train with it. Engineer Eli Lucas and Fireman Jesse Johns stuck to their posts on the engine. The engineer was killed, and the fireman injured, but it was thought that he would recover. Three passengers, Mrs Young and her little daughter, of nearby Summerfield; and Henry Brown, of Beallsville, were also killed and Mrs Young's husband Nathan was seriously injured. Railroad officials stated that the trestle had been in perfect condition and its collapse had been solely the result of the derailment of the train.[990]

As westbound Cleveland Cincinnati Chicago and St Louis (Big Four) Railroad freight train No. 61, consisting of an engine and tender and 33 cars and a caboose, was approaching a bridge a mile west of Lawrenceburg Junction, Indiana, United States, at 10 p.m. on Friday 31st January 1896, its engineer saw that the second car was off the track. Knowing that the derailment was likely to result in the collapse of the bridge he opened his throttle wide and succeeded in getting his engine and the leading car across the structure. The derailed second car struck the side of the bridge and the remainder of the train piled up behind it, causing the bridge to collapse just as the engineer had expected and almost the whole of the remainder of the train went down with the structure, going down 40 feet to the creek below. The rearmost waggon and the caboose remained on the line short of the bridge. The engineer, fireman, brakeman and a fourth man who were riding on the engine and a second brakeman riding in the caboose thus all escaped injury. Fifteen of the cars which went down were empties, the five which followed were loaded with coal and the remainder with general merchandise. There were also a number of

---

987 *Connellsville Courier*, Connellsville, Pennsylvania, United States, 23rd November 1894
988 *Norwalk Daily Reflector*, Norwalk, Ohio, United States, 20th November 1894
989 *Defiance Daily Crescent*, Defiance, Ohio, United States, 20th November 1894
990 *Columbus Daily Herald*, Columbus, Indiana, United States, 6th April 1895

tramps riding the train, probably 12 but possibly as many as 17. The bodies of two of the train riders were removed from the wreck shortly after the accident and a third man, who was badly hurt, died after being taken to Lawrenceburg. Three tramps were unhurt and six were unaccounted for. Different accounts had it that five, eight, nine or even 14 tramps had died. Strangely one of the tramps killed, named Truitt, had three passes in his pocket from Aurora to Indianapolis for himself, his mother and his sister. It was a mystery therefore why he had chosen to steal a ride.[991] The derailment was apparently caused by a broken wheel.[992]

Southbound passenger train No. 2 of the Houston and Texas Central Railroad, in charge of Conductor John West with Engineer Charles Clark, was derailed on Monday 26th September 1898 near Corsicana, Texas, United States, and its chair car and sleeping car fell from Chambers Creek trestle, about 15 feet high, into a slough of dirty water. About a dozen passengers, Superintendent L.A. Daffan, General Freight Agent R.D. Berry and Porter Will Walling (coloured) were injured. The worst hurt was Judge G.W.L. Davis, of Oak Cliff, who was crushed under a seat. Miss Mae Moss, of Denison, bound for Austin, was credited with saving the life of Freight Agent Berry by holding his head above water until help arrived. There were fortunately only a porter and a newsboy in the sleeping car; they escaped through a window. All the other coaches of the train overturned except the mail car.

An injured passenger, Mr Caswell, described his experiences. "I saw that we were going to turn over and I tried to brace myself, but got knocked loose from my hold once or twice and rolled over with the car. We went down the embankment a few feet from the trestle and it made my hair stand on end when I saw how near we were to that dangerous part of the road."

The sleeping car, at the rear of the train, had been the first to leave the track at a point where it was under repair, and had run along off the rails until it reached the trestle, from which it had plunged, dragging the chair car with it, and half pulling off the day coach or ladies' car.

A relief train with surgeons was sent to the scene from Corsicana, and another from Houston.[993]

Judge Davis died of his injuries two days after the accident.[994] A second passenger possibly died later.

---

[991] *The Batesville Tribune*, Batesville, Indiana, United States, 6th February
[992] *Silverton Standard*, Silverton, Colorado, United States, 8th February 1896
[993] *Galveston Daily News*, Galveston, Texas, United States, 27th November 1898
[994] Ibid, 29th November 1898

CHAPTER 11

TUNNEL ACCIDENTS

As mentioned in the previous volume of this series and remarked upon by the writer L. T. C. Rolt in his classic work on British railway accidents, *Red for Danger*, Victorian railway passengers often felt an acute trepidation as they were passing through a tunnel, and made repeated pleas to the railway authorities that all tunnels should be artificially illuminated, pleas which, probably for practical reasons if nothing else, fell upon deaf ears. Rolt goes on to point out that these fears were by no means unjustified, especially in the period before the introduction of the absolute block system, when trains were merely spaced by the time interval system. If a train should chance to break down in the open there was a good chance that the crew of a following train might be able to see it in time to pull up, even if there was insufficient time for the guard of the first train to go back and protect its rear. In the smoke- and steam-filled darkness of a tunnel there was no such hope. Furthermore, should a serious collision occur in a tunnel the confined space would be likely to result in the wreckage being hideously compressed, and should a fire break out in a long tunnel provided with ventilating shafts, the latter could act as chimneys. In addition, the rescue work would be hampered by the darkness and the difficulty of working in a confined space where the use of breakdown cranes was impossible, and the only resort was to drag the wreckage bodily out of the tunnel. Because of the special characteristics of tunnel accidents this book will devote a separate chapter to them, though in fact up until the beginning of the 20th century only one major disaster occurred within a tunnel, in the Clayton Tunnel in England back in 1861.

A head-on collision in the single-track tunnel just north of the station at Salem, Massachusetts, United States, on the Eastern Railroad on 5th October 1881, had an unusual cause. There were signal towers (boxes) at each end of the tunnel, each provided with three ball signals, the latter supplemented by lights during the hours of darkness. The display of only one ball or light permitted a train to enter the tunnel; three, two or no balls or lights meant that the line was blocked. The signals were mechanically worked but the signalman at the north end of the tunnel was guided by bell signals and indications sent electrically from Salem station. On the evening of the day of the accident the engineer of a southbound passenger train found three lights showing at the north end of the tunnel and the train had been almost brought to a stand when the signals cleared and it proceeded into the tunnel, the north end signalman having received the usual bell signal and indication that the tunnel was clear. A few moments later the passenger train collided head-on with a northbound work train which had also entered the tunnel under clear signals. Fortunately the crew of the work train had become aware that the signals were not working properly and had brought their train to a stand before the collision. No one was hurt and only minor damage resulted. It was subsequently discovered that the branch of a tree had fallen onto a telegraph wire and pressing it against a signal wire had caused a false clear signal. Ironically on 29th November of the same year there was a rear collision in the same tunnel in which a brakesman lost his life. Even though over 10 years had elapsed since the Revere disaster of August 1871 the Eastern Railroad was still not using the electric telegraph to space following trains.[995]

A signalman's error, or rather series of errors, was the cause of a serious accident in Canonbury Tunnel, England, on 10th December 1881. This accident involved no less than four trains, that is it was a triple collision. Simple collisions were common, double collisions were rare and triple collisions very rare, while this accident is probably unique in being a triple collision in a tunnel. The 544 yard tunnel was on a short branch running from the Great Northern Railway main line to the North London Railway, west of that company's Canonbury station. The signal boxes most directly involved were the Great Northern box, Finsbury Park No. 1, at the Finsbury Park (north) end of the tunnel and Canonbury Junction box where the branch joined the North London line, the latter box being of course a North London one. All the trains involved were North London trains, bound for London Broad St terminus. The accident, or series of accidents, was serious, resulting in the death of three passengers and a guard, serious injuries to another five passengers and slight injuries to 10 more.

The first witness to give evidence at the subsequent Board of Trade inquiry into the accident was Signalman Henry Hill, the North London man at Canonbury Junction. He said that a Great Northern train had passed his box at 8.49, he gave "Line clear" back to Finsbury Park No. 1 (henceforth No. 1) box at that time and it was cleared back to him from Canonbury Station box at 9.1. He was next offered a Chalk Farm train by the North London Railway's Highbury box at 8.55, and offered it to Canonbury station, but it was refused. About the same time No. 1 box had offered him another train from the Great Northern line, but he refused it and gave the "positive block" bell signal, six beats, to both No. 1 box and Highbury. Highbury acknowledged the six beats, but No. 1 box did not. Canonbury Station box accepted the Chalk Farm train at 9.1. Meanwhile he had received two beats from No. 1 box, to which he had replied with seven beats, which was not acknowledged. His block indicator to No. 1 box remained at "Train on line". As the Chalk Farm train was passing the junction at 9.2. he saw a train emerge from Canonbury tunnel. He gave No. 1 box seven more beats, which were not acknowledged, and two minutes later No. 1 box again gave him two more beats. He gave No. 1 box seven beats again, was quickly given two beats again and gave seven beats a fourth time. About 9.5 he heard the second train run into the back of the first. He called the first train on by a green flag at 9.7 when Canonbury station cleared the Chalk Farm train. He knew nothing of the further collisions. Signalman Hill explained

[995] *A History of Railroad Accidents, Safety Precautions and Operating Practices*, pg. 261

that when he was offered trains from the Great Northern and Highbury simultaneously, he gave priority to the Highbury train as the distance from Highbury box was less than that from Finsbury Park No. 1 box. He gave the six beats to both No. 1 box and Highbury because both signalmen had offered him trains and Canonbury Station had not cleared back the previous train. He had given seven beasts to No. 1 box as the Great Northern signalman there did not acknowledge his six beats. He understood both to be signals indicating that the line was obstructed, the seven indicating that the obstruction would not last long. Hill's evidence was more or less confirmed by his 14-year-old booking boy, Arthur Long.[996]

On duty since 6 a.m. at Finsbury Park No. 1 box had been Relief Signalman William Hovey. He related that he had given the "Be ready" signal for a North London up passenger train to Canonbury Junction and had it accepted at 8.47, his needle being put to "Line clear". He had given the "Train entering section" signal for this train at 8.48 and Canonbury Junction put the needle to "Train on line". Canonbury Junction cleared that train back at 8.49 with three beats. At 8.54 he had given the junction box the "Be ready" for another train, which was acknowledged by one beat. At 8.56 he had offered it again and had received seven beats in reply at 8.59. The seven beats was a strange signal to him but looking in the North London rule book he found that it meant "permissive block". He had then cleared his starting signal and let the train go, despite his block needle from Canonbury Junction remaining at "Train on line". He gave "Line clear" back to Finsbury Park No. 4 box at 9 o'clock and was given the "Be ready" for a second train. He gave one beat on the gong to Canonbury Junction, received seven beats back again, exhibited a green flag from the box, shouted to the driver to, "Go at caution," and cleared his starting signal at 9.5. At the same time No. 4 box offered another train and he received "Train on line" for it at 9.6. The previous train having only gone one minute he delayed asking to Canonbury Junction till 9.10, when he again received seven beats, the block indicator stubbornly remaining at "Train on line". He cleared his home signal, gave the driver a green flag, cautioned the driver, pulled off his starting signal and the train proceeded. He cleared the train back to No. 4 box at 9.10 and was offered a fourth train. This stopped at his up home signal at 9.13 and the process was repeated; he gave Canonbury Junction one beat, received seven beats back, cleared his home signal, spoke to the driver, cleared his starting signal and the train went on.

Signalman Hovey admitted that he had not received "Line clear" from Canonbury Junction for any of the four trains. He had never had the seven beats bell signal before. There was no speaking telegraph to the North London box and he had not time to read the North London rules but had simply assumed that the seven bells sanctioned permissive block working as carried out on the Great Northern Railway, whereby one train was permitted to follow another into an occupied block section after having been cautioned. He denied having received six, as opposed to seven beats from Canonbury Junction. He had been appointed as relief signalman on the basics of his experience and certified as competent to work any box in the London district.

Having gone into some detail on how the mistakes were made which led to the accident, it is now time to consider what actually happened to the trains. The first train involved in the collisions was the 8.25 a.m. from High Barnet, which left Finsbury Park at 8.52. Its driver, William Dawkins, said that his train had been stopped four minutes at

[996] Board of Trade report

the No. 1 box starting signal, which had then been cleared and his train had entered the tunnel. At the far end of the tunnel the Canonbury Junction home signal was at danger. He had stood there seven minutes, seeing the Chalk Farm train pass as he drew up to the signal. He whistled three times while waiting and after about six minutes had felt another train run into the rear of his, shortly after which the Canonbury Junction signalman had called him forward with a green flag. Guard James Cody and Underguard Harry Ayling of the Barnet train confirmed their driver's evidence and indicated that the first collision had only been a slight one.

Driver Henry Harvey had left Finsbury Park with the 8.58 a.m. train for Broad Street about 9.3. He said that he had been checked at the No. 1 box home signal but denied having been shown a green flag by the signalman or spoken to. He had then proceeded into the tunnel, the advanced starting signal being off. The repeating signal in the tunnel for the Canonbury Junction up branch home had been at danger and he had applied the brakes. The next thing he saw was the red tail lamp of a train in front of him and he had run into it at about two miles an hour. The first train then went forward, and he was drawing his train out of the tunnel when it was run into by the third train, which he had not realised was following.

Underguard Henry Catherall was acting as head guard of the 8.58 train. He heard his driver whistle several times, but he already had the brakes on. His train stopped in the tunnel, but he did not realise that it had run into the train in front. It then began to draw forward out of the tunnel when he heard another train enter the tunnel behind them. He got down from his van and was making his way back to No. 1 box but had to wait for a down train to pass. He turned his light to red and showed it to the driver of the oncoming train, but doubted whether, in the smoke and steam, it could be seen. The third train, travelling at less than 10 miles an hour, then collided with the rear of the second. Guard Catherall said that he continued to go back as the passengers in the third train, who were in darkness, began to scream out, and he told them to get out and stand against the tunnel wall, as another down train might be coming. His progress impeded by an open carriage door, he had got back about three carriage lengths beyond the brake van of the third train when he heard yet another train coming. He managed to get a detonator down on the rail before this train reached him, then tripped over a sleeper and fell into the four foot and his lamp went out. The fourth train was travelling at a good speed, about 18 miles an hour, but he managed to roll into the six foot out of its way. Clearly, he had a narrow escape. Picking himself up, the guard continued back to No. 1 box, and found the signals off for a fifth train, which he could hear approaching from Finsbury Park. Guard Catherall said he told the No. 1 box of the collisions and the fifth train was stopped at the home signal.[997]

Driver Charles Ruffit stated that he had been in charge of the engine of the 8.43 a.m. train from Enfield, which had left Finsbury Park at 9.6. He had been checked at No. 1 box, but the home signal had been cleared as he approached it and the starting signal was clear. He saw no green flag from the box and heard no warning from the signalman. He entered the tunnel and seeing the Canonbury Junction repeater in the tunnel had whistled to alert his guard. He saw nothing of the train in front and was running at about 8 miles an hour when the collision occurred at 9.10 or 9.11. Two or three minutes later another train ran into the back of his.

[997] Ibid.

Underguard Frederick Adams of the third train was badly stunned and could not really tell the inquiry very much. Head Guard West of this train was one of those killed.

Driver Edward Harvey stated that he had left Finsbury Park at 9.13, five minutes late. All the No. 1 box signals had been off except the distant and like the other drivers he had neither seen the signalman's green flag nor heard him shout. His train had entered the tunnel travelling at about 20 miles an hour and he had seen nothing of the third train before he ran into it. It was this final collision which caused the casualties in the accident, four passengers as well as guard West being killed outright, five passengers seriously injured and well over a hundred less badly hurt.

Colonel W. Yolland investigated the accident for the Board of Trade and concluded, not surprisingly, that it had resulted largely from a great mistake made by Signalman Hovey in introducing an unauthorised mode of working. Without taking time to read the relevant clause (No. 7) of the North London block regulations, which would have taken him less than half a minute, he had undertaken to work the line under a permissive block, a system not used on the North London line or on Great Northern passenger lines. The inspecting officer stated that 'there is not a syllable in these instructions which points out or prescribes how the "permissive block" is to be worked.' In fairness to signalman Hovey this was true, but in the North London's code of signals for the bells and gongs (reproduced in the accident report) seven beats was misleadingly described as 'Obstruction of line to station in rear (Permissive block)' this table of bell codes probably being what had misled signalman Hovey. The inspecting officer pointed out, though, that Hovey had worked No. 1 box on no less than 18 occasions in 1880 and 1881, surely enough time to have familiarised himself with the box instructions, and to have queried anything he did not understand with either the signal inspector or one of the regular signalmen. Nevertheless Colonel Yolland recommended that 'seven beats, permissive block' be struck out of the North London instructions.[998]

The verdict of the jury at the inquest into the deaths of the victims concluded that the accident had resulted from the unauthorised mode of working adopted by Signalman Hovey, but, doubtless to his relief, they stopped short of finding him guilty of manslaughter, stating only that he had been guilty of a grave error of judgement. Following Colonel Yolland's lead they recommended that the phrase "permissive block" be immediately expunged from the North London signalling regulations as misleading and that a uniform and less complicated system of signalling be adopted.[999]

Park Avenue Tunnel, New York, United States, suffered a serious accident on 22nd September 1882. The New York Central and Hudson River Railroad tunnel, leading to New York Grand Central terminus, already had so much traffic by this date that it had four lines through it. On the morning of the 22nd a mishap apparently occurred to the driving wheel of a locomotive on one of the central main lines through the tunnel and blocking of the main lines by the disabled engine resulted in all the traffic having to be worked over the local lines. The 7.53 a.m. New York New Haven and Hartford Railroad train was standing in the tunnel at the 86th Street station when it was run into from behind by a New York Central and Hudson River Railroad Harlem local train, which demolished

998 Ibid.

999 *The Week's News*, London, England, 7th January 1882

its rear two cars. Mrs Eugenie Aubert, a teacher at the 12th Street public school, was killed instantly and a number of other passengers were injured. The seven most seriously injured passengers were taken to the Grand Union Hotel, where a Mr William Howe, of Mount Vernon, died from his injuries the same afternoon, and a third passenger died later. The great majority of the passengers on the Harlem train were schoolgirls on their way to the Normal college. Some of them became hysterical, the newspaper reported, after the collision, and jumped out of their carriages in the dark tunnel, landing in six inches of mud and water. They had then wandered around in the darkness, unable to find their way out, until rescued by a Professor Dundon from the college and some of the New York fireman called to the accident.[1000]

The day after the accident William Rawson, the operator at the 98th Street station, was placed under arrest for having neglected to telegraph ahead the Port Chester special train (the Mount Vernon train) and thus having permitted the Harlem train to follow it. He told a reporter, 'All I can say is that the New Haven train passed into the tunnel at 8.49, and the Harlem train was not permitted to enter the tunnel until 9.03 and was ordered to proceed very slowly and keep a close watch for danger signals.' A passenger on the New Haven train complained that it had been standing at the 86th Street station at least six minutes when the collision occurred, perhaps awaiting a platform at Grand Central station.[1001]

The coroner's jury which inquired into the deaths of the victims of the Park Avenue Tunnel accident (confusingly called the Fourth Avenue Tunnel accident by the newspapers) reached its verdict on Thursday 5th October. Their findings were comprehensive. They found Telegraph Operator Rawson guilty of culpable negligence in failing to signal the New Haven train but did not stop there. They found New Haven conductor George Rood guilty of gross and criminal negligence for running his train with insufficient brakesmen and without sufficient signalling equipment, for accepting an unqualified person of immature age and inadequate experience in place of the regular brakesman, and 'in other respects'. They found Brakeman Robert Robbins (presumably the immature inexperienced brakesman referred to) guilty of culpable negligence in not following the rules relating to the running of trains – did he fail to go back and protect his train? The railway company was found guilty of negligence for having insufficient brakesmen. The jury urged more railroad safety legislation, condemned the railroad company for sacrificing safety for economy, censured Abraham Close, the New Haven engineer, for leaving his train in an emergency, criticised the lack of a block system on the rapid transit lines, and recommended a more perfect system of tunnel lighting, a rearrangement of the ditches and roadbed in the tunnel to make them safe to walk over and improved tunnel ventilation.[1002]

The little Mediterranean island of Malta had an accident on its sole seven-mile-long railway line on 20th March 1889 when the boiler of engine No. 4, which was reportedly in poor condition, exploded inside Floriana Tunnel, without it would seem any very

---

[1000] *The Fitchburg Sentinel*, Fitchburg, Massachusetts, United States, 23rd September 1882
[1001] Ibid, 25th September 1882
[1002] Ibid, 6th September 1882

serious consequences.[1003]

Considerably more serious than the Malta mishap was an accident six months later, on 29th September 1889, in Pranerottolo Tunnel, Italy, on the line between Naples and Foggia. Two trains, one of them one of them possibly a troop train, collided head-on in what was described as a long and narrow tunnel near Ariano station. The train from Foggia, which was the one carrying the soldiers, consisted of 25 vehicles, the foremost of which were cattle waggons. Three people, two of them soldiers, were killed in the accident and 27 injured. Those injured had to wait for a considerable time within the darkness of the tunnel before assistance arrived from Ariano station. The accident was allegedly due to a signalling error and a signalman named Urbani was arrested. A day or so later the *Times* was able to erasure its readers that Messrs Thomas Cook and Son, the travel agents, had learnt by telegram from their Naples agent that no British subjects were among the casualties.[1004]

Park Avenue Tunnel, New York, United States, on the approach to the New York Central and Hudson River Railroad's Grand Central station, already the scene of an accident in 1882 suffered another early on the morning of Friday 20th February 1891. A shop train of nine empty New York New Haven and Hartford Railroad passenger cars was en route from Grand Central station to the yards at Mott Haven, running slowly through thick fog. Riding on the train were 20 or 30 car cleaners, including one woman, also a young news boy. When the train was partly through the tunnel it was run into from behind by New Haven train No.10, a local working in charge of Conductor George Hutchinson and Engineer Louis Fowler, which was running at full speed. The rear two cars of the first train were telescoped and the engine of the local train was badly smashed in and battered. Immediately after the collision the telescoped cars caught fire from the car stoves and two of the cleaners were burnt to death before they could be released. The New Haven was the only railroad company working into New York which still used stoves for heating by this date. A New York Central light engine on the down line also ran into the wreckage. Six of the cleaners in all lost their lives and another cleaner was seriously injured.[1005] Those killed were Helen Supple, William Anton Zelnor, John Hencke, Michael Mullane, James Flynn and John Murray.

After an inquest lasting six days the coroner's jury found that the accident had been caused by Engineer Fowler overrunning a red signal at 72nd Street in the fog and his failing to hear, if it had rung, an associated warning gong. The jury found the officers and directors of the New Haven Company responsible for the deaths of the victims and recommended that additional openings be made in the tunnel to improve the ventilation and lighting, that anthracite coal be used for the engines, that the signal boxes at each end of the tunnel be double-manned at all times, that each train passing through the tunnel be advised of the identity of the preceding train and its headway and that trains be ordered to run at a uniform speed through the bore.[1006]

1003 *The Malta Railway* by Joseph Bonnici and Michael Cassar, pgs. 52 and 136
1004 *The Times*, London, England, 1st and 2nd October 1889
1005 *Brooklyn Daily Eagle*, Brooklyn New York, United States, 20th February 1891
1006 *New York World*, New York, New York, United States, 6th March 1891

Another New York, United States tunnel was the scene of an accident in 1893; this time it was a derailment near Parkville Tunnel, beneath the Coney Island Boulevard on the New York Brooklyn and Manhattan Beach Railway. On the evening of Tuesday 20th June 1893 a train which had left Sheepshead Bay at 5.45 p.m. was approaching the tunnel. Its six cars, the front and rear closed and the others open, were crowded with about 1,000 men women and children who had watched "The Suburban" race at the course at Sheepshead Bay. Every seat on the train was taken and men were standing in the aisles of the closed cars. The train was running at full speed when it became derailed close to the entrance to the tunnel. The engine and leading two cars remained on the track and the remainder of the train was dragged to the mouth of the tunnel, where it became divided, the engine and first two cars passing through the tunnel to come to a stand 25 feet beyond it and part of the rear of the train coming to a stand within the tunnel itself. The train being so crowded, there were a large number of casualties. Two passengers were killed outright, no less than seven died after being removed to hospital and another hundred were injured.[1007]

It was clear that the train had become derailed at a series of facing trap points protecting the point where the line became single before passing through the bore. These points had either been defective or had been moved while the train was passing over them. At the inquest Long Island Railroad Train Master Dwight Baker testified that he had been riding on the train and had gone to the adjoining tower controlling the points and said to the signalman, Peter McGarry, "You threw that switch before the train passed." McGarry had replied, "Yes, I thought that the train was all over." This would seem to explain how the accident had happened. Before moving the points McGarry would have had to replace the distant and home signals and release the facing point lock, making four lever movements in all. There should also have been a detector bar which would have prevented the points being moved while a train was standing on it, but the bar had been removed in January and was not replaced until after the accident. Allegations made at a previous hearing that McGarry had dozed off just before the accident and had perhaps reversed the points while still half asleep, were refuted by three other towermen who spoke of having conversed with him over the telegraph only five minutes previously. At the conclusion of the inquest the jury returned a verdict finding the Long Island Company (who leased the Manhattan Beach line) guilty of gross negligence in not proving a detector bar and exonerated signalman McGarry from all blame.[1008]

The accident was also investigated by the State Board of Railroad Commissioners, who concluded that in fact Switchman McGarry had been dozing just before the accident and being suddenly awakened had grabbed the levers and reversed the road, derailing the train. Also that there had been negligence on the part of Signal Foreman John Sanderson, who had become aware of the missing detector bar on 15th May and had ordered an assistant to replace it, which had not been done. The bar had originally been removed to replace the rail beneath it. The commissioners recommended that all open cars be closed off on both sides to at least halfway up from the floor with some form of wire netting, that the long side steps be removed, doors provided at the ends of the open cars and an

---

1007 *Newark Daily Advocate*, Newark, Ohio, United States, 21st June 1893

1008 *The Brooklyn Daily Eagle*, Brooklyn, New York, United States, 30th June 1893 aisle down the centre of them.[1009]

The 4.8-mile-long single-track Hoosac Tunnel, in Massachusetts, United States, the longest tunnel in North America up to the end of the 19th century, was the location of an accident on 8th September 1894. Extra freight train No. 245 left North Adams, at the west end of the tunnel, at 9 p.m. that Saturday evening. It passed through the tunnel to a point about 1,000 feet east of its central shaft when the spring hanger of its engine broke and its engineer stopped the train to make a repair. The engine was detached from its train, jacked up and after about 15 minutes' work the job was done, the engine was let down again and the train was ready to depart. Meanwhile another freight train, No. 22, came up to the west portal of the tunnel. To ensure safety only one train was allowed to be in the tunnel at any one time and there were signal towers at each end of it with protecting signals. The signal at the North Adams portal was at danger, but on the arrival of train No. 22 the operator immediately cleared it to white and the train entered the still occupied tunnel. It was travelling at full speed when it collided with the caboose of the train in front, killing Brakeman George Minnick, who had been in the caboose, and William Turpenty, who at the moment of the collision had been making his way between the engine tender and the leading car. Engineer Walter Clapp and Fireman Charles Frazer of the colliding train were seriously injured. The accident blocked the tunnel for at least a day.[1010]

The collision was clearly the fault of one of the two operators. H.W. Hodgkins, the North Adams operator, was arrested the following day and according to one newspaper refused to make any statement regarding the accident. Another newspaper stated that he had alleged that he had received the O.K. signal from the east portal operator.[1011] At that time. A.M. Crosier (or Crozier) the East Portal operator, was, according to one source, still at liberty, though a warrant had been issued for his arrest.

At the inquest into the accident on 13th September, Operator Crosier gave evidence that on telling Operator Hodgkins that he, Crosier, had not been offered extra 245 east, Hodgkins had replied, 'I am in it, but I don't care so much if no one is hurt.' [1012]

The conclusion of the inquest was reportedly that Hodgkins had in fact failed to report the entry of either of the trains involved in the collision to his colleague at the east portal.[1013] No further details were given for his double failure. None of the three brakemen with the train seem to have considered going back to protect it; perhaps the company rules did not require them to do so.

Poor ventilation in a single-line tunnel was the principal factor a disastrous runaway accident in Italy on the evening of Thursday 11th August 1898. An unusually heavy goods train was provided with three engines to climb the severe gradients leading to the single-track Giori tunnel. With poor quality coal even three locomotives nevertheless proved insufficient and the train slowed down and then stalled in the tunnel. Sulphurous fumes

---

1009 *Syracuse Evening Herald*, Syracuse, New York, United States, 12th July 1893
1010 *Bangor Daily Whig and Courier*, Bangor, Maine, United States, 10th September 1894
1011 *Sandusky Register*, Sandusky, Ohio, United States, 10th September 1894
1012 *The Fitchburg Sentinel*, Fitchburg, Massachusetts, United States, 14th September 1894
1013 Ibid, 18th September 1894

(probably a mixture of carbon monoxide, carbon dioxide and sulphur dioxide) asphyxiated the engine crews and probably the guards, and the train then ran back through the tunnel out of control to collide with a passenger train near Ponte Decimo. Both trains were wrecked, and nine railwaymen and two passengers lost their lives.[1014]

A follow-up report in *The Times* from an Italian correspondent spoke of the many similar single-line tunnels on steep gradients in Italy, with for example 52 of them on the main line across the Apennines between Bologna and Florence, most of them on a gradient of 1 in 40 and some, such as the Apennino tunnel, of considerable length. The problem of fumes was especially bad when there was a wind blowing at the lower end of the tunnel, pushing the smoke and steam in the same direction as the train. The correspondent referred to an incident four or five years earlier when a heavy train of sleeping and dining cars carrying one of the crowded heads of Europe and suite had emerged from the 2,727-metre-long Appennino tunnel into the open air at Pracchia station and its driver and fireman were found lying unconscious on the footplate, overcome by the fumes. On other occasions troop trains had stalled in tunnels and the soldiers had been seriously affected. The matter had come to the attention of the Italian Government and an engineer, a Signor Saccardo, had come up with a solution in the form of a powerful ventilating fan placed at the higher end of the tunnel which blew air through it, greatly improving conditions in the tunnel. Giori tunnel had unfortunately yet to be provided with such a fan.[1015]

The penultimate tunnel accident of the 19th century occurred at Charing Cross, Glasgow, Scotland, on the North British Railway's low-level line on Wednesday 28th March 1900. Once again it was a rear collision, the result of a signalman's error. Two workmen's trains were involved, and seven passengers were killed outright or died shortly afterwards, and 31 others and a guard were injured, some of them very seriously.

Signalling technology had progressed markedly since the early days of the railways, and this line was protected by what was regarded as the safest form of signalling then available, namely Sykes' Lock and Block, but an accident happened, nonetheless. The lock and block system was meant to ensure that a starting signal could not be lowered for a train to proceed until it had been accepted by the box in advance, and that once a train was in the section it could not be cleared back and a following train accepted until it had released a lineside treadle, thus ensuring that a signalman could not mistakenly clear out a train which had not actually passed him. The Achilles heel of the system was the need to provide a key to free the locking to cope with cases where a train did not proceed, or where the instrument had failed. The signalmen were required to report all uses of the release key and warned that any tampering with the instruments would render them liable to instant dismissal.

On the morning of the accident Signalman James Milne at Queen Street West box was offered and accepted the 5.45 a.m. workmen's train from Bridgeton Cross by the Queen Street East box at 5.57. He received the "Train entering section" signal for it at 6.1, offered it and had it accepted by Charing Cross box at the same time and it passed him at 6.3, not being booked to stop at Queen Street. At 6.6 the East box offered him, and he

[1014] *The Times*, London, England, 13th August 1898
[1015] Ibid, 16th August 1898

accepted the 6 a.m. workmen's train from Bridgeton Cross. He waited a short time, expecting to receive the "Train out of section" for the 5.45 train from Charing Cross. It did not come so he rang up Charing Cross on the telephone and asked the signalman what he was stopping the traffic for. "Did I not clear?" was the reply. "No," he had said, and Charing Cross at once gave the "Train out of section" signal and accepted the 6 a.m. train, the time being 6.7. Signalman Milne said that he had at once lowered his signals for the train, which like the first one had been checked, and gave the "Train entering section" signal to Charing Cross as the train passed him at 6.9. Shortly afterwards the Charing Cross signalman telephoned him and asked if it was the 6 a.m. train approaching. Told that it was, he had replied, "My God," or possibly something ruder, "there's another train in the tunnel." A few minutes later a surfaceman came to his box and told him to stop all traffic. The working that morning had been perfectly normal up to the time of the accident.

On duty at Charing Cross was Relief Signalman James Rentoul. According to his account he had accepted the 5.45 a.m. workman's train at 6.3 from Queen Street West and had received the "Train entering section" signal at the same time. Previously, at 5.57, he had accepted the 5.20 a.m. workman's train from Springburn from Queen Street West, it had been accepted by Finnieston, the next box in advance, at 5.59, had passed his box at 6.3 and had been cleared back by Finnieston at 6.4.

At about 6.9 the Queen Street West signalman had rung him on the telephone and asked him if he had not cleared back, or words to that effect. He had replied, "Oh, haven't I?"

"No," came the reply. He had then at once inserted his release key into the instrument so that he could fully replace his starting signal lever and accept another train.

Directly afterwards at about 6.10 he had accepted the 6 a.m. Bridgeton Cross train from Queen Street West. Immediately after he had accepted the 6 a.m. train a doubt had come into his mind and he had rung Signalman Milne at Queen Street West and asked him what train was coming. Told that it was the one due at Charing Cross at 6.10 he realised that he had made a mistake and that there would be a smash.[1016]

Signalman Rentoul stated that the electric treadle ahead of his starting signal had not been working properly the previous Saturday night, and on the Monday morning he had found that the treadle was again not working properly and was not freed by the release key, so that he had had to manually lift the lock which dropped into the tappet to release the mechanical locking on his starting signal and his block instrument. On the Tuesday the apparatus had worked correctly and, on the Wednesday, the day of the accident, it had also been working correctly, but on receiving the telephone call from Signalman Milne he had jumped to the conclusion that it had failed again and had used the release key. He had given the "Train entering section" signal to Finnieston for the 5.45 train at 6.4, immediately Finnieston had accepted it and he had cleared his signals before the train had reached him so that he could get on with eating his breakfast without interruption. He had been still eating his breakfast when Queen Street East had enquired why he had not cleared the 5.45 and he had assumed that that train must have passed him unnoticed while he was eating. It had not occurred to him to check with Finnieston regarding the 5.45 train but just as he had accepted the 6 a.m. train the Finnieston signalman had telephoned

[1016] Board of Trade report

him and asked if the 5.45 was not coming. It was at that point that he had realised what he had done. Signalman Rentoul pleaded that he had been afraid of delaying traffic and explained that he lived some distance from Charing Cross station and that his only way of getting to work for early turn was to walk, which took him an hour.

Driver James Brown, in charge of the engine of the 5.45 a.m. train, explained at the Board of Trade inquiry into the accident that he had had some trouble with the Westinghouse brakes on leaving Gallowgate station and at College station had asked a porter to look and see if any of the brake blocks were rubbing. The porter had indicated that they did not seem to be. On leaving Queen Street station the brakes had come on hard and had stopped the train in the tunnel between there and Charing Cross. Alighting from his engine with a lamp he had gone back and found the brakes of two carriages to be hard on and had released them. Returning to his engine, he had set off with his train again but only covered about 10 or 12 yards before it stopped a second time, and he had discovered that another train had run into the back of them. They had been stopped, he thought, six or seven minutes in the tunnel.[1017]

Guard Michael Grant of the 5.45 a.m. train was seriously injured in the accident (his right foot had to be amputated) and his evidence had to be taken in the infirmary. He remembered the signalman at College telling him of a problem with the brakes binding and his calling to a porter to release them (it will be noted that driver Brown had said that *he* called a porter, so perhaps they both did). They had subsequently come to a stand in the tunnel between Queen Street and Charing Cross at about 6.4. He had not left his van, not knowing what was wrong and expecting his train to move at any moment. The train had restarted at 6.8 but came to a stand a second time. He had then alighted onto the track and was going forward to find what was wrong when he had heard another train coming. Running back, shouting and whistling, he had just reached his van when his train began to move for a third time, so he had jumped into his van to avoid being left behind just as the collision occurred. His right foot had been jammed in the wreckage and he had been trapped there, suspended by his leg, until released. Guard Grant said that his side and tail lamps had been burning brightly but that the tunnel had been full of steam and smoke. He had not taken any steps to protect his train as he had considered that it was well under the cover of the Charing Cross distant signal.

Driver William Lamond stated that he had been in charge of the engine of the 6 a.m. workman's train from Bridgeton Cross to Kilbowie. He had not stopped after leaving Bridgeton Cross until his train had run into the rear of the 5.45 one in the tunnel. He had seen nothing of the train ahead before the collision. His train had been travelling at the time at 10 to 15 miles an hour. He was only slightly hurt and his engine little damaged.

Lt-Col H.A. Yorke of the Board of Trade placed the responsibility for what he termed 'this terrible calamity' on Signalman Rentoul for the latter's rash and thoughtless use of the release key. Driver Brown and Guard Grant of the 5.45 train also came in for criticism for their failure to look further into the cause of the brakes problem at College station, a criticism which seems very unfair given that the cause of the problem could not be discovered after the accident, though the locomotive superintendent was of the opinion that there was no problem, but that Driver Brown had mishandled the brakes. More justifiably, one might think, Guard Grant was found fault with for not going forward to

---

[1017] Ibid.

enquire of his driver what was wrong when his train first stopped in the tunnel, and for not going back to protect his train in accordance with Rule 217.

The inspecting officer was also critical of the fact that the lock and block apparatus had not been cased in, even after it had been in use for 18 months, so that any unauthorised person could interfere with it, as Signalman Rentoul had admitted doing on the Monday before the accident. In addition failures of the apparatus had not been reported to the Superintendent of the Line and the Telegraph Superintendent on the special forms provided, while the number of alleged failures was excessive and indicated ‘a want of intelligent and careful maintenance of the apparatus’ by the linesman, John Moran. (Charing Cross stationmaster Stevenson had stated that there were 27 cases of treadle failures recorded in the key release book, others which were not there recorded and he had written 20 letters to the District Superintendent reporting various defects in the apparatus.) Finally there were discrepancies in the times recorded in the Charing Cross train register book as compared to those at Queen Street West and Finnieston showing either that Signalman Rentoul had been careless in making his entries or otherwise that he had deliberately made wrong entries to cover up other irregularities.[1018]

Fairmount Park Tunnel, Philadelphia, Pennsylvania, United States, on the Baltimore and Ohio Railroad, was the location of a serious collision shortly before midnight on Friday 11th May 1900. The trains involved were two sections of fast New York bound express freight train No. 1023, the first section having 45 cars and the second 39. The line climbed through the tunnel 80 feet to the mile (1 in 66) and in order to more easily work the heavy first section up the hill Engineer Craig on Engine No. 1032 elected to divide it within the tunnel, going forward with the front 28 cars and leaving 17 behind. The train having thus been split Brakeman Cooney went back through the tunnel with his lamp to protect the rear portion of the train, which should also have been protected by the block system in force between the towers at either end of the tunnel. He had not proceeded very far when he saw the headlight of engine No. 684 pulling the second section coming towards him. He waved his lamp and shouted, but the engineer of the advancing train apparently did not hear him, though the brakeman on the eighth car said later that he did. Brakeman Cooney just had time to dive to the side of the tunnel as the collision occurred. The second section of the train had entered the tunnel running at some speed and ran into the rear of the back portion of the first section in the middle of the 3,811-foot-long bore. Engineer George Laub and Fireman William Hinchman were crushed to death in the collision and the cars piled up in a heap and caught fire. The rear car of those of the first section left in the tunnel was a naphtha rank waggon which exploded, throwing burning naphtha in all directions and setting the rest of the wreckage ablaze. Four oil cars, each carrying 4,000 gallons of fuel, were also involved and they exploded in turn.

After the collision Brakeman Cooney struggled through the smoke and flames and finally managed to reach the end of the tunnel and fresh air. Meanwhile the crew of the first section of the train were unaware of the collision and having disposed of the front of his train re-entered the tunnel with engine No. 1032 and managed to draw out the remainder of their train, which they found to be on fire, despite the terrible heat in the tunnel. Brakeman Cooney eventually made his way back to Race Street tower, which he

[1018] Ibid.

found to be deserted, and he telephoned to the adjoining towers to stop all traffic. The Baltimore and Ohio Railroad tower operator at Race Street, William Lantell, had failed to stop the second section, which was following 10 minutes behind the first, having not restored his signal to danger after the first section had passed while awaiting line clear from Park Junction tower at the far end of the tunnel. A search was made for him after the collision, but he could not be found. He had disappeared, presumably with vague ideas of an escape as he feared a long gaol sentence for having caused the accident, but subsequently he gave himself up to Baltimore and Ohio Superintendent Bent at 3 p.m. on the Saturday afternoon. He then confessed that the accident had been his fault, saying, "I did it. The white signal was not changed, and the train went through. I was asleep or dozing, and why I cannot say. I am ready to stand the censure however." He was arrested, pending the coroner's inquest.[1019]

Half an hour after the accident a Baltimore and Ohio engineer with a freight engine and a lot of nerve managed to back into the tunnel and pull out 17 cars of the second train. Two of these at the end of the train had contained gunpowder and there had been grave fears that should the fire reach them the resulting explosion would have demolished the whole tunnel. The other cars withdrawn from this end of the tunnel were livestock cars. Above the tunnel the noise of the collision and explosions had sounded like an earthquake and it was not until Brakeman Cooney had emerged from the tunnel that it was realised what had happened. An alarm of fire was then raised, and group of residents led by Common Councilman John Daly, who lived over the tunnel, went into the tunnel from the northern portal, but half suffocated by the fumes from the burning oil found that they could do nothing and had to retreat. After the firemen had arrived, they in turn made repeated attempts to enter the tunnel to fight the conflagration, but each time they were driven back. Eventually it was decided by Fire Chief Baxter to take the drastic step of making a hole in the roof of the tunnel to let out the smoke and flames, with the aid of a gang of labourers and some dynamite. This was done, and the fire was finally extinguished at 6 p.m. on Saturday after 18 hours.

Pending the reopening of the line Baltimore and Ohio trains were diverted over the Pennsylvania Railroad tracks between Eastwick and Belmont.[1020]

It was rumoured that there had also been five tramps stealing a ride on one of the freight trains, who had perished in the accident, a brakeman having seen seven tramps board the train and only two get off of it prior to the accident, though the railroad company denied that they could have been on the train and no traces of their remains were found. In addition to the deaths of the two enginemen and the possible deaths of the tramps about a dozen firemen were seriously injured in their attempts to control the blaze and had to be taken to hospital, while another 20 were slightly hurt.[1021] A conservative estimate of the damage caused by the accident was $175,000.[1022]

The coroner's jury found Tower Operator Lantell criminally negligent and censured the conductor of the train for not having sent back his flagman. The jury commented, 'We believe keeping operators or signalmen at work for twelve hours consecutively is inviting disaster, especially when alone in the tower, and we recommend that the hours of

[1019] *Philadelphia Inquirer*, Philadelphia, Pennsylvania, United States, 13th May 1900
[1020] *Philadelphia Times*, Philadelphia, Pennsylvania, 12th May 1900
[1021] *Boston Daily Globe*, Boston, Massachusetts, United States, 12th May 1900
[1022] *Philadelphia Inquirer*, Philadelphia, Pennsylvania, United States, 13th May 1900

service of such employees should be reduced to eight hours per shift. We suggest that a more modern system of signals be installed, and that the tunnel be made a more absolute block by electric devices. We would also suggest that the tunnel should be lighted and ventilated as an additional safeguard.' Lantell was committed to await the action of the grand jury.[1023] They found that he had a case to answer and he stood trial at the beginning of December, when Judge Audenried directed that his prosecution be abandoned as it had not been proved that he had been guilty of neglect.[1024]

1023 *The Philadelphia Times*, Philadelphia, Pennsylvania, United States, 2nd June 1900
1024 *Philadelphia Inquirer*, Philadelphia, Pennsylvania, United States, 7th December 1900

CHAPTER 12

LEVEL CROSSING COLLISIONS

Wherever roads and railways intersect they must do so either by means of a bridge or on the level. Bridges have the advantage of allowing road and rail traffic to operate entirely without interference from one another, but they are much more expensive to provide, especially in flat countryside, due to the need to provide extensive earthworks for approach embankments or cuttings as well as the cost of providing the span itself. Where underbridges are provided there is often also the need to build retaining walls and there can be problems with drainage.[1025] The provision of level crossings, especially if, as in many parts of the world, they are unstaffed, is the much cheaper alternative, but where they exist, they introduce an additional source of danger to both road and rail users. Robert B. Shaw in his outstanding book on American railway accidents *A History of Railroad Accidents, Safety Precautions and Operating Practices*, writes that he has excluded ordinary grade (i.e. level) crossing collisions from consideration on the grounds that they are not proper train accidents, inasmuch trains usually suffer much less harm from such accidents than the road vehicle involved. This series of books, on the other hand, will attempt to include, amongst others, a selection of such accidents that have features of interest even if none of the casualties in them, if any, occurred to train users. The internal combustion engine was invented during the period covered by this volume, in 1884, and the first motor car appeared in 1885 or 1886.[1026] However, as far as the author has been able to discover the first level crossing collision involving a motor vehicle did not occur until 1st May 1904, in France,[1027] and thus all the accidents that we will be concerned with in this volume will involve the collisions of trains with vehicles drawn by animals (usually horses) or powered by electricity or perhaps steam.

A tragedy occurred, and a more serious accident was only narrowly averted, at Weeton, near Harrogate, England, on Friday 6th May 1881. A lad was leading two horses pulling a heavy road roller across the North Eastern Railway main line when the 11.35

[1025] *Backtrack* Volume 24, pg. 88.
[1026] Wikipedia – History of the Internal Combustion Engine
[1027] Wikipedia – Liste des accidents ferroviaires en France au XX siècle

a.m. up express to Leeds caught them. The train struck and killed the boy and the horses but missed the road roller itself, which was still on the down line, thus preventing potentially more serious consequences to the train crew and passengers.[1028]

Four people died in an accident on 19th August 1882 at Anoka, Minnesota, United States. The victims were two young men and two young women who were returning with a horse and carriage from a Saturday night dance. The horse was also killed.[1029]

Philadelphia. Pennsylvania, United States, saw a level crossing collision between a train and a horse-drawn streetcar (tram) on Tuesday 7th November 1882. The accident, on the Frankford Road crossing, involved a Pennsylvania Railroad train carrying emigrants from New York and streetcar No. 74 of the Second and Third Streets Railway. It happened at about 10.30 p.m. The conductor of the streetcar alighted from it and went forward to the ungated level crossing to check if any trains were in sight, in accordance with the rules of the streetcar company. Seeing, as he said later, neither a train nor a signal light he called his driver on. The streetcar was midway over the line when its driver saw the train bearing down upon him and close at hand. He immediately whipped up his horses, but it was too late, and the engine caught the rear end of the streetcar, demolishing it. Enoch Cubberly, the engineer of the train, said that in the darkness he had only seen the streetcar at the last moment, when he had whistled for brakes to the brakeman and applied the hand brake on the engine. The only occupant of the streetcar, a Miss Annie Schmidt, aged 27, had been sitting near the door close to where the engine struck, and was killed instantly.[1030]

Another collision between a train and a streetcar occurred in Pennsylvania only two months later, on the morning of 12th January 1883, in somewhat different circumstances and this time without any deaths. This accident, in Anderson Street, Allegheny City, just outside of Pittsburgh, took place at a gated level crossing. The streetcar, owned by the People's Railway Company, became unmanageable coming down the hill to the crossing and crashed through the gates in front of a Pennsylvania Railroad Chicago express. The engineer of the express was fortunately able to check his train and reduce the force of the collision, the train striking the rear platform of the car and pushing it out of the way. A man riding on the platform was seriously hurt and two women passengers fainted from fright, but the actions of the driver and conductor of the crowded car in keeping both its doors closed saved the remainder of the occupants.[1031] The newspaper report does not mention any horses pulling the street car, possibly it was powered by a steam dummy or tram engine (electric street cars did not come into use anywhere until a couple of years later).[1032]

[1028] *The Times*, London, England, 9th May 1881

[1029] *Stevens Point Journal*, Stevens Point, Wisconsin, United States, 26th August 1882

[1030] *Williamsport Gazette Bulletin*, Williamsport, Pennsylvania, United States, 10th November 1882

[1031] Ibid, 13th January 1883

[1032] Wikipedia – Street cars in North America

The growth of American cities in the late 19th century led to a corresponding growth in the number of streetcar lines, of level crossings of such routes with railway lines and of collisions on the latter. Thus a third encounter between a steam train and a streetcar happened in June 1883, this time in Chicago, where a Wabash Railroad express bound for Kansas City collided with a car on the night of the 30th of that month at the Root Street crossing, near the then southern edge of the city. The collision was a violent one, the car was reportedly 'smashed to fragments' and the lamps in the car exploded and set it on fire. Of its 13 occupants only one escaped injury, several were badly burned, and it was feared that two or three of those most severely hurt would die. The driver of the streetcar had misunderstood a signal from the man stationed at the crossing.[1033]

The run of collisions with streetcars in the United States in 1883 ended with one on 5th October, again in Philadelphia, Pennsylvania. This time an incoming Philadelphia and Reading Railroad company train running on its North Penn branch and a Union line streetcar were the protagonists in a meeting at the junction of Susquehanna Avenue and American Street. As usual it was the streetcar which came off worse from the encounter. Ten of the most severely hurt of them were taken to the Episcopal hospital where two, James Devine and John McKeown, had died by the end of the day, McKeown leaving a wife and seven children. The railroad and streetcar companies were apparently both to blame for the accident. There were no gates at the crossing, though the city council had directed that the Philadelphia and Reading Railroad provide them a year earlier. For its part the streetcar company was operating the car with just a driver and no conductor, so that the driver was expected to guide his team of horses, see that the passengers deposited their fares in a box in the car and watch for trains at the crossing. The driver was apparently inside the car at the moment the accident occurred, the horses having been left in charge of a boy. The car had stopped and waited at the crossing for a northbound train to pass but had then proceeded to cross without the boy guiding the team noticing a southbound one, and the engine of that train, travelling at about 25 miles an hour, had struck and wrecked the car.[1034]

Rashness in trying to beat a train over a level crossing was (and is) of course a worldwide phenomenon and on Sunday 14th June 1885 an example of such foolhardiness was manifest near Vienna, Austria, when an omnibus driver attempted to beat a Südbahn train in the village of Perchtoldsdorf. The train caught the rear of the omnibus, killing one omnibus passenger outright and seriously injuring a number of others, including five children. The errant driver escaped unhurt.[1035]

If humans could be rash and foolhardy, then horses could also get out of control and cause accidents. Black Rock Crossing, near New Britain, Connecticut, United States, was the scene of an accident caused by horses which became unmanageable on 1st October 1885. Charles Dyson, aged 18, had set off on the morning of that day in his father's omnibus to go to school in the town, accompanied by his two sisters, Minnie, aged 15,

[1033] *Decatur Daily Republican*, Decatur, Illinois, United States, 2nd July 1883
[1034] *The Helena Independent*, Helena, Montana, United States, 6th October 1883
[1035] *The Times*, London, England, 17th June 1885

and Charlotte, aged 13. As they neared the New York and New England Railroad level crossing the horses somehow took fright and dashed across the line directly in front of a passenger train running at full speed. The train struck the vehicle, squarely dashing it to pieces. Minnie was thrown 50 feet and was picked up dead, her brother and sister were thrown 100 feet and were alive when found but both died after being taken home. The newspaper reporter commented that of John Dyson's seven offspring only four boys remained. Two men had lost their lives at the same crossing some time previously.[1036]

In a variation on previous train-streetcar collisions one in St Louis, Missouri, United States, on 26th January 1886 involved a car being drawn by mules rather than horses. The animals were pulling a Broadway line streetcar when they were struck by a St Louis Iron Mountain and Southern Railroad Texas express on the Seventh Street level crossing. The engineer of the train was almost able to pull up and although the car was partly lifted up and overturned none of its dozen passengers was hurt. One of the unlucky mules, though, was killed.[1037]

Princess Lieven, described as 'a member of a noble Russian family' had a narrow escape from death on 7th January 1889. She was on her way from Charkoff (now Kharkiv) on Christmas Day (celebrated on 7th January by Russian Orthodox Christians) to visit her father, who lived in a village near Krasnopavlovka station on the Koursk (now Kursk) – Azoff (now Avov) line. As the sledge in which the princess was riding approached the railway it was seen that a train of 12 carriages was being set back towards the crossing. The coachman, noting that crossing barrier remained raised, nevertheless decided to continue across the line, and the sledge was struck by the train, which was moving at some speed. The horses were thrown to one side, the coachman was run over by the train and killed, and the princess fell between the rails, where seven carriages passed over her. The engine driver had whistled and shouted to the coachman to warn him of the danger and stopped his train as quickly as possible, when it was found that the princess, although she had temporarily lost consciousness (fainted?) was actually unhurt, and she soon recovered with the administration of restoratives. It was then discovered that the woman crossing keeper was lying insensible by the track. She was apparently suffering from carbon monoxide poisoning from the smoke from a charcoal fire, and although she had gone out and attempted to lower the barrier her strength had failed her, and she had collapsed.[1038]

A streetcar collision in Cincinnati, Ohio, United States, on 6th February 1889 involved, like that in St Louis in January 1886, a car drawn by mules. Car No. 4 of the Covington and Cincinnati streetcar company, drawn by two mules, was crossing the Chesapeake and Ohio Railroad tracks at Pike Street at about 9 o'clock on the evening of that day when it was run into by a Kentucky Central Railroad train, the latter company presumably having running powers over the Chesapeake and Ohio line. Most of the passengers on the car escaped with a few bruises and a shaking but a Mrs Loebker, a

[1036] *The Fitchburg Sentinel*, Fitchburg, Massachusetts, United States, 2nd October 1885
[1037] *The Trenton Times*, Trenton, New Jersey, United States, 27th January 1887
[1038] *The Times*, London, England, 8th and 9th January 1889

widow, had a leg torn off at the thigh, though a child on her lap escaped unhurt. The two mules were killed. There were the usual recriminations after the accident, the passengers on the car swearing that the engineer of the train had not rung his bell or given any warning of his approach. There was no watchman at the crossing, which street cars used every 10 minutes, and the newspaper wondered that more accidents had not happened there.[1039]

Six deaths occurred in an accident at Kalamazoo, Michigan, United States, on 8th May 1889. A freight train required assistance at Ostend, the first station west of Kalamazoo, and a switch engine was despatched from there with to run as fast as possible to Ostend to prevent further delay. At the same time a streetcar was approaching the line along West Main street. The level crossing was equipped with barriers, but these had not been lowered and the streetcar driver, seeing that the gates were raised and hearing no warning whistle from the engine, naturally assumed that it was safe to cross. The engine struck the car squarely and carried it some distance before it could be brought to a halt. Of the dozen passengers on the car five were killed instantly and a sixth died shortly afterwards. All of those killed were women, and they included the wives of a former Lake Shore and Michigan Southern superintendent and of a mathematics professor at Kalamazoo Baptist College, whilst the wife of the owner of the Kalamazoo paper mill was among those injured. One passenger who had been riding inside the car, a boy riding on its platform and the car driver escaped injury. Witnesses stated that the engine had been travelling at 25 miles an hour, a speed in excess of that permitted within city limits. The gatekeeper stated that he was not required to attend the crossing between 6 and 6.45 p.m. as no trains were scheduled to run at that time. He was presumably away having his tea. The newspaper added that 'Great excitement prevails in the city, and thousands of people have visited the scene of the accident.' [1040]

A few days later it was reported that Engineer Michael Wellham and Conductor John Bush had been charged with the manslaughter of a Sarah Hadlock, one of the women killed in the accident.[1041]

In an article entitled 'The Railroad Monster' the *Trenton Times* newspaper described an accident at a level crossing near Little Silver station, New Jersey, United States, on Friday 5th July 1889. The people involved in the accident were the aged mother of a Joseph A. Ward, a New York collar and shirt manufacturer; Mr Keating, her son-in-law, who was a New York hat manufacturer; Mr Keating's little daughter Elsie and Kate, the child's Irish nurse. The party were out on a pleasure ride in a horse and carriage. They reached the level crossing just as a Central of New Jersey Railroad train was pulling out of the New York and Long Branch station. That train having passed Mr Keating drove onto the crossing with the double track line right in front of the 3.30 p.m. express train from New York, which ran, the newspaper report added, fast from New Jersey to Long Branch, stopping only at Red Bank, and was in charge of Edward Lock, 'one of the oldest and best conductors of the road.' What was more to the point was that the train was

1039 *Newark Daily Advocate*, Newark, Ohio, United States, 7th February 1889
1040 *The Marion Daily Star*, Marion, Ohio, United States, 9th May 1889
1041 *The Salem Daily News*, Salem, Ohio, United States, 16th May 1889

travelling at 40 miles an hour as it reached the crossing and the pleasure seekers had no chance. The carriage was described as being ‘smashed into kindling wood’ and the poor horse was ‘torn to pieces’. The four human victims were carried various distances down the line. Several people at the station witnessed the accident and rushed down the line to help after the train had passed. They found that the three adults had been killed instantly and were dreadfully mutilated. The little girl, Elsie, was still alive, but died from her injuries before she could be moved to her father’s summer house, half a mile away. There was only a minute’s difference between the times of the two trains at the crossing.[1042]

On Tuesday 2nd September 1890 a fire broke out at carpentry shop and dwelling house of a Mr Nathaniel Hemenway in the Maplewood neighbourhood of Malden, Massachusetts, United States. The alarm was raised, and a horse-drawn ‘new chemical fire engine’ was sent out. Hurrying to the blaze the fire engine was struck by a train at the level crossing adjoining the Boston and Maine Railroad’s Faulkner station. George Phillips, the driver of the fire tender, was killed outright and third assistant engineer Samuel Kingston badly hurt. One of the horses was also killed. Mr Hemenway was seriously burned endeavouring to save some of his property from the fire.[1043]

A driver in charge of a waggon carrying gunpowder might be expected to be especially careful in view of his dangerous load but on 7th November 1890 a train collided with one as it was crossing the line near Bochum, Westphalia, Germany. The force of the collision caused the gunpowder to explode, blowing the waggon and horses a considerable distance. The waggon driver, the engine driver and a large number of the train passengers were injured, but surprisingly no one seem to have been killed. The noise of the detonation could be heard several kilometres away and large numbers of spectators soon appeared at the scene of the accident.[1044]

More typical than the Westphalia accident was one near Camden, Indiana, United States, on Sunday 27th December 1890. A party of four young people were out on an evening pleasure ride in a two-seater conveyance. Arriving at a level crossing they presumably failed to notice in time a fast-approaching Vandalia Railroad train, which struck the vehicle, wrecking it and throwing them in every direction. A James McKain was killed outright when he was thrown against a telegraph pole and a Miss Lizzie Snowberger died of her injuries soon afterwards. The other two were little hurt. The accident happened at about 8 p.m., so it would have been pitch dark and the train was perhaps hard to see.[1045] [1046]

On Friday 23rd January 1891 an elderly couple, Mr James E. Owens, aged 79, and his wife Nancy, aged 74, were crossing the Baltimore and Ohio Railroad just outside of the

1042 *The Trenton Times*, Trenton, New Jersey, United States, 6th July 1889
1043 *The Indiana Gazette*, Indiana, Pennsylvania, United States, 4th September 1890
1044 *The Times*, London, England, 8th November 1890
1045 *Fort Wayne Sentinel*, Fort Wayne, Indiana, United States, 30th December 1890
1046 *The Trenton Times*, Trenton, New Jersey, United States, 29th December 1890

city of Washington, D.C., United States, with a horse or horses pulling a milk waggon. It was a cold day and their faces were well covered up for warmth. With their ears thus insulated from outside sounds as well they failed to hear the whistling of a train bound for the capital and the train could not be seen from any great distance. Both were killed instantly.[1047]

Were the weather foggy of course using a level crossing was extra hazardous. Thus on 18th February 1891 during a dense fog a Great Eastern Railway train dashed into a waggon pulled by a team of four horses at Beccles, England. The train did not escape either in this case, one waggon and the guard's van being thrown off the track. The guard of the train, named Newburn, was seriously injured in the derailment, having both arms fractured. The gatekeeper, holding the crossing gate open for the waggon to pass, was also seriously hurt but the man in charge of the waggon was not injured. One of the horses was killed and the permanent way was much damaged.[1048]

The mistake which had cost four lives at Little River in 1889 cost another at Nashua, New Hampshire, United States, on 26th August 1891. Mr W.H. Campbell, 'a well-known paper box manufacturer', his wife and daughter were riding in a horse-drawn carriage when it was struck by a Boston and Maine Railroad train at the Temple Street crossing. Mrs Campbell was fatally injured, and her husband suffered severe scalp wounds. Mr Campbell had waited for an up train to pass and then set off across the line into the path of a down one.[1049]

An accident which could have proved much more serious occurred at Rockville, Maryland, United States, exactly a month after the Nashua tragedy. On Saturday 26th September 1891 a waggon loaded with stone was struck on the Baltimore street crossing by a fast-running eastbound Baltimore and Ohio Railroad express. The rear of the waggon was knocked to pieces, but its driver and the horse escaped. The driver, a Mr Uriah Rickets, had been walking alongside the waggon when he saw the train close at hand. Whipping the horses, he had leapt from the track in time to save himself. The safety gates had been raised and several witnesses who were stood close at hand claimed that the engineer of the train had failed to whistle at the usual spot and had given no warning until nearly opposite the depot and within a few yards of the crossing. The railroad company promised to investigate the matter.[1050]

Possibly the first level crossing involving an electric streetcar was that at Flushing, New York, United States, on 12th October 1891. A Long Island Railroad freight train collided with the car at 8.30 a.m. on the Bridge street crossing. Of the 10 passengers on the car a Miss Mary Hill, principal of the College Points school, and Miss Annie Van Nostrand, an

---

1047 *The Marion Star*, Marion, Ohio, United States, 24th January 1891
1048 *The Times*, London, England, 19th February 1891
1049 *The Fitchburg Sentinel*, Fitchburg, Massachusetts, United States, 26th August 1891
1050 *The News*, Frederick, Maryland, United States, 29th September 1891

attendant at the same school, were badly injured, Miss Hill possibly fatally so.[1051]

The colony of Victoria, which in a few years would become a part of the Commonwealth of Australia, witnessed a level crossing accident on 12th February 1892 on the then still to be completed Curdie's River line. Although the line was not quite finished it was ready enough for its contractors, Messrs Buscombe Chappel and Bell, to operate passenger as well as goods trains over it. As one of these passenger trains was approaching a level crossing named Smith's Crossing it was in collision with a drag (a type of carriage) pulled by four horses and being driven by a Mr William M'Donald. The drag was en route to pick up the family of a Mr Stansmore, of the Camperdown livery stables, the owner of the drag, from a Mr Frank Madden's (sheep?) station and bring them back to Camperdown on their way back to Melbourne. The two leading horses almost got clear of the line before the train reached the crossing and were only slightly injured, but the two pole horses (those harnessed to the shaft of the carriage) were killed outright and the drag itself was smashed to pieces. Mr M'Donald and a lad who was accompanying him were thrown clear and escaped unhurt, and none of the several passengers on the train was injured. The tender of the engine was thrown off the line by the force of the impact and the engine itself so badly damaged that it could not proceed. It was intended that the crossing be protected by gates, but with the construction of the line not yet finished these had still to be hung.[1052]

A bad accident occurred at Elmira, New York, United States, on the evening of Thursday 1st December 1892. The victims of the accident were returning, in what was described simply as a waggon, from the theatre, when it was struck by Erie Railroad train No. 5 at the unguarded Pennsylvania Avenue crossing. Mr Conklin, who was driving the waggon, was the only one of the party not to be killed. He had apparently failed to see the train and drove onto the line directly in front of it. The single horse had just cleared the line when the engine of the train struck the waggon about its centre. Freed from the demolished waggon, the terrified horse ran away down the street, and Mr Conklin, who still had hold of the reins, was pulled away from the line and probable death, though he was badly injured. Killed instantly were Mrs Conklin, Edward Blanchard and his wife, and a Maggie Pitts, while Mr Conklin's son was badly injured.

Engineer Cable New on the locomotive knew nothing of the accident until he pulled up at Elmira station, where his attention was drawn to the remains of the bodies of three of the victims caught on the engine pilot. He told the subsequent inquest that his booked time at the station was 11.6 p.m. His train had been a heavy one and was 15 minutes late at Susquehanna, but he had made up six minutes on the journey. He was particular about his engine's headlight (all American engines carried headlights, unlike British ones which at this time did not) and had a good one that night. He had been travelling at 15 miles an hour at the Pennsylvania Avenue crossing, the speed of trains being limited to that over guarded crossings within the city limits (the newspaper report had spoken of the train continuing from the crossing to the station 'at a terrific speed'). He had sounded his whistle and the automatic bell of the engine was ringing. As he got to the crossing, he had

[1051] *Middletown Daily Press*, Middletown, New York, United States, 12th October 1891
[1052] *The Argus*, Melbourne, Victoria, 13th February 1892

taken his eyes off the track for two seconds to drink some cold tea. The fireman had been firing. He had felt no jerk at the crossing. His long-distance vision was as good as it had ever been. He did not suppose that the engine whistle could be heard at a very great distance because of the weather.[1053]

Asked by a lawyer if it was a company rule not to pass over unguarded crossings at more than four miles an hour, Engineer New replied that he thought that such was the case.

Before December 1892 was out four more people had been killed in a level crossing collision with a streetcar in Chicago. The accident happened on Thursday 29th, during the morning rush hour when traffic was at its heaviest. Streetcar No. 216 was thus very crowded as it approached the tracks of the Pittsburgh Fort Wayne and Chicago Railroad at the junction of 57th Street and Stewart Avenue. At the same moment an engine was backing down the tracks at a speed, according to the railwaymen, of 15 miles an hour. For some reason the crew of the engine did not see the streetcar nearing the line, while the passengers in the car, unable to see much out of its heavily frost-coated windows, did not see the engine bearing down upon them. They thus remained unaware of their impending fate until there was a sudden violent impact. The engine dragged the car about 150 yards down the line. The horses pulling it had been across the track and escaped injury – released by the collision they galloped away, while nearby residents, alerted by the cries of the injured passengers, rushed to their assistance. The injured victims were carried to nearby houses and at least one, a George Blake, to a saloon, where he very shortly expired. Five police patrol waggons then arrived, and the officers took charge of the scene. Another critically injured passenger, a Mrs Carson, was loaded into a police waggon to be taken to the Mercy hospital, but she died before it reached there, and the waggon was diverted to the city morgue. The bodies of the two other fatal victims were discovered under the tender of the engine.

In addition to dealing with the dead and injured the police set about arresting those whom they believed might be responsible for the accident. First, they arrested Conductor O'Connor and Driver Stanley, of the streetcar, and Engineer Rosscup and Fireman Meager, of the engine. Then they turned their attention to the two railwaymen who were in charge of the crossing; Peter Schwartz, manning a signal tower at the crossing, and his groundman, John Allbright. The two men denied being responsible for the accident but were placed in custody with the other four. Allbright was the same watchman who had been at the crossing when a previous accident with a streetcar, which had cost two lives, had occurred on 10th February. The newspaper reporter commented that it could not be doubted that the accident was 'due to the grossest carelessness'. The men had not been attending to their duty and neither knew of the likelihood of an accident until it had happened. Both had been in Allbright's shanty whereas Schwartz should have been up in his tower watching for approaching trains. There was also considerable doubt as to how fast the engine had been running, and although the engineer and fireman insisted that there had been a red light on the tender the driver and conductor of the streetcar both denied having seen one. The hour of the accident was not stated in the newspaper report

1053 *Hornellsville Weekly Tribune,* Hornellsville, New York, United States, 9th December 1892 but clearly it must have happened in the pre-dawn darkness.[1054]

A sleigh, drawn by two horses and carrying a party of six women and several children, who were returning home to Passaic City, New Jersey, United States, was struck by Erie Railroad train No. 14, the Wells-Fargo Express, at a level crossing with Monroe Street, in that town, half an hour after midnight on Tuesday 24th January 1893. Mamie (or Mary) Ryan, 22, was killed instantly, Maggie Splain (or Spillane), 21, died of her injuries within a few hours, and a Mrs Thomas O'Brien, the divorced wife of a West Street, New York, saloon keeper, and John Moore, the coloured driver of the sleigh, were fatally injured. The two horses were also killed but the other two persons presumably escaped. The newspaper noted that 'There was no flagman nor gate at the crossing.' The victims were all stated to be quite prominent people, and were all members of a straw riding party.[1055] [1056] (A straw or hayride was a traditional American and Canadian activity consisting of a recreational ride in a waggon or sleigh loaded with straw or hay for comfortable seating.[1057])

Another bad accident involving a streetcar in Chicago, Illinois, United States, occurred on 17th July 1893. This time it was the 49th Street crossing with the Grand Trunk Western Railroad. Again it was a crowded morning horse-car (the accident happened at 7.30 a.m.), owned by the Halstead Street Company. There were 46 men and women on board the vehicle with some of the men being forced to ride on the footboards because there was no room inside. The newspaper account noted that 49th Street here passed over a network of tracks and that the crossing had always been regarded as a dangerous one. Like that at 57th Street and Stewart Avenue concerned in the accident the previous December, it was provided with gates (barriers) worked from an adjoining signal tower.

As the horse-car was nearing the crossing a long freight train going west passed over it. That train having cleared the crossing, George Barmen, the towerman, raised the crossing barriers. Driver Charles Stalnecker naturally assumed that this meant that it was safe to cross, whipped up his horse and began to cross. His colleague, Conductor Frank Barnett, had also alighted from the car and walked ahead to check that all was clear, but could see nothing other than the rear of the departing freight train. Suddenly the towerman saw an advancing Great Northern Railroad passenger train and at once began to lower the gates again, but too late to arrest the progress of the streetcar. The men who had been riding on the footboards of the car were the lucky ones, for they were able to jump off in time and escape, but the passengers inside had no chance of escaping death or injury as the train, travelling moderately fast, demolished the car, throwing it 30 feet through the air. Three or possibly four horse-car passengers lost their lives, one or two men and two women, and many more were injured. The bodies of the dead were taken to the city morgue and carriages took the injured to hospital.

Engineer Jones of the passenger train stated that he did not see the streetcar until the horses were on the track and he had then been too close to be able to pull up.

1054 *Sandusky Daily Register*, Sandusky, Ohio, United States, 30th December 1892

1055 *The Daily Review*, Decatur, Illinois, United States, 24th January 1893

1056 *Middletown Daily News*, Middletown, New York, United States, 24th January 1893

1057 Wikipedia – Hayride

Nevertheless, Police Captain Ward arrested the engineer, his fireman, James Campbell, and a man named Henry Hughes who had also been riding on the engine. They were to be held at the police station pending the coroner's inquest. The newspaper report made no mention of the towerman being arrested.[1058]

Mutual negligence, the *Olean Democrat* newspaper suggested, was the cause of a 'Railway crossing horror' at State street crossing, North Le Roy, New York, United States, on 20th August 1893. The accident claimed as its victims three members of the Bovee family, 'one of the richest and most prominent in Genesee county' together with a Miss Nancy Wicks, another resident of Le Roy, and a Miss Emma Bowden, who had been visiting the Bovees. The part of five were on their way to church on this Sunday morning, riding in a carriage drawn by a 'spirited team' of two horses. They reached the crossing at the same time as the eastbound Lehigh Valley Railroad flyer. Mr Bovee, who was driving, failed to spot the train until the last moment, when it was too late to turn around. The train was running at 60 miles an hour. Its engineer whistled loudly, and Mr Bovee whipped his horses in a frantic attempt to get clear, but the engine struck the forward part of the carriage. All five of its occupants and the two horses were killed and their bodies badly mangled; the victims could only be recognised by their clothing. After the accident the train backed to the crossing and the bodies were picked up and conveyed to the Bovee's house, three miles from Le Roy. It is not clear from its report why the newspaper thought the railway was in any way to blame for the accident. The same weekend an August Kasten and John Frey, of Syracuse, New York, returning from a fishing excursion, were killed by a Rome Watertown and Ogdensburg Railroad train near Woodard station. They had been seated in a platform waggon driven by a Joseph Gardner, whom it appears tried to cross the line in front of the train. Gardner and the horse pulling the waggon escaped unhurt, but Gardner's mind was stated to have been affected by the accident. Kasten left a widow and seven children.[1059]

The derailment of a streetcar on a level crossing in Decatur, Illinois, United States, led in turn to its being struck by a Wabash Railroad train on Tuesday 14th August 1894. The southbound electric car left its track at the Edwards Street crossing. The rails of the streetcar line were necessarily short pieces of steel at the point where they crossed the Wabash track and it was here that the derailment occurred. Standing on the far side of the crossing at the Cerro Gordo Street switch, waiting for the southbound car to pass, was Motorman Harry Douglas in charge of a northbound car. Seeing what had happened, he moved forward with his car to try and help re-rail the southbound one. The two cars had just been coupled together when eastbound Wabash freight train No. 94 was seen approaching. Seeing that it was no use attempting to re-rail the disabled car in the short time left Motorman Douglas uncoupled his car again and beat a quick retreat to the Cerro Gordo switch, while the sole passenger on the derailed car, a Mr C.J. Shores, of Kankakee, ran down the track towards the approaching freight train in the hope of stopping it, but as it was 8.30 p.m. it was dark and he had no lantern with him, so all that he could do was shout, and his efforts were unavailing. When the Wabash engineer was

1058 *Decatur Daily Republican*, Decatur, Illinois, United States, 18th July 1893
1059 *The Olean Democrat*, Olean, New York, United States, 22nd August 1893

finally close enough to see the obstruction, he reversed his engine, but the collision followed very shortly afterwards, and the front of the car was demolished. Witnesses related, however, that worse damage to the car was done later when, in the attempt to move it clear so that the freight train could get past, its rear end was badly damaged by coming into contact with a telegraph pole.

No one was hurt but rumours circulated, as often happens in such cases, that there had been a disastrous wreck at the crossing. A breakdown train was sent out, only to find when it reached the scene that there was nothing for it to do, and it was sent back again. Engine No. 389 of the freight train was undamaged, and eventually continued with its train, followed by the St Louis accommodation passenger train, delayed half an hour by the mishap.

Wabash divisional engineer John Brant stated afterwards that he did not see how the railway could be blamed in any way for the mishap. The streetcar had been blocking the line and its employees had not thought to exhibit any danger signals to the engineer of the freight train. Superintendent Ferguson of the streetcar line stated that the conductor had done all he could to save the car. The damage to the car would cost between $600 and $800.

Concluding its report on the accident the newspaper wrote that, 'The Edward Street crossing is an exceedingly dangerous one. No watchman is there at night. When these rapidly moving trains pass so often it is surprising that frequent collisions do not occur.' [1060]

A tragic accident happened near Chartham, England, on the South Eastern Railway, on 9th October 1894. It involved a waggon drawn by a single horse carrying 21 hop pickers and their children. At this date, when many working-class London families were too poor to afford any other sort of holiday, they could 'enjoy' a working holiday, exchanging the back streets of the metropolis for the hop-fields of Kent. The waggon had set off at about 6 a.m. from Canterbury for the Horton Chapel Farm of a Mr Corrie, reached from the main Canterbury-Ashford road by a private lane which crossed the railway at an occupation crossing, protected by a pair of stout wooden gates which opened outwards, away from the line. The morning was a very foggy one. As the waggon got to the crossing the driver of the waggon, Thomas Finch, who was an employee of the farm, asked two boys named Bertie Ward and James Draper, two of the hop pickers, to open the gates. Ward climbed down from the waggon and opened the near gate whereupon Finch proceeded to take the waggon onto the crossing without stopping at all to check whether or not a train was coming, despite the very limited visibility. It was a fatal mistake.

The 4.15 a.m. goods train from Ashford, in charge of Driver Edward Sweetman, and running two hours late, was bearing down upon the crossing. Driver Sweetman told the Board of Trade inquiry into the accident that he had passed through Chartham station at about 35 miles an hour and had whistled at the whistle board 400 or 500 yards in the rear of Horton's Crossing. His engine had only been about 30 yards from the crossing when a horse's head had suddenly appeared over the left rail of the down line (that which his train was on). He had immediately whistled, shut off steam and applied the vacuum brake

[1060] *The Daily Review*, Decatur, Illinois, United States, 15th August 1894 with full force, but the left buffer of his engine had struck the obstruction, he thought, between the horse and the front of the waggon. The horse was thrown over to the up line, but parts of the waggon were caught beneath his engine and were still there when he had been able to stop a quarter-mile down the line.

Driver Finch had not instructed the boys Ward and Draper to listen out for any trains or to look to see if there was anything coming. Draper had not had time to get over the line and open the far gate before Finch drove onto the crossing. Finch apparently saw the train at the last moment and as he could not go forward tried to go back but failed. He was one of those killed in the accident. Four of the hop pickers, one a 1½-year-old child, were also killed outright, two more, one a child of 2½, died later in Canterbury hospital, and nine others were injured.

Major General C.S. Hutchinson, who inquired into the accident, found no blame attached to Driver Sweetman, his fireman or the guard of the train and that the accident was entirely due to the imprudence of Driver Finch. Why, during a thick fog, he had not stopped even for a moment to listen for the sound of an approaching train, could of course never be known as he had paid for his mistake with his life. The inspecting officer noted that over the 6½ years up to 30th June 1894 there had been 137 fatal accidents, or 21 per annum, at occupation crossings, as opposed to 191, or 29 per annum, at public road level crossings. Increased safety might be provided by the provision of a crossing keeper, and if necessary, protecting signals, or by treadles which activated a bell at the crossing until the train had passed. On new lines one or other arrangement had been adopted when thought necessary by an inspecting officer, but the Board of Trade had no statutory powers to require such measures on an existing railway. As regards this crossing he recommended the cutting down of some bushes on the Chartham side of it to improve visibility in clear weather.[1061] [1062]

By 15th October a relief fund begun by the Mayor of Canterbury had raised upwards of £150 for the families of the victims.[1063]

An accident near Hoyleton, South Australia, on 18th January 1895, had a luckier outcome. Mr John Saunders and his son Willie were bringing wheat into town in a waggon drawn by a four-horse team.

Approaching a level crossing Mr Saunders alleged that the driver of a light special train had failed to whistle for the ungated crossing, though he admitted that he had failed to notice the approach of the train until he was within a chain (22 yards) of the crossing. He had then jumped off and attempted to catch the heads of the leaders (the front two horses) 'but this made them all the more eager to cross'. Striking the shaft of the waggon, the train lifted it clear of the line, smashing a front wheel and the forepart of the body of the vehicle. The horses were thrown against a fence and the shafters slightly injured. Wheat was scattered over the ground and a few bags carried some distance by the engine of the train. Mr Saunders and his son were unhurt, though the accident gave Mrs Saunders and their other children some anxious moments as they had witnessed it from a short distance away. The damage was estimated at £20 or £30. Mr Saunders stated that he

[1061] Board of Trade report
[1062] *The Times*, London, England, 10th October 1894
[1063] Ibid, 15th October 1894

had previously had several narrow escapes at the crossing and complained that trains often failed to whistle. The line passed through a small cutting to the north of the crossing, restricting the view in that direction.[1064]

Less fortunate than Mr Saunders and his son were a Mr George Warren and Mr L.S. Burton, whose horse and hooded buggy or trap was hit by a train at Campbell's Crossing (also known as Kerr's Crossing and the 29 Mile Crossing), South Australia, on 23rd February 1895, killing both. The men, who were old friends, had been returning to Gawler. In this case the crossing was under the protection of a woman crossing keeper, but strangely was not provided with gates. As the mid-day train from Roseworthy was nearing the crossing its driver saw the buggy about to cross the line. He gave a continuous blast on his whistle, while Mrs Cullen the crossing keeper and her little boy motioned for the driver of the buggy with a flag to stop. This action appears to have confused Mr Warren, the driver of the buggy, for according to witnesses in reply to the warning shouts he had replied, "What are you holding up that flag for?" and in a dying statement had said that he had understood the waving of the flag to indicate that it was all right for him to proceed. It was not and the train struck the hind wheel of the buggy, throwing its occupants a distance of 15 feet. They were both fatally injured; Mr Burton died in a few minutes; Mr Warren lingered until the following day. At the inquest there was an attempt to prove that the crossing keeper had failed to keep a proper lookout and the railway company was criticised for permitting the view at the level crossing to be obstructed by a cottage (presumably the crossing keeper's cottage), a shed and some trees, especially as it was approached as far as the railway was concerned down a steep gradient. There was also the usual claim that the speed of the train (it had been travelling at about 35 miles an hour at that point) had been too high. In the end the inquest jury returned an open verdict. No comment appears to have been made as to the lack of gates at the crossing. The newspaper commented that, 'Habitual users of the crossing have been too apt to treat the presence of the keeper as an idle formality, and to cross the line in spite of the greatest efforts to prevent them,' and that, 'A second more would have enabled Mr Warren to do what others have probably done, that is to have escaped a collision by a hair's breadth.' [1065] [1066]

Another accident with a streetcar, which originated with a mishap to the latter vehicle, occurred near the Sharpsburg Bridge in Pittsburgh, Pennsylvania, United States, half an hour after midnight on Tuesday 25th June 1895. The efforts of the engineer of a West Pennsylvania Railroad fast meat train prevented the accident having worse consequences. A Sharpsburg and Citizen's line electric car had just got onto the level crossing with the West Pennsylvania line when its trolley pole slipped off the line, leaving it stuck on the crossing and in darkness. As Conductor Lindsay was working to replace the trolley pole a freight train was seen to be approaching at considerable speed. The streetcar passengers made a mad scramble for the exit and most of them succeeded in getting off before the collision, while Engineer Black, of the freight train, seeing the trolley, reversed his

[1064] *Adelaide Observer*, Adelaide, South Australia, 26th January 1895
[1065] *South Australian Chronicle*, Adelaide, South Australia, 2nd March 1895
[1066] *The Advertiser*, Adelaide, South Australia, 5th March 1895

engine and applied the air brakes and was able to much reduce the speed of his train. Daniel Nolan, the motorman of the streetcar, and four passengers, all residents of Sharpsburg, were injured. Motorman Nolan had struck to his post in the cab of the streetcar and was severely, if not fatally, hurt.[1067] [1068]

On the evening of Monday 16th September 1895 Southern Railroad fast mail train, No. 35 from Washington D.C. to Atlanta, Georgia, collided with a horse and waggon on a level crossing at Lawyers station, Virginia, United States. The train did not stop at Lawyers and was probably travelling at high speed at the time of the collision. The man driving the waggon, supposedly a Joseph Callahan, who lived near Rustberg, two women and two small children were killed outright and a sixth person riding in the waggon, a girl of about 16, was fatally injured and died soon afterwards. The horse was also killed.[1069] A newspaper report commented, 'The railroad officials are at a loss to understand how the accident occurred at the place named as they say the track is visible for a distance of two hundred yards or more, and the occupants of the vehicle should have been able to see the approach of the train.' [1070]

Encounters between trains and massive road vehicles such as transporters and large articulated lorries would not happen until well into the next century, but on 26th September 1895, near Howden, England, a North Eastern Railway train met a fairly substantial obstruction in the form of a traction engine pulling a thrashing machine. Like the Chartham tragedy the previous year the accident occurred on an occupation crossing on a very foggy morning. The train involved was the 5.30 a.m. passenger train from Hull to Leeds and the collision occurred at Brind Crossing, 1,474 yards west of Howden station. The crossing was provided with gates. The railway was dead straight for some distance on either side of the it but the road approaches to the line were below the level of the railway, and trees and hedges meant that it was necessary to actually get onto the crossing to get much of a view of a coming train.

The traction engine was the property of a Mr John Shaw, of Howden, and it was being used to tow a thrashing machine from Howden to a Prickett Hill farm. Both the starting point for the traction engine and the farm lay on the south side of the railway line, but the only practicable route between them with the engine was over the public road level crossing at Howden station, along that road through Brind village, and then down the private road that went over Brind Crossing. The traction engine was accompanied by a driver and fireman, a flagman (The Locomotives on Highways Act of 1865 required all mechanical road vehicles to be preceded by a flagman. It was repealed in 1896. [1071]), and a man following it to close gates and suchlike. When the traction engine reached the crossing at about 6.20 a.m. the flagman, William Stainton, went ahead to see if the line was clear, and was joined by the fireman, Ellis Morritt. The fireman asked Stainton if he could hear anything and the latter replied that a luggage train had just passed on the down

[1067] *Fort Wayne News*, Fort Wayne, Indiana, United States, 25th June 1895
[1068] *Fort Wayne Sentinel*, Fort Wayne, Indiana, United States, 25th June 1895
[1069] *Millersburg Grit*, Millersburg, Indiana, United States, 19th September 1895
[1070] *Brooklyn Daily Eagle*, Brooklyn, New York, United States, 17th September 1895
[1071] Webpage – Car History 4 U

line, but he could hear nothing else. It was too foggy to see very much. Reassured by the flagman the fireman called for the driver to, "Come on." The traction engine had just cleared the line, but the thrashing machine had not, when the passenger train, which none of the four had heard or seen, suddenly loomed out of the mist and hit the machine, smashing it to pieces. No one was quite sure where the fireman was at the moment the collision occurred, but afterwards he was found lying on the ground, unconscious, on the up side of the line, and he died a few hours later.

Driver Robert Wilcockson on engine No. 167 told the Board of Trade inquiry into the accident that he had stopped at Howden and had left there at about 6.20, two minutes late, for the next station and stop at Wressle. Running at normal speed through the thick fog he had suddenly glimpsed an obstruction in front of his engine and the collision had taken place almost at once. His engine had been knocked about a bit but not derailed. The brake pipe had been severed in the collision and the train came to a stand down the line. It had been a very foggy morning and he could not see more than 10 yards. He had never seen a traction engine on that crossing before. He always whistled for public road level crossings, but not for occupation crossings. He had asked the men with the traction engine how the accident had happened, and they had told him that they had not heard his train coming, and that the traction engine fireman had been killed.

The accident was investigated by Lt-Col G.W. Addison, who commented that there was nothing in the surroundings of Brind Crossing to make it a dangerous one, the railway line being both level and straight, but that he considered that a crossing frequently used by traction engines should be regarded as being of more importance than an ordinary occupation crossing. Mr Shaw informed him that his engine went to Pricketts Farm six times a year, making therefore 12 annual crossings of the line, and carts used the crossing more frequently, if not daily. Referring to General Hutchinson's suggestion with reference to the Chartham accident of the installation of warning bells at crossings, the inspecting officer said that he was told by Mr Shaw that traction engines took about two minutes to cross the line. This implied that two minutes' warning of the approach of a train would be required, which would mean treadles a long way off, and he feared that most users of the crossing would disregard bells ringing for so long. Instead he was of the opinion that common-sense suggested that the local railway staff should be informed when it was intended to take a traction engine over the railway so that the most suitable times to cross could be arranged and trains either held back or their drivers cautioned to take special care.[1072]

An unusual level crossing accident occurred at the Grove street Crossing, Jersey City, New Jersey, United States, on 20th February 1896. Involved in the accident were North Hudson line horse-car No. 4 and a caboose (guard's van). The horse-car was on its way from Jersey City to Hoboken. Drill (shunting?) engine No. 867 in charge of engineer James Harding passed over the crossing just before the horse-car reached it, towing caboose No. 20417, occupied by Conductor Milton Hannis. When the engine and caboose got to Erie Street, a block beyond the Grove Street crossing, the engineer reversed his engine and began to set back, and half a block short of the crossing he cut the caboose loose to make a flying switch (a practice termed fly shunting in Britain – the

[1072] Board of Trade report

engine is uncoupled from the waggons it is pushing or pulling and the points are reversed between the engine and its vehicles so that the waggons go along one line and the engine along another). When Conductor Hannis in the caboose saw the horse-car he applied the brakes but could not stop it before it hit the car. The line over the crossing was very busy, with trains passing to and fro at all hours of the day, and the rules of the horse-car company required its conductors to walk over the crossing in front of the car to ensure that it was safe to cross. This the horse-car conductor, Henry Goelet, had done, and he had been assured by the two flagmen stationed there, George Dixon and Michael McIntyre, that it was safe.

The impact with the car lifted it from the track, smashed in its sides and drove it down the line. The cries of the frightened and injured passengers led to an immediate call for the police and doctors, and ambulances were sent from the Christ, St Francis and City hospitals, but only six passengers were injured and but two, a Mrs Bernard and her son, were badly enough hurt to be taken to St Francis hospital, the others being carried to their homes. The flagmen, Dixon and McIntyre, said afterwards that they had never known the caboose to be fly shunted before and they had expected the engineer to set back into the siding with the caboose attached as usual, the vehicle being stabled in the siding on the far side of the crossing overnight. Engineer Harding and Conductor Hannis could not be found after the accident.[1073]

Eleven months after the Brind Crossing collision there was another encounter between a train and a traction engine drawing a thrashing machine, this time on a field (accommodation) crossing near Etchingham station (England) on the South Eastern Railway. The train concerned was an express, the 3.40 p.m. departure from London Charing Cross to Hastings. This time no one was hurt in the accident.

The first person to give evidence at the subsequent Board of Trade inquiry into the accident was Mr Walter Baldwin. He said that on Tuesday 28th July 1896, the day before the accident, he had taken the traction engine and thrashing machine from Cottendon Farm, the home of his employer, Mr Caffyn, via the field crossing to a Mr Pearson's Forge Farm. He had worked there on Tuesday and Wednesday, and then on Wednesday evening had to take the engine and machine via several fields and the crossing to Mr Pearson's other farm, Court Lodge. He had arrived at the crossing at about 4.45 p.m., one of his men had unlocked the padlocked crossing gates, he had checked that the signals worked from the signal box at Etchingham station were at danger and had started across. He had got as far as the down line when the leading wheel of the engine struck a rail, it slewed around and stuck. Finding that he could go neither forward nor get the engine to go backwards up the incline to the upper crossing gate he had sent one of his men to inform the Etchingham stationmaster but did not send anyone in the opposite direction as there were some platelayers working near the down distant signal who must have realised what had happened. He then busied himself in trying to remove the engine from the line. He knew there was a down train which passed about 5 p.m. but had there been no problem he could easily have got across the line before that train arrived. He had uncoupled the thrashing machine and was close to moving the engine when he heard the whistle of the down express. The train was pulling up when it struck the engine at eight

1073 *Trenton Evening Times*, Trenton, New Jersey, United States, 21st February 1896 or ten miles an hour. Mr Baldwin said he had never taken a traction engine across a railway line before. He had not given any notice to the railway that he needed to take the engine over the line. His employer, Mr Caffyn, said that he had never thought to inform the railway authorities either. Mr Pearson said that he had not known a traction engine to use the crossing before but that thrashing machines drawn by horses had always used it.

Signalman Charles Wolgar in Etchingham box said that the Hastings express had been signalled to him from Ticehurst Road box at 5.2 p.m. and he had had the "Train entering section" signal for it at 5.7. Shortly afterwards, as he was closing the gates of the public road level crossing at the station, Porter Fernar had come running up and shouted, "Caffyn's engine is across the line, keep all the signals up." He had therefore not cleared his down signals but could do nothing more to stop the express, which had already been off Ticehurst Road two minutes. He could not see the accommodation crossing from his box.

Ganger William Brett said that on the 29th July he and his men had been clearing hedges near the Etchingham down distant signal. He had seen the traction engine in the field, taking water from a ditch, and then 10 or 12 minutes later one of his men had remarked that he thought the engine was stuck on the line. He had sent the man to go and check and then climbed to the top of the cutting, from where it was plain that the engine was definitely stuck. He had then sent one of his men, Platelayer Weston, up the line to stop the train, and the latter had covered 200 or 300 yards when the train appeared, and Weston had stood in the four foot with his arms raised signalling for the driver to stop. Ganger Brett said that as his work that day had nothing to do with the permanent way, he had left his detonators and red flag back in the yard at the station.

Also running up the line to warn the express was Brakesman Edward Pinnock of the 2.30 p.m. up goods train from Hastings, which had been shunting at the station. He did have detonators and a red flag with him and managed to get one fog signal down and was waving the flag when the express passed him.

Driver Edward Pope, in charge of engine No. 84 hauling the down express, related that he had had clear signals at Ticehurst Road and had been running at about 60 miles an hour when he found the Etchingham distant at "Danger". Shutting off steam, he had whistled and when the signal was not cleared had applied the brakes. Coming nearer the signal he had seen a platelayer holding his hands up, then a guard putting down a detonator and he had made a full brake application. Just past the distant signal he saw the obstruction itself, and he had then reversed his engine and his fireman had applied sand. He thought that he had only been travelling at two or three miles an hour when he hit the traction engine. The front bogie wheels of his engine were derailed.[1074]

Major F.A. Marindin investigated the accident for the Board of Trade. He said that he did not blame Baldwin for the accident as the man had sent someone to the station as soon as he found he was stuck and had pluckily stuck to the traction engine trying to shift it until the express was dangerously close. The engine had clearly not stalled due to lack of steam, but its leading wheels had become embedded in the gravel ballast. According to a strict interpretation of the rule book Driver Pope should have stopped at the distant signal and then proceeded at caution but this rule was notoriously a dead letter and drivers were not expected to obey it. Such a rule was worse than useless, was unfair to

[1074] Board of Trade report

the company servants and should be modified.

The inspecting officer was critical of the ganger, whom he thought should have been more alive to the danger, as the man sent by Baldwin had had time to cover the 500 yards to the station and return before the man sent out by the ganger had got very far back. He was also less than impressed with the ganger not having had his detonators and flag with him. More important though than any fault of the railwaymen involved was the danger posed by the use of traction engines on these crossings, as the presence of a fixed signal and of the platelayers near the crossing was a mere chance. It was, as he understood, Major Marindin said, very doubtful if the railway companies had any power to prevent such field crossings being used as the owners liked, and it was absolutely necessary that the travelling public should be safeguarded. An obligation should therefore be imposed (presumably by legislation) on the owners of traction engines to give use to the nearest station of their need to use an accommodation or occupation crossing so that the railway authorities could take measures to protect trains, while the railway companies should be given the power (subject to appeal) to absolutely prohibit the use by traction engines of unsuitable level crossings.[1075]

Findlay, Ohio, United States, saw another level crossing collision with the horse-drawn chemical waggon of a fire department on 3rd December 1896, two days after a man in nearby Lima in charge of a waggon and horses owned by the Solar Refining Company had been struck on a level crossing by a Chicago and Erie Railroad train and killed.[1076] The Findlay accident was less serious, the firemen concerned having a narrow escape, though two horses were killed. The Findlay fire department were responding to an alarm in the north part of the city, with six men (five firemen and a reporter for the *Findlay Republican* newspaper) and a team of four horses pulling the waggon. As the waggon approached the Main Street level crossing the adjoining Arlington Hotel restricted the view of the Lake Erie and Western Railroad track to the east of the crossing. Just as the crossing was reached Erie train No. 3 loomed up in front of the firemen. Any attempt at stopping the team was out of the question, though Fireman William Earl braked hard and Fireman William Turner managed to turn the leading pair of horses to the west. Fortunately for the lives of the fireman the engine of the train reached the crossing a few seconds before the horses, so that the waggon crashed into the side of the baggage car behind the engine. The head pair of horses escaped but of the second pair one was killed outright and the other so badly injured that it was immediately shot to put it out of its misery. Both the chemical waggon and the baggage car were damaged but none of the firemen was thrown off by the sudden stop and none were hurt. Engineer Chas Nutting driving the train had not seen the horses and knew nothing of the impending accident before it happened.[1077]

On the afternoon of Wednesday 24th February 1897 seven members of the Woodward family lost their lives when a covered waggon in which they were riding was hit by a Southern Railroad train at Avondale Crossing, Sherman Heights, three miles outside of

---

[1075] Ibid.
[1076] *The Daily Herald*, Delphos, Ohio, United States, 3rd December 1896
[1077] *The Times Democrat*, Lima, Ohio, United States, 4th December 1896

Chattanooga, Tennessee, United States. Possibly because it occurred in a part of the country where few newspapers were published the accident was little reported in the press despite its relative magnitude. Incoming passenger train No. 7 struck the waggon dead centre as it was attempting to cross the line in front of the train, throwing its occupants into the air. The victims almost all suffered fractured skulls. They were the family of a well-known farmer, living near King's Point. The accident presumably occurred at an ungated level crossing.[1078]

A similar if slightly less serious accident to the Avondale one in terms of casualties occurred just over five months later on about 5th August 1897 at the Hollywood Crossing, New York, United States. John Mauhck, a prosperous local farmer, was taking his family for a drive in a light three-seater waggon. Mrs Mauhck had spoken to her husband as they neared the level crossing, as to her dread of it, situated as it was at the foot of a steep gradient and at the end of a sharp curve on the railway. He, however, had assured her that her fears were groundless, and that their horse could be depended on to clear the track, even if a train was already in sight. They had not observed a combined locomotive and observation car of the New York Ontario and Western Railroad until it was close upon them when Mr Mauhck tried to hurry across. He almost made it, but the pilot of the engine just caught the hind wheels of the waggon as they were leaving the rails and waggon and occupants were hurled 40 feet down a rocky embankment to the bed of the Willowemoc creek. One of the family's children was killed outright, a second child died soon afterwards, and fears were entertained for the lives of the other four children and their parents.[1079]

Thomas Laws, the son of the editor of the *Auckland Star* newspaper, was one of three people who lost their lives when a train collided with a horse-drawn omnibus on a level crossing with Custom street in Auckland, New Zealand. The omnibus passengers had just left a circus near the railway, where they had been watching a performance of the *Flying Jordans'*, famous trapeze artists who performed worldwide.[1080] The omnibus driver, W. Ness, apparently failed to hear the whistling of the train, which collided with the front of the omnibus. The 'bus was dragged a short distance down the line before the train could be stopped and the horses pulling it were also killed. Driver Ness, who had only just come out of hospital, and a youth named Charles Akast were the other people to die and several more passengers who had been sat at the front were injured, while those sat at the back escaped unhurt. The driver of the 10.10 p.m. departure from Auckland station had seen nothing of the omnibus until it had suddenly shot onto the line in front of his engine.[1081] [1082]

Three days after the Auckland accident, on 4th October 1897 another serious one occurred at the perhaps all too appropriately named Dead Man's Cut, three miles north of

1078 *Lorain County Reporter*, Elyria, Ohio, United States, 27th February 1897
1079 *Hornellsville Weekly Tribune*, Hornellsville, New York, United States, 6th August 1897
1080 Webpage – Sage Journals: Nineteenth Century Theatre and Film
1081 *The Telegraph*, Brisbane, Queensland, 4th October 1897
1082 *The Leader*, Melbourne, Victoria, 9th October 1897

Willow Springs, Missouri, United States. The train had, according to its conductor, given the proper signals for the crossing but the driver of the waggon had failed to heed them. The crossing, situated on a curve and approached by a heavy grade, was however considered to be a particularly dangerous one. Six of the seven occupants of the waggon were killed outright, and the survivor was thought to be fatally injured. They were mostly members of the Wooton family, from Texas County, Missouri, and had been on their way to Arkansas to pick cotton.[1083]

The next major level crossing collision appears to have been on Saturday 26th February 1898, at the Western Avenue crossing, a mile and a half south of Blue Island, Illinois, United States. The seven people who died were all teenagers or young adults, aged between 14 and 20. They had been on their way from the village of West Harvey to a surprise birthday party for their friend, the daughter of Ferdinad Landgraf, the proprietor of the Union house in Blue Island. Driving the bus was George Sauerbier, whose 14-year-old son was seated beside him and was one of those killed; after the accident, although injured himself, Sauerbier had continued to Blue Island to convey the sad news of the affair. As the omnibus was coming to the level crossing with the Grand Trunk Western Railroad line around a bend on the country road a northbound train could be seen approaching but the road leading to the crossing was down a slight slope, and with the surface slippery from snow and ice the driver was unable to pull up, especially as the frightened horses, seeing the train, had darted forward across the track. The train struck the middle of the waggon, the bodies of two of the victims being carried down the line on the front of the engine, and not discovered until the train had backed to the crossing. After the accident doctors were summoned by the fairly recently invented telephone and they treated the injured with the help of willing members of the public.[1084]

A level crossing collision at Chiltern, Victoria, on 18th April 1898, had some interesting features. A two-horse coach owned by Messrs Crawford and Co. of Beechworth was following a buggy and pair being driven by a Mrs M'Vean, of Howlong. As the two vehicles came up to the American Crossing, near Chiltern station the woman pulled her horses into the fence, having been warned by a boy of the arrival at the crossing of the 9.5 a.m. goods train from Benalla. The coach driver failed to realise why the woman had pulled up and simply overtook her buggy and continued towards the crossing, until his horses saw the train and becoming frightened swerved towards the station. There was then no time either to get over the line in front of the train or to turn the coach around, and the next moment the train dashed into it. The driver, Nicholas Downing, and his two passengers, Mr G. Moye, a Chiltern auctioneer, and Mr James Cunningham, a music teacher from Bright, were somehow thrown clear and were lucky to escape with relatively minor injuries. One of the horses was killed and the coach smashed to pieces.

There were the usual conflicting statements after the accident. The two passengers on the coach swore that the train had given no warning of its approach. The railwaymen at Chiltern station, on the other hand, stated that the engine whistle had been sounded twice,

---

[1083] *The News*, Frederick, Maryland, United States, 5th October 1897
[1084] *Fort Wayne News*, Fort Wayne, Indiana, United States, 28th February 1898

and they were supported by a permanent way inspector who happened to be a riding on the train. Whatever was the case the crossing was regarded as dangerous, and it had previously been provided with gates and a crossing keeper. The gates had subsequently been taken away and the crossing keeper removed, no doubt to save money. The local council had drawn the attention of the railway department to the danger of leaving the crossing unguarded, but no notice had been taken of them. The view along the railway from the road was interfered with in one direction by a house and a garden containing several large pine trees, and in the other direction by the old gatehouse. Though this was the first serious accident at the crossing there had been several near misses, and it was hoped that the crossing keeper would be re-instated.[1085]

On 5th September 1898 what was probably the worst level crossing accident anywhere in the world since the Burlington disaster, 43 years earlier, occurred at Cohoes, New York, United States. It was a collision between a Night Boat special train of the Delaware and Hudson Railroad and a trolley car (electrically powered streetcar) of the Troy City Company. The streetcar line ran from Cohoes via a bridge over the Hudson River to Lansingburg and crossed the railway on the Cohoes side of the bridge. Because of the frequent rail traffic it was the practice for all trolley cars to be brought to a stand at the crossing and for the conductor to alight and go forward to check that all was clear before signalling the motorman across. It was unclear initially after the accident whether the conductor had done so on this occasion, but at any rate the trolley car was halfway across when the train came up, travelling a high speed. Because of the procedure of the trolley car conductors checking the line the train engineers felt safe driving over the crossing at high speed. The engineer in this case only saw the trolley car when his train was a short distance from the crossing and it was impossible for him to pull up, though he did his best to reduce speed while the motorman of the trolley car attempted to accelerate out of the way. The car was carrying 35 passengers, most of them young people. Eighteen of them were killed when the train struck the front of the trolley, and it was feared that at least 10 more would die. The pilot of the engine was smashed but the train passengers were not hurt. The injured passengers were taken to the Cohoes hospital and some to the Continental Knitting mill as there were too many for the hospital to cope with.[1086] The following day Walter Congdon, the trolley car conductor, was arrested and lodged in Cohoes jail, it being alleged that he had signalled for the motorman to cross with the passenger train already in sight.[1087]

Fatal carelessness of the type so often evident in level crossing collisions was the cause of one at Allenwood, New Jersey, United States, on 16th December 1898. A party of eight persons was returning after a drive to Asbury Park and were within a stone's throw of their homes when the rig in which they were riding was struck by Pennsylvania Railroad train, the vehicle having entered the crossing when the train was less than 50 feet away. Allenwood was an unstaffed flag station and having no one wishing to alight and there being no passenger at the station signalling for the train to stop, the engineer

---

[1085] *The Argus*, Melbourne, Victoria, 19th April 1898
[1086] *Middletown Daily Argus*, Middletown, New York, 6th September 1898
[1087] *The Sandusky Star*, Sandusky, Ohio, United States, 7th September 1898

was proceeding at some speed. At a grocery store a few yards from the crossing Thomas Allen was serving his customers. When they heard the sound of the crash, all rushed outside and Allen stumbled over the body of his wife, while a few feet away was the body of his cousin. A little later he discovered that his mother, brother and baby son were injured and two of them likely to die. He was said to be crazed with grief. In the 21st century he would doubtless have been offered grief counselling but that was not an option in 1898 of course; one wonders how he coped. The newspaper report on the accident added that the whole of Allentown was shrouded in gloom over the tragedy and that the Allens were the principal family of the hamlet, which was named after them. Driving the carriage had been another Allen, William, who lost consciousness after the collision but apparently was not badly hurt.

There was another harrowing story in the newspaper report relating to one Joseph Cranmer. He was one of several local residents who had run from their houses with lanterns to assist after the accident. The train, which had been brought to a stand half a mile down the track, was then set back to the crossing, and Cranmer went to speak to the engineer. Getting to the front of the locomotive he found the crushed and lifeless body of his daughter, Jeanne, wedged in the pilot of the engine.

Having done what they could at the scene the train crew went on, with instructions to hurry doctors to Allenwood from Manasquan. Several medical men set off at once to the scene upon getting the news.[1088]

London, Ontario, Canada, had a level crossing collision involving a trolley car on 1st February 1899, though it was only a minor one. The motorman in charge of the car saw an express pulling out of the Grand Trunk Railway station as he was crossing the tracks a short distance away on Richmond street. He applied full power to get clear and damage to the car was confined to a crushed rear step.[1089]

Considerably more serious than the Ontario mishap was an accident across the border in Pennsylvania, United States, two days later. Five men and a woman were riding on a two-horse waggon loaded with furniture on 3rd February 1899 when it was struck by a Baltimore and Ohio Railroad train at Riverton station, near McKeesport. Four of the men were killed instantly and the other man and the woman died within a few hours.[1090]

A Sunday afternoon drive ended in tragedy for William Reinhardt and his family on 9th July 1899 at Columbus, Ohio, United States. Mr Reinhardt was described as a prosperous German merchant living at 256, Donaldson Street, Columbus. That Sunday afternoon they had gone out for a ride and ironically had stopped off to offer their condolences to the family of Jacob Huifman, who had been run down and killed by some cars the previous Thursday. After a short visit they had started for home. As they got to the Woodlawn Avenue crossing with the Big Four (Cleveland Cincinnati Chicago and St Louis) Railroad there were no signals to indicate any danger and they drove onto the crossing just as a

---

[1088] *Middletown Daily Argus*, Middletown, New York, United States, 17th December 1898
[1089] *Manitoba Morning Free Press*, Winnipeg, Manitoba, Canada, 3rd February 1899
[1090] *Stevens Point Journal*, Stevens Point, Wisconsin, United States, 11th February 1899

westbound express came around the corner. There was, the newspaper reporter wrote, 'a shriek of horror', and the engine of the express struck the carriage, carrying it nearly 300 yards down the line and crushing the horse. Mother, father and two of the children were killed outright and two more children died later. There was one survivor, Clarence, 14, now an orphan, who was taken to hospital with a broken collarbone.[1091]

At about 5 p.m. on Sunday 12th August 1900 an omnibus, driven by a man named Paters and owned by Henry Bittner of Slatington, Pennsylvania, United States, was returning thereto from Cherrysville, where its passengers had been attending the funeral of a Sophia Schoeffer. The passengers were almost all relatives of the deceased woman. At a point about three miles outside of Slatington the road along which the omnibus was travelling crossed the line of the Lehigh and New England Railroad at Benniger's Crossing. The crossing was approached around a sharp curve in the road and the omnibus came along at a good rate of speed, its occupants unaware of any danger until suddenly, as it neared the line, a special train, consisting of just an engine and a single coach, came into view only a short distance away. Unable to pull up short of the crossing the omnibus driver whipped up his four horses in a desperate attempt to get across the line in front of the train. He failed to do so, and the train struck the middle of the omnibus, throwing its passengers in all directions. Fifteen of them were killed outright or died shortly afterwards and another 10 were seriously hurt, several it was feared fatally. Doctors were sent for, together with a special train, and the injured were taken to South Bethlehem. The crossing where the accident occurred was not provided with a watchman and locals complained that it was impossible to hear an approaching train. The four horses escaped injury as they had cleared the line at the moment of impact.[1092]

In an all too familiar scenario a party riding in a carriage on a Sunday evening pleasure trip attempted to drive across a railway line at an unguarded level crossing without taking sufficient care and their vehicle was struck by a train at a location the newspapers called Oswego Junction, four miles west of Syracuse, New York, United States, on 2nd September 1900. The horse or horses and carriage were struck by what was probably a New York Central and Hudson River Railroad eastbound fast mail train travelling at 40 miles an hour. Patrick Foley of Syracuse, his sister Ellen Foley of Split Rock and Josephine Blanchard of Syracuse were killed, and a Michael Maloney was injured. There was no night flagman at the crossing and another train standing by the crossing had blocked the view of the mail train approaching on an inside track.[1093] [1094]

The Oswego Junction accident turned out to be the last significant level crossing collision of the 19th century. With the new century and the rise of the motor car, the motor lorry and the motor coach there would be many thousands more, in practically every country in the world, as will be seen in volumes to come.

1091 *Fort Wayne Sentinel*, Fort Wayne, Indiana, United States, 10th July 1899
1092 *Trenton Times*, Trenton, New Jersey, United States, 13th August 1900
1093 *Sandusky Daily Star*, Sandusky, Ohio, United States, 3rd September 1900
1094 *The News*, Frederick, Maryland, United States, 3rd September 1900

CHAPTER 13

RAIL/RAIL CROSSING COLLISIONS

At locations where railways and roads intersected the idea of having them cross on the level had obvious advantages in terms of economics over the cost of building an over or underbridge, especially if the level crossing was unmanned and the construction of a bridge would require the building of substantial approach embankments or possibly approach cuttings leading to an underbridge. With road users being limited in the early days to pedestrians, cyclists and horse-drawn vehicles, the dangers to a train from a collision on such a level crossing were also likely to be minimal. For railways crossing other railways on the level though, the situation was very different. A collision with another train on a crossing was likely to be far more serious than a collision with, say, a horse and cart, and furthermore a train driver approaching a level crossing and seeing another train nearing or on it could not pull up in anything like the distance that a road user could. It is no surprise therefore that in most countries, where the railway system was either built directly by the state or under state supervision, that rail/rail crossings were not provided. It was only in countries where the railways were built by competing private companies without state interference that such crossings existed, most especially in the United States, where lines were often built as cheaply as possible and the economy of not needing to construct a bridge was doubtless an important consideration. Thus, the United States had many rail/rail level crossings, Great Britain about 20 or so at one time and its colonies, such as Ireland and Canada, had a handful. One or two did also exist elsewhere, such as at Istravoz (later Hilal), Turkey, where the Oriental Railway crossed the Smyrna – Cassaba Railway.[1095] As has been pointed out the American companies also seemed to have a perverse proclivity for building rail/rail crossings where they could easily have been avoided, perhaps because in the early days train speeds were low, trains infrequent and night operation unusual so that the dangers were not fully appreciated.[1096] In contrast to Great Britain, where from early on all rail/rail crossings were protected by fixed signals and signalmen, many such crossings in the United States were not so protected, especially on lines with light traffic, no doubt to save the expense of having to

[1095] Wikipedia – Hilal railway

[1096] *A History of Railroad Accidents, Safety Precautions and Operating Practices* by Robert B. Shaw, pg. 99

provide the men and equipment to do so. As a result rail/rail crossing accidents in Great Britain were practically unknown by the 1880s, whilst they continued to occur quite frequently in the United States.

On 8th October 1881 a serious accident occurred at Fulton Junction (later renamed East Clinton), Illinois, United States. A Chicago Milwaukee and St Paul Railroad passenger train from Rock Island to Racine, Wisconsin, ran into a heavily loaded Chicago and North Western Railroad as the latter was passing over the crossing of the two railroads. The trucks of the freight train were thrown onto the platform of the depot there and hit several persons. Recently appointed station agent J.J. Flanagan and a 13-year-old boy, Arthur Cuddy, the son of a labourer, both of Fulton, were killed outright, and Lester Fisk, the son of Clark Fisk, suffered probably fatal internal injuries. L. Rosenblatt, a travelling man from Beloit, was badly but not dangerously bruised and H.B. Suther, of Clinton, Iowa, suffered a crushed foot and ankle. The passengers on the Milwaukee train suffered a shaking. The trucks of the freight train and the engine of the passenger train were piled up in a wreck.[1097] The failure of the air brakes of the passenger train to work was given as the cause of the accident.[1098]

A Missouri Pacific Railroad construction train and a Burlington and Missouri Railroad freight train collided on the crossing at Falls River, Nebraska, United States, on Saturday 11th February 1882, wrecking three box cars and an engine though no one was hurt. Each engineer supposedly considered that he had the right of way and that the other would stop.[1099]

Fog was a factor in a collision between freight trains of the Wabash St Louis and Pacific and Chicago Cleveland and Columbus (name given in newspaper report but possibly not correct) Railroads at the crossing of the two lines at an unstated location 12 miles from Chicago, Illinois, United States, early on Saturday 2nd September 1882. The engineer and fireman of the Columbus train were badly but not fatally hurt, and an engine and six cars were wrecked, the damage amounting to $8,000. The accident was said to have been the result of carelessness.[1100]

As the number of independent railroad companies building new lines in the United States increased towards the end of the 19th century so did the number of rail/rail crossings between such lines and the number of collisions on them. Thus, another accident occurred on 1st November 1882, this time at Diamond Crossing, near Scranton, Pennsylvania. A Delaware Lackawanna and Western Railroad train and a Lehigh and Susquehanna Railroad train collided. Watchman John Glynn was killed, and several railwaymen were injured. Twenty-five cars were thrown from the track and an engine

---

1097 *The Stevens Point Journal*, Stevens Point, Wisconsin, United States, 15th October 1881
1098 *Daily Kennebec Journal*, Augusta, Maine, United States, 10th October 1881
1099 *The Globe*, Atchison, Kansas, United States, 13th February 1882
1100 *The Burlington Weekly Hawkeye*, Burlington, Iowa, United States, 7th September 1882

demolished.[1101] Then on 25th November 1882 there was an accident at Lynn, Indiana. Freight train No. 12 of the Grand Rapids and Indiana Railroad ran into an Indiana Bloomington and Western Railroad train at the point where the lines intersected, without it would seem any serious consequences.[1102]

Trains of the Jeffersonville Madison and Indiana and Ohio and Mississippi Railroads collided at Seymour, Indiana, United States, on about 21st February 1883. 'No lives lost, but a large loss of property.' [1103] Again on Sunday 25th February 1883 Wabash St Louis and Pacific Railroad and Cleveland Cincinnati Chicago and St Louis Railroad freight trains came together on the crossing at Sheldon, Illinois, United States. Both trains were badly damaged, and several cars ditched but no one was hurt. After being blocked for several hours the lines were open again by midnight.[1104]

On about 13th August 1883 a waggon containing 400 kegs of blasting powder, consigned to H.P. Mason & Co, was left on a Kentucky Central Railroad siding at Winchester, Kentucky, United States, where the Kentucky Central line was crossed by the trunk line of the Chesapeake and Ohio Railroad. On the morning of 17th August, a Chesapeake and Ohio mixed train from Mount Sterling to Lexington, consisting of a smoking and baggage cars, two passenger cars and numerous freight cars arrived at Winchester. This train stopped east of the crossing with the Kentucky Central line and its engine and three leading freight cars were detached and ran forward to clear a set of points leading to the siding in which the waggons were to be placed. At the same time shunting operations were also in progress on the Kentucky Central line and that company's engine and trucks were blocking the crossing. Seeing this the engineer of the Chesapeake and Ohio train, which also had to foul the crossing before it could be set back into the siding, reversed his engine but the rear portion of his train, which had not been properly braked, had meanwhile started to run forward and colliding with the front portion pushed it onto the crossing. By an unlucky chance the Chesapeake and Ohio locomotive struck the truck loaded with the blasting powder. The collision was severe enough to break open several of the kegs and sparks from the engine then caused an explosion which shook every house in Winchester to its foundations.[1105]

Conductor Randolph Martin of the Kentucky Central train was stood right alongside of the waggon loaded with explosives and was killed instantly, while Conductor Charles McMichael of the Chesapeake and Ohio train was grievously injured and died a few hours later. Upon his engineer whistling for brakes he had been running along the top of the cars to apply the brakes on the front one. Several other persons who had been near the crossing were more or less badly hurt.

Chesapeake and Ohio Engineer Ben Schultza (or Dennis Schiller according to another report) ran off after the explosion and had not been seen since. His fireman, named Baxter, also ran off, badly frightened, but later returned. The engine of the mixed train

[1101] *Port Jervis Evening Gazette*, Port Jervis, New York, 2nd November 1882
[1102] *Fort Wayne Daily Gazette*, Fort Wayne, Indiana, United States, 28th November 1882
[1103] *Newark Daily Advocate*, Newark, Ohio, United States, 23rd February 1883
[1104] *Fort Wayne Weekly Sentinel*, Fort Wayne, Indiana, United States, 28th February 1883
[1105] The Advertiser Courier, Hermann, Missouri, United States, 22nd August 1883

was wrecked, and the Kentucky Central freight cars were 'blown to atoms'. The depot and the adjoining Railway Hotel were damaged. C.D. Emmons, the agent at Winchester for both companies, was held responsible by Winchester residents for the accident as a Kentucky law forbade the storing of powder on a siding for more than 24 hours at a stretch. One newspaper report blamed the layout of the crossing and switches at Winchester for the accident.[1106] [1107]

At Owatonna, Minnesota, United States, on 11th October 1883, Eastbound train No. 14 of the Chicago Milwaukee and St Paul Railroad was struck on the crossing of the two lines by westbound Winona and St Paul Railroad freight train No 15. The engine of the Winona train and three flat cars of the Milwaukee one were wrecked. The crew of the Winona train said that they had divided it before reaching the crossing, leaving all the brakesmen with the rear portion. On the steep descent to the crossing the engine had then been unable to control the balance of the cars. 'It is quite a loss to both companies.' [1108]

A serious collision occurred on 28th October 1883 at what was mysteriously termed in a newspaper report the 'Joliet crossing', which appears to have been at Hartsdale, four miles east of Dyer, Indiana, United States. A northbound Pittsburgh Cincinnati Chicago and St Louis Railroad (Panhandle) passenger train collided on the crossing with an 'outgoing' Michigan Central Railroad freight train. Both engines were wrecked and fell on and set fire to the flagman's box, where the flagman Peter Callie, had retreated upon seeing that a collision was about to happen. He was killed, as were Engineer William Morris of the Panhandle train and Head Brakesman John Kellogg of the Michigan Central train. Engineer Donegan of the Michigan Central train, 'severely injured and crazed by burns and fright', ran all the way to Dyer, 'where he arrived a lunatic.' His fireman, John Gorman, was also hurt but the passengers escaped with a severe shaking. The newspaper report concluded, 'it is impossible to locate the responsibility for the accident.' [1109]

A train organised by the Union Sunday school organisation of Streator, reportedly the largest excursion train ever run by them and consisting of 15 well-filled coaches, was involved in a serious accident on 21st August 1884 at Kankakee, Illinois, United States, which was also its destination. The New York Central and Hudson River Railroad station was actually on the Streator side of a crossing with the Illinois Central Railroad line, but the train was heading for the pleasure grounds, which presumably were on the far side of the crossing. As the train was passing over the crossing it was struck by a southbound Illinois Central passenger train which cut through the excursion train, derailing itself and three cars of the excursion. Real estate agent John Penn, probably the father of one of the children, was killed and several others were injured, a couple possibly fatally. The Illinois Central engineer alleged that his air brakes had failed to work. The accident happened at

---

[1106] *Ottumwa Daily Democrat*, Ottumwa, Iowa, United States, 19th August 1883
[1107] *The Advertiser Courier*, Hermann, Missouri, United States, 22nd August 1883
[1108] *The Standard*, Albert Lea, Minnesota, United States, 18th October 1883
[1109] *Reno Evening Gazette*, Reno, Nevada, United States, 29th October 1883

about 11 a.m.[1110]

An expensive accident occurred south of Chicago, Illinois, United States, on 13th October 1884 when a Pittsburgh Cincinnati Chicago and St Louis Railroad train and a Baltimore and Ohio Railroad one collided at the crossing of the two roads. Five meat-laden refrigerator cars were wrecked and burned. The loss was $15,000.[1111]

The strangely named Know Nothing Crossing at Greenfield, Massachusetts, United States, was the scene of a collision on 30th March 1885. A slowly moving eastbound Fitchburg Railroad freight train was on the crossing when it was run into by a northbound Connecticut River Railroad passenger train. Fitchburg Engineer Ed Bemis was scalded on one arm but stuck to his damaged engine and by shutting off steam was able to avoid collided with a westbound train a short distance beyond the crossing. His head brakesman, George Reynolds, suffered a broken jaw and was badly cut and bruised about the head. Engineer Jim Clancy of the Connecticut River train was slightly scalded but none of its passengers was hurt. Conductor J.G. Doble was in charge of the freight train. It was claimed that the freight train had the signal to proceed over the crossing and this seemed likely to be true.[1112]

A Pittsburgh Cincinnati Chicago and St Louis Railroad train was in a collision on a rail/rail crossing near Chicago. Illinois, United States on 10th January 1886 when there was a collision between a switch engine and an empty train at the Ashland Avenue crossing, the other railroad being the Chicago Milwaukee and St Paul. The newspaper report did not make it clear which train was on which road. Trainmen Thomas King and Dennis O'Connell were badly hurt, the latter probably fatally.[1113]

The Pittsburgh Cincinnati Chicago and St Louis Railroad seem to have been particularly unlucky at this period as one of their trains was involved in yet another collision on 25th August 1886, this time with a Cleveland Cincinnati Chicago and St Louis Railroad train at Milford Center, Ohio, United States. The northbound passenger train of the latter railroad collided with a Panhandle freight train. Two passengers were killed, Assistant Sergeant of Arms Landon, of the Ohio House of Representatives – a policeman, and a Charles Phillips.[1114] [1115]

A relatively minor collision at Milford Junction, Indiana, United States on 25th November 1886 had a rather unusual cause. A westbound Baltimore and Ohio Railroad freight train ran into a Cincinnati Wabash and Michigan Railroad train which was

---

1110 *Decatur Morning Review*, Decatur, Illinois, United States, 23rd August 1884
1111 *The Daily Gazette*, Fort Wayne, Indiana, United States, 14th October 1884
1112 *The Fitchburg Sentinel*, Fitchburg, Massachusetts, United States, 31st March 1885
1113 *The Semi Weekly Age*, Coshocton, Ohio, United States, 12th January 1886
1114 *Decatur Morning Review*, Decatur, Illinois, United States, 26th August 1886
1115 *Galveston Daily News*, Galveston, Texas, United States, 27th August 1886

standing on the crossing of the two roads. No one was hurt but a caboose and several cars were wrecked, and the Baltimore and Ohio engine was thrown off the line and damaged. The Baltimore and Ohio engineer explained afterwards that his whistle had become defective and so he had been unable to whistle for brakes.[1116]

A section of the New South Wales Government's Illawarra line had only been open for two days when it was temporarily blocked by a collision on 23rd June 1887 at North Wollongong between a contractor's locomotive (named *Gladstone*) returning from Clifton with empty ballast waggons and No. 2 engine of the Mount Pleasant Coal Company at the crossing of the two lines. The drivers of both engines reportedly whistled but neither heard the other. The driver of the Mount Pleasant engine, who suffered a cut arm and a slight shaking, was the only casualty but both engines were much damaged, the cost of repairs to the Mount Pleasant one being estimated at £1,800 and to the contractor's engine £300. The Mount Pleasant Company stated that it intended to claim compensation from the Government for having dispensed with the signalman at the crossing since they took over the line.[1117] [1118] [1119]

After the run of minor or relatively minor rail/rail crossings which had occurred in the United States around this period its neighbour Canada suffered a major disaster on 15th July 1887 at St Thomas, Ontario.

The Michigan Central Railroad, running from east to west, crossed the London and Port Stanley Railroad running from north to south almost in the centre of St Thomas. Two semaphore signals on the south side of the London and Port Stanley line protected it from that direction though perhaps confusingly at the time of the accident they were preceded by a third signal, not yet brought in use and standing at "All clear". The regulations also dictated that all London and Port Stanley trains must stop dead before entering the crossing, but this rule was apparently something of a dead letter.

On the evening of the day of the accident a Grand Trunk Railroad excursion train left Port Stanley at 6.40 p.m., consisting of nine coaches and a baggage car and packed with almost 400 passengers, many of them children who had been on church picnics. Because there were insufficient seats a proportion of the passengers were standing in the aisles or on the platforms between the cars. In charge of the train was Conductor Richard Spettigue with Engineer Henry Donnelly, Fireman Harry Angles and Brakesmen Allan Davis and George Mason. George McMullen, a friend of Engineer Donnelly, was also riding on the engine.[1120]

The excursion train was travelling at 40 miles an hour as it neared St Thomas. Meanwhile a Michigan Central Railroad freight train, consisting of a mixture of box cars and tank waggons holding 3,000 gallons of crude oil, was also nearing the crossing. If the yet to be brought into use semaphore was off the two other semaphores on the Port Stanley line were plainly standing at "Danger". The excursion train slowed, but not enough.

1116 *The Marion Daily Star*, Marion, Ohio, United States, 26th November 1886
1117 *Freeman's Journal*, Sydney, New South Wales, 25th June 1887
1118 *The Daily Telegraph*, Sydney, New South Wales, 25th June 1887
1119 Webpage – Railway History in Illawarra, New South Wales, by C.C. Singleton
1120 *Wreck!* By Hugh A. Halliday, pgs. 64-70

Engineer Donnelly applied the air brakes, but they failed to work. He whistled for brakes and reversed his engine as Fireman Angles and Mr McMullen jumped from the engine. The two brakesmen, surrounded by children on the platforms, attempted to screw down the hand brakes but it was too late and at about 7 p.m. the excursion train, still travelling at between 12 and 25 miles an hour, ploughed into the Michigan Central freight train, as bad luck would have it striking one of the oil tank cars, so that burning oil immediately poured over the engine, baggage car and front coaches of the excursion train.

After the collision uninjured passengers and townspeople rushed forward and attempted to rescue trapped victims, ignoring or ignorant of the danger posed by a second tank car adjoining the one which had been struck. Ten minutes later this waggon exploded from the heat, killing a fireman, sending flames 100 feet into the air, causing horses to stampede through the streets, which ran down and killed a woman, and shattering windows four blocks away. Many onlookers were scorched and several badly burned and they then fled in a panic. It took the firemen two hours to get the blaze under control.[1121]

Seventeen people died in all, including 13 passengers and Engineer Donnelly of the excursion train. Most of the dead passengers came from just four families. Rumours spread that Engineer Donnelly had been drunk but there was no real evidence to suggest that he had been. He was probably guilty of approaching the crossing at too high a speed and possibly was misled by the out-of-use signal, though the inquest jury paid little attention to this possibility, or to the basic danger of having the two lines crossing on the level. There was more concern about the failure of the air brakes. Neither Conductor Spettigue nor Fireman Angles had checked them before leaving Port Stanley. There was talk that that the air brake on the leading car was markedly different from those on the cars behind it, and of a possible kink in the brake pipe, but there was no proof of anything. The jury returned a verdict finding Conductor Spettigue guilty of manslaughter for having failed to carry out the brake test, declared that Donnelly had been unfit to drive the engine without stating why, condemned the overcrowding of the excursion and the excessive speeds of trains through St Thomas.

On 30th September the Elgin County Assizes Grand Jury threw out the manslaughter charge against the conductor, and there the matter rested.[1122] [1123] [1124]

What was reportedly a narrow escape from a more serious accident occurred on the 16th Street Crossing, Chicago, Illinois, United States, on about 26th July 1888. A Chicago and Eastern Illinois Railroad train was stated to have had the right of way, the Chicago Burlington and Quincy Railroad train being blocked by a gate. The Chicago and Eastern Illinois engine was badly damaged, and four cars of the Burlington train were derailed. The cost of the damage was estimated at $10,000. A Grand Trunk Railroad train on a parallel line narrowly escaped being struck by the overturning Eastern Illinois engine. The accident was reported to have been entirely the fault of the Burlington trainmen, none of whom was at the rear of their train. Possibly it was being propelled over the

---

1121 Ibid.

1122 Ibid.

1123 *Bismark Daily Tribune*, Bismark, Dakota Territory, United States, 16th July 1887

1124 *Manitoba Daily Free Press*, Winnipeg, Manitoba, Canada, 17th July 1887 crossing, though the newspaper report does not specifically say so.[1125] Less than a month later, on 17th August, a Chicago Burlington and Quincy Railroad passenger train due in Chicago at 10.30 p.m. cut through a Pittsburgh Cincinnati Chicago and St Louis (Panhandle) Railroad freight train on the Western Avenue crossing. Five freight cars were demolished and the Burlington engine damaged, the passengers getting away with a bad fright and a shaking. This accident was blamed on a failure of the air brakes.[1126]

A failure of the air brakes to work was likewise blamed for a collision at Lima, Ohio, United States on 4th September 1888. Pittsburgh Fort Wayne and Chicago Railroad passenger train No. 9 was running an hour late as it approached Lima when its air brakes allegedly failed to work, and it cut clean through the baggage car of Lake Erie and Western Railroad train No. 5 which was going over the crossing of the two lines. The only person in the baggage car was baggagemaster Van Norman, who was very lucky to escape with a few bruises. Behind the baggage car the smoking car was full of passengers returning from an Odd Fellows' reunion at Bluffton. They were badly shaken up, but none sustained more than cuts and bruises. The engine of the Fort Wayne train, which consisted of two baggage cars and six coaches, was derailed. Had the Lake Erie train reached the crossing a moment sooner it could have been a major disaster.[1127]

It was not a failure of the air brakes but excessively slippery rails due to snow and perhaps the misjudgement of the engineer which caused a bad collision at Hastings, Nebraska, United States, on 21st December 1889. A St Joseph and Grand Island Railroad morning freight train was unable to pull up in time due to the slippery rails and ran into the side of a Burlington and Missouri River Railroad passenger train on the crossing. Fireman James Caspill and Brakesman James Martin of the St Joseph train were fatally injured. The engineer of the passenger train was seriously injured but the passengers escaped with a severe shaking up.[1128]

For some reason another accident in Lima, Ohio, United States, on 2nd May 1890 was barely mentioned in the newspapers despite the fact that it involved three fatalities. A Pittsburgh Akron and Western Railroad company train was run into by a Cincinnati Hamilton and Dayton Railroad one and one of the engineers and two tramps were killed.[1129]

As the Cleveland Canton and Southern Railroad Coshocton express was approaching Newburgh station, near Cleveland, Ohio, United States, at 9 a.m. on 13th June 1890, it was run into by the engine of a Cleveland and Pittsburgh Railroad freight train. One passenger car was almost cut in two but no one in it was killed, W.R. Carver of Kent, Ohio, being considered the most seriously injured, and his injuries were not considered

---

1125 *Newark Daily Advocate*, Newark, Ohio, United States, 27th July 1888
1126 Ibid, 18th August 1888
1127 Ibid, 6th September 1888
1128 *Morning Organiser*, Portland, Oregon, United States, 22nd December 1889
1129 *Newark Daily Advocate*, Newark, Ohio, United States, 3rd May 1890

dangerous. About a dozen others were slightly cut and bruised. The Pittsburgh train was fortunately moving very slowly.[1130]

Valley Crossing, nine miles south of Columbus, Ohio, United States, was the scene of a collision on 15th February 1891. A southbound Columbus Hocking Valley and Toledo Railroad passenger train collided with the engine of a northbound Scioto Valley Railroad freight train. Both engines were badly wrecked, and a number of people seriously injured, including Engineer Charles Bothwell, of Portsmouth, who was brought to Columbus. The passengers injured included a Mrs Robert McCune, of Columbus, a little daughter of A.B. Cheney, of Canal Winchester, who was thrown against a stove and badly burned, and Mrs Gen Grosvenor, of Athens, and her three daughters, who were slightly hurt. Two young men named Johnson, who were stealing a ride, were perhaps fatally injured.[1131]

What was described as one of the worst collisions that had ever occurred there happened on the Union Crossing, Elizabeth, New Jersey, United States, at about 5 p.m. on the afternoon of Friday 29th May 1891. A Philadelphia fast train of the Central of New Jersey Railroad collided with a Pennsylvania Railroad freight train. Engineer William Copson saw the freight train on the crossing and applied his air brakes but 'they failed to work'. Seeing the inevitability of a collision his fireman then jumped and escaped unhurt, but the engineer stuck to his post as it hit the moving freight cars, throwing them to one side. The engine was thrown over on its side, with Engineer Copson beneath it. At first it was feared that he had been crushed to death, but he was released after a few minutes and taken to hospital, terribly scalded. The train passengers were of course greatly excited and a few of them sustained cuts and bruises. None of the crew of the freight train was hurt. Both roads were blocked for nearly three hours.[1132]

On about 30th August 1891 the rear car of a Baltimore and Ohio Railroad excursion train, containing 28 passengers, was struck by a New York Lake Erie and Western Railroad coal train at the crossing of the two roads at Lodi, Ohio, United States. The car was thrown on its side and was described as having been totally demolished but none of the passengers was killed and only three were seriously injured.[1133]

The crossing was replaced by an overbridge at an unknown date.

A Lake Shore and Michigan Southern Railroad and a Grand Rapids and Indiana Railroad train collided on the crossing near Kendalville, Indiana, United States, on 10th October 1891. One train was a passenger and the other a freight, but the newspaper reports do not clarify which company's train was which. Several passengers were injured in the early morning accident and there was considerable damage to property.[1134]

1130 *Fort Wayne Sentinel*, Fort Wayne, Indiana, United States, 13th June 1890
1131 *Newark Daily Advocate*, Newark, Ohio, United States, 16th February 1891
1132 *Marion Daily Star,* Marion, Ohio, United States, 30th May 1891
1133 *Mitchell Daily Republican*, Mitchell, South Dakota, United States, 1st September 1891
1134 *Sioux Valley News*, Correctionville, Iowa, United States, 15th October 1891

Taopi, Minnesota, in the Mid-West of the United States, was the location of an accident on 29th October 1891. Chicago St Paul and Kansas City Railroad train No. 1 from Chicago, due at St Paul at 7.30 a.m., had just started up and had entered the crossing when it was struck by a Chicago Milwaukee and St Paul Railroad freight train. Engineer Chambers of the Milwaukee train and a tramp riding under the forward end of the baggage car of the Kansas City train were killed, Fireman Lynch of the Kansas City train was fatally injured and Fireman Thorston of the Milwaukee train seriously hurt. The Milwaukee trainmen claimed that the Kansas City train ran into them. Both locomotives were derailed.[1135]

On 12th November 1891 a sad accident occurred at Waco, Texas, United States, at the crossing of the Missouri Kansas Texas (Katy) Railway and the San Antonio and Aransas Pass Railway (most American companies called themselves railroad but for some reason these two preferred railway). At 9 a.m. a Katy switch engine was pushing a string of freight cars towards the Brazos River bridge when the leading car struck the car of a San Antonio train on the right angle crossing of the two lines at South First Street. The San Antonio truck was broken up and Brakesman J.M. Brown buried in the wreckage and fatally hurt, with a broken thigh and internal injuries, though he lingered in great agony until three in the afternoon. He was a stranger to Waco, and it was only his third day in the job.[1136] Whether the accident was in any way related to his inexperience was not stated.

The crossing at Clyde, Ohio, United States, saw an encounter between freight trains of the Cleveland Cincinnati Chicago and St Louis (Big Four) and Lake Shore and Michigan Southern Railroads on 1st July 1892. The Big Four train was backing over the crossing at a time when the Lake Shore train was due. The target protecting the crossing was thrown to stop the backing movement when the Lake Shore train was seen approaching but too late, as the Big Four train was by then very close to the crossing, and its caboose was struck and overturned by the Lake Shore train, which also knocked the trucks from under a coal car. Both trains were moving slowly, and no one was hurt. The damage was only slight, and the wreckage was cleared by a Big Four crew in 40 minutes. Meanwhile Lake Shore fast mail and passenger train was diverted via Sandusky.[1137]

Careless shunting was probably also the cause of an accident near Sandusky itself on 30th October 1893. In this accident eastbound Lake Erie and Western Railroad passenger train No. 4, due in Sandusky at 1.40 p.m., had slowed up for the crossing of the Sandusky and Columbus Short Line Railroad and was on the crossing when its smoking car was struck by three or four Short Line cars which were being shunted and which were thrown off the track. As it happened only the brakesman and one passenger, travelling engineer Thomas Cafferty, of Lafayette, Indiana, were in the car and Cafferty suffered two broken

---

1135 *Herald Despatch*, Decatur, Illinois, United States, 31st October 1891
1136 *Galveston Daily News*, Galveston, Texas, United States, 13th November 1891
1137 *Sandusky Daily Register*, Sandusky, Ohio, United States, 2nd July 1892

ribs, cuts and bruises, the brakesman being unhurt. Cafferty was treated by a surgeon and went home by the 1.15 a.m. train. The freight car involved in the collision was badly broken up and its load of salt was scattered about the wreck. A small fire also broke out among some newspapers being carried for sale, but Conductor William Christy soon put it out. The Lake Erie line was blocked, and its trains had to use the Lake Shore tracks. As might be expected the Lake Erie trainmen blamed the Lake Shore men for the accident.[1138]

Still in Ohio, on 17th November 1893 several railwaymen were badly hurt when a Cleveland Canton and Southern Railroad train and one of the Lake Erie Alliance and Southern Railroad collided on the crossing at Minerva. Both train crews claimed that the target at the crossing had given them the right of way.[1139]

Wabash Railroad train No. 5, a Company's premier heavyweight passenger train between St Louis, Missouri and Chicago, Illinois, was delayed two hours when it collided with an Indiana Illinois and Iowa Railroad freight train at the crossing with the Kankakee and Seneca Railroad at Essex, Indiana, United States, at about 12.30 p.m. on 9th December 1893. The freight train was badly damaged, but no one was hurt.[1140] [1141]

The collision of another premier American train, the vestibule limited of the Richmond and Danville Railroad, at 1 a.m. on 17th January 1894, was a little more serious than the Indiana one, though there were no fatalities. It happened at Chester Crossing, South Carolina, where the Richmond and Danville and the Georgia Carolina and Northern Railroads intersected in a shallow cutting. State law required that all trains stop short of the crossing and the conductors of both trains claimed that theirs had done so. The passenger train appears to have reached the crossing first, for when its engineer, A.E. Williamson, became aware of the danger he opened his throttle and almost succeeded in clearing the crossing. Meanwhile the engineer of the freight train had applied his brakes, but his engine struck the middle of the last sleeper, pushing it against the side of the cutting, the engine then coming to a stand on the Danville track. Several passengers and Pullman Conductor Davis were slightly hurt and were treated by Dr Davega of Chester. The engineer and conductor of the freight train jumped before the collision, were slightly hurt and had reportedly disappeared.[1142]

The crossing at Last Chance, Missouri, United States, witnessed a serious accident at 3.10 a.m. on 30th June 1894. A southbound Kansas City Fort Scott and Memphis Railroad train was going over the crossing when it was run into by a northbound Kansas City Pittsburgh and Gulf Railroad stock train travelling at full speed. The engine of the Pittsburgh train struck the fifth car of the Fort Scott train, derailing six cars of the latter

---

1138 *Lima Times Democrat*, Lima, Ohio, United States, 31st October 1893
1139 *The Salem Daily News*, Salem, Ohio, United States, 18th November 1893
1140 *The Daily Review*, Decatur, Illinois, United States, 10th December 1893
1141 Webpage – Classic Streamlines: The Banner Limited
1142 *The Constitution*, Atlanta, Georgia, United States, 18th January 1894

and four cars of the stock train. Fireman Jack Robbins of Kansas City fell under the engine and was crushed to death, and about 25 head of cattle were killed. The wrecked cars were burned in order to clear the lines. The Fort Scott train was said to have had right of priority.[1143] Given the fact that the Pittsburgh train apparently made no attempt to stop and that the accident occurred in the early hours it seems probable that its engineer had fallen asleep and its fireman had either also dozed off or was engrossed in firing.

A collision at Clymers, Indiana, United States on 23rd August 1894 had a rather different cause from the previous one. A Terre Haute and Logansport (Vandalia) Railroad freight train was run into on the crossing by Wabash Railroad freight train No. 64. The engine and seven cars of the Wabash train and four cars of the Vandalia train were derailed and both lines blocked. Engineer Davis of the Wabash train jumped and sprained an ankle. Wabash train No. 42 was diverted via the Monon and Panhandle roads and delayed five hours. Engineer Davis told a subsequent investigation that he had stopped his train short of the crossing but (unknown to him?) it had broken in two and the rear portion catching up and colliding with the front portion had forced it forward into the path of the Vandalia train.[1144]

The crossing of the Louisville and Nashville and New Orleans and Northeastern Railroads about two miles from New Orleans was one of a number in the United States where all trains were required to come to a stop before entering the crossing, an obviously cheaper if inferior means of ensuring safety to the provision of fixed signals worked from a signal box, such as was required in Great Britain by this date. On 14th October 1894 Engineer Hanley of a Louisville and Nashville "across the lakes" excursion train consisting of eight well-filled coaches duly came to a stand and whistled before proceeding over the crossing. As it was part way over the crossing the excursion's passengers, many of whom were women and children, saw a train bearing down upon them on the other line and made a rush to escape, but before any of them could reach the doors at either end of the car it had been struck by the other train and was thrown into a ditch. No one was killed outright but an estimated 19 passengers on the Louisville and Nashville train were injured, some of them seriously. Engineer Hanley had tried to get his train across in time and opened the throttle. In the collision his train divided, the engine and seven cars stopping some distance down the line from the struck last car.

The colliding train belonged to the East Louisiana Railroad, which had trackage rights (the equivalent of British running powers) over the New Orleans and Northeastern line into New Orleans. Like the Nashville train it was carrying excursionists for 'the lake coast resorts of the pine lands of St Tammany parish.' None of its passengers was hurt. After the accident there was 'great indignation' against Engineer T.P. Simpson of the colliding train who disappeared into the swamps. Later he re-appeared and was arrested on a charge of criminal negligence. He alleged that his air brakes had failed to work and when he realised that he could not avoid a collision he and his fireman had jumped.[1145] [1146]

[1143] *The Weekly Herald Despatch*, Decatur, Illinois, United States, 7th July 1894

[1144] The Fort Wayne Sentinel, Fort Wayne, Indiana, United States, 23rd and 25th August 1894

[1145] *Sandusky Register*, Sandusky, Ohio, United States, 15th October 1894

Failure of the air brakes was again cited as the reason for a collision on Christmas Day 1894 between a Houston and Texas Central Railroad train and one of the Missouri Kansas Texas Railway at 6 p.m., a mile west of Waxahachie, Texas, United States. A westbound Houston passenger train was passing over the crossing when, as in the New Orleans accident, it was its rear car which was struck by a Katy passenger train. A moment after the collision an overturned stove started a fire, but a line of hose which was stretched from the Houston locomotive was used to extinguish the blaze before it could cause more damage. After half an hour all the injured passengers had been released from the wreckage and they were carried on improvised stretchers to neighbouring farmhouses for medical assistance. Two passengers, both from Waxahachie, were seriously and possibly fatally injured. The Katy train was in charge of Conductor Littlefield and Engineer Mays. The line was cleared by midnight. Some of the injured passengers were able to continue their journeys but the majority were too badly hurt to travel further for several days.[1147]

A log train on the Chicago and North Western Railroad (referred to in the newspaper report as the Lake Shore road) was going over the crossing at Medina Junction, Wisconsin, United States, on 5th January 1895, when it was run into by a Wisconsin Central Railroad train in charge of Conductor Palmer, with Engineer Martin Moore and Fireman Robert Beck. Moore, Beck, and Head Brakesman Bat Gardner, who was also on the engine, were very lucky to escape with slight injuries, and probably only did so because the engine had struck the ends of two cars. The Central engine was badly damaged and two North Western cars and 12 Central cars were 'practically ruined'. The Central train had been travelling at full speed at the time of the collision. Again there was a state law requiring all trains to come to a dead stand at the crossing but this had been routinely disregarded at Medina Junction by the engineers of heavy southbound and eastbound trains on account of a hill beyond the crossing which they needed to take a run at to avoid having to "double up" the train (take it up in two portions). The newspaper commented that Engineer Moore would doubtless lose his job, 'and if he does, scores of friends in this city {Stevens Point} and all along the line will deeply sympathize with him, as there isn't a more popular young man on the road.' [1148]

At 5.30 p.m. on 3rd December 1895 it was the turn of Alexandria, Indiana, United States, to have a crossing collision, albeit a fairly minor one. A Cleveland Cincinnati Chicago and St Louis (Big Four) train going over the crossing was run into by train No. 124 of the Lake Erie and Western Railroad in charge of Conductor J.M. Rainer and Engineer Joab Bowsher, both of Lima, Ohio. The Lake Erie engine was derailed, and damaged and two Big Four box cars were derailed. No one was hurt. As to the cause of the collision the newspaper report stated, 'it is supposed that the Big Four train had the crossing and that the L.E.& W. train could not be brought to a stop, Alexandria being one

---

[1146] *Bangor Daily Whig and Courier*, Bangor, Maine, United States, 15th October 1894
[1147] *Galveston Daily News*, Galveston, Texas, United States, 26th December 1894
[1148] *Stevens Point Gazette*, Stevens Point, Wisconsin, United States, 9th January 1895

of the worst places on the road at which to handle a train, especially to make a stop.' [1149]

The worst rail/rail crossing collision in the United States, or indeed anywhere else in the world, occurred on 30th July 1896 two miles from Atlantic City, New Jersey. The popular seaside resort of Atlantic City was linked to the city of Philadelphia by the West Jersey and Seashore (controlled by the Pennsylvania) and Atlantic City (controlled by the Philadelphia and Reading) Railroads, which during the 1880s and 1890s ran highly competitive services between the resort and the metropolis. Trains of the rival companies often raced one another on their lines, which ran closely parallel for some distance across the marshes. It was therefore especially unfortunate that the two roads crossed one another not far from their respective termini in Atlantic City. Unusually for an American location the crossing was, however, controlled from the continually manned Meadow tower, working interlocked signals in each direction, with both stop signals close to the crossing and distant signals 2,000 feet from it. The ringing of a bell informed the towermen when a train was a mile and a half distant and once the signals had been cleared for one road they could not be replaced until the train had passed.

On the evening of 30th July a special West Jersey and Seashore train (No. 700) left the South Carolina Avenue station in Atlantic City at about 6.40, its departure slightly delayed from the timetabled 6.30 p.m. Its 14 cars were filled with over 1,000 happy excursionists from Bridgeton, Millville, Vineland and other South Jersey towns who had travelled to Atlantic City under the auspices of the Improved Order of Red Men (a fraternal organisation dating from 1834). Darkness was falling as the train approached the crossing, and Engineer John Kreiner could see an Atlantic City train coming from the opposite direction, but his signals were clear, and he kept going. By the time that he had realised that the Atlantic City train was not pulling up it was too late to do anything. The West Shore engine and the leading baggage car cleared the crossing, but the first passenger car was smashed by the Atlantic City engine and the next two completely telescoped into one another. The colliding engine then overturned in a ditch, burying its engineer. No less than 59 passengers in the front coaches were killed outright or died of their injuries, though the father of one child saved its life by throwing it from the train a moment before he himself was killed. The collision occurred at about 6.48 p.m. Fire broke out in the wreckage but a fire engine from Atlantic City was soon on the spot and put it out.

As usual the newspapers gave eye-witness accounts of the accident, including a long one from a Dr Smith, one of the rescuers, who told a reporter, "It was an awful sight. Only four or five bodies had been taken from the wreck when I arrived. The Reading {Atlantic City} express struck the excursion train on the Pennsylvania {West Jersey and Seashore} road, cutting it clean in two. The Reading engine struck the second coach of the excursion train with terrific force. The coach was smashed into bits, there was nothing left of it except a tangled mass of splinters, mixed up with which were the mangled and torn bodies of men, women and children. Death had come instantaneous, as from a lightning stroke, in most of them. There was no semblance of a railroad car in that heap of kindling wood. The third coach had not received the full force of the blow, but it was ripped up bodily and lay overturned with one side torn off. The engine and baggage car

1149 *The Times Democrat*, Lima, Ohio, United States, 4th December 1895 of the excursion train stood on the rails unharmed, as did the cars at the other end of the train. The train had been cut in two, and in the luckless car which opposed the great Reading engine rushing on at nearly a mile a minute nearly every passenger was killed or severely injured. I saw the remains of the Reading engine in a heap of scrap iron and rubbish piled up with the smashed car. The engine was crushed so completely that it was simply a scrap heap. I could not see whether the engineer and fireman were killed, but they could not have escaped. The wreckage caught fire from the ruins of the engine and a fire engine from Atlantic City was brought out. The flames were soon subdued, although I could not tell whether any of the victims had been burned to death. I saw one little boy laid out on the bank with his clothes burned quite off, but he might have been dead before the flames reached him. The Pullman car on the rear of the Reading express did not leave the track with the cars in front of it, and the inmates suffered little harm beyond bruises, cuts and sprains. Those who had been spared on both trains, stood about dazed and helpless until the screams of the wounded and dying brought them to their senses. The physicians who had been summoned from Atlantic City and brought by special train over the Pennsylvania road came in thick and fast. I assisted them in temporarily dressing wounds in the field, but it seemed as if almost every victim pulled from the wreck had been killed at the instant of the collision. I saw the bodies of five children lying side by side on the grass as they had been lifted out. The sights all about me were heartbreaking and horrible. Mothers were looking for children and husbands for wives and the air was filled with groans and cries." [1150]

The colliding train was Atlantic City train No. 23, which had left Kaighn's Point Ferry, Camden, across the Delaware River from Philadelphia, at 5.40 p.m. in charge of Conductor Pinkerton and Engineer Edward Farr. It soon became clear after the accident that it was Engineer Farr, who was also killed, who was responsible for it, though Towermen George Hauser and William Thurlow were initially arrested and lodged in Atlantic City jail. Farr's mistake was inexplicable as he was an experienced engineer, but every witness, including his own conductor, testified that the signals had been against him. Strangely the trip would have been his last one even had there been no accident as he had just tendered his resignation." [1151]

Fireman O'Halligan testified that he had not seen the signal until they were within 25 feet of it because he had been occupied firing. He had saved his life by running back into the tender.

Towerman Hauser was criticised for having given the excursion train the right of way over an express but said that he had given priority to the train which he saw first. Possibly Engineer Farr, who would have known from the working timetable that there were no regular trains at that hour, had realised that it must be an excursion over which his express should have priority. Or he might have misidentified the excursion train as an opposing train on the Atlantic City line, given that the two roads ran close together. Yet another possibility was that he was engrossed in racing an Atlantic City bound West Shore express. Whatever the reason he had clearly run through the signals, though he clearly realised his mistake at the last moment, reducing the speed of his train from 45 to 15 miles an hour by reversing his engine and applying the air brakes, and had

1150 *The Fort Wayne Gazette*, Fort Wayne, Indiana, United States, 1st August 1896

1151 *A History of Railroad Accidents Safety Precautions and Operating Practices* by Robert B. Shaw, pgs. 103-106 stuck to his post. [1152]

On 7th August the coroner's jury ruled that Engineer Farr was mainly responsible for the accident by his failing to obey the Meadow signals but a minority of them also censured Towerman Hauser for using poor judgement in giving priority to the excursion train and Engineer Kreiner (or Greiner) for not exercising greater care.[1153]

Robert B. Shaw points out that the management of the two railway managements could also be blamed for having a flat crossing on such busy lines, scheduling expresses to arrive and depart simultaneously and allowing or perhaps even encouraging competition between the two roads. A contemporary editorial in the Railway Gazette suggested the obvious solution of the two railroads exchanging termini but reflected that the railway executives reply would be that "only the impracticable view of one unacquainted with the situation would suggest such a thing." [1154]

Dundalk station, Ireland, the scene of a rail/rail crossing accident in 1868, suffered another one on 3rd October 1896. By this date the former Dublin and Belfast Junction and Irish North Western Companies of the 1868 accident had both become part of the Great Northern (Ireland) Railway. The rail/rail crossing with the Dublin – Belfast main line and the single-track branch from the Enniskillen line to Barrack Street goods station was controlled by fixed signals worked from Dundalk South box, while the single line was worked by the electric train staff between Dundalk West box and Windmill Road Junction box, the latter adjoining the goods station. Dundalk South box was not a staff station; it had bell communication with the West box, but not with Windmill Road. From the west the crossing was protected by the West box starting signal, 450 yards from the crossing, and the South box home signal, 32 yards from the crossing. However, the procedure for westbound trains was for the West box signalman to ring the South box, but if the signalman in the South box did not accept the train it was still allowed to draw past the West box starter at danger as far as the South box home signal.[1155]

On the day of the accident a special goods train of 12 empty coal waggons and a brake van arrived at the West box at 12.40 a.m. but was not accepted by the South box as a light engine was using the crossing en route to the engine shed sidings. Not having had the train accepted, Signalman Patrick Coleman gave the driver, Francis Warnock, the train staff and told him that he might draw down to the crossing. Signalman Coleman could see that the South box home was at "Danger". The train failed to stop at the protecting signal and collided with the tender of the light engine on the crossing.

After the collision the train stood at the crossing for rather more than half an hour, when Locomotive Foreman William Fenton told Driver Warnock to take his train on to the goods station. Driver Warnock set off but on arriving there found that he had left his train behind. He then set off back towards the crossing and collided with the coal waggons a quarter mile from the crossing. Guard Patrick Keowa tried to stop the driver coming back too far, but the waggons had come to a stand on a river bridge with no

---

[1152] *Trenton Evening Times*, Trenton, New Jersey, United States, 5th August 1896
[1153] Ibid, 8th August 1896
[1154] *A History of Railroad Accidents Safety Precautions and Operating Practices* by Robert B. Shaw, pgs. 103-106
[1155] Board of Trade report

footway and the guard was unable to get to the front of his train.

Driver Warnock claimed that the South box home signal had been off but the Board of Trade inspector, Lt-Col G.W Addison, pointed out that his statement was not only contradicted by other witnesses, but the interlocking would have prevented Signalman James Sharkey in the South box from clearing the signal. Both accidents were due, the inspecting officer stated, to the carelessness of the enginemen of the goods train, from a cause which to the great credit of railway servants was as uncommon as to be almost unknown, namely Driver Warnock was, if not actually drunk, under the influence of alcohol, and his fireman, John Hodges, certainly was drunk. It seemed likely that the men had twice gone into a pub at Castleblaney to get drink. Both men had been dismissed.

Notwithstanding the failure of the trainmen, Lt-Col Addison was not happy with the arrangements at the crossing in several respects. He recommended that no trains should be allowed to pass the West box starter at danger, communication should be proved between the South and Windmill Road boxes and the South box home signal should be moved further back from the crossing, or an outer home provided. He was also unhappy that there was no block working through Dundalk station yard, although no exemption from it had been granted by the Board of Trade.[1156]

Fog was a factor in a collision at Decatur, Illinois, United States, on 30th December 1896 at the crossing of the Illinois Central and Wabash Railroads, the scene of at least one previous collision. Southbound Illinois Central local freight No 193 was stopped taking water at the crane south of the depot at 10.30 a.m., with its rear cars protruding over the crossing. There was a dense fog. A Wabash Railroad Springfield local freight composed of 25 jumbo coal cars, only one of which was fitted with air brakes, was approaching the crossing. Engineer Whalen shut off steam at Water street and began to brake at Morgan Street, but the heavy train could not be stopped, and it smashed the gates protecting the crossing and struck the Illinois Central train. Engineer Whalen and Fireman Ralston stuck to their engine and were not hurt, and the damage done was not great.[1157]

Another Wabash Railroad train was involved in a collision at Newton, Indiana United States, on 17th March 1897. A westbound Wabash freight train stopped short of the crossing, its engine was detached and ran over the crossing with the Chicago and Erie Railroad to a coal dock. As it was returning to its train a westbound Chicago and Erie express was approaching the crossing. The crossing was controlled from a signal tower but apparently there was some problem with the equipment, the newspaper report stating that 'the interlocker refused to work'. In the evening darkness the towerman tried to signal to the express with a lantern but it struck the Wabash engine, and both were badly damaged, while three cars of the express were derailed. Erie engineer Joe Glass was caught under his engine and seriously cut and bruised but he was the only casualty. The lines were blocked until the next morning.[1158]

[1156] Ibid.
[1157] *The Daily Review*, Decatur, Illinois, United States, 31st December 1896
[1158] *Fort Wayne Gazette*, Fort Wayne, Indiana, United States, 19th March 1897

A westbound Lake Erie and Western Railroad freight train collided with a Pennsylvania Railroad pay train at Burgoon, Ohio, United States, on 14th April 1897. A misunderstanding of the target protecting the crossing resulted in both trains attempting to go over it at the same moment. The two engines collided,[1159] and both overturned into a ditch, but the engineers and firemen jumped and escaped injury.

On 20th August 1897 an excursion train was run by Joe Bennett, proprietor of the People's Tea Store, Lima, Ohio, United States, from Lima to Toledo and back via the Detroit and Lima Northern Railroad. It was almost back in Lima, running as regular southbound passenger train No. 41, when at 10.40 p.m. it was in collision with an eastbound Lake Erie and Western Railroad freight train at the crossing a mile and a half from the city. The excursion train was in charge of Conductor Harper and Engineer Lenhart, and the freight train was in charge of Conductor T.E. Davis, with Engineer Charles Hartman and Fireman William McLellan, plus Brakemen Whelan and Davis. Like many American crossings no signalling was provided but all trains were ordered to stop before proceeding over it. Engineer Lenhart did in fact bring his train to a stand before proceeding over the crossing at about 6 or 8 miles an hour when its fifth car was run into by the freight train and it was cut in two. One passenger, Mrs John Burnham, was probably fatally injured, suffering a fractured skull and internal injuries, while another 10 passengers were seriously and about 20 less seriously injured.

It would seem that Engineer Hartman of the Lake Erie train was at fault. His heavy train of 40 loaded and six empty waggons had left the yards on the south side of the city at 10.17. He had closed the throttle and reduced speed as he approached the crossing but did not come to a stand. Then as the way appeared to be clear he had opened the throttle for an incline beyond the crossing, when the passenger train suddenly appeared from his left. He then reversed his engine while Engineer Lenhart opened the throttle of his locomotive in an attempt to get clear. Fireman McLellan was somewhat hurt, and Conductor Davis and Brakeman Durbin were shaken up. A cut of cars on the "Y" connecting the two roads had prevented Engineer Lenhart from seeing the Lake Erie freight train earlier, or Engineer Hartman from seeing the passenger train.[1160]

Minor rail/rail collisions continued to occur at regular intervals in the United States with further examples on 26th October 1897 at Lerna, Illinois; on 21st February 1898 at New Washington, Ohio; 27th July 1898, at Mattoon, Illinois; 2nd December 1898, at Clyde, Ohio; and 17th December 1898 at West Point, Mississippi. More serious was the one at Tuscola, Illinois, on 31st December 1898. The Chicago and Eastern Illinois, Illinois Central and Indiana Decatur and Western Railroads all intersected with a space of 20 feet at Tuscola. A Chicago and Eastern Illinois passenger train was setting back over the crossing with the Illinois Central to reach its turntable before departing eastwards to Danville, when it was struck by a southbound Illinois Central fast freight train consisting of 30 waggons loaded with crushed stone. Some lumber sheds obscured the view of the engineer of the passenger train of the freight train, but he told a newspaper reporter that

[1159] *The Times Democrat*, Lima, Ohio, United States, 15th April 1897
[1160] Ibid, 21st August 1897

he had whistled for the crossing before proceeding and his whistle had been unanswered. It was theorised that the freight train had experienced a problem with its air brakes. Fireman George Thomas of the freight train was killed, Fireman James Hopton of the passenger train was thought to be fatally injured and a brakeman was missing, believed buried beneath one of the engines. Engineer Hart of the freight train jumped and escaped with a severe cut over his eye. There were only half a dozen passengers, who were unhurt. The crossing had previously been the scene of several narrow escapes and the railroad companies were in the process of erecting an interlocking tower there, which was almost ready.[1161]

In early November 1899 members of the 20th Kansas Volunteer Regiment were returning home from their participation in the Spanish-American War. On 3rd November a Missouri Pacific Railroad troop train carrying the men of Company F was involved in a collision with a Missouri Kansas Texas Railway freight train on the crossing at Trilby, Kansas, United States, 10 miles north of Blue Mound. The soldiers escaped with a severe shaking, but the Missouri Pacific fireman was injured and the engine badly damaged.[1162]

Early on 8th December 1899 Chicago Milwaukee and St Paul Railroad fast mail train No. 57 collided with Wisconsin Central Railroad freight train No. 26 on the crossing at Schleisingerville (renamed Slinger 1921), Wisconsin, United States. Both engines were much damaged, and it took about three hours to clear the lines. Passenger G.F. Maynard of St Paul, Minnesota, was hurt about the head and Mail Agent McCann was slightly hurt.[1163]

Westbound Central of Georgia Railroad and northbound Georgia and Alabama Railroad freight trains collided on the crossing at Dawson, Georgia, United States, on 29th March 1900. No one was hurt. Thick woodland obscured the view of each train from the other and each engineer blamed the other for the accident. The accident occurred during the morning and by the end of the day wrecking trains had cleared the lines.[1164]

On Friday 13th April 1900 what was described as 'an expensive wreck' occurred at Stillwater Junction, Ohio, United States, at the intersection of the Cincinnati Hamilton and Dayton Railroad and the jointly owned Dayton & Union and Pittsburgh Cincinnati Chicago and St Louis (Panhandle) Railroads. The Cincinnati Hamilton and Dayton local train, in charge of Conductor Clement of Delphos, was on the crossing when the other train ran into it. No one was hurt but the Dayton & Union and Panhandle engine was badly damaged, and a car-load of tobacco scattered over the right of way.[1165]

[1161] *Daily Review*, Decatur, Illinois, United States, 1st January 1898
[1162] *The Daily Northwestern*, Oshkosh, Wisconsin, United States, 3rd November 1899
[1163] Ibid, 8th December 1899
[1164] The Atlanta Constitution, Atlanta, Georgia, United States, 30th March 1900
[1165] *The Daily Herald*, Delphos, Ohio, United States, 14th April 1900

A Pittsburgh Fort Wayne and Chicago Railroad train was in trouble again on 20th June 1900 when double headed passenger train No. 15 with 14 cars crashed into a Lake Erie and Western Railroad freight train on a crossing in Lima, Ohio, United States, 'the air brake apparatus refusing to work'. No one was hurt and this time it was cement which was scattered all over the tracks.[1166] Finally, on about 24th June a Panhandle express from Cleveland to Pittsburgh collided with the engine of an Alliance and Northern Railroad train at the intersection at Alliance, Ohio, United States. This third accident to involve a Panhandle train was more serious than the other two as Fireman Conrad Weaver of the Alliance and Northern train was killed, while his driver, Albert Dunlap was injured. Property loss was estimated at $25,000.

At about noon on 5th November 1900 a westbound Green Bay and Western Railroad passenger train was going over the crossing with the Wisconsin Central Railroad at Plover, Wisconsin, United States when a freight train on the latter road backed into it. The engine and tender of the passenger train were struck by the caboose and a car of the freight train, both of which vehicles were derailed. Brakeman George Schilling jumped from the freight car as it went over but sustained a broken thigh and a badly sprained ankle.[1167]

[1166] Ibid, 20th and 21st June 1900

[1167] *The Stevens Point Journal*, Stevens Point, Wisconsin, United States, 5th November 1900

CHAPTER 14

FIRES

Fires often broke out on trains following derailments or collisions, especially perhaps in the more northerly areas of North America where trains were commonly heated in the early years by wood or coal stoves, and a stove overturned in an accident would more likely than not set the wooden coach ablaze. Stoves for train heating gave way towards the end of the 19th century to the use of high-pressure steam from the locomotive, which eliminated the danger of fires from upset stoves but introduced a different hazard should the steam pipes be damaged in an accident. (This danger was itself eventually eliminated following the invention of the "Vapor" car heating system using low-pressure steam by American Egbert Gold in 1903.[1168]) In the same way the substitution of gas for oil for use in train lighting removed the danger of fires from broken oil lamps though this introduced a new danger in the form of fires from escaping gas, and a number of accidents were made much more severe by fires started in that way. This chapter will, however, not concern itself with fires breaking out as secondary events, but will consider only primary train fires, that is accidents where there was no precipitating derailment or collision but merely a fire. Causes of train fires other than sources of heat or light included matches carelessly discarded by smokers, sparks from steam engines setting fire to goods and external sources such as forest fires. The first electric railways were also in operation before 1901, but no serious fires from this form of traction seem to have occurred this early.

A fire which was not properly extinguished at the first attempt or which broke out again could have resulted in a disastrous explosion in the colony of Victoria at the end of June 1882. On the evening of 29th June 1882 the ordinary goods train due in Sandhurst (renamed Bendigo 1891) at 6 p.m. was halfway between Woodend and Sandhurst when a waggon with a load of chaff (the husks of cereal crops used as animal fodder) was discovered to be on fire. Alarm as to the fire was greatly increased by the fact that it was marshalled next to a waggon containing three tons of blasting powder. As soon as the train reached the next station the waggon as shunted into a siding and the fire apparently

[1168] Wikipedia – Steam generator (railroad)

put out. The offending waggon was then reunited with the remainder of the train and it continued to Sandhurst. At 2 a.m. the next morning, as the night shunters at Sandhurst were just about to go home to bed, they noticed a glow coming from the goods yard. Hurrying over there they found that the waggon of chaff was on fire again, and furthermore that it was still marshalled next to the waggon containing the blasting powder, presumably either dynamite or lithofracteur (an explosive compound of nitro-glycerine). The fire had fortunately not progressed too far and was easily extinguished. The newspaper reporter commented that, 'There is undoubtedly someone to blame in this matter – in the first instance for placing a truck of inflammable matter next to one containing blasting material, and secondly, when the fire was discovered and presumably extinguished, that precautions were not taken to prevent it again breaking out, or even separating the two trucks on arriving at the Sandhurst station.''[1169]

The 1.15 p.m. Pennsylvania Railroad passenger train from Jersey City on Sunday 22nd October 1882 was as usual crowded, with an estimated 680 passengers in its 10 coaches. As it reached the Bergen Cut travelling at about 35 miles an hour its leading smoking car was crowded with passengers, many of them standing in the aisle. Suddenly a dense cloud of smoke, mixed with flames, came through the open front door of the car, facing the engine, and the surprised passengers sprang up to shut it. The next moment it was thrown open again, Engineer Joseph Sieg and his fireman staggered into the car and it was seen that the engine cab was a mass of flames. The train was now running away, and the engineer told his fireman to go to the closet at the rear of the car and apply the brake from there, but the fireman found his way completely blocked by a mass of excited passengers with no way through. A passenger at the front asked the engineer, "What is to be done?" and the engineer (according to the somewhat melodramatic, but perhaps true, newspaper account) gritted his teeth, turned and went back through the door to the engine tender, then disappeared into the flames beyond. Shortly afterwards the train began to slow up and it came to a stand on the bridge over the Hackensack River. Then Engineer Sieg reappeared above the tender, weak and half unconscious, his clothes burned off him and badly burnt himself. Passengers helped him into the car, the fire in the engine cab was extinguished with buckets of water from the river, and Conductor Whelpley, who had been in the third car at the time, ordered up an engine and freight car with which to take the injured engineer to Jersey City, from whence he was taken to St Francis' Hospital. The fireman took the train on to Newark. The sister at the hospital later told a newspaper reporter that the engineer was badly hurt but she thought he would survive. The fire had been caused by flames from the open firebox door igniting the woodwork of the cab, fanned by a strong wind.[1170]

The Midland Railway overnight Scotch express, which left London St Pancras station for Glasgow St Enoch at 9.15 p.m., caught fire in the early hours of Sunday 29th October 1882, and a passenger was burnt to death, the only passenger fatality in England and indeed Great Britain in a primary train fire in the whole of the 19th century. The

[1169] *Adelaide Observer*, Adelaide, South Australia, 8th July 1882 from *Bendigo Advertiser* 29th June

[1170] *Bangor Daily Whig and Courier*, Bangor, Maine, United States, 26th October 1882 signalman at Rothwell Haigh box saw an unusual light from the train but thought it must be a reflection from the open firebox. His colleague just down the line at Hunslet South Junction cabin, however, saw that the train was on fire and sent the "Stop and Examine" signal to Hunslet Station cabin, but the train pulled up 100 yards beyond his box. Driver Frank Carlisle, in charge of the engine of the train, said that the whistle of the communication cord had sounded as he approached Hunslet South Junction and looking back along his train he had seen a light and had brought his train to a stand. It is not very clear from the Board of Trade report whether the fire was first noticed by Conductor Robert Donaldson, in charge of the Pullman car *Enterprise*, the vehicle in which the blaze originated, or by Conductor Andrew Baillie, in charge of the Pullman car *Excelsior*, but at any rate both of them pulled the communication cord and roused their sleeping passengers; four men in the *Enterprise* and an unstated number of women in the *Excelsior*. *The Times* reported that three of the four male passengers in the *Enterprise* were rushing frantically about in their nightclothes as the fire spread rapidly and when the train stopped the conductor and three of the passengers jumped from the burning coach, one of the passengers through a window as he could not open the door.

Even at that early hour of the morning the burning train and shouts of the passengers attracted attention and soon a number of railwaymen and police were on the scene. The train was uncoupled, and the driver pulled the front portion of it, including the burning *Enterprise*, a short distance ahead and railwaymen from Hunslet sidings threw buckets of water at it. The decision was then made, according to *The Times* by a Police Sergeant Vincent, to pull the burning car half a mile down the line to Hunslet station, where there was a water crane. This was done but then it was found that the Pullman car was too high up to go under the crane and the fire was finally extinguished using a hose from Messrs Nicholson's chemical works. The Board of Trade inspector who investigated the accident, Colonel W. Yolland, thought that the decision to move the train forward was an error of judgement which probably cost the dead passenger his life. After the train had been divided Conductor Donaldson attempted to re-enter the *Enterprise* to check that all four men had got off, but the smoke was too thick for him to do so. After the fire had been extinguished the badly burnt body of a Dr Arthur, who had been returning from working for the Government medical service in Ceylon to his home near Aberdeen, was found.[1171]

Colonel Yolland made a very careful examination of how the fire could have originated and decided that it must have been caused by the lighted reading lamp used by a Mr Cranston in Berth 8, either from it igniting something whilst burning or due to a spark from it as it was being blown out. He did not believe Conductor Donaldson's story of what had happened and thought that the conductor had probably been breaking the company's rules by smoking and that, panicking when the fire broke out, he had failed to ensure that Dr Arthur, whom it would seem was probably considerably the worse for drink, had left the car. The inspecting officer also pointed out that the Midland Company was continuing to use the Harrison's cord communication system, no longer sanctioned by the Board of Trade, and criticised the rule which stated that when the cord was pulled the train should not be stopped until it was within the protection of fixed signals, which meant that in the present accident the train was not brought to a halt as quickly as it might have been, when in the case of a fire every moment was vital. Finally he considered that

---

[1171] *The Times*, London, England, 30th October 1882

the use of reading lamps should not be permitted in sleeping berths.[1172] The verdict of the jury at the inquest into the death of Dr Arthur contained a string of recommendations very similar to those in the Board of Trade report. They had been handed a list of questions to answer by the coroner when giving their verdict, drawn up perhaps by Colonel Yolland. From the evidence at the inquest it is possible that Dr Arthur was under the influence of opiates taken to help him sleep rather than drunk or perhaps he was suffering from the combined effects of alcohol and drugs, both of which he was probably addicted to. Dr Arthur had also suffered from sunstroke and had sustained a head injury.[1173 1174]

In what was almost a repeat of the Hunslet tragedy in the United States, a Pullman sleeping car attached to the Washington express caught fire near Wilmore, Pennsylvania, on 13th March 1883. The passengers were able to escape unhurt but lost all their clothes, valises, jewellery and so on as the car was burnt out. Among them were General Pearson, Pension Agent Dudley, Colonel Ben Wilson, a former Çongressman from West Virginia; ex-Congressman Best from Virginia, General Cockens, of Indiana, Colonel Davidson, a local from Beaver, Pennsylvania, Colonel Porter, Connellsville, Pennsylvania, and Postmaster and ex-Congressman Calkins of Indiana. The burning sleeping car was detached from the remainder of the train and most of its passengers were forced to walk half-dressed and barefooted over half a mile of frozen ground to a replacement sleeping car at about 4 a.m.[1175] The loss to the Pullman Car Company was estimated at $20,000.[1176] The cause of the fire was not stated.

A fire in a ladies' compartment of a Great Western Railway train between Wrexham and Ruabon, Wales, on 24th May 1883, had an unusual cause. The train was stopped, presumably after the communication cord had been pulled, and shunted into a colliery siding where the fire was put out. It was discovered that a careless workman, engaged in constructing an iron overbridge across the railway, had dropped a red-hot bolt onto the train as it was passing beneath him.[1177]

On 10th September 1883 westbound Wabash Railroad freight train first No. 91 was brought to a stand about four miles east of Jacksonville, Illinois, United States. The train included from a stock car containing two fine horses in charge of a groom, supposed to be one Scott Pickerell. As the train was proceeding towards Jacksonville the brakeman in the car behind the stock car smelt burning and saw that the stock car was on fire. At the same time the person in charge of the horses appeared at the manhole of the stock car waving his coat and hat as a signal for the train to be stopped. The brakeman duly stopped the train and saw the groom disappear and then reappear. Supposing that the groom could take care of himself the brakeman rushed forward to the engine as soon as

---

[1172] Board of Trade report
[1173] *The Times*, London, England, 8th and 9th November 1882
[1174] *The Railway Magazine* November 1963 – December 1964, pgs. 251-3
[1175] *The Olean Democrat*, Olean, New York, United States, 20th March 1883
[1176] *Chester Times*, Chester, Pennsylvania, United States, 14th March 1883
[1177] *The Times*, London, England, 25th May 1883

the train had been brought to a stand for water to fight the fire. The blaze had got too good a hold, however, and after a fruitless fight to put it out the firefighting efforts had to be abandoned and the car left to burn itself out. In the remains of the car was found the charred body of the groom, along with those of the two horses. He had either been overcome by the smoke or injured by one of the horses and had burnt to death in the car along with the two animals. The horses had been on their way from Buffalo, New York, to Kansas City for exhibition in a fair there. It was supposed that the groom had been smoking, and that sparks from his pipe had set fire to the straw in the car.[1178 1179]

The American state of Pennsylvania had had an oil industry from the 1850s [1180] and it was a tank of this oil which caused a bad train fire on the 15th January 1884 on the narrow gauge Bradford Bordell and Kinzua Railroad. On a hill overlooking the railroad line was a large oil storage tank belonging to the Anchor Oil Company, and it was from this tank that a quantity of oil escaped and ran down the hill to form a pool across the track about midway between Tayport and Bradford stations. An oil company employee saw that the oil had escaped and noticing a passenger train approaching ran to try and warn it, but because of deep snow he was unable to get to the line in time.[1181] When Engineer Patsy Sexton of the crowded 9 a.m. train from Wellesville to Bradford spotted the oil he reversed his engine and applied the air brakes, but it was too late to stop the train. [1182 1183] Almost as soon as the train reached the oil it burst into flames, which enveloped it. The intense heat cracked the coach windows and the car filed with flames. Panicking passengers rushed for the train doors but most of them were forced to jump through the windows of the single coach. The engineer and fireman of train No. 2 stuck to their posts as long as they could before they too were forced to jump and the train ran away down a steep descent, driverless, and then became derailed on a curve after about half a mile. Most of the approximately 40 passengers suffered more or less severe burns but the thick snow by the lineside cushioned them as they jumped from the train.

As soon as news of the accident became known a relief train with surgeons and cots was despatched to the scene. Here it found the engine lying on its back, the passenger and baggage cars a smoking mass of ruins and injured passengers lying moaning with pain by the sides of the line. Three elderly women who had it was supposed had been too frightened to jump from the train were, as the contemporary newspaper account put it 'burnt to a crisp'.[1184] It was thought likely that some of those injured might also die.

A mysterious affair took place on the Midland Railway on 9th April 1884. As a fast train was about to depart from Bristol Temple Meads station a respectably dressed man hurried up to the guard and asked him to place a small package in his van. As there was no address on the package the guard was reluctant to do so but eventually consented to do

[1178] *The Daily Miner Butte*, Montana, United States, 12th September 1883
[1179] *Saturday Herald*, Decatur, Illinois, United States, 15th November 1883
[1180] Wikipedia – Pennsylvania Oil Rush
[1181] *A History of Railroad Accidents Safety Precautions and Operating Practices* by Robert B. Shaw, pg. 341
[1182] *The Indiana Democrat*, Pennsylvania, United States, 24th January 1884
[1183] *Gettysburg Compiler*, Gettysburg, Pennsylvania, 23rd January 1884
[1184] *Chester Times*, Chester, Pennsylvania, United States, 16th January 1884

so and the man took a seat in a carriage some way back along the train. The train was passing Charfield, England, 20 miles from Bristol, when smoke was seen to be issuing from the van, and the driver, on a signal from the guard, stopped the train at Berkeley Road, the next station. The van was then found to be full of smoke and some of its woodwork was charred, whilst the guard suffered burns to the back of his head and his coat was singed. The railway was reluctant to say too much about the affair to newspaper reporters, but it was believed that a quantity of chemicals in the package handed to the guard had exploded, setting it on fire. Several detectives had apparently followed the suspect who handed over the package and traced him to Liverpool.[1185]

In the months preceding this incident several Fenian bombs had exploded, some of them on railway property. *The Times* at any rate evidently believed this to be another attempted Fenian outrage.

A horrific fire occurred on a Greely Salt Lake and Pacific Railroad train carrying Orton's Anglo-American Circus which was en route from Fort Collins to Golden near Windsor, Colorado, United States, on 29th August 1884. As the train neared Windsor, a small station near Greeley, Engineer Collepriest noticed that its leading car was on fire, reversed his engine and whistled for brakes. This car was where the roustabouts or casual circus labourers slept, and there were 60 of them in the coach, sleeping in three tiers of bunks on either side. The circus performers occupied a different car, further back the 17-vehicle train. The layout of the roustabout's car made their escape much more difficult. There were two side doors but the forward one was closed with men in bunks sleeping up against it. The rear side door was also closed and those attempting to escape that way found that the empty berths next to it contained rubbish on fire which filled the car with smoke and cut off escape in that direction. The only means of escape left was via a small side window between the car and the engine, while the suffocating smoke made it difficult to arouse the sleeping men. When the victims could be got out the injured were laid out on the cactus-strewn prairie by the side of the train and the contemporary account described the screams of those unable to escape, the glare of the flames illuminating the injured writhing in agony on the cactus beds and the terrified howls of the circus animals in the adjoining car 'making an appalling scene'. The performers from the rear of the train gazed at the scene with white faces while Albert Lake, in charge of the animals, and his friend Kent walked over the cacti in their bare feet to pour buckets of oil on the burns of the victims and wrap them in blankets. Engineer Collepriest took his machine into Greeley and returned with Dr Jesse Howes, President of the Colorado Medical Association.[1186]

Many of those injured suffered not only burns but broken limbs and dislocated joints through being forced through the small window in their escape. The bodies and parts of bodies of the dead were loaded onto a flat car and taken into Greeley at daylight, where they were buried together in one huge coffin, 10 feet long and seven feet wide. It was impossible to obtain a complete list of the fatalities as many had been taken on only a day or so before and their names were unknown. There were believed to be at least 10 dead. Three of them were never identified. An inquest was opened the same day but could not

---

1185 *The Times*, London, Pennsylvania, England, 11th April 1884

1186 *The Atchison Globe*, Atchison, Kansas, United States, 30th August 1884 proceed very far as the manager and other officials of the company had gone to an engagement at Golden.[1187] A number of the survivors agreed, however, that there had been two barrels of naptha in the fateful car which had been set alight either by a spark from the engine or, as indicated at the resumed inquest, by a torch put down by one of the men when he began to play cards. It was the highly inflammable naptha which had made the fire spread so quickly. The local *Greely Tribune* newspaper was also critical of the circus manager, who immediately continued to Denver with the train, leaving the county to pay for the burial of the dead men. Despite the disaster the circus had performed as usual at Denver, its amended destination, the next day. The circus also tried to remove the injured from hospital and take them forward on the train to save the hospital bills, but Dr Hawes refused to release them.[1188]

As the 12.20 p.m. up goods train from Woodend to Macedon, Victoria, was descending Forest Creek bank on 14th November 1885, its guard noticed that the second waggon from the engine, containing straw, was on fire. He had the train stopped without delay when it was found that the fire had extended to five waggons, three filled with straw, one chaff and the other granite. Despite every effort to extinguish the fire the contents of seven waggons were burnt and five waggons were destroyed and two more damaged. The cost of the damage was estimated at £500.

A spark from the engine was believed to have caused the fire. Commenting on this, the newspaper wrote, 'The engine, which is of the heavy express type, is fitted with a "gridiron" spark-arrester, placed over the firebox, this style being regarded by the department as the best of all they have tried yet. The arrester consists of a strong wire mesh, those made in the colony having a gauge of three-sixteenths of an inch. The one on the engine in question was, however, fitted in England, and has a much coarser gauge, whilst it was not very carefully fitted at the sides, thus allowing sparks to escape. The arresters made in the colony are said to be very effective.'[1189]

A tantalisingly brief report from the *Times* correspondent in Egypt told of the arrival there on the evening of 17th July 1886 of Sir H.D. Wolff, Nubar Pasha and Messrs Walpole and Bruce and added that their train had been late 'owing to the fact that the carriages had taken fire – an event of almost daily occurrence on the Egyptian railways.'[1190] It would be interesting to know why train fires were so frequent in Egypt.

The presence of mind of a passenger prevented what might have been a serious train fire in Queensland on 8th September 1886. Mr Pugh, a police magistrate travelling to Hendon on official business, was the only passenger in the first-class compartment of the evening down train from Warwick. The train had not gone very far when the oil lamp in the compartment began to flicker and burn abnormally, and between Deucher's Crossing and Hendon the lamp suddenly burst, and burning oil spread all over the floor of the

[1187] Ibid.
[1188] Webpage – Circus Train Fire of 1884; One of Weld County's Worst Disasters
[1189] The Telegraph, Brisbane, Queensland, 21st November 1885
[1190] *The Times*, London, England, c19th July 1886

compartment and was soon blazing fiercely. Mr Pugh dealt with the situation by taking off his overcoat and using it to beat out the flames, which he managed to do after a hard struggle. He then tried to attract the attention of the guard, but was unable to do so, even after getting out onto the footboard. (Queensland passenger trains were clearly not provided with communication cords, even as late as 1886.) Soon afterwards the train reached Hendon anyway. Very little damage was done, except to Mr Pugh's coat, which was both burnt and smeared with unburnt oil which he had wiped up with it.[1191]

The *Oregon* Pullman sleeping car, making its first westward trip on the second section of Pennsylvania Railroad express No. 9 from New York, caught fire about 4 a.m. on 13th November 1888 near Conemaugh, Pennsylvania, United States. A boy passenger, looking from between the curtains of his berth, found that the front of the car was on fire. He cried out in alarm and the sleeping passengers jumped from their berths in fright. One pulled the alarm cord and a porter used a fire extinguisher with good effect. The train was stopped, and the passengers jumped down from it, clad only in their nightclothes, so that they suffered considerably in the cold mountain air. By the time the fire was extinguished half of the car was destroyed, so it had presumably made its first and last journey. Almost all the passengers lost some or all of their possessions, some arriving at Pittsburgh wrapped only in blankets. The train was delayed several hours. The fire was blamed on an overheated stove in the ladies' toilet.[1192] Fires due car stoves would soon belong to history for measures to replace them with steam heating were well under way. By the end of 1890 an English newspaper was able to report that there was a legal prohibition on their use in New York State, and the railroads of New England and Canada were fast following suit in adopting steam heating, though those of the western American states were lagging behind.[1193]

Brief reports in British press mentioned a serious accident involving a train of 38 waggons laden with petrol in Russia on about 27th March 1890. According to a report from Tiflis (now Tbilisi), Georgia, the train caught fire near Belagora station and was totally destroyed, the driver, fireman and guard losing their lives.[1194 1195]

A forest fire was raging fiercely on Sunday 10th May 1891 around Austin, Pennsylvania, United States, and a work train carrying Superintendent Badger of the Sinnemahoning Valley Railroad and 75 men steamed out of town and up into the woods to fight the fire, which was threatening to destroy the town. It had proceeded about five miles and was deep in the forest when the intense heat and blinding smoke prevented any further progress. A decision was then made to reverse the train as the forest fire was closing in on every side. A newspaper reporter at Coudersport wrote, 'It was finally decided to dash past the burning skidway {meaning of this word not apparent}, and the engineer and fireman, with their faces covered with dampened cloths, and their hands and

---

1191 *Warwick Examiner and Times*, Warwick, Queensland, 11th September 1886

1192 *Newark Daily Advocate*, Newark, Ohio, United States, 14th November 1888

1193 *The Times*, London, England, 19th November 1890

1194 *The Times*, London, England, 29th March 1890

1195 *The Railway Press*, London, 5th April 1890

arms wrapped in wool, mounted the little engine and pulled out through the wall of fire. The 75 exhausted men gathered in groups on the flats for protection or lay on their faces on the floor. As the blazing furnace of logs was approached the heat became unbearable, and the smoke so blinding and stifling the men were obliged to cover their mouths with clothes. Just opposite the millions of feet of burning logs, where the heat and smoke and flames were the greatest, a terrible thing occurred. The engineer had forgotten that such great heat would surely spread the rails, and he pulled the throttle wider in the hope of sooner escaping from the torment of heat and smoke. Then there was a lurch, an ominous heaving, and a shriek of despair as the train toppled over into the hail of fire beneath. A scene ensued never to be forgotten by those who escaped, though every man will bear to his grave a mark of that awful moment. The cars caught fire like so many paper playthings, and the men within, half blinded and scarcely realising anything except that they were being slowly roasted to death, struggled fearfully to regain the track, where safety lay, for a time at least. Those uninjured from the fall and only smarting from the pain of the intense heat, bravely turned their burned, blackened hands to aid their more unfortunate fellows. At this hour it is impossible to secure details, though enough is known of the scene that followed the hurling of the struggling mass of men into the furnace of flame, to say its like had never occurred before. Supt. Badger, of the Sinnamahoning Valley Road, in charge of the relief train, had worked the hardest of all to save the properties of others, when the train ditched and rolled over so suddenly, he must have been injured so as to be unable to help himself, and owing to the smoke and panic he was not found until too late-jammed in the wreck; he had evidently slowly burned to death. At this time – 9.30 – it is known that six others also miserably perished at once, or died soon after, and 30 others of the party were badly burned, many probably fatally, owing to the fears that they inhaled the flames that seemed to spring fairly into their faces. Seven others of the party are missing, and their fate is unknown, though they are likely in the charred wood of the logs or train. Relief parties started for the scene as soon as the fearful news spread, many male relatives of the men injured insisting on accompanying the train, though they will hardly be able to reach the place of the wreck unless the fires have burned themselves out. Owing to the great devastation done to everything in the way of the fire, communication is badly interrupted, and it is impossible to learn the names of the men burned, or those still missing.’’[1196]

Another report from Bradford stated, ‘The train was pulled into the forest several miles, and before the crew realized it the train caught fire and was destroyed. The men on board leaped from the cars when they caught fire and sought safety in a stream that flowed nearby. The water in the creek was boiling hot and those who jumped into it were severely scalded. There they crouched while the tempest of flame rolled over them, the intense heat blistering and burning the terror-stricken men in the most shocking manner. Badger did not take to the stream, he seemed dazed and ran ahead of the blaze. He was overtaken by the flames, and his lifeless body burned to a crisp, was found to-day. Four companions have not been found.’[1197]

A fire on 27th August 1891 on the 9.17 a.m. Great Northern Railway suburban train from Barnet to Moorgate Street near Finchley (later Finchley Church Road), England,

---

[1196] *Bloomington Weekly Leader*, Bloomington, Illinois, United States, 15th May 1891

[1197] Bradford Era, Bradford, Pennsylvania, United States, 12th May 1891 arose from a new cause, for the coaches of the train were lit by electricity. The current for the electricity was supplied by a dynamo in a special compartment in the guard's van. Soon after leaving Finchley station the guard discovered flames coming from the dynamo compartment and that the fire had taken a considerable hold of the woodwork. He stopped the train and porters from the station used water to get the fire under control. An electrician met the train at Finsbury Park and found that the dynamo was undamaged and could still provide electricity for lighting the train through the tunnels to Moorgate street. 'It is conjectured that undue friction in connection with the dynamo had set light to the woodwork of the chamber.' [1198]

On Friday night, 5th February 1892, fire broke out in a first-class sleeping car of an express train between Bucharest and Braila in Romania. In what tended to be standard procedure in sleeping cars in various countries to prevent unauthorised entry, the doors of the carriage were locked, and 18 sleeping passengers were burnt to death or suffocated. Two other passengers, M. Danielescu, from Galatz, and 'a Captain' escaped with their lives. The staff of the sleeping car had failed to notice the fire in time. It had been caused by 'a defect in the heating apparatus'. [1199] [1200] [1201]

A report from Vienna stated, 'The interior of the carriage somehow took fire as the train was speeding along. The fire must have spread with awful rapidity, for as soon as the cries and screams of the imprisoned passengers attracted attention the flames seemed to be bursting from the windows. The unfortunates shouted to have the door of their fiery furnace opened, while at the same time they struggled to break it open. But the smoke and flame soon overcame them, and when rescue came all were dead but two, some having been burned to death and others suffocated.' [1202]

What was probably the most disastrous train fire of the 19th century happened in Russia on 11th January 1893, to a train carrying about 200 military recruits who were on their way to join their regiments. The train was travelling at speed on the Slatoust-Samara Railway when the blaze broke out in the forward portion of the long coach in which the soldiers were travelling. Fanned by the wind, due to the speed of the train, the fire spread from one end of the coach to the other very quickly. The train was stopped as soon as possible and the coach, which was by now completely enveloped in fire, was detached from the remainder of the train. The young soldiers immediately made a rush for the door and some jumped through the windows, most of those who survived escaping in this manner. In their desperation to escape several soldiers got wedged together in the narrow passage leading to the door and they were thus trapped. Seven were dragged from the mass of struggling bodies badly burnt, one of whom died later the same day. No less than 40 others were burnt to death and eight more died from smoke inhalation making a total in all of 49 fatalities. The cause of the fire was not stated, and possibly was not

[1198] *The Times*, London, England, 28th August 1891
[1199] *The Standard*, London, England, 9th February 1892
[1200] *Sydney Evening News*, Sydney, New South Wales, 10th February 1892
[1201] *Benton Harbor Weekly Palladium*, Benton Harbor, Michigan, United States, 12th February 1892
[1202] *The Standard*, Albert Lea, Minnesota, United States, 17th February 1892

known.[1203]

A combination of hazardous luggage being carried by one passenger and the carelessness of a second passenger led to a fire on the Kentucky Central branch of the Louisville and Nashville Railroad near Talbott, Kentucky, United States, on 27th February 1893. The train concerned was an early morning one which had passed through nearby Paris at 7.58 a.m. One of its passengers, riding in the smoking car, was a plumber who had placed a can of gasoline (petrol) by the door leading to the baggage car. Another passenger, M.H. Applegate, a newsboy, lit up a cigarette and carelessly threw away the still burning match, which chanced to set fire to some spilt oil on the outside of the can. Seeing the obvious danger William Tarr, a distiller, who was sat near the coach door, opened it and attempted to kick the can out of the car, but instead his kick knocked the stopper out of the can and its whole contents then caught alight, enveloping the car in flames and forcing the passengers to run for the doors or jump from its windows. It was described as a miracle that none of the passengers died in the ensuing blaze, which destroyed the car. As it was, Tarr, Applegate and five other passengers were severely burned, one of them in helping to remove the corpse of a Mrs Florence Longmore Letton from the baggage car so that she could be buried as intended rather than being cremated.[1204]

The town of Hinckley, Minnesota, United States, was affected by a catastrophic forest fire on 1st September 1894 which destroyed the town and killed an estimated 500 people. The town was served by two different lines, the Eastern Railroad of Minnesota and the St Paul and Duluth Railroad. By mid-afternoon the fire had set light to half the town and its residents knew that their only hope was to flee by one or other of the railroads. The first train to leave that afternoon was on the Eastern Railroad. Northbound local freight train No. 23 had arrived there at 2.40 p.m. but could not leave before the arrival of southbound passenger train No. 4 at 3.25. When that train arrived it was decided that to continue southwards was impossible and that the duty of the train crews was to save as many of the town's residents as possible. Part of freight train No. 23, with its engine leading, was coupled onto the rear of the passenger train No. 4 and about 400 people climbed aboard the train. When some of that number had done so they were urging the train crews to go but Engineers William Best of the passenger train and Ed Barry of the freight train waited as long as they dared for others to get aboard. Finally at 4.15 p.m. they crossed the Grindstone River bridge, whose sleepers were smoking, waited five minutes on the far side for any further townspeople to join them and then ran as hard as they could through the fire, crossing en route 19 wooden bridges, many of them on fire. The telegraph poles were also on fire and by delaying so long at Hinckley they were also running on the time of an opposing southbound train, which had, however, been abandoned back the line. When they finally pulled into West Superior an exhausted Engineer Barry had to be carried from his engine.[1205]

On the St Paul and Duluth Railroad the southbound Limited was due in Hinckley at

[1203] *The Times*, London, England, c19th January 1893
[1204] *Hamilton Daily Democrat*, Hamilton, Ohio, United States, 28th February 1893
[1205] *Reno Evening Gazette*, Reno, Nevada, United States, 4th September 1894

4.15 p.m. About a mile and a half outside of town it was flagged down by 200 fleeing townspeople. Engineer James Root briefly considered dashing through the fire, then realised that that was impossible and instead set back at speed to a marsh known as Skunk Lake four miles to the north. Before he had got there the forest fire, its flames driven at hurricane speed, had smashed coach windows and set them ablaze. A dozen passengers, maddened by the heat, reportedly threw themselves from the train. The engine's regulator was hot enough to blister Root's hands, the engine curtains were ablaze, and the seat was scorchingly hot, while the engineer could only stay at his post because his fireman was dousing him with water, having climbed into the water tank. Finally they reached Skunk Lake and the 125-odd passengers on the train and 200 townspeople took shelter in the shallow water of the marsh, where they waited for four hours for the fire to pass. Two Chinese passengers for some reason refused to leave the train and were burnt to death. The whole train was consumed by the fire. [1206] [1207]

According to Shaw another St Paul and Duluth train, the northbound Limited, rescued more citizens from Hinckley, returning southwards with them. A report from Mora, Minnesota on Monday 3rd September also mentions another Eastern Minnesota train which left St Paul at 1.5 p.m., picked up 300 people from Hinckley as late as 6 p.m. (though the fire was so intense that people were falling dead in the street at 4) and retreated southwards towards St Cloud, but which had not been heard of since and was believed to have burned with all on board, there being no chance of the people on it being alive unless they could have found a stream or slough. If 300 people died on this train it would be the greatest death toll in any train accident in United States history, but further details have not been found. The same Mora report mentions St Paul and Duluth Railroad freight train No. 46 going into the ditch one and a half miles west of Pokegama, Wisconsin (near Superior) presumably because of the collapse of a fire-damaged bridge, and that it had fire all around it and 95 people in its caboose, who must have perished unless rescued. Finally the report mentions fears for an eastbound passenger train which left Duluth that Saturday afternoon with a journey through 200 miles of fire-swept country, and which had not appeared in Marquette, Michigan.

About 19th December 1895, while a passenger train on the Siberian Railway in Russia was running at high speed, a carriage occupied by Count Golovine caught fire. The count jumped from the moving train and saved his life, but the countess and their two children were burnt to death.[1208]

The railway from Vryburg in the Cape Colony to Bulawayo in Rhodesia, which formed part of Cecil Rhodes' ambitious plan for a Cape-Cairo railway, opened on about 4th November 1897. A special train from Johannesburg-Eastern Provinces was run in connection with the opening and met with an accident at Gaberones (now Gaborone) in Bechuanaland when a fire broke out in its electric light dynamo van. The electrician and

---

[1206] *A History of Railroad Accidents Safety Precautions and Operating Practices* by Robert B. Shaw, pg.342
[1207] *Fort Wayne Gazette*, Fort Wayne, Indiana, United States, 4th September 1894
[1208] *Salem Daily News*, Salem, Ohio, United States, 24th December 1895

his assistants suffered severe burns but none of the passengers was hurt.[1209]

What was described as the Western mail train caught fire 27 miles short of Bourke, New South Wales, on Monday 28th November 1898. A blaze thought to have been due to a spark from the engine developed in a waggon carrying chaff and fanned by the wind rapidly spread to one carrying furniture. The train was stopped and there were fears that it would all be consumed by fire as the waggons were soon a mass of flame. The burning trucks were separated from the vehicles in front of and behind them but there seemed to be nothing at hand to get them off the running line into a siding. Fortunately a Mr Dwyer, who drove up from a homestead close by, was able to offer the use of a stout wire rope which was attached at great risk to the underside of the burning waggons. The railwaymen and about 50 train passengers hauled at the rope to try and shift the waggons. The first attempt was unsuccessful, and they had to wait until the fire had consumed much of the waggons and their contents before the trucks could be shifted at a second attempt. The train was then reunited, the two waggons being left to burn themselves out, and it finally reached Bourke three hours late just after 9 p.m.[1210]

1209 *The Times*, London, England, c6th November 1897

1210 *The Daily Telegraph*, Sydney, New South Wales, 29th November 1898

CHAPTER 15

CHEMICAL EXPLOSIONS

Railway accidents involving explosions were of two types; those which might be defined as chemical explosions and locomotive boiler explosions. Accidental chemical explosions, such as those due to mishaps with blasting powder or munitions, were rare, whereas locomotive boiler explosions were, at this date, considerably more common. This short chapter will deal with chemical explosions.

Arab and Islamic opposition to growing European influence in Egypt led to a war between Britain, aided by France, and Egypt in the summer of 1882 and the defeat of the Egyptian army at Tel-elKebir on 13th September.[1211] Just over a fortnight later at 4 p.m. on 28th September 1882 at 4 p.m. there was a massive explosion involving an ammunition train at Cairo station, followed by a series of smaller explosions. The series of explosions continued for over three hours, setting the station on fire, and a *Times* correspondent found 'a crowd of terrified Arabs hastening with shrieks from the station', while, 'Owing to the energetic efforts of the military, who had to contend with great difficulties, the pumping material being scanty, the fire was localized and finally got under control by 9 o'clock. The station was saved, but all the goods sheds, containing provisions for the Army for 10 days, as well as about 100 trucks of ammunition, was destroyed.' It was at first feared that 30 people had been killed; later the number of dead was given as five, with 20 wounded, including a surgeon of the 60th Rifles. Initially it was thought that the explosions were due to foul play, but this idea was later discredited, though arsonists were arrested for afterwards setting fire to two waggons of hay and trying to set fire to an adjoining portion of the city.[1212]

At the beginning of October *The Times* added that the War Office in London and officials of the Royal Arsenal were unable to account for the explosion. The ammunition was tested to be able to withstand a temperature of 150°F, proof against the most excessive heat experienced in the tropics, and all theories of spontaneous combustion

[1211] Wikipedia – History of Egypt under the British

[1212] *The Times*, London, England, 29th and 30th September and 1st October 1882 were also dismissed as untenable. One theory was that 'some of the loosely-made and badly packed Egyptian cartridges captured at Tel-el-Kebir had got mixed up with the English ammunition'. Another possibility was the explosion of a shell which had had its safety pin removed prior to use in battle, had not been used and then been returned to the limber box in an unsafe condition.[1213]

It was not munitions but commercial explosives which were involved in a major disaster at Kings Mill, Ohio, United States, on the Panhandle (Chicago St Louis and Pittsburgh) Railroad (Strictly on the Little Miami Railroad which was leased by the Panhandle) on 15th July 1890. A freight train had a waggon containing 500 kegs of powder to deliver to the siding serving Peter's cartridge factory, close to King's Mills station. Instead of setting back with the waggon into the siding the latter was uncoupled a little below King's Mill station and pushed into the siding. Gaining momentum the loose waggon struck two waggons loaded with soda ash in the siding and these in turn ran against another waggon loaded with 800 kegs of powder, whereupon the whole lot exploded. Fire broke out following the explosion and the cartridge factory and King's Mill station were burnt down. The freight train itself was stopped 300 feet from the scene and escaped destruction though it was seriously damaged.[1214] Twelve people were reportedly killed outright, and a number of others injured, some seriously. Most of the dead were railwaymen, and employees of the cartridge factory, the injured were mainly cartridge factory employees. No trace remained of Brakesman William Franley, who had been standing on an empty car when the explosion occurred. A house close by was also demolished and a mother and child in it killed. Some of those injured were also expected to die.[1215] The accident occurred between 4 and 5 p.m. but at midnight the fire was still burning strongly and had consumed everything around, very dry weather having made everything combustible. A crowd of 400 or 500 people had gathered and were busy fighting the flames. The damage caused was estimated at about $30,000. The King's Powder Mills themselves were a quarter mile away and so were not involved in the explosion.[1216]

It is difficult to believe that such loose shunting of waggons containing explosives was permitted; certainly it would not have been in Great Britain by this time; the powder waggons would have been labelled 'Not to be loose shunted'.

Only 10 months after the King's Mills disaster, on 19th May 1891, another major explosion occurred in the United States, this time near Tarrytown, New York, on the New York Central and Hudson River Railroad. Dynamite, used for blasting during track construction, was the cause of the explosion. The dynamite was carried in the first of the four flat cars which formed the train, 24 cases of it each containing 50 pounds of explosive, 1,200 pounds in all. The force of the explosion was said to have been terrific, the ground trembled like an earthquake, the walls of several houses in Tarrytown were shaken and cracked and showers of window glass fell into the street. The flat waggons

1213 *The Times*, London, England, 2nd October 1882

1214 *Chillicothe Morning Constitution*, Chillicothe, Missouri, United States, 16th July 1890

1215 *The Daily Northwestern*, Oshkosh, Wisconsin, United States, 16th July 1890

1216 *Chillicothe Morning Constitution*, Chillicothe, Missouri, United States, 16th July 1890 were blown to pieces, the engine was mangled, and its tender blown into the river. The known victims killed included Timekeeper John McCarthy, Frank Morrissey, who had charge of the dynamite, Brakeman John Smith and 15 Italian labourers, whose names were not known as they were only known by numbers. A large number of others, including at least a dozen Italian labourers, were injured. Many of the bodies were badly mangled, reminding veterans of a scene from the civil war. Two men who had been fishing on the riverbank at the time were supposed to have been blown into the river and drowned. The track was torn up over a distance of 600 feet and the ground littered with debris. Five of the bodies had been blown into the river and not yet recovered at the time of the report. A panic was caused at the nearby Tarrytown high school where teachers and pupils thought that there had been an earthquake and rushed out into the street. There had been nearly 50 labourers on the train.[1217]

Initially the cause of the explosion was thought to be the ignition of the dynamite by a spark from a passing engine but a couple of days later Edward Finnegan, the foreman of the workmen, explained from his hospital bed that he had given the labourers strict orders not to smoke. The dynamite cartridges were not capped when they left the manufacturers (the blasting caps contained a small amount of primary explosive which set off the main charge. The dynamite itself was safer to handle before the blasting caps were added, hence they were not added in the factory.[1218]) At the accident scene a man was sat on a box adding the caps. He had accidentally dropped one and that caused the explosion.[1219]

The explosion of a powder car between New Hartford and Winsted, Connecticut, United States, on 20th September 1892 was much less disastrous than the previous two accidents detailed, though 12 other waggons of the Philadelphia and Reading Railroad train were piled up in a wreck and two brakemen were hurt.[1220]

Again on 10th October 1893 a waggon carrying over 200 kegs of powder blew up an eastbound Pittsburgh Fort Wayne and Chicago Railroad train shortly after it had passed through North Lawrence, Ohio, United States, fatally injuring Engineer Colvin and Fireman McCowan. A crater 20 feet deep was caused by the blast and the track was damaged over a length of 150 yards. Twenty cars were wrecked, the damage was estimated at $25,000 and traffic was suspended for several hours. The cause of the explosion was not known.[1221]

The deaths of an unknown number of train passengers in Santander, Spain, on 3rd November 1893, were a relatively small part of a much larger catastrophe. Santander is a port on the north coast of the country and on the day of the disaster the cargo steamer *Cabo Machichaco* was being unloaded. The steamer was carrying a general cargo of iron, petroleum, flour, wine and so on, also five hundred cases of dynamite, a fact which its

---

[1217] *Salem Daily News*, Salem, Ohio, United States, 20th May 1891

[1218] Wikipedia – Detonator

[1219] *Newark Daily Advocate*, Newark, Ohio, United States, 23rd May 1891

[1220] *Middletown Daily Press*, Middletown, New York, United States, 21st September 1892

[1221] *The Trenton Times*, Trenton, New Jersey, United States, 10th October 1893 captain had not declared to the port authorities. Had he made it known that his ship was carrying such a large quantity of explosives the steamer would not have been permitted to enter the port. At 3 p.m. a fire was declared in one of the coal bunkers and the 20 cases of dynamite which the captain had ventured to declare were moved a safe distance away. It was soon realised that the fire had taken too strong a hold for it to be extinguished and the decision was made to procure a tug to tow the steamer away from the quay and if possible, out to sea. The operation of getting a tug to the scene took time and meanwhile a large crowd of spectators had gathered to watch the proceedings, some of them scrambling onto the tug.

At 4.30 p.m., as the tug had arrived and was attempting to move the steamer away from her moorings, the fire, which had continued to spread rapidly, reached the petroleum being carried on the steamer, and there were a series of violent explosions, as first the petroleum, then the ship's boilers and finally the 480 cases of dynamite blew up. Steamer and tug disappeared and everyone in the vicinity was killed instantly. As chance would have it a well-filled train had just arrived at Santander station when the explosion occurred, and the train was wrecked and set ablaze and many of its passengers killed.[1222] Six hundred people in all died in the explosion, it is not known how many of them had been on the train.[1223]

The supposed placing in the stove of the smoking car of a Missouri Kansas Texas of Texas Railroad train as it was leaving San Marcos, Texas, United States, of a dynamite cartridge was blamed in a report from Atlanta, Georgia, for causing an explosion which shattered half the car. The blast, on 24th January 1894, instantly killed Ed Binding, a travelling agent for the San Antonio city brewery, and J.C. Heldelhelmer, a prominent merchant from Austin. The other six passengers in the car were seriously injured.[1224] If the explosion was due to dynamite it was not known who was responsible for the outrage or their motives. However, a report from closer to the scene (Galveston, Texas) blamed the occurrence, probably more reliably, on a burst steam pipe.[1225]

A disastrous explosion occurred at Vredendorp (contemporary spelling in *The Times*, now Vrededorp), a poor densely populated suburb of Johannesburg, in the Transvaal, at 3.30 p.m. on the afternoon of Wednesday 19th February 1896. Eight waggons loaded with dynamite were being shunted on the Nederlandsche Zuid-Afrikaansche Spoorwegmaatschappij (Netherlands-South African Railway) when the whole consignment exploded, tearing a hole 30ft deep, 200ft long and 80ft wide. Every building within half a mile of the explosion was destroyed and the damage extended to the principal streets of Johannesburg, some distance off, where there was wholesale destruction of glass and the wrecking of numerous solid masonry structures. The suburb where the explosion took place was part of the 'Malay Camp'. 'Its residents are, or rather were, of the lowest class of coloured people and Boers, who, too poor to live

---

[1222] *The St James' Gazette*, London, England, 6th November 1893
[1223] Webpage – Remembering the Machichaco Tragedy of 1893
[1224] *The Atlanta Constitution*, Atlanta, Georgia, United States, 25th January 1894
[1225] *Galveston Daily News*, Galveston, Texas, United States, 25th January 1894

in a house as we know the term, generally erect a shelter for themselves. It may be of corrugated iron or mud bricks, but mostly of the tin linings of packing cases.' In modern language it was a slum settlement.[1226]

As the explosion had occurred in the middle of a highly populated area the death toll was high, and by 22nd February 67 bodies had been recovered, in addition to which there were two coffins filled with 'remains' – presumably bits of bodies.[1227] In addition to those killed there were many more injured. All the beds at Johannesburg hospital had been filled by the early arrivals, those brought in later had to be lain on the floor and after that an overflow infirmary was organised, while during the evening 100 of the slightly injured had been treated and discharged. The corpses of the dead were laid out in rows in the Wanderers' Athletic Club.[1228]

On 21st February the *Times* correspondent in Johannesburg wrote that a truck load of detonators had not been originally sent with the dynamite trucks but had been shunted alongside them. He added that, 'The Sanitary Board have strict regulations against the keeping of any quantity of explosives in the town, and frequent appeals have been made to the railway company to prevent large quantities of dynamite from being detained in the sidings.' [1229] The Cape Town correspondent added that, 'The dynamite had been imported in bulk and made up into cartridges in Pretoria, passing thenceforward as Transvaal-made dynamite. There were 55 tons on the trucks, with 90 cases of detonators. It was probably the detonators separately exploding which led people to believe that the boiler of the locomotive burst. The Netherlands Company's own regulations are shown to have been broken in several respects.' [1230]

The following day the Cape Town correspondent wrote, 'It is impossible yet to say what is the number of the victims of the Johannesburg explosion, as search parties continue to excavate bodies from a large area. The Cape Government has contributed £1,000 to the relief fund, which now amounts to over £104,000. A great amount will be necessary, as at least 1,000 families are already known to be homeless. A commission of inquiry has been appointed by the Transvaal Government.' [1231]

On 27th April 1896 a nasty accident happened in India, when an explosion occurred in a third-class carriage of the up Bombay mail train between miles 925 and 926, between Sikandarabad and Dadri stations, south of Ghaziabad. The sides and floor of the carriage were blown out by the blast and some of the passengers hurled out of it. The crowded carriage and a postal van were also set on fire by the explosion, forcing 'natives' to jump from the train with their clothes on fire while it was travelling at high speed. The fire also damaged the communication cord, and the train ran a further two miles before it was stopped. One passenger was killed on the spot, he had been the person responsible for the explosion, as he had surreptitiously brought onto the train 'some native bombs, such as are used in the display of fireworks on the occasion of

---

1226 *The Times*, London, England, 22nd February 1896
1227 Ibid, 24th February 1896
1228 Ibid, 21st February 1896
1229 Ibid, 22nd February 1896
1230 Ibid.
1231 Ibid, 24th February 1896

marriage festivals'. Fourteen other passengers were seriously injured, three of whom died before reaching Ghaziabad and seven more in hospital there. Four other injured passengers recovered. 'General Maitland, R.E., who has arrived in India on a special mission connected with recruitment for the Military Works Department, was amongst the passengers.' [1232] [1233]

[1232] Ibid, 30th April 1896

[1233] Administration Report on the Railways of India for 1896-97

CHAPTER 16

BOILER EXPLOSIONS

The railways of the 19th century were powered almost entirely using steam engines, and locomotive boiler explosions were therefore much more common than chemical explosions. Although the explosion of a locomotive boiler could often prove fatal to the engine crew, it was rare for train passengers to suffer and probably because of this, information regarding boiler explosions in countries outside of Britain and the United States has been hard to obtain.

The boiler of a Missouri Pacific Railroad engine exploded on 21st September 1881 between Chetopa and Parsons, Kansas, United States. The very powerful blast wrecked the engine and 10 or 12 cars and killed Engineer George Adams, Fireman Simon Bailey, a John Denny 'and a stranger named O'Neill'. All four were in the engine cab and were blown between 100 and 200 yards and terribly mangled. 'Bailey's head was blown completely off and could not be found.'[1234]

The North Eastern Railway suffered a boiler explosion on 26th December 1881. Although it was Boxing Day goods as well as passenger trains were running and at about 4.30 p.m. engine No. 204, pulling three waggons and a brake van, was drawing up behind a mineral train on the up mineral line at Thornaby Iron Works Junction, half a mile east of South Stockton station, when its boiler exploded violently. The boiler and firebox were thrown forward and landed on the fifth and sixth waggons from the rear of the mineral train, while the brake van of the mineral train was set alight. The driver and fireman of No. 204 were killed outright and the guard and a waggon inspector in the brake van fatally injured.

Major-General C.S. Hutchinson investigated and concluded that the explosion had resulted from the firebox crown sheet having overheated due to lack of water. Thus weakened, it had been unable to resist the steam pressure when water was admitted to the

[1234] Ibid, October 1881

boiler as the engine pulled up behind the mineral train. Whether the lack of water was due to carelessness on the part of the enginemen or to a failure of the single water gauge, it was impossible to say. The water gauge had been checked by a fitter only four days earlier and was then working correctly so it seemed unlikely that it had not been working correctly, possibly the engine crew had mistaken an empty gauge for a full one. The lead plug had fused but not melted due to a hard incrustation on its bottom surface. The inspecting officer thought that there could be no grave objection to the use of such plugs provided that they were renewed at frequent intervals.[1235]

The 4 p.m. Illinois Central Railroad suburban train, consisting of a baggage car and four coaches drawn by engine No. 167, and due in Chicago, Illinois, United States, at 4.20, was running 10 minutes late at 43rd Street on the afternoon of 9th January 1882. When it was a very short distance north of that station the boiler of the engine exploded. A Professor Leslie Lewis told a reporter, "It was running along smoothly, when we heard a dull thud; hot water and steam and several pieces of iron came flying into the car. The occupants made a rush for the door and crushed each other till it seemed they would crowd each other off the platform." The train ran off as if nothing had occurred, and we stopped at 39th Street. The whole boiler was blown off the running gear, which remained undamaged. Pieces of flue and other parts of the boiler were thrown great distances. Some dropped through the carpenters connected with the new building of Mr John Bordens, others landed in the yard of Mr Goodwin, and the smokestack was thrown in the lake. The engine cab remained untouched. Engineer John Glover was slightly scalded and suffered a cut foot and his face was slightly cut by a little glass. Fireman Edward Scanlan was slightly scalded. The door of a car was blown off and the passengers were also slightly scalded, and a brakeman jumped off and hit his head against a brick wall sustaining slight injuries. A popcorn pedlar named Isaac Perkins was badly but not fatally scalded. Everyone in the baggage car was well blackened with soot. The cost of repairs to the engine was estimated at $10,000 and to the car $500.[1236]

A serious accident occurred at Dunbar, Scotland, on the North British Railway, on 1st September 1882. The 8.5 p.m. up goods train from Leith Walk, Edinburgh to Berwick had stopped at the up advanced starting signal where it had taken water and a banking engine had joined its rear for the climb to Grantshouse. It had then set off again and had proceeded about a quarter mile when the boiler of engine No. 465 at the front of the 40 waggon train had exploded violently. A piece of its boiler fell onto the down line and was then run into by the 7.50 p.m. down goods train from Berwick to Edinburgh which was derailed. Driver John Hamilton, Fireman Andrew Melrose and a third man riding on the engine, described variously as a publican,[1237] and a wine and spirit merchant, were all killed.

Major-General C.S. Hutchinson stated that, 'The cause of this disastrous boiler explosion is involved in much obscurity.' There was no sign of corrosion in the boiler, no other defects with it, the lead plug was intact, and it had not run out of water. Driver Hamilton, the inspecting officer remarked, 'had committed two grave offences' in

---

1235 Board of Trade report

1236 *The Locomotive*, February 1882

1237 *The Times*, London, England, 4th September 1882

carrying an unauthorised passenger and in failing to stop at Portobello to take up a full load of waggons, probably to avoid the unauthorised passenger being noticed, but that of course did not explain how the explosion had occurred. Major-General Hutchinson was forced to put forward the 'sudden generation of high pressure steam theory' [1238] which Christian Hewison in his book rejects as meaningless and suggests that there was a problem with the two small safety valves failing to operate properly and being unable to cope with the large amount of steam produced in preparation for the ascent of the bank to Grantshouse.[1239]

The boiler of the engine of a Central Iowa Railroad train exploded on 16th October 1883 as it was climbing a bank four miles north of Oskeloosa, Iowa, United States, killing Engineer Edward Bailey instantly and causing probably fatal injuries to Fireman Charles and Brakesmen Elmer and Blanchard. The boiler was carried forward 150 ft and the wreckage of the train piled 30 ft high.[1240]

The Victorian Railways had a bad start to April 1884. On 2nd April there was the head-on collision on the single line between Little River and Werribee (described in the Single Line Collisions chapter) and the following day there was a boiler explosion near Sunbury. A goods train which had left Melbourne at 2 a.m. on 3rd April 1884 with engine No. 49 and 24 empty waggons was approaching Jackson's Creek viaduct, three-quarters of a mile beyond Sunbury station, when the outer shell of the firebox burst, killing Fireman R. Hutchin and seriously injuring Driver James Harris.[1241]

An investigation into the explosion by a board appointed by the Railway Commissioners concluded that there was no evidence that the steam pressure had exceeded 130 lb/sq. in or that the boiler had been low on water and that the explosion had been due to severe corrosion, furthermore that 'most probably this explosion would not have occurred had there been a regular and systematic method of inspecting locomotive boilers on the Victorian Railways such as we suggest under the head of recommendations.' Other recommendations were that the thickness of boilers should be such that the calculated bursting pressure was five or six times the maximum working pressure, for regular hydraulic testing, for fusible plugs placed so as to melt when an inch of water remained and made of a proper fusible metal, not lead, and for the use of a third safety valve inaccessible to the engine crew.[1242]

On 5th July 1884 the 4 p.m. North British Railway goods train from Balloch to Kipps was just setting off from Balloch, Scotland, and was passing the up platform when the boiler of its engine, No. 417, exploded. The driver and fireman were thrown some distance, the driver being severely scalded, a man named Jack standing on the up platform was fatally injured and died shortly afterwards and another man named William Livingstone stood nearby was struck by a fragment of the boiler and knocked

---

1238 Board of Trade report

1239 *Locomotive Boiler Explosions* by Christian Hewison, pgs. 99-101

1240 *The Locomotive*, December 1883

1241 *The Australasian*, Melbourne, Victoria, 12th April 1884

1242 *The Age*, Melbourne, Victoria, 10th May 1884

unconscious.[1243] The collision was stated not to have been a violent one but the two last named men had been stood close to the locomotive.

Major F.A. Marindin of the Board of Trade reported that an examination of the boiler showed clearly that it had failed along the longitudinal seam or joint at the top, where there was a groove or flaw of long standing. 'The plate was in many places holding by a mere skin, and it is surprising that it had not given way sooner, but from its appearance I have very little doubt but that the flaw would have been detected if the boiler had been subjected to an hydraulic test when it was last in the shops for thorough repair.' Major Marindin added that he understood that new boilers made by the company had a better type of joint and every boiler re-tubed in the last two years had been hydraulically tested. He recommended that all the old boilers should also be tested as soon as possible, and all boilers tested periodically thereafter.[1244]

On 25th July 1884, a Lehigh Valley Railroad shifting engine was running from Whitehaven to Nescopeck Siding, Pennsylvania, United States, when its boiler exploded, killing all four men on the engine; three of their bodies were blown to pieces and could not be found. As no flagman was left to warn an approaching freight train it ran into the wreckage and several of its cars were demolished. The victims were Engineer J.H. Hassell, Fireman Armbruster, Brakeman Hassell (the engineer's son) and Telegraph Operator R.E. Smith.[1245]

What was described as 'an awful accident' happened at the Jersey Shore, Pennsylvania, United States station of the Beech Creek Railroad, 14 miles above Williamsport, on 9th December 1886. Four men lost their lives when Beech Creek engine No. 4, which had been in the workshops at Jersey Shore for repairs, blew up. It had been taken out for a trial trip and run up and down the road between the shops and the Pine Creek station at Jersey Shore, a distance of three-quarters of a mile. Four men were riding on the engine; Engineer P.M. Knight, Fireman Allen Ramsey, Fireman James Wearne and Shop Hand J.C. Field. The engine had stopped on the track midway between the junction and Jersey Shore station when the boiler suddenly burst, and the engine was 'blown to atoms'. Ramsey, Wearne and Field were blown 100 ft into the air and fragments of their bodies were found 800 ft away. Knight was blown 1,500 ft to the right of the track into a field, and part of his body was found against a tree, while the other part was found in a swamp nearby. Not enough remained of the engine for the cause of the explosion to be discovered. Knight, Wearne and Ramsey had families. All lived in Jersey Shore.[1246]

An up-freight train on the eastern division of the Canadian Pacific Railway was involved in accident two miles east of Nepigon, Ontario, Canada, on 28th December 1886. Engineer Ramsay and Fireman Harry Brunell were accompanied on engine No.

[1243] *The Times*, London, England, 7th July 1884
[1244] Board of Trade report
[1245] *The Locomotive*, September 1884
[1246] Ibid, February 1887

269 by Brakeman Frederick and a carpenter named Gordon from Port Arthur. At the point mentioned the crown sheet collapsed and the boiler burst. A fearful rush of steam blew all but the engineer out onto the track, where they must lie in the cold, with the temperature 40 degrees below zero F. All were terribly scalded and frozen. Gordon died two hours later with a broken back. Brunell had one leg broken, and was badly frozen on the face, hands and feet, while Frederick was badly scalded and smashed about the face but able to walk. Engineer Ramsay escaped injury.[1247]

The boiler of engine No. 253 of the Philadelphia and Reading Railroad exploded at Haak's Switch, eight miles west of Tamaqua, Pennsylvania, United States, on 3rd December 1887. Five of the crew of six of the freight train being pulled by No. 253 were on the engine when it exploded, and three of them were killed instantly. While the other two were so badly hurt that their recovery was thought unlikely. Engineer Alexander Walker and Brakemen Albert Gudner and Wallace Ettinger were killed, Fireman Joseph Reifsnyder suffered serious head and internal injuries and Conductor David Pfleger was badly scalded and broke his shoulder. The body of Gudner was found in the woods 85 yards from the engine; he had struck a tree which had crushed the entire back of his head but was probably already dead from the effect of the explosion. All the bodies were found 20 to 50 yards from the wreck. The engine was left more a pile of scrap iron than a locomotive. Gudner and Walker left large families. Ettinger was only 20 years old and had only been on the railroad a short time. The injured men were both married. The cause of the explosion was unknown. The locomotive was an old camel-back which had been remodelled and at the time of the accident was pushing a train of empty coal cars. It was said to be in an extremely bad state of repair.[1248]

Two days after the Haak's Switch boiler explosion, on 5th December 1887, there was one at Stellarton, Nova Scotia, Canada, on the Intercolonial Railway, which killed William Eastwood, Daniel Robertson and Alexander Frazer and seriously injured Fireman Alexander Murray. The three men who were killed had been on their way to work. The conductor and engineer had stepped into the telegraph office only two minutes before the explosion happened and thus escaped injury. The engine was wrecked, large sections of it being thrown a good distance. The station building was badly damaged. The noise of the explosion could be heard over 20 miles away.[1249]

In a spectacular boiler explosion at Strømmen, Norway, on 22nd December 1888, engine No. 11, the train engine of a double-headed goods train, was thrown up into the air by the force of the blast and landed upside down on the cab and tender of the pilot engine of the train, No. 36. Five men were slightly injured in the accident. The implied cause of the explosion was the jamming of the safety valve with a block of wood by the engine driver in order to obtain a higher steam pressure to haul the train up a gradient. The safety valves were easily accessible on the class of engine involved (Litra A). The engines had been taking water when the explosion took place. Engine No. 36 was little damaged in

[1247] Ibid.
[1248] Ibid, February 1888
[1249] Ibid.

the accident and it was towed to Kristiania (now Oslo) for repairs, still with engine No. 11 on top of it. It continued in traffic for many years, while No. 11 was scrapped after the accident and used for spare parts.[1250] [1251]

As already noted in the Tunnel Accidents chapter, the little Mediterranean island of Malta had an accident on its sole seven-mile-long railway line on 20th March 1889 when the boiler of engine No. 4, which was reportedly in poor condition, exploded inside Floriana Tunnel, without it would seem any very serious consequences.[1252]

At the beginning of 1890 a new branch was being built from Warragul to Neerim, Victoria, and at about 5 p.m. on the 7th March an engine owned by Mr Ahern, the contractor, brought a train of about 30 workmen along the line. The trucks carrying the workmen were then detached and the engine proceeded light to Muddy Creek station. It had been standing there for about a quarter of an hour when its boiler exploded, throwing its dome, weighing two and a half tons, 300 yards, and its outside tanks, weighing two tons, 50 yards. No one was killed outright but Mr Ahern himself, a lad named David Chautem who had been standing nearby, Fireman Wallace Williamson and an employee named F.M Andrew were all scalded, and pieces of iron broke the windows and fell onto the roof of a government inspector who lived in a gatehouse close by. The driver of the engine, John Mundy, had been stood about 50 paces from the locomotive and escaped unhurt.[1253] Chautem, the most severely scalded of the victims of the accidents, died from his injuries on Sunday 16th.[1254]

At 1015 a.m. on the morning of 2nd November 1890 the boiler of engine No. 44 of the Werra Eisenbahn exploded at Schleusingen, Germany, as it was about to work a train to Themar, injuring three railwaymen and damaging several buildings. The cause of the explosion was not determined but it was thought likely to have been due to faulty material or construction of the boiler.[1255]

A boiler explosion occurred near Oxmoor, Alabama, United States, 17 miles west of Birmingham, on 17th January 1892, on a line owned by the DeBardeleben Coal & Iron Company and leased by McNamara Brothers. An engine ran out of water when running between Eureka mines and Oxmoor, and the engine crew tried to reach a tank half a mile away as quickly as possible. The boiler exploded while the engine was running at full speed, wrecking it and 'blowing Engineer Joseph Hunt and Fireman Bradford into a thousand pieces'. The track was torn up and a large hole made in the ground.[1256]

1250 *NSB Teknikk Årgang* 5 1979 No. I, pgs 63-4
1251 Webpage – Jernbane.net
1252 *The Malta Railway* by Joseph Bonnici and Michael Cassar, pgs. 52 and 136
1253 *Traralgon Record*, Traralgon, Victoria, 11th March 1890
1254 *Great Southern Advocate*, Korumburra, Victoria, 21st March 1890
1255 Wikipedia – Bahnhof Schleusingen
1256 *The Locomotive*, March 1892

On 28th January 1892 a boiler explosion occurred near St Clair, Pennsylvania, United States. The locomotive was demolished and fragments of it thrown in every direction. Joseph Zeigler, Napoleon Paul, Jacob Turner, William Winterstein and Harry Sands were all killed. Paul's remains were found 300 ft away in a creek. Turner's body was 500 ft away across the valley; his head was crushed.

Winterstein was found after several hours' search halfway up a mountain, 400 ft above the wreck. Zeigler and Sands were thrown 200 ft up the bank along with the firebox. Trees were uprooted and the hillsides were lined with debris.[1257]

On 17th September 1892 engine No. 2219 of the Kaiserlich-Königlichen Staatsbahnen (Austrian State Railway) blew up between Kalwang and Wald, Austria, due to low water, killing the engine crew. The engine was thrown sideways eight metres.[1258]

Philadelphia and Reading Railroad engine No. 563, pulling a heavy train of empty coal cars, blew up at Conner's Crossing, near Schuylkill Haven, Pennsylvania, United States, on 14th November 1892, killing five men and fatally injuring a sixth. Those killed were Engineer Henry C. Allison, Fireman William Mackey, William Cowley, William Kendrick and William Moyer. Brakeman Michael Dobbins was badly scalded, and it was thought would probably die. The engine had been running northbound. William Cowley, with his crew, had brought up a train of empties and after running them into the Cressona yards had boarded engine No. 563 at the Mine Hill Crossing to reach their homes, and, as was customary, had ridden in the engine cab. Dobbins, the only one who had escaped instant death, had been crowded out of the cab and had taken a seat on the tender. The men had barely been on the engine two minutes when the explosion occurred. They were thrown into the air in all directions and killed instantly. The boiler and firebox were blown off the trucks and the line was so badly damaged that traffic was considerably delayed. The north and southbound midnight Buffalo trains were compelled to run via the Little Schuylkill branch from Tamaqua to Port Clinton.[1259]

Four persons died, and several were injured when the boiler of Philadelphia and Reading Railroad engine No. 948 exploded at Lebanon, Pennsylvania, United States, on 13th May 1893. Those who died were Levy Yocum, John Yocum, George Solliday and Willie Loudon. William Anspach was fatally injured, and his daughter Annie, who had just brought her father's dinner, was seriously and possibly also fatally injured. George Seller and Isaac Beard were also seriously hurt and several people half a block away received minor injuries.[1260]

A boiler explosion on the Marietta and North Georgia Railroad at Hiawassa station, Tennessee, United States, on 20th June 1894, killed Fireman James Devers outright, and badly injured Engineer J.C. Devers, baggagemaster J.C. Sanger and Brakeman A.D.

[1257] Ibid.
[1258] Wikipedia – Kesselzerknall
[1259] *The Locomotive*, January 1893
[1260] Ibid, July 1893

Bentley, Sanger being fatally hurt. 'The accident occurred at the foot of a mountain, where the large pushing engine of which J.C. Devers was engineer was coupled up to a southbound train to pull it over the mountain. The engineer of the train had detached his engine and gone over the mountain, and young Devers had come off the siding and backed the pushing engine onto the train. Bentley and the elder Devers were on the ground making the coupling, and Sanger was in the baggage car. Young Devers was hurled into the car, the front end of which was torn away by the explosion, and without doubt was killed instantly.' Sanger was fearfully scalded and died the next day. The elder Devers was thought to be fatally injured but was still alive at the time of the report. One of the boiler sheets was supposedly known to be cracked.[1261]

The boiler of a Canadian Pacific Railway consolidation engine exploded near Field, British Columbia, Canada, on 30th July 1894, while it was pushing a freight train up a steep grade known as "the hill". The locomotive was blown to pieces, killing instantly Engineer Wheatley and Fireman Hunt. Brakeman George Kemp on the rear car was fatally injured by flying fragments of the boiler and died on 1st August.[1262]

Kaiserlich-Königlichen Staatsbahnen (Austrian State Railway) engine No. 5668 suffered a boiler explosion at Prinzersdorf, Austria, Austria-Hungary on 21st March 1895, due, like the 1892 accident, to low water, and again the engine crew were killed. The derailed engine continued for 60 metres along the track before ending up against the side of the cutting.[1263]

The boiler of a Lehigh and Hudson Railroad locomotive exploded on 10th November 1895 between Sugar Loaf and Lake station, near Warwick, New York, United States, while the train it was pulling was in motion. The main part of the boiler was thrown 50 ft into the air and fell 25 ft from the track, leaving the engine trucks uninjured, so that the train ran about a mile and a half before it could be stopped. Conductor Martin O'Neill and Engineer William Cooper were dead when found, and Brakeman James L. Sloane was unconscious and died very shortly afterwards. Fireman Herbert Beetner was found walking along the line, with most of his clothing torn off. He was dazed, and said that he remembered going through the air, and found himself in a field, 50 yards from the track. He lost consciousness soon after he was found and died a few hours later.[1264]

Two days after Atchison Topeka and Santa Fe Railroad eastbound train No. 2 had been attacked by train robbers in New Mexico, its engine suffered a boiler explosion at Peterton, near Osage City, Kansas, United States, on 4th October 1896, and was wrecked, resulting in seven deaths. Killed as a direct result of the explosion were Engineer George Strunk or Strump, Fireman Henry Holister, William McAdams, Artur Smith and two other unidentified men. McAdams, Smith and the unidentified victims were all tramps

1261 Ibid, August 1894
1262 Ibid, September 1894
1263 Wikipedia – Kesselzerknall
1264 *The Locomotive*, January 1896

who had been riding on the baggage car. M. Purcell, J.B. Stone, H.S. Fowler, WF.F. Evans, F.A. Tiffany, Mrs Edna Maxwell, Wilford Burns and James Coleman, presumably all passengers, were injured. A newspaper report stated that, 'The wreck was marked by scenes of the wildest confusion among the terrified passengers. The nerves of many were at a high pitch as a result of their experience with the road agents in New Mexico, and when the crash came the first impression of nearly all was that the train had been attacked again by robbers.' The accident proved too much for one passenger, a William Beckler, bound for Chicago, who in front of a coach full of passengers drew a pistol from his pocket and shot himself dead. He had been drinking heavily.[1265] Fears by the passengers that the wreck was due to dynamite having been placed on the track were refuted by the railway officials, who insisted that it was a boiler explosion, pure and simple.[1266] The explosion was blamed on the engineer, who should have stopped for water at Osage City but had decided to push on to the next water tank because he was running late.

Eastbound Chicago and Alton Railroad freight train No. 101 had stopped at First Street and Grand Avenue, Kansas City, Missouri, United States, on 4th May 1897 when its boiler exploded, killing instantly Engineer H.H. Roberts, and injuring Fireman Charles Pardoner so severely that he died shortly afterwards. A dozen men standing on the platform of the station had narrow escapes from death. The boiler had failed by the blowing of the crown sheet. At the coroner's inquest on 6th May a number of experts testified that they had examined the wrecked boiler and had found that most of the stay bolts were badly corroded. The verdict was that the deaths were caused by "defective parts of the boiler". [1267]

The boiler of the locomotive of a train from St Johns to Harbor Grace exploded near Freshwater Valley station, Newfoundland, on 12th June 1897. Engineer Glasgow was killed instantly, and Fireman Byrne was badly injured. The explosion occurred while the train was in motion, breaking the coupling which connected the engine with the cars. Glasgow's body was thrown some distance into the air. Little damage was done to the cars, which did not leave the track, and none of the passengers was injured.[1268]

On 21st June 1897 the boiler of a freight engine on the Ferrocarril Internacional Méxicano (Mexican International Railroad) exploded three miles outside of Ciudad Porfirio Diaz, Mexico, demolishing the engine and killing Engineer James Griffin and Fireman Jacob Mann. The big mogul engine was thrown 50 ft and turned completely over. The train was coming slowly across a bridge, and the engine had just reached the other side when the explosion occurred. The cause of the explosion was unknown. The body of Griffin was found about 100 ft from the track. The fireman was blown into the coal tender. No one else was hurt, and only one car was wrecked, the rest of the train remaining standing on the bridge.[1269]

[1265] *Fort Wayne News*, Fort Wayne, Indiana, United States, 5th October 1896
[1266] *The Locomotive*, January 1897
[1267] Ibid, July 1897
[1268] Ibid, August 1897
[1269] Ibid.

As the 3 p.m. Londonderry to Belfast train of the Belfast and Northern Counties Railway was approaching Antrim, Ireland, on 7th August 1897 at about 5.58 p.m. one side of the firebox of engine No. 58 collapsed and there was a rush of steam and water onto the footplate. Fireman Cornelius McAllister was thrown off the engine a quarter mile from the station and killed and Driver James Turner was found dead on the footplate when the train stopped. No one else was hurt and the guard stopped the train with the automatic brake when he found that it was running past the platform.

Major F.A. Marindin, in a long report, delayed by his illness, found that the accident was due there being a large number of broken copper stays in the firebox, 19 of 30 of which in one group on the right-hand side of the firebox were broken. He concluded his report by stating, 'I think that it can hardly be doubted that a critical examination of the fire-box, such as is made by the boiler examiner, would have detected the faulty condition of the right-hand side, if it had been made shortly, or even some considerable time, before the accident took place. The only way to guard against such accidents is, evidently, to have frequent examinations, more worthy of the name than the present superficial ones, and it is satisfactory to learn that on the Belfast and Northern Counties line the boiler examiner is now ordered to make a proper examination of all boilers on every shed day – that is, once a week.' [1270]

The boiler of Canadian Pacific Railway locomotive No. 354 exploded at White Creek bridge, 12 miles west of North Bend, near Vancouver, British Columbia, Canada, on 30th September 1897 while the engine was running at full speed. Brakeman George Elson was thrown down a steep embankment and killed almost immediately. The engineer and fireman were scalded but escaped death.[1271]

The boiler of engine No. 70, engaged in shunting at Brisbane Roma Street, Queensland, exploded shortly after noon on 29th August 1898. Driver Maclaren and Fireman Easton escaped with their lives but were badly scalded. The engine was severely damaged, and a shed roof was also damaged.[1272] The inspection of the wrecked engine disclosed an old crack in the boiler plate, the failure of which was presumably the cause of the explosion.[1273]

Another boiler explosion occurred at Brisbane Roma Street, Queensland, on 2nd December 1898. This time involving engine No. 62, which had just come from the engine shed and was about to pick up the carriages to form a train to Gladstone. Driver John Marshall was thrown back onto the tender but escaped with bruises and shock and Fireman Alfred Webb was unhurt. The greater part of the top of the boiler was blown completely off, with pieces of iron flying in various directions, some severing the telegraph wires. Locomotive Engineer Horniblow examined the engine and was 'of

[1270] Board of Trade report
[1271] *The Locomotive*, November 1897
[1272] *Telegraph*, Brisbane, Queensland, 30th August 1898
[1273] *The Week*, Brisbane, Queensland, 2nd September 1898

opinion that the accident was due, as in the previous case, to a fracture in the boiler plate. There was no indication of corrosion in the iron to suggest that the explosion was due to a worn condition of the plate. He could not say when the boiler was last inspected, but full inquiries would be made as to that, and to all the circumstances of the accident.' [1274]

A subsequent report by the Railways Commission found that both boilers had exploded because of longitudinal grooving along the lap joints coupled with a failure to undertake regular boiler inspections by a railway which was stretched by heavy traffic and insufficient resources.[1275]

Driver David Biddelph and Fireman Thomas Marra were killed in the explosion of the boiler of an engine at Zeehan, Tasmania, on 16th May 1899. A board of inquiry appointed to investigate the cause of the explosion came to the conclusion that it was caused by 'the presence of some explosive substance in the fire-box or ash pan of the boiler', and thus, not a proper boiler explosion at all, but this conclusion was questioned in a letter to a newspaper from 'LUCEM ALBUM' who suggested that it was more likely due to a faulty safety valve,[1276] and another newspaper wrote that the verdict of the board had been ridiculed by local engineers.[1277]

A disastrous boiler explosion occurred at the Cardenas roundhouse, near Tampico, Mexico, on 15th August 1899, killing Engineers Felix Limer, L. Fitzgerald, J. Hussey and William Gibson, together with three fireman and wood passers whose names were not given. An engineer named Lockhart was blown 90 ft and fatally hurt and two Mexican shop employees were described as also being fatally hurt, making a total of 10 fatalities.[1278]

The 7.15 a.m. down Great Eastern Railway goods train from Ipswich to Felixstowe had just drawn up at the home signal at Westerfield, England, on 25th September 1900 after dropping off a waggon when there was a violent explosion and its boiler was thrown onto the down platform, 40 yards from the engine, after knocking down a telegraph pole, a signal post and a gatekeeper's hut. Driver John Barnard and Fireman Macdonald were killed and a policeman and a newspaper boy standing nearby were slightly hurt.[1279]

Lt-Col P.G. von Donop investigated the accident assisted by Mr Thomas Carlton of the Board of Trade Marine Department. They found that the stays of the copper firebox had failed, as an experiment these were of bronze rather than copper and had not been properly maintained by the boilersmiths at Ipswich, unused to dealing with stays made of such material, and who had not been given proper instructions.[1280] [1281]

[1274] *The Brisbane Courier*, Brisbane, Queensland, 3rd December 1898
[1275] Ibid, 26th January 1899
[1276] *Launceston Examiner*, Launceston, Tasmania, 27th June 1899
[1277] *The Argus*, Melbourne, Victoria, 1st June 1899
[1278] *The Locomotive*, January 1900
[1279] *The Times*, London, England, 26th September 1900
[1280] Board of Trade report
[1281] *Locomotive Boiler Explosions* by Christian Hewison, pgs. 107-9

CHAPTER 17

VANDALISM

'Accidents' due to vandalism and sabotage have always been a problem for the railways. They are obviously very closely related, inasmuch as both represent deliberate attempts to cause trains to crash, as opposed to accidents caused through non-deliberate carelessness or negligence or purely accidentally. Here they will, however, be treated separately, as they are slightly different. This separation is perhaps a little artificial, and if the perpetrator of a given act is not caught or his or her motives not sufficiently obvious it may be difficult to decide which it was, but it has nevertheless been felt that the distinction is worth making.

Vandalism is usually carried out by either children or young adults and its results are most commonly, though not always, less serious than those following from acts of sabotage. The motives for train accidents resulting from vandalism are usually either simple curiosity, in the case of younger children – 'I wanted to see what would happen' – or in the case of older children and teenagers a reflection of general anti-social behaviour; so that such individuals who attempt to derail trains also carry out other acts of 'mindless vandalism', and perhaps also commit burglaries or carry out arson attacks. If the person carrying out an act of vandalism is an adult they are often found to be of low intelligence or mentally deficient.

Sabotage on the other hand is usually undertaken by adults, who employ methods more likely to cause death and injury, so that while a vandal might for example just place an object upon the track or jam a stone into a set of points, a saboteur would be more likely to employ methods such as blowing up a section of track using explosives.

The 6.39 p.m. up departure from Guildford, England, on Saturday 16th May 1881 was passing Worplesdon signal box when it struck some obstructions on the track. The London and South Western Railway passenger train was not derailed but its carriages were jolted sufficiently violently for some of the perhaps more nervous passengers to scream out in alarm. The train was brought to a halt and large pieces of iron were found to have been placed at intervals along the line.[1282] It clearly did not take the police long to find the culprit for on the following Saturday William Steere, aged 12, appeared at the Guildford County Court, where Driver Joseph Back gave evidence of running into the obstructions. He said that having stopped and examined his engine he had continued to Woking, where he had reported the matter. Henry Glover, the Woking stationmaster, produced in the court three large pieces of iron – portions of disused rails – found at the spot named by the driver and which bore distinct marks of wheels having passed over them. Mr Barker, the deputy chief constable of Surrey, said that the prisoner, on being told what he was being charged with, had said after some hesitation that, "I put iron on the metals and waited in the hedge to see what would happen." The boy was committed for trial.

Cases of trains running into obstructions placed on the line seem to have occurred with remarkable frequency, at least in Great Britain. Few accounts of such vandalism have come to light from other countries, almost certainly not because there were none but because they were not serious enough to be reported internationally. Another typical act of vandalism occurred in England on 14th August 1882, when the engine driver of the 9.15 p.m. Great Western Railway train from Dudley to Old Hill reported to the stationmaster at the latter place that he had felt a bump as he was crossing a bridge near Wright's Lane. Proceeding to the spot, the stationmaster discovered a plank, which had clearly been struck by the engine, and several bricks lying across the metals. The stationmaster removed the obstructions before the arrival of the next train, which was almost due. 'At present,' the *Times* reporter continued, 'three little boys are suspected, and it is expected that proceedings will be taken against them.' [1283]

When a train ran over two fishplates near Hanley, England, on 29th June 1884, a lenient view was taken of the matter because the perpetrators were three children aged only between five and seven. It was decided that they had been only playing and had no malicious intent. Seven was in any case the legal age of criminal responsibility.[1284] They were, however, 'brought before the Bench and severely admonished.' [1285]

A man named George Wilson was committed for trial by Nottingham magistrates on a charge of having placed no fewer than 64 separate obstructions on the railway over a distance of half a mile at Elton and Orston, England, on the Great Northern Railway, on 17th February 1886. The obstructions included a sleeper and masses of stone and brick. A goods train ran into them, was severely shaken and its engine was damaged. He confessed to the offence.[1286] At the Nottingham Quarter Sessions on 5th April Wilson, a butcher aged 25 was sentenced to six months' imprisonment. His defence counsel stated that he was 'of weak intellect'. [1287]

[1282] *The Times*, London, England, 18th May 1881
[1283] Ibid, 23rd May 1881
[1284] Webpage – English Legal History
[1285] *The Times*, London, England, 1st July 1884
[1286] Ibid, 26th February 1886
[1287] Ibid, 6th April 1886

A Mr John Gunn, standing in the garden of his house at Hounslow, England, shortly after 5 p.m. on Monday 2nd May 1892, saw four youths playing about near a nearby level crossing known as Murray's Crossing. He then saw them place something, he could not make out what, on the line in front of the 5.33 and 5.40 London and South Western Railway trains, and heard those trains crush the objects as they passed over them, whereupon he at once informed the Hounslow stationmaster, who proceeded down the line and found the youths busy placing further objects on the rails. The objects were later found to be brick-bats and clumps of concrete and cement which had been left lying about the lineside where a new siding had just been put in. A 'watchman's' (probably a fogsignalman's) fire pail was also put on the line. George Cooper, 17; Harry Stevens, 16; Edward Newman, 14; and Albert Cooke, 14, were arrested and committed for trial at Brentford Police Court two days later. When arrested the elder prisoners attempted to throw the blame on to the younger two boys.[1288]

What was described by *The Times* as 'A boy's freak' ended in a very painful punishment for nine-year-old Harry Perkins of Bedford, England. He interfered with a set of points at the London and North Western Railway station on 12th August 1897, derailing several waggons of a goods train, damaging two, and caused a brakesman and a shunter to jump for their lives. The Bedford magistrates ordered him to receive six strokes with the birch.[1289]

'Train Wrecked for Revenge' was the headline in the *Philadelphia Times* of 17th May 1899. Train No. 4 of the Philadelphia and Reading Railroad, consisting of an engine, combination car and day coach, had been travelling at high speed about three miles from Tamaqua, Pennsylvania, United States, when it encountered a spike placed on the rail on Longacres Curve. The train was thrown from the track just as a northbound coal train was passing on the other line and the derailed passenger engine struck the coal train, throwing several coal trucks off the line in turn. Engineer Samuel Grier of the passenger train was killed instantly, and Fireman John Short of the coal train was fatally injured. Fireman Lott Kerstetter of the passenger train was frightfully scalded, but it was hoped that he would recover. Two passengers were also injured. Suspicion soon fell on an Oliver Ohl; he was arrested and confessed to having put the spike on the line. According to the *Philadelphia Times* Ohl was 35 years old and said that he had committed the deed as an act of revenge against the railroad company. However, other newspapers stated that he was in fact a youth of only 17, and had placed the spike not out of maliciousness, but just to see it flattened by the train. He had been working in a timber yard adjoining the line. The *Lebanon Daily News* commented, 'Placing nails or spikes on the railroad is a boy's trick, but a very bad one. No one should ever do it. This boy's thoughtless act killed the engineer, maimed two firemen, injured others and wrecked two engines and several cars. Unintentionally he brought sorrow into many homes. The engineer leaves a widow and seven children.' Whether this incident was vandalism or sabotage clearly depends on the

---

1288 Ibid, 4th May 1892
1289 17th August 1897

age and motive of the person involved.[1290] [1291] [1292]

Finally on 16th September 1900 the little Malta Railway seems to have suffered its only recorded act of vandalism when the 7.55 p.m. train from Notabile ran into a pile of stones heaped onto the rails near San Salvatore station. The train was not derailed, and no one was hurt, but £3 12s worth of damage was caused. A reward of £50 was offered for information leading to the arrest of the culprits but no one was ever apprehended.[1293]

1290 *Philadelphia Times*, Philadelphia, Pennsylvania, United States, 17th May 1899
1291 *Lebanon Daily News*, Lebanon, Pennsylvania, United States, 17th May 1899
1292 *Train Wrecks of Eastern Pennsylvania* by Charles J. Adams III and David J. Seibold, pgs. 139-40
1293 *The Malta Railway* by Joseph Bonnici and Michael Cassar, pg. 171

CHAPTER 18

SABOTAGE

Four distinct motives for sabotaging trains may be distinguished (after Robert B. Shaw):

a) As an aid to intended robbery, either of the train passengers or to steal goods being carried by the train for example bullion.

b) To satisfy a personal grudge against the railway company, such as by an ex-employee who feels that he has been wrongly dismissed

c) To hinder the operation of the railway, as for example during a strike

d) As an act of guerrilla warfare or terrorism, perhaps against an occupying enemy power. Sabotage against the railways was a feature of most 19th-century conflicts from the American Civil War onwards. In the period under consideration in this volume this type of sabotage was employed fairly extensively in the campaign for the liberation of Cuba from Spanish rule and a little later during the Second Boer War in Southern Africa, where it was used against the British.

Tunisia, in North Africa, was formerly a province of the Ottoman Empire, but in early 1881 it was invaded by France, who established what was nominally a protectorate, but was effectively a colony,[1294] this being part of the European Scramble for Africa, as the 19th-century colonisation of the continent came to be known. As might be expected there was resistance to the French occupation, and on 30th September 1881 insurgents attacked the station staff at Oued Zergha after trains from Tunis and Ghardimaou had crossed there, killing the stationmaster and 10 other staff (who were in fact mostly Maltese and Italians rather than French) and burning down the station and adjoining houses. When the afternoon train from Tunis arrived at Medjez-el-Bab, the preceding station, smoke rising from the area of Oued Zergha and the fact that the telegraph wires were down indicated that something serious had happened down the line, and the train was shunted into a siding there. Meanwhile an opposing train from Ghardimaou had passed through Beja, the last station before Oued Zergha from the western side, and was continuing its journey.

[1294] Wikipedia – French Conquest of Tunisia

When the latter train arrived within a few kilometres of Oued Zergha, travelling slowly, it was found that the rails had been lifted at different points. The damage was repaired, and the train continued its journey, as 'no Arabs were anywhere to be seen'. The track was broken up again further on, and this time the train was derailed, despite its low speed. No one was hurt and the train was then abandoned, and the passengers, railwaymen and the train's escort walked to Oued Zergha. There 'a frightful spectacle met their eyes. The stationmaster had been burned alive and the railway officials and servants had been massacred.' The party continued along the line to Medjez-el-Bab and the same night the insurgents blew up the engine of the abandoned train and burnt its carriages.[1295]

Upon receiving news of the massacre at Oued Zergha the French General Logerot despatched troops to the area. A train set out from Beja for Medjez-el-Bab and 'on quitting the first tunnel from Beja they found the summit of the defile covered by Arabs, who commenced firing into the train. The rails were found to be covered with stones, and the woodwork of all the bridges was in flames. The French soldiers steadily returned the Arab fire till they arrived at kilometre 95, when the train left the rails amid the shouts of the insurgents, who began rushing towards it. Colonel de Bord then ordered Commandant Quelfucci to attack the heights, and the Arabs were successfully dislodged from their position with some loss.' The fighting continued until dusk, when the Arabs stopped firing, and the French soldiers withdrew to Beja.[1296]

Robbery was the supposed motive for the derailment of a train near Woodsville, New Hampshire, United States, on the Boston, Concord and Montreal Railroad 21st August 1882. The fast *White Mountain Express* was running around a curve at Dark Hollow on a 20-foot-high embankment when engineer George Peebles saw an obstruction on the track. He braked and reversed his engine, but the locomotive struck the object, a chain, and the train went down the embankment. The engine and cars were badly damaged, and several people were said to be 'severely but not seriously' hurt.[1297]

The derailment of a troop train between Picasco and Inca in Chile on 6th October 1882 took place during the War of the Pacific which was in progress between Chile and an alliance of Peru and Bolivia. 'Torpedoes' laid by Montoneros (cavalry militia) derailed part of the train and killed seven or eight soldiers. Also riding on the train was the Italian consul at Picasco, who was captured by the Montoneros, who demanded $100,000 for his release. An Italian man-of-war was reportedly dispatched in an attempt to secure his release.[1298]

On Tuesday 13th November 1883 Ferrocarril Nacional de México (Mexican National Railroad) train No. 10 left Saltillo for Nuevo Laredo on time. All went well until just after 7 p.m. when it came to a sudden halt about three miles north of La Jarita. Going forward from the rear coach to find what had happened Conductor Burt found that the

[1295] *The Times*, London, England, 5th October 1881
[1296] Ibid, 8th October 1881
[1297] *The Oshkosh Northwestern*, Oshkosh, Wisconsin, United States, 22nd August 1882
[1298] *The Gazette*, Earlville, Illinois, United States,14th October 1882

engine was lying in the ditch badly damaged, and that the whole train other than the rear car had left the track. The conductor found that the fishplates had been removed and the rails shifted. He reached the front of the train and found the engineer, Madden, who said that he was badly hurt and that the fireman, Charles Festler, had been killed. Some passengers had by now alighted and also came forward to see what had happened when 15 or 20 masked horsemen appeared out of the bush and ordered everyone to get back onto the train. Having got the passengers to re-embark the horsemen ransacked the baggage car, then called for the conductor to open the safe, or they would kill every goddam American on the train. Mr Burt was unable to do so, even with an axe. Eventually the robbers decided to take the safe, which was small, with them, and searching for the train for further safes they found some silver bullion. They also took the small amount of money that the conductor had on him, but missed his watch, which he had hidden in his shoe. After about three quarters of an hour they fled southwards, having stolen about $8,000. Leaving the passengers with Brakeman Beikerch, the conductor and two other men then set off for help and reaching Sanchez station flagged down a train coming towards them, which turned out to be carrying railwaymen and soldiers. The explanation for the appearance of the second train was that train No. 10 had been reported as being on time at La Jarita and when the telegraph had failed 10 minutes later it was suspected that it had been attacked, especially as the authorities had discovered that an attack was planned on the pay train, due to go out the next day. The second train had thus been despatched to investigate. The leader of the gang of robbers, who spoke very good English, was believed to be an American.[1299]

Less than a fortnight after the Mexican sabotage a Southern Pacific Railroad train was attacked near Gage, west of Deeming, New Mexico, United States, on 24th November 1883. Again the motive was robbery. The robbers, described variously as three masked men and as cowboys, removed a fishplate and spread the rails, throwing the engine, mail car, a coach and half of a sleeper car from the track. They then fired about 20 shots at the train, and Engineer Webster was shot dead. The fireman was also shot several times but ran and hid in the sage bush. The 19 passengers hid their valuables but were not robbed, except one who left the train, who was relieved of $150, while the conductor was robbed of $200. Shots were fired at the mail car, and the mail was ransacked but nothing was taken. In all the robbery only netted the gang between $600 and $1,000. Had they stopped the train the night before they could have secured nearly $10,000. The robbers remained at the scene about an hour, until it was dark, before making their getaway. The rear brakesman ran back five miles to Gage station, from where he telegraphed to Deming, and a special train carrying 100 citizens was despatched from Deming and returned with the 19 passengers, the mail and the body of the dead engineer.[1300]

Large posses set off in pursuit of the robbers and the Wells Fargo Express Company and the railroad company offered a reward of $2,500 for the capture of each robber or $10,000 for the whole gang. 'It is said John Price, the notorious Mexican desperado, was

---

[1299] *Laredo Times*, Laredo, Texas, United States, 18th November 1883

[1300] *Galveston Daily News*, Galveston, Texas, United States, 27th November 1883 chief of the gang.'[1301]

The Little Miami division of the Panhandle (Pittsburgh Cincinnati Chicago and St Louis) Railroad suffered a serious act of sabotage at the Miami River bridge, eight miles east of Dayton, Ohio, United States on 10th April 1884, described as 'the most complete wreck that has occurred here in years.' A rail was displaced at the west end of the bridge. The engine and tender went through the bridge into the river and the baggage car, two coaches and a sleeper rolled down the 20-foot-high approach embankment. Engineer John Thomas and Fireman John O'Connell of Xenia went down with the engine. The fireman was killed outright, the engineer had died by the next day and baggagemaster George Cooper and Express Messenger Smith were said to be fatally injured. Robbery or a grudge against the railway company were put forward as motives. It was believed that a 35-year-old man seen lingering about Beaver station, near the bridge, was the culprit. He had broken open a tool shed to obtain equipment to damage the track, the tools later being found hidden by leaves in a woodshed.[1302]

A very serious act of railway sabotage, if sabotage it was, occurred in Spain on 27th April 1884, on the Badajoz line in Spain. A passenger train from Badajoz was derailed on or at the approach to a bridge over the Alcudia River between Chillon and Almadenejos stations, and almost the whole train fell into the river. The initial report spoke of twenty dead;[1303] the final figure was 59 dead and a similar number injured. Most of the passengers on the train had been soldiers on furlough, but there were also some peasants. The accident took place in an almost uninhabited district. Whether the train was derailed deliberately or whether it was an accident was the subject of much controversy. The *Times* correspondent in Madrid wrote on 28th April that 'the positive statements made last night by the Madrid Press...that the catastrophe...was undoubtedly due to the criminal work of political partisans, hostile to the existing form of government, afford another example of the disposition to exaggerate and jump to conclusions... up to the present, no absolute proof has been received that such a horrible and useless crime as the intentional destruction of a bridge over which a train full of harmless people was to pass had been contemplated or carried out...'[1304] The Madrid correspondent continued in the same vein in a piece written on 1st May 'The bridge had no strategic importance, it was known to be in a bad state, previous to the late heavy rains, and it is clearly proved that it was not cut or blown up. No insurgents have, previously or since, been seen in the district...'[1305] but on the same day added that insurgent bands had destroyed a railway bridge between Gerona and Figueras, while on 3rd May it was reported that insurgents had 'pulled up the metals on the Fluvià railway bridge in the same manner as was done at the Alcudia bridge.'[1306] Finally an official report from the Minister of Public Works following an examination of the scene by an engineer stated that 'the engineer assures me that he has found indications that the rail on the side upon which the train fell was either removed

1301 *Newark Daily Advocate*, Newark, Ohio, United States, 27th November 1883
1302 *The Daily Gazette,* Colorado Springs, Colorado, United States,12th April 1884
1303 *The Times*, London, England, 28th April 1884
1304 *The Times*, London, England, 29th April 1884
1305 Ibid, 2nd May 1884
1306 Ibid, 5th May 1884

from its place, or was without spikes, which might be the origin of the train leaving the line.'

The violent rape of a Mrs Jennings, the elderly wife of a prominent white man, led to an act of sabotage near Albany, Georgia, United States. The rapist, named George Clarke, was apprehended and sentenced to death but for reasons which are not entirely clear he was regarded as a martyr by many blacks, who vowed to thwart his execution. Two companies of the Albany Guards were sent to Dawson, where the execution was to take place, to keep order, but their special train was derailed a mile and a half from Albany, on probably 24th August 1884, by obstructions placed on the line. Three cars were derailed but no one was hurt. Another train was procured, and the military then continued to Dawson, where the hanging took place without incident on the 25th.[1307] [1308]

Clear Creek station, several miles south of Hempstead, Texas, United States, on the Houston and Texas Central Railroad, was the location of a major act of sabotage on 14th November 1884. A section house was broken into to obtain the necessary tools to interfere with the track, and fishplates were unfastened, and spikes pulled out so weakening the track that it gave way when a train passed onto it. Robbery was apparently the motive and it was believed that the wreckers had intended to derail and plunder a freight train, but a northbound passenger train reached the scene first. The engine and tender passed over the damaged track safely, but the following two baggage cars, mail and express cars and two Pullman sleeping cars went down a 30 foot embankment into a creek. It was initially reported that 10 people had died and 15 were more or less seriously injured.[1309] The dead included a German woman and her two children, immigrants bound for Austin; three sons escaped unhurt.[1310]

Robbery was also the motive for the deliberate derailing of a fast passenger train near Purvis, Mississippi, United States, early on 23rd December 1884 on the New Orleans and Northeastern Railroad. Engineer Robert Turner and his fireman were fatally crushed between the tender of their locomotive. The train had been travelling at 40 miles an hour when it ran into an obstruction placed on the line. Railroad Detective Tom Gill investigated the affair and soon had reason to suspect three men, who were arrested and placed in the jail at Purvis. The three were charged with the crime and eventually confessed, saying that they had expected a freight train which they wished to wreck in order to steal what they wanted for Christmas. They had not intended to derail a passenger train, or to hurt anyone, and were sorry for what had happened. They had wrecked a freight train, they said, at the same location about a year previously for the same purpose.

News of the arrest of the three men was supposedly kept very quiet, but soon became common knowledge in the railroad community. At 12.30 a.m. on 27th December 50

1307 *San Antonio Light*, San Antonio, Texas, United States, 26th August 1884

1308 *The New York Times*, New York, New York, United States, 26th August 1884

1309 *The Daily Gazette*, Fort Wayne, Indiana, United States, 15th November 1884

1310 *Galveston Daily News*, Galveston, Texas, United States, 15th November 1884 railwaymen; conductors, engineers, firemen, brakemen and section men boarded the southbound train at Meridian for Purvis, where they went straight to the jail. Battering down the jail doors with axes and sledgehammers they led out the three men, who were piteously pleading for mercy, and placed ropes around their necks. The three were dragged to a spot near where the derailment had taken place and despite their cries for quarter they were hanged from trees, where their bodies were found still dangling in the morning. The work of lynching had been completed by 2 a.m., and the lynching party then stole quietly away.[1311]

The derailing of trains by sabotage seemed to occur almost every week in the United States in late 1884, usually for motives, where apparent, of robbery or a grudge against the railway company. On the evening of 31st December 1884 Baltimore and Ohio Railroad express passenger train No. 6 due in Cumberland, Maryland, at 7 p.m. was derailed near Grafton, West Virginia, by an iron bar wedged in the frog of a switch. Engineer Carl Hebron was killed outright and Fireman George Dawson terribly scalded and died within a few hours. Two passengers were injured.[1312] It transpired that the motive for this act of sabotage was revenge. A man named Mike Burke had been sacked by the Baltimore and Ohio Company some months previously for unstated reasons. He had then begun stealing from freight cars, had been eventually caught and jailed. Mrs Burke had then carried out the act of sabotage, unbeknown to her husband, as an act of revenge. Mr Burke said he was overwhelmed by his wife's action and refused to see her. She made a full confession of her crime.[1313] Her fate is unknown.

On the same day as the Grafton incident a Wabash St Louis and Pacific Railroad eastbound passenger train was wrecked at Prairie Switch, Indiana, United States. The train was late and was running at about 50 miles an hour as its engineer attempted to make up time before reaching Fort Wayne. The switch lamp indicated "All clear" but the switch had been opened for the siding. The engine, tender and mail car were derailed, and the tender and mail car struck two freight cars stabled in the siding. Locking arms, the engineer and fireman had gone down with their machine and somehow escaped without serious injury. As was often the case the sudden shock broke the coupling behind the mail car and the remainder of the train did not leave the track. The motive in this case was suspected to be robbery, and within two or three days the local sheriff had arrested a man named George Scott, of Whitehouse, Ohio, Scott confessed that the motive for the sabotage had indeed been robbery, the intention being to wreck the train and steal from the dead or dying passengers. When the saboteurs had seen that their attempt had failed, they had fled. Scott agreed to turn state's evidence and at the time of the newspaper report the search was on for his fellow saboteurs. Investigation showed that the pin had been taken out of the bottom of the switch stand and the strap removed, so that the switch could be moved without altering the switch light.[1314]

[1311] *Indiana State Sentinel*, Indianapolis, Indiana, United States, 31st December 1884
[1312] *Cumberland Daily Times*, Cumberland, Maryland, United States, 2nd January 1885
[1313] Lebanon Daily News, Lebanon, Pennsylvania, United States, 6th January 1885
[1314] *Huntington Democrat*, Huntington, Indiana, United States, 8th January 1885

Barely a week after the above two acts of sabotage, on Thursday 8th January 1885, another occurred at Bethpage Junction, New York, United States, on the Long Island Railroad. The accommodation passenger train which had left Long Island City at 5.30 p.m. for West Deer Park was running through a switch when its engine was derailed and overturned at 6.50 p.m. Engineer Henry Burchell was killed outright, and Fireman Jacob Dodsall was badly scalded by escaping steam. He was removed to St Catherine's Hospital, Brooklyn, but died the morning after admission. 'Investigation revealed that the accident was due to train wrecking.' [1315] The train had been in charge of Conductor James McNamara and consisted of only an engine, a combination baggage and smoking car and one passenger car. Both cars left the track and the few passengers were badly shaken up but not hurt. One of the switch rails had been moved so as to throw the locomotive from the track. The act was clearly one of a person with some knowledge of railway working, it was supposed a discharged employee.[1316]

The frequency of such acts of sabotage in the United States seemed to ease off later in 1885 but in 1886 there was another major strike, the Great Southwest Strike, during which acts of sabotage were committed by the strikers in an attempt to stop the running of trains. Thus, for example on 26th April a Missouri Pacific Railroad train was wrecked at 4 a.m. between Wyandotte and Kansas City, Missouri, United States, near the Kaw River Bridge. Spikes had been removed from the rails on a curve and the engine and four waggons went down an embankment. Of the strike-breaking crew Fireman Ben Horton and Brakeman George Carlisle were killed and engineer J. H. Fowler was seriously injured.[1317] Six Knights of Labour union members were later charged with causing the derailment on the evidence of an informer.[1318]

A very brief report in an American newspaper stated that a dispatch from Vienna on 3rd November 1886 had referred to the wrecking of a transport train, presumably due to an act of sabotage, between Nicoliaeff and Odessa, in Russia, in which 40 soldiers had died.[1319]

The Missouri Pacific Railroad train from Omaha to Kansas City was derailed by sabotage at 11.40 p.m. on Tuesday 11th January 1887 at Dunbar, Nebraska, United States, killing Engineer James De Witt. A pair of rails had been displaced using stolen track maintenance tools. The train fell down a 40-foot-high embankment and a sleeper, chair car and smoker ended up in a creek. The engine overturned on top of the embankment, crushing the engineer beneath it. None of the passengers was killed but several were badly hurt, as was Express Messenger H. Chenowith, whose life was despaired of for some time. Within a couple of days two men, David W. Hoffman and James W. Bell, had been arrested in connection with the crime and Bell confessed that they had been planning the job for some weeks, as they were both broke and wanted to leave the city.

[1315] *Biddaford Daily Journal*, Biddaford, Maine, United States, 10th January 1885
[1316] *Philadelphia Times*, Philadelphia, Pennsylvania, United States, 10th January 1885
[1317] *Decatur Review*, Decatur, Illinois, United States, 27th April 1886
[1318] Wikipedia – Great Southwestern Railroad Strike of 1886
[1319] *New Era*, Humeston, Iowa, United States, 11th November 1886

They found that the train was carrying $16,000 of gold and $20,000 of silver on the night that they carried out the attack. They had stolen the tools and damaged the track, when Bell said he became frightened and urged Hoffman to come away, lest someone was killed. Hoffman had replied that, 'he didn't care a damn how many were killed, he wanted that money.' Bell had then gone away, followed by Hoffman. When the train shortly left the track, they had returned, but had been too frightened to go near the baggage car where the money was stored. Hoffman was hanged at Nebraska City on 22nd July, Bell, having turned state's evidence, received 10 years in the penitentiary.[1320]

An act of railway sabotage occurred in Lower Burma in July 1888. On 3rd of that month a band of 300 dacoits from the Tharrawaddy district cut the railway line and telegraph wires 200 miles south of Prome. When news of the assembling of the dacoits reaching Rangoon a train carrying a body of military police was despatched to Zeegon station, which was under attack from the dacoits. At the moment that the train reached there the dacoits were actually engaged in destroying the line, and the pilot engine preceding the train was derailed. Subsequent to the attack military police occupied the stations near the site of the attack and all mail trains were preceded by pilot engines.[1321]

The *Times* correspondent in Rangoon speaks of the attackers as dacoits, implying that they were simply professional armed robbers or bandits, but it would appear that they were far more likely part of the resistance movement against British rule which had sprung up after the British took control of the whole of the country after the Third Burmese War of 1885,[1322] especially as in the same report the correspondent stated that the Tharrawaddy district had for some time been in a disturbed condition, and that 'a very heavy punitive police – tax has been imposed on all the inhabitants of the district. This tax is so heavy that it is stated on good authority to amount practically to a confiscation of the property of the poorer classes. The local newspapers report that in consequence of the people of Tharrawaddy being made desperate by recent taxation an attack was made on the line. The dacoits who attacked the line are reported to have declared that they only wanted to injure government property and to attack public officials.' [1323]

On 14th July the Rangoon correspondent added that a Rangoon resident had been discovered with documents apparently from the Mengwoon Prince with 'a remarkable proclamation calling on all Burmese residing in Lower Burmah in the districts traversed by the iron enemy – the Burmah State Railways – to unite and make up for the want of foresight previously displayed and assail the only vulnerable place in the power of the great enemy. The people living on the Prome and Toungoo lines are exhorted to destroy the parts of the railway in their immediate neighbourhood. Special mention is made of mail trains, which are desired to be overturned and destroyed utterly.' [1324]

About 9th August a band of saboteurs intended to tear up the railway between Pegu and Pyuntaza but were foiled by the arrival of a special train carrying reinforcements of

---

1320 *Weekly Nebraska State Journal*, Lincoln, Nebraska, United States, 29th July
1321 *The Times*, London, England, c5th July 1888
1322 Wikipedia – Burmese Resistance Movement 1885-95
1323 *The Times*, London, England, c6th July 1888
1324 Ibid, 16th July 1888

military police.[1325] On 18th August the *Times* correspondent in Rangoon effectively admitted that the attacks on the railway were political in nature. 'The most serious feature in connection with the recent attack on the railway is that nearly all the people concerned are villagers, hitherto known as respectable men and unconnected with dacoits or bad characters. They appear to have been driven to make the attack by the indiscriminate imposition of punitive taxes by the Deputy Commissioner.' [1326]

Presumably unrelated to the Burma sabotage, which occurred elsewhere in what might be loosely termed Greater India, the Calcutta (now Kolkata) to Bombay (now Mumbai) mail train was derailed on 23rd July 1888 near a bridge over the Godavery (Godavari) River near Najik, a rail having been removed. The engine of the train crossed the gap but its tender and the following coaches were thrown off. Sixty yards of rail had been removed and the fishplates and bolts were found beneath the bridge. No one was hurt.[1327]

A brief report from St Petersburg, Russia, stated that a train had been derailed in a tunnel at an unstated location on the Trans-Caspian Railway on 10th March 1889. The rails had been deliberately removed, and there were heavy casualties, amounting to 50 dead and injured. The band of robbers responsible had been captured.[1328] The idea of derailing a train to rob it in a tunnel seems strange, as surely access to the derailed train would then be difficult.

Derailments due to train wreckers, as they were termed, continued to occur regularly in the United States during the later 1880s and a nasty one involving an oil train occurred on 25th October 1889, near Kokomo, Indiana, on the Lake Erie and Western Railroad. Following the derailment, caused by an obstruction jammed into a set of points, 12 oil tank waggons exploded and burned. Engineer Mehl, Fireman Burnett and Head-Brakesman Spellman were thrown beneath the burning tanks and 'burned to a crisp'. The report on the sabotage added that, 'This has been the third attempt at train wrecking in this vicinity within the last two months.' [1329]

On about 6th August 1890 men working on the permanent way between Havana and Matanzas, Cuba, were surprised by an armed gang who ordered them to tear up the rails. A freight train was thereby derailed, and the gang shot dead the trains conductor, a man named Rodriguez. Troops were subsequently despatched to the scene from Matanzas.[1330] The gang was led by a Manuel García (also known as The King of the Fields of Cuba) [1331] and the attack can be seen as a preliminary to the war of independence fought against the occupying Spanish, though the war proper did not begin until 1895.

1325 Ibid, 13th August 1888
1326 Ibid, 21st August 1888
1327 Ibid, c28th July 1888
1328 *The Atlanta Constitution*, Atlanta, Georgia, United States, 11th March 1889
1329 *Reno Evening Gazette*, Reno, Nevada, United States, 25th October 1889
1330 *The Times*, London, England, 11th August 1890
1331 Wikipedia – Manuel García Ponce

A further act of sabotage occurred in Burma on 5th October 1890. The engine and 14 carriages of the down mail from Rangoon ran off the line about 55 miles from there. About 20 passengers were injured and one was killed. 'The line had been tampered with.'[1332]

As with the Purvis derailment of 1884 robbery was the motive for the derailing of a Rio Grande Railroad passenger train between Brownsville and Port Isabel, Texas, United States, on 19th January 1891. The saboteurs removed the fishplates and spikes from one rail, tied a rope of hide to the rail and concealed themselves nearby. When the train reached the spot the rope was pulled, the rail shifted out of place and the train derailed. The engineer was badly scalded, and the fireman suffered a broken arm and leg and cuts to his head and the two enginemen were groaning and appealing for help, but the robbers ignored them and proceeded to loot the train. The passengers and railwaymen were then blindfolded and locked into a box car. There they might have burnt to death, for the derailed coaches caught fire from the overturned locomotive, but they were released by some local farmers. A sheriff and posse set off in pursuit of the perpetrators.[1333] The robbers did not profit as much as they had hoped from their crime, for the train often carried very heavy shipments of Mexican silver, between $100,000 and $125,000 worth every time a steamer was due to leave Brownsville, the silver being carried to New York, where it was sold for a large profit. Clearly this had been the robber's target.[1334]

Once the robbers had gone and he could get out of the train the conductor cut the telegraph wires and rigging up a temporary instrument was able to telegraph to Brownsville for assistance. A relief train arrived after about an hour. The Government offered a reward of $1,000 for the apprehension of the saboteurs.[1335]

It was the famous Orient Express which was initially reported to have been wrecked by sabotage on 30th May 1891 near Tcherkesskeui in Turkey. A later report stated that in fact it was not the named train but just an ordinary express. A gang of about 30 brigands, led by a noted robber named Anastasius, at first seized the watchman at Tcherkesskeui and then proceeded to remove some rails. The engine, tender baggage van and a second- and a third-class carriage were derailed, the engine driver having had no warning of danger. No one was killed or seriously injured in the derailment, which had happened just before midnight, though a few were bruised and shaken. Immediately after it had occurred the brigands boarded the train, brandishing revolvers and yelling in chorus. They first seized the conductor of the train, whom they made to act as interpreter, then went along the train relieving the passengers of their money and watches, except the few third-class passengers, whom they left alone. The brigands then chose the five wealthiest looking passengers, who happened to all be from a party of German tourists, to hold for ransom, and marched off with the five together with the Swiss engine driver, named Freundinger. The engine driver was chosen to act as a replacement translator for the

[1332] Ibid, 8th October 1890

[1333] *Kokomo Saturday Tribune*, Kokomo, Indiana, United States, 23rd January 1891

[1334] *Oil City Derrick*, Oil City, Pennsylvania, United States, 26th January 1891

[1335] *San Antonio Light*, San Antonio, Texas, United States, 23rd January 1891 conductor, who was ordered by the brigands to return to Constantinople with a note demanding 200,000f ransom, or else the hostages would be killed. The ransom was eventually paid, and the hostages were released after nine days of captivity.[1336]

At about 9.45 p.m. on Tuesday 16th June 1891 the Omaha and Chicago express of the Chicago Milwaukee and St Paul Railroad was wrecked half a mile east of Coon Rapids, Iowa, United States. As his locomotive jumped the track Engineer William Liddell reversed it and applied the brakes but two hundred feet from the point where the engine left the rails was a covered trestle bridge across a stream and the speed of the train could not be much reduced in such a short distance. The engine and mail car fell 50 feet from the bridge to the stream below and the baggage car, day coaches, chair car and one sleeping car fell from the side of the trestle and overturning in their fall landed upside down. The coaches were electrically lit so there was no fire. Engineer Liddell was unhurt, and Fireman Albert Rouse escaped with slight injuries but Express Messenger Aldroyd and passenger Henry Candon of Perry, Iowa, were killed outright and passenger John McCarthy of Dunkirk, N.Y. was fatally injured. Conductor Charles R. Cornelius suffered a broken arm and a baggagemaster, two brakemen, a postal clerk and about a dozen passengers were more or less hurt.

Uninjured passengers began the rescue work in a storm of heavy rain and lightning with the aid of lanterns brought from Coon Rapids station and neighbouring farmhouses. Local doctors were summoned and a special train with physicians and workmen was sent from Perry. It was at first theorised that the recent and continuing heavy rain had caused the roadbed to subside and a rail to break but it was then discovered that the derailment had been caused by a sleeper which had been fixed to the approach trestlework of the bridge. The motive for the act of sabotage was believed to have been robbery.[1337]

The Bostian Bridge over Third Creek (Shaw wrongly states the accident occurred at the bridge over the Catawba River, which is in fact crossed a short distance further on by the Norman bridge) about two miles west of Statesville, North Carolina, United States, was the scene of a major disaster on 27th August 1891. At about 2 a.m. on the morning of that day a Richmond and Danville Railroad train travelling at about 30 miles an hour left the track on the immediate approach to the bridge and fell over 80 feet to the creek below. Engineer William West, Fireman Warren Fry, baggagemaster Hugh Linster and about 18 passengers were killed and about 25 more of the 85 passengers on the train were injured. Three of the fatalities were Asheville firemen returning from the state tournament at Durham. A harrowing story was told of a Miss Luellen Pool who had held her mother's head out of the water until her strength was completely exhausted, she had to let her mother's head drop and the woman drowned.[1338] Miss Pool then related to rescuers how the water was rising from her knees to her chin, but at that point she was pulled from the wreckage and escaped the fate of her mother. The water in the creek had been rising because the wreckage was acting as a dam. The same newspaper was able to

1336 *The Times*, London, England, 2nd, 3rd, 4th, 5th, 6th, 8th, 10th, 12th and 15th June 1891
1337 *Benton Harbor Weekly Palladium*, Benton Harbor, Michigan, United States, 19th June 1891
1338 *The Salem Daily News*, Salem, Ohio, United States, 28th August 1891

record that, 'The behaviour of the coloured people at the wreck was most praiseworthy.'[1339]

As was usual at this time an inquest was quickly held and on 31st August the coroner's jury returned a verdict that 'the said wrecking of the train {was} caused by a loose rail, the bolts and spikes of the same having been taken out by some person or persons unknown to the jury, with tools or implements belonging to said railroad company (the Western North Carolina, which was now leased and worked by the Richmond and Danville), which said tools or implements were by gross negligence on the part of said railroad company left in an open shed accessible to every passer-by.' The jury added that, 'We also find that several of the cross-ties at and near the break to the said railroad track, where said rail was displaced, were unsound and should have been replaced, and that the superstruction {sic} on the bridge was in part defective and unsafe and further that the high rate of speed maintained in running trains over this bridge deserves and has the censure and condemnation of this jury.' The jury's verdict would seem to imply that they were not really sure that the accident was the result of sabotage, or was due to track defects and perhaps excessive speed.[1340] The local *Landmark* newspaper rejected the theory that the accident was due to sabotage and advanced a third theory, that the light and empty rear car, the *Daisy*, had first become derailed and had pulled the remainder of the train with it. The railroad company, predictably, preferred to ascribe the accident to sabotage, perhaps by someone with a grudge against the company.[1341]

Revenge on the part of an engine driver was the apparent motive for a collision on the line between Klingnau, in the Swiss canton of Aargau, and Waldshut, in Baden, Germany, on 1st December 1891. The driver had been given notice of his pending dismissal at the end of the year for an unstated reason and was only entrusted with what was termed a 'supplementary service' to Waldshut. Driving a light engine, he arrived at the station at Döttingen, Switzerland, where the signals were against him. Despite this he failed to stop, running through the station at full speed, ignoring the whistling of the station staff. At the far end of the station his engine collided head-on with the locomotive of an approaching passenger train. The driver and fireman of the light engine (some reports say of the passenger train) were killed and three passengers possibly fatally injured.[1342] [1343]

In South America Chile saw a civil war in 1891 between its president, José Balmaceda, supported by the Chilean Army, and Congress, supported by the Chilean Navy. The civil war ended in September 1891 with the defeat of the army and the suicide of Balmaceda.[1344] The festivities in connection with the installation of the new president, Jorge Montt, on 26th December 1891, were however marred by an attack on a train carrying a number of naval officers from Valparaiso to the capital Santiago, supposedly

[1339] *Landmark*, Statesville, North Carolina, United States, 31st August 1891
[1340] Ibid.
[1341] Ibid, 3rd September 1891
[1342] *The Times*, London, England, 2nd December 1891
[1343] *The Philadelphia Inquirer*, Philadelphia, Pennsylvania, United States, 2nd December 1891
[1344] Wikipedia – Chilean Civil War of 1891

carried out by supporters of the late President Balmaceda. The train was derailed by obstructions placed on the line but there was no loss of life.[1345]

A report in the French press spoke of a failed attempt on the life of the Czar of Russia by Nihilists at an unspecified date, but probably in early October 1892, at Skiersnimice station in Russian Poland. A dynamite bomb was exploded under a train supposed to that of the Czar, killing five people and wounding 14. The Czar had entered the station by another line and was unhurt. 'Forty arrests have been made.' [1346]

The state of Tasmania was the perhaps rather unlikely location for a serious sabotage attempt on 20th September 1893. The up express from Launceston left that place at 3 p.m. in charge of Guard Harry Reynolds and Driver George Jones. The train ran normally until it reached the Horseshoe Bridge, between Bridgewater and Brighton. The bridge was a wooden trestle structure on stone piers, about 150 ft long and with a maximum height of about 65 ft. The train was derailed on the immediate approach to the bridge and ran onto the structure off of the rails, but most fortunately did not fall from the structure, though the engine and leading carriages came to a stand leaning over the side of it. The whole of the train was derailed except the rear guard's van. Examination of the line showed that fishplates had been removed and the ends of two rails lifted, in addition to which a number of obstructions had been placed at intervals along the line, consisting of rails, stones, platelayer's tools and so on. The cowcatcher of the engine had pushed off the first two obstructions, but the train had become derailed at the point where the track itself had been interfered with. It was thought that if the train had travelled 10 feet more before it could be stopped the whole of it would have fallen from the bridge to the creek below. Only the immediate application of the Westinghouse brake by the driver had saved the train from disaster. After the accident the guard crawled along the bridge and warned the passengers to keep their seats lest they should alight on the viaduct and fall over the edge of it in the darkness.[1347]

Ironically the Minister of Railways had only recently alluded in a speech to parliament upon the freedom of the Tasmanian railways from accident. The government offered a reward of £500 for the apprehension of the perpetrators of the crime.[1348]

Shortly after the accident a brown felt hat with a red lining was found near the accident scene and was traced to 27-year-old Charles Briggs, a former railway labourer. Other evidence, such as his reluctance to discuss the accident and his apparent nervousness after it had happened weighed against him and he was charged with two counts of placing obstructions on the railway and interfering with the track with intent to commit murder and to endanger the safety of railway passengers. He stood trial twice, in December 1893 and February 1894, and each time the jury failed to reach a unanimous verdict. After the second mistrial the case against him was dropped. [1349] [1350] If Briggs was

---

[1345] *The Times*, London, England, 28th December 1891
[1346] *Auckland Star*, Auckland, New Zealand, 13th October 1892
[1347] *The Mercury*, Hobart, Tasmania, 21st September 1893
[1348] Ibid.
[1349] Ibid, 9th October and 13th and 14th December 1893
[1350] *Tasmanian News*, Hobart, Tasmania, 27th and 28th February and 1st March 1894

indeed the culprit his motive was presumably a grudge against the railway company, given that he was a former employee. It has not been possible to find a mention of why he left his job on the railway or if he had been sacked; did he feel perhaps that he had been unfairly dismissed?

Robbery was the motive for the derailment of the northbound Southern Pacific Railroad overland express just after midnight on 16th February 1894 at Roscoe, California, United States. As the train approached Roscoe its engineer saw a man standing with a flaring torch at a wide open switch. The distance was too short to pull up in and the train was derailed as a fusillade of bullets was fired at the engine. Engineer Thomas jumped and escaped, but Fireman Masters was caught between the engine and tender and badly crushed. Two men, one on either side of the line, then fired at the cars of the train and demanded access to the express car. Upon the express messenger refusing they blew the car open with dynamite, tearing the side of the car off. The robbers then looted the car and although they were unable to get into the safe, they escaped with a large amount of loot. Fireman Masters lingered in agony for an hour, begging for a pistol to shoot himself with, and died before he could be released. One of two tramps riding on the engine was also killed, the other having a miraculous escape.[1351]

At the end of April 1894, a rebellion known as the Revolution of the 44 broke out in the Central American state of El Salvador against the government of General Carlos Ezeta.[1352] On 3rd May 1894 the president attempted to proceed by rail from Acajutla to Sonsonate with 1,500 troops en route to the assistance of Government forces operating against the rebels at Santa Ana. The rebels, having become aware of the movement of this troop train removed some of the rails on a steep gradient, wrecking the train, eight coaches being telescoped. Two hundred (some accounts say three hundred) soldiers were reportedly killed and 122 more wounded.[1353] President Ezeta was deposed in June and fled to Panama.[1354]

At about 7 a.m. on Monday 2nd July 1894 an eastbound Canadian Pacific Railway Halifax express, which had left Montreal at 8 p.m. on Sunday, fell from a trestle over the west outlet to Moose Head Lake, between Askwith and Moosehead stations, Maine, United States. The train was travelling at its usual speed when the engine and tender, baggage, mail, and first and second-class cars fell 25 feet into a rocky gorge and the Pullman sleeping car was overturned. The trestle was a wooden structure with an iron bridge in its centre; 40 feet of its western end collapsed under the train. The first three cars were telescoped, their tops torn off, timbers wrenched to pieces, 70-pound rails twisted like twigs and baggage and boxes mixed up and broken. The passengers only numbered 23. They were greatly frightened and four of them were killed outright, along with Engineer Fred Leavitt, who was crushed beneath his locomotive. The passengers killed were Fred Foss, Canadian Pacific station agent at Greenville Junction; mail clerk

1351 *The Standard*, Ogden, Utah, United States, 17th February 1894
1352 Wikipedia – Revolution of the 44
1353 *The Times*, London, England, 17th May and 4th June 1894
1354 Wikipedia – Carlos Ezeta

Walter Starkey, G.C. Hoyt and a man who could not be immediately identified. In addition, Fireman Angus Macdonald was fatally injured and Charles Grant, station agent at Jackman, Maine, was probably fatally hurt.

Investigations made after the accident determined that the train had been derailed by an obstruction placed on the track by some person or persons. The company offered a reward of $4,000 for the arrest and conviction of the perpetrators and it was thought that the state authorities would also offer a reward.[1355] [1356] No motive for the sabotage was suggested.

A nationwide Pullman railroad strike began in the United States in May 1894 following a decision by the Pullman Palace Car Company to cut wages during a recession.[1357] Like the 1877 strike, and as often the case with such industrial disputes in this country, the strike soon became bitter and troops were called in to supress it. Sacramento, California, was a strong union area and when an attempt was made to run train No. 4 from there to San Francisco on 11th July the strikers decided to stop it. Two and a half miles west of Sacramento the strikers concealed themselves by the line and when the train reached them. they opened fire on it with Winchester rifles. They had also sawn through the timbers of the wooden Yolo trestle, leaving the rails intact, so that as soon as the train passed onto the trestle it was derailed.[1358] Engineer Clark and four soldiers died in the wreck. By the next day a striker named Salter Worden had been arrested charged with causing it. [1359] Worden was tried and sentenced to death for the crime. In February 1894 he confessed to having taken part in the sabotage in the hope of winning a commutation of his death sentence from California Governor Budd.[1360] The sentence was indeed eventually commuted to life imprisonment in June 1898 after a board had decided that Worden was insane at the time that he committed the offence.[1361]

Chicago Rock Island and Pacific Railroad train No. 8, known as the "Fort Worth accommodation" was wrecked at Lincoln, Nebraska, United States, on 9th August 1894 at 10 p.m. at a point where the line crossed Salt Creek, and the Union Pacific and Chicago Burlington and Quincy lines by a high trestle. As the train reached the start of the trestle the rails spread, and the engine and coaches ran for 50 feet over the sleepers and then fell 40 feet to the creek below. Spilt coals from the engine ignited the bone-dry timbers of the wooden trestle and very quickly it became a mass of flames, while the wooden coaches of the train were also set ablaze. Some of the passengers from the chair car who did escape were too dazed or frightened to lend any assistance to try and rescue other passengers while it was reported that a party of three Russians, man, woman and child were so paralysed with fear that they could not be induced to leave their burning coach and so burnt to death. On the other hand, a J.R. Brinker, a Denver travelling salesman, was

---

1355 *Winnipeg Free Press*, Winnipeg, Manitoba, Canada, 4th July 1894
1356 Ibid, 12th July 1894
1357 Wikipedia – Pullman Strike
1358 *Reno Evening Gazette*, Reno, Nevada, United States, 11th July 1894
1359 Ibid, 12th July 1894
1360 Ibid, 9th February 1898
1361 *Idaho Daily Statesman*, 18th June 1898

thought to have perished in the wreck but turned up alive and well the next day, having missed his train at Beatrice.[1362] Because the bodies of the victims were so consumed by the fire it was difficult to say exactly how many had died, some accounts give a figure of 11 deaths, including Conductor C.D. Stannard, Engineer Ike Depew and Fireman William Craig, others put the number of dead as high as 24.[1363]

There was no question that the wreck was the result of an act of sabotage as the marks of a wrench on a loosened rail and of a crowbar on sleepers were plainly visible. There were deep dents in the sleepers where a crowbar had been used and loose spikes were lying all around. Section man J.D. Glover said that he saw three well-dressed men jump on a westbound freight train around the time of the accident, each carrying a long, new satchel [1364] but it was a local black man, George Washington Davis, who was arrested and tried for causing the wreck. At the first trial the jury refused to find him guilty of first-degree murder. He was tried again, and this time he was convicted of second-degree murder, but he never admitted to the crime and there was much doubt as to his guilt. After 10 years in gaol Governor Mickey of Nebraska granted him parole on the grounds of an absence of motive and lack of evidence as to his guilt. No one else was ever tried and who really caused the accident and why remains a mystery.

The Cuban war of independence from Spain began in February 1895 and was characterised by guerrilla tactics, including attacks on trains. One of the first of these was on Sunday 11th August, when a train of volunteers had left Havana to fight the insurgents. It was blown up by dynamite at Bolondron and the *Reuter* news agency reported that, 'Only a few of those in the train are said to have escaped.' [1365] On about 3rd September the rebels threw a train off of the rails at Cienfugos, injuring two passengers [1366] and on 10th September a mine exploded under the last carriage of a train conveying a battalion of the Luchana regiment to Puerto de Guantanamo, killing two soldiers and wounding seven.[1367] The next day (11th September) a train carrying both soldiers and civilians was thrown off the line near Puerto de Baga, Neuvitas by a band of rebels who lay hidden close by. The rebels then opened fire, killing five soldiers and two civilians in the first discharge. The remainder were taken prisoner but were released after 12 hours, the rebels keeping the soldiers' arms.[1368] On 8th October a fireman was killed when rebels derailed a train near Santo Domingo.[1369]

About 22nd November 1895 bandits derailed a train on the Pisco – Ica railway in Peru. The motive was robbery, mail and the passengers' baggage being taken.[1370]

[1362] *Fort Wayne Sentinel*, Fort Wayne, Indiana, United States, 10th August 1894
[1363] *Newark Daily Advocate*, Newark, Ohio, United States, 11th August 1894
[1364] *Fort Wayne Sentinel*, Fort Wayne, Indiana, United States, 10th August 1894
[1365] *The Times*, London, England, 17th August 1895
[1366] Ibid, 9th September 1895
[1367] Ibid, 12th and 13th September 1895
[1368] Ibid, 19th September 1895
[1369] Ibid, 14th October 1895
[1370] *Galveston Daily News*, Galveston, Texas, United States, 24th November 1895

The rebellion in Cuba continued with a major act of sabotage on 25th November 1895. Insurgents commanded by Leonico Vidal wrecked a train carrying 200 Spanish soldiers, together with arms and ammunition, near Ceinfuegos. A dynamite bomb exploded under the front of the train, wrecking the engine and leading car and killing the engineer, fireman and 13 soldiers. Eighteen more soldiers died when the rear coaches fell down a steep embankment and about 50 soldiers were wounded, some fatally. The remaining troops were so disheartened that they surrendered without resistance and the rebels gained thousands worth of dollars of arms and ammunition.[1371] In further attacks an engine driver and two fireman were killed in the derailment of a mixed train between Nuevitas and Puerto Principe on about 27th November [1372] and a driver and conductor died when a goods train was derailed at Jicotea on about 7th December. Another attack on a train carrying 300 passengers, many of them soldiers, on about 2nd January 1896, between Nuevitas and Puerto Principe killed a fireman.

On 13th July 1896 an explosion wrecked a passenger train between Cidra and Sabanilla in Matanzas province, Cuba. Six people were killed and 18 injured.[1373] [1374]

The 1878 Congress of Berlin which followed the defeat of Turkey by Russia resulted in the independence of Montenegro, Romania and Serbia but the province of Macedonia was restored to the Ottoman Empire. The struggle for the independence of Macedonia from the Turks continued, however, with Bulgaria, Serbia and Greece all competing for control of the area.[1375] It was assumed that it was in connection with this struggle that a Turkish military train was derailed on the Salonica – Nish line between Demirkapu and Krivolak stations, Macedonia, at 8 p.m. on 22nd October 1896. The engine driver and two officers were killed, and the fireman and four guards were injured. An ambulance train with doctors was sent to the scene.[1376] [1377]

Attacks on trains on the London Underground had been carried out by Fenians in October 1883 (at Praed Street and Charing Cross) and in January 1885 at Gower Street. No one had been killed in any of these bombings, though 62 passengers were injured in the Praed Street attack.[1378] The last such attack on the London Underground in the 19th century occurred on 26th April 1897, and this time there was a fatality. At 7.1 p.m. on that day a "Circle" line train from Farringdon Street had just come to a stand at Aldersgate Street station when there was an explosion in a compartment of a first-class carriage. The platform was at the time crowded with people bound from the City to the East End 'and the wildest confusion prevailed'. Three passengers who had been sat in the affected

---

[1371] *Landmark*, Statesville, North Carolina, United States, 3rd December 1895
[1372] *The Times*, London, England, 2nd December 1895
[1373] *The North Adams Daily Transcript*, North Adams, Massachusetts, United States, 14th July 1896
[1374] *The Times*, London, England, 16th July 1896
[1375] Wikipedia – Macedonia
[1376] *London Echo*, London, England, 23rd October 1896
[1377] *The Daily Mail*, London, England, 24th October 1896
[1378] Wikipedia – Attacks on the London Underground

compartment were seriously hurt and a medical man who shortly attended the accident advised the removal of 10 of the most badly hurt to St Bartholomew's Hospital, which was situated less than two hundred yards from the station. In all, 60 people were wounded and one, a Henry Pitts, of Tottenham, subsequently succumbed to his wounds. Initially the accident was blamed on the explosion of a Pintsch gas cylinder beneath the carriage, used to supply gas for lighting, but it soon became apparent that the cylinder had been damaged by the explosion rather than having caused it, and the explosion was in fact due to a bomb. The jury at the inquest on Mr Pitts returned a verdict of murder against a person or persons unknown. This attack was carried out not by the Fenians but by an anarchist group and was perhaps connected to Queen Victoria's upcoming Diamond Jubilee in June.[1379] [1380]

In March 1897 Greece and Turkey were on the verge of war over the status of Crete, then an Ottoman province but which its Greek population wanted to be unified with Greece. In preparation for the coming conflict Turkey was running troop trains through Macedonia towards the Greek frontier and it was one of these trains which was the target for an act of sabotage as it was crossing a bridge over the Varduari River, near Salonika (Thessaloniki from 1912), on about 10th March. As the train, filled with a reported 3,000 soldiers, was on the bridge the structure was blown up with dynamite. Several carriages were thrown into the river and many of their occupants, unable to get out, drowned before they could be rescued. Other soldiers managed to escape from the carriages but were weighed down by their equipment, and also drowned. The exact number killed was unknown, but it was believed to be large. The transport of troops to the Greek frontier had to be suspended until the bridge could be repaired. The sabotage was blamed on 'Macedonian malcontents' sympathetic to the Greeks.[1381] [1382] [1383]

On 9th June 1897 the daily 6 a.m. passenger train from Havana to Matanzas, Cuba, was dynamited near Campo Florido, dynamite bombs placed in a culvert on a curve being detonated electrically. Two engineers, one fireman, one conductor and seven soldiers were killed, three passengers seriously injured and 30 less seriously so. Coincident with the explosions the rebels opened fire on the train, wounding a captain. The insurgents then advanced on the train with the supposed intention of plundering the $6,000 in the express car but were driven off by the train's military escort. One rebel was killed, apparently by the shock of the explosions.[1384]

In a similar attack to that of 9th June an 'exploring train' as it was termed by a newspaper was wrecked by rebels at a point 110 kilometres from Havana, Cuba, on 26th June 1897. A culvert was destroyed by dynamite between Taco Taco and Bacanagua. The explosion was detonated as the train from Vuelta Alajo was passing over the culvert and

---

[1379] *The Standard*, London, England, 27th April 1897
[1380] *The Times*, London, England, 27th and 28th April 1897
[1381] *Boston Sunday Globe*, Boston, Massachusetts, United States, 14th March 1897
[1382] *The Liverpool Courier*, Liverpool, England, 14th March 1897
[1383] Wikipedia – Greco-Turkish War (1897)
[1384] *Newport Daily News*, Newport, Rhode Island, United States, 10th June 1897

the engine fell into the gap left by the destroyed culvert. Sixteen soldiers were killed outright and the engineer, fireman, Captain Suarez of the artillery volunteers and five soldiers of the Battalion of Asturias were seriously wounded. Two of them died later from their injuries. Another eight soldiers were slightly wounded. All the injured were sent to Palacios. There were also civilian passengers on the train, none of whom were hurt but who were, not surprisingly, much frightened. They were eventually transferred to another train.[1385]

In another major action on about 11th November 1897 the Cuban rebels dynamited and derailed a train between Nuevitas and Puerto Principe. The explosion overturned the engine and destroyed several coaches and killed the engineer, fireman and 10 other persons and injured 27 soldiers and passengers.[1386]

About 4th February 1898 a train was blown up by a dynamite bomb placed on the track between Bonita and San Vincent, in Santiago de Cuba province, Cuba. The cars were destroyed and according to one account four Spanish soldiers killed and 16 wounded. Another gave the casualties as five dead and 27 wounded.[1387] [1388] This seems to have been the last railway sabotage of the Cuban war of independence. The Spanish-American War broke out in April, which Spain soon lost, and by the end of the year the Caribbean island had gained its independence.

As well as giving up Cuba at the end of the Spanish-American War, Spain transferred control of the Philippines to the United States. The Filipinos, too, wanted their independence and this led to the Philippine-American War, which was in progress over very much the same period as the much better known (Second) Boer War. Two acts of railway sabotage were recorded during the war, on 17th August and 22nd September 1899, both near Angeles. In the first incident the Americans had occupied Angeles and the Filipino forces attempted to recapture the place, using a train hauled by two locomotives, but the American soldiers foiled the attack by derailing the train.[1389] In the second the 'insurgents' themselves derailed a train near Angeles and fired on the soldiers guarding it. One soldier, a Private Charles Ziemans, and 'an unknown civilian' were killed before the attackers were driven off, leaving six dead. 'Troops were immediately sent in pursuit.' [1390] The Filipinos lost the war but finally gained their independence after the Second World War in 1946.

The guerrilla war with numerous attacks on trains which had so characterised the Cuban war of independence was to be shortly repeated in southern Africa where what is usually known as the Boer War, but which was more correctly the Second Boer War, had

1385 *Jeffersonville News*, Jeffersonville, Indiana, United States, 28th June 1897
1386 *Pontiac Sentinel*, Pontiac, Illinois, United States, 19th November 1897
1387 *Fort Wayne News*, Fort Wayne, Indiana, United States, 5th February 1898
1388 *The Times*, London, England, 8th February 1898
1389 *Fort Wayne News*, Fort Wayne, Indiana, United States, 17th August 1899
1390 *Reno Evening Gazette*, Reno, Nevada, United States, 23rd September 1899

broken out on 11th October 1899 between the British and the Boer republics of the Transvaal and the Orange Free State.[1391] The use of guerrilla war tactics by the Boers really only began in earnest in 1900 but an armoured train from Cape Town to Mafeking was derailed and captured by the Boers the day after the conflict had broken out. The train, called the *Mosquito*, and said to be one of the smaller armoured trains, left the track at Kraaipan, Cape Colony, a rail having been removed. The Boers then shelled it with artillery for half an hour before capturing it, together with a Lieutenant Nesbitt and 15 men. The engine driver and a 'native' escaped.[1392]

Two days later a hospital train, consisting of an engine and one carriage flying a white flag, was started for Maribogo, just short of Kraaipan, with the intention of visiting the Boer camp and asking for permission to remove the wounded to Kimberly. It got no closer than half a mile south of Maribogo station when it too was derailed, its driver having reduced speed when he saw that the track had been disturbed. Some apparently hostile Boers then appeared, and the train had to beat a retreat at full speed to Vryburg.[1393]

In another early attack on 15th November 1899 the Boers derailed an armoured train sent out from Estcourt on a reconnaissance trip to Chieveley. It had just set off on its return trip when it was attacked near Frere, Natal. Three miles north of Frere two waggons were derailed by an explosion of dynamite and the Boers opened up a heavy fire on the train with 7 pounder shells. Accompanying the train were 90 men of the Dublin Fusiliers, 80 men of the Durham Light Infantry Volunteers, 10 sailors under a Captain Haldane, Lieutenant Franklin and a British newspaper correspondent for *The Morning Post*, a Mr Winston Churchill, later to become the Prime Minister who saw Britain through the Second World War. An engagement with the Boers lasting over an hour then took place. Eventually the derailed waggons were cleared from the line under Boer fire and the engine returned to Estcourt with 10 wounded soldiers, leaving the remaining troops to make their way back on foot. Churchill accompanied the engine as far as Frere station, then returned on foot to care for other wounded. He was taken prisoner by the Boers, but later escaped. The Pietermaritzburg correspondent for the *Times* recorded that 'Mr Winston Churchill behaved most gallantly and distinguished himself by his courageous conduct.' One soldier died in the derailment and two or three were killed in the fighting.[1394]

The next reported sabotage was not until the morning of 24th February 1900 when a train carrying part of the City of London Volunteers to the front was derailed at Milders Vliet, Cape Colony, 15 miles from Cape Town, several lengths of track having been removed. 'The damage was not serious.'[1395]

[1391] Wikipedia – Second Boer War

[1392] *The Times*, London, England, 16th October 1899

[1393] Ibid, 16th October 1899

[1394] Ibid, 15th, 17th and 18th November 1899

[1395] *Evening Democrat*, Warren, Pennsylvania, United States, 24th February 1900

After defeats by the Boers at the beginning of the Second Boer War the British poured in reinforcements and went over to the offensive. The Towns of Ladysmith and Mafeking were relived on 27th February and 17th May 1900 and the Boer cities of Johannesburg and Pretoria captured on 31st May and 6th June 1900. British observers considered that the war was then all but over, but the Boers chose to continue fighting, employing guerrilla tactics aimed at British supply and communication lines.

On 30th July 1900 the British troops suffered a major loss in the derailment of a train near Frederickstad on the Krugersdorp-Potcheefstroom Railway, in the Boer republic of the Transvaal. A special patrol had been ordered to ride along the railway to check for possible sabotage but for some reason this order was not carried out, and a supply train, escorted by a detachment of the Shropshire Light Infantry, went off the line, killing the engine driver and 15 men.[1396]

Three days later another train was derailed by Theron's Scouts near Honing Spruit, Orange River Colony, at 1 a.m., its sleeping passengers being awakened by the 'rattle of derailed waggons and rifle shots'. Two men were killed in the derailment. The *Times* correspondent in Pretoria added that, 'Theron (the Boer commandant) was polite and at once liberated the American Consul, his friends, and Lord Algernon Lennox. The soldiers were taken prisoner but were released after a five mile march over the veldt.' [1397]

On 12th September 1900 the *Times* correspondent in Pretoria wrote that the Boers were making incessant attacks on the railway lines, 'which although they only dislocate traffic for a short time, throw an enormous strain on the staff of the Imperial military railways.' In the week up to 7th September a supply train had been derailed and captured near Klipp River, Transvaal, on 1st. The line south of Kronstad was cut and an ordnance stores train captured, the line was cut at America Siding and a train was fired upon near Heidelberg on the 2nd, the stoker being killed, and the driver injured. On 3rd the line was cut south of Heidelberg, and the rails were torn up and an engine derailed on the Pietersburg (Transvaal) line. On 4th the line was cut early in the morning at Hector Spruit, on 5th the Krugersdorp line was cut. Near Balmoral, Transvaal, an engine was blown up and five trucks derailed by dynamite and on the 7th the railway was blown up near Roodeval.[1398] An impressive catalogue of sabotage for just one week.

Near Pan, Transvaal on 1st October 1900 the Boers derailed and fired on a train carrying the 2nd Coldstream Guards, killing five men and wounding a number of others and Lieutenant C. Heywood. It was reported that, 'The men behaved extremely well, and everyone kept their heads, or in all probability the losses would have been heavier, as the Boers opened fire within 60 yards {of} where the train was derailed.' [1399] The fatalities were from Boer fire and not from the initial derailment.

1396 *The Times*, London, England, 2nd August 1900
1397 Ibid, 6th, 7th and 8th August 1900
1398 Ibid, 14th September 1900
1399 Ibid, 4th October 1900

The final act of Boer sabotage before the end of the century occurred near Val, Transvaal, on 29th December 1900, when they removed four lengths of rail and derailed the first of several trains from Standerton, which had left there at 5.30 a.m. Several trucks were wrecked, and the Boers made off with five waggon-loads of provisions, setting fire to the supplies they could not carry away.[1400]

The war dragged on until the end of May 1902 when the Boers were finally defeated after the British had built a chain of blockhouses to protect their supply lines, linked by barbed wire, and adopted a scorched earth policy, burning homesteads and interning civilians in concentration camps.[1401]

1400 Ibid, 1st January 1901
1401 Wikipedia – The Second Boer War

CHAPTER 19

MISCELLANEOUS ACCIDENTS

The vast majority of railway accidents fall into one of a small number of categories such as derailments, collisions, bridge failures and so on. A few, though, do not easily fall into any obvious category, hence this short chapter on miscellaneous accidents. Many of the accidents listed in this chapter were fairly minor ones which have been included because of their unusual features.

Just under three years after the 1880 Rimutaka accident in New Zealand when a train was blown over by the wind, it was eclipsed in magnitude by a disaster of a more or less similar type in the United States. On 21st August 1883 a Chicago and North Western Railroad train left Rochester, Minnesota, at about 4 p.m. for Zumbrota, on the Rochester and Northern division of the company. The train was travelling at high speed at an unstated location when it was struck by what the newspapers of the time called a cyclone, but which would now presumably be termed a tornado. News of the disaster was limited because the telegraph wires were down and no information could be gained until a relief train arrived at Owatonna, Minnesota, with 35 injured persons. A witness who had been to the scene then told a newspaper reporter that every car in the train had been wrecked and almost literally shattered to pieces by the sudden stop caused by the train being blown from the rails. 'The gentleman stated that nine dead bodies had been taken from the ruins and a large number of those who were seriously injured were taken to Rochester and Owatonna. At the time he left the work of extricating the unfortunate victims was still progressing and the number of killed would reach a score when the final summing up was made.' In the town of Rochester itself the storm had demolished three hundred houses, damaged two hundred more and killed a number of its citizens.[1402]

Collisions with animals which have strayed onto the line have always been fairly commonplace. In countries with temperate climates the animals are usually cattle or perhaps sheep, in tropical countries they might perhaps be elephants, tigers or leopards. A

[1402] *The Terre Haute Express*, Terre Haute, Indiana, United States, 23rd August 1883 collision with an elephant in a country such as Canada would naturally be unusual, but one did occur on 15th September 1885, near St Thomas, Ontario, and the elephant was no ordinary beast but the famous *Jumbo*, the animal star of P.T. Barnum's *Greatest Show on Earth*. The circus had finished its shows at St Thomas and at 9.30 p.m. keeper Matthew Scott was busy loading *Jumbo* and a smaller elephant named *Tom Thumb* into their cars for transport to the next destination. Between the siding on which the elephant's cars were stabled and the circus grounds ran the Grand Trunk Railway main line. The sidings could be reached via a level crossing a short distance away, but the railway company had, as it turned out unwisely, given the circus permission to take down a fence so that the elephants could be taken more directly to their car. Just as the keeper was taking the animals across, he was surprised to see a special westbound freight train approaching. He first tried to get the elephants to go down the lineside embankment out of the way but *Jumbo*, sensing no danger, refused to budge. He then tried to get them across to the siding, but they were only part of the way across when the engine of the freight train struck first *Tom Thumb* and then *Jumbo. Jumbo* was carried about 100 yards down the line and his body derailed the engine. He died within a few minutes and *Tom Thumb* also expired a few hours later. Keeper Scott was heartbroken at *Jumbo's* death while Barnum tried to capitalise on the accident by making up a story that *Jumbo* had died saving the keeper's life.[1403] [1404]

Canada also experienced a rather strange accident on 7th June 1886 at Petewawa Bridge, Ontario. A Canadian Pacific Railway freight train crossing the bridge 10 miles west of Pembroke included a car loaded with a steam shovel for Smith's Falls. As the car was crossing the bridge some part of the shovel caught the framework of the bridge and the bridge collapsed, causing the train to fall 40 feet to the river below. Conductor Williams was killed instantly, and a number of others were dangerously injured. The engine and tender escaped and proceed to Pembroke with all the injured who could be found.[1405]

Continuous heavy rain caused a bad accident on the line from Marseilles Saint Charles to Lyon Perrache via Grenoble, in France, on Friday 12th November 1886. A passenger train from Marseilles had passed Peypin station and was two kilometres from that at Sisteron, travelling at full speed, when a terrific mass of earth and stones fell from Mont Gerves onto the locomotive and the leading first-class carriage. The engine was overturned and the first-class carriage 'smashed to atoms'. Three people were killed more or less outright and three more died of their injuries before they could be taken to any house, the dead comprising four passengers, the engine driver and a telegraph clerk. The landslide reportedly covered the line with 300,000 cubic metres of earth and gradually covered the whole train. The accident happened at 11.50 a.m. Assistance was sent from Sisteron and the injured passengers dug out of the earth, while the body of the driver was found crushed beneath his engine. A further landslide at 5 p.m. did no more harm. A relief train returned with 20 injured passengers to Digne

[1403] Webpage – Jumbo the Circus Elephant and his Tragic Death
[1404] *Boston Daily Globe*, Boston, Massachusetts, United States, 17th September 1885
[1405] *Warren Daily Mirror*, Warren, Pennsylvania, United States, 8th June 1886

at 7 p.m. Two more of the passengers and the conductor of the train died of their injuries on Sunday 14th, two days after the accident, and the life of the fireman was despaired of. Also, on Sunday M. Millaud, the new Minister of Public Works, visited the scene of the accident. [1406] [1407]

The stoppage of a steam-powered express train at a station at an intermediate station always incurred a considerable loss of time not only in the time actually spent standing at the platform but in the deceleration and re-acceleration of the train. There was therefore an incentive for express trains to make as few stops as possible in order to reduce the overall time for the journey. To cater for passengers wishing to alight at stations at which the express did not call the idea was early adopted in Great Britain, though not elsewhere, of having slip coaches which could be detached at the appropriate point along the journey to drop passengers at intermediate stations without the whole train needing to stop. The idea was widely adopted in Great Britain in the 19th century, especially by the Great Western Railway. It did, however, have a number of disadvantages, for example after the introduction of corridor trains passengers riding in the slip coach were unable to access the restaurant car. Nor was the reverse procedure of attaching a coach from an intermediate station to a non-stop express feasible. There was also the possibility of the slip coach accidentally becoming detached prematurely, while once the slip coach had been detached, even at the correct point, there was then an infringement of the absolute block system, inasmuch as there were then two 'trains' within the same block section at the same time. It is doubtless a tribute to the care of the staff involved that there were no serious accidents involving slip coach use, but there were a few more minor ones, one of which will be next described.

On 14th February 1887 the Great Northern Railway 9 a.m. down express from London King's Cross had a slip coach for Stamford attached at Peterborough for the run of about 12 miles to Essendine, the junction for the branch to Stamford, to prevent the express having to stop again at Essendine or passengers for Stamford having to change at Essendine for Stamford. Unfortunately as the train was passing Werrington Junction signal box, 3¼ miles north of Peterborough, the slip coach accidentally became detached and Signalman Charles Tuffnail sent the "Train divided" signal to Helpstone cabin. Slip Coach Guard William Johnson also realised what was happening when he saw the main part of the train pulling away from him and he held out a red flag, first at one side and then at the other side, to attract the attention of the Head Guard, William Payne. His flag was seen by Alfred Shore, the foreman of a gang of platelayers working on the line just north of Werrington Junction, who held up his arms and showed a red flag to the driver of the express, Edward Platt. Seeing this, Driver Platt applied the continuous brakes and quickly brought the express to a stand. The problem was that the train brakes, though continuous, were still of the non-automatic type, and did not therefore apply on the slip coach. Having waved his red flag Guard Johnson belatedly applied his hand brake but could not stop the slip coach before it ran into the back of the main train. Eight passengers and three guards, including Guard Johnson, were injured or shaken.

[1406] *The Standard*, London, 15th November 1886

[1407] Wikipedia – Liste des accidents ferroviaires en France au XIX siècle

Major F.A. Marindin, the Board of Trade inspecting officer, considered that Guard Johnson had been 'exceedingly injudicious' in waving his flag before he applied the hand brake, though he pointed out that had the train been fitted with automatic brakes the collision would have been avoided. He was probably being hard on the guard, who said that he had not expected the express to stop before Helpstone, two miles from Werrington Junction. The cause of the coach having become detached was unclear.[1408 1409]

Four years before the Werrington Junction mishap a slip coach accident had occurred at Halesowen in August 1883 and there were two more before 1901, at Reading in May 1894 and Warwick in December 1895, all three on the Great Western Railway. None of them resulted in a fatality.[1410]

Powerful winds blew a passenger train from the rails near Fiume, Hungary, on about 5th January 1889. Three persons lost their lives, and many were hurt.[1411]

The same number of deaths as in the Fiume accident and 89 cases of injury resulted when a train of 19 carriages and waggons was blown over an embankment near Madhangur, India, on the Northern Bengal Railway on 15th April 1889.[1412]

Eastbound Southern Pacific Railroad passenger train No. 19 encountered a waterspout at Johnson Siding, near Del Rio, Texas, United States, on 11th September 1890. The engineer of the train, seeing the flooded track ahead, had stopped his train and with his fireman walked forward to check the safety of a bridge crossing a stream. The water continued to rise with great rapidity and the two men were swept away by the floodwater but fortunately were able to swim ashore downstream. The floodwater destroyed two small bridges and several hundred yards of track while some reports stated that the waterspout, 40 feet wide, also struck the train itself, overturning the engine, baggage and mail cars though the passengers escaped. Three years earlier a waterspout had reportedly wrecked a freight train near Anacacho, just east of Del Rio.[1413 1414]

An unusual, though not unique accident, occurred on the Jervois Bridge at Port Adelaide, South Australia, on the morning of Tuesday 16th December 1890, when a train crossing the bridge was struck by a ship. The *Formosa*, a full-rigged sailing ship from Melbourne, had been moored at the Copper Company's wharf with her boom out and as the train was passing the martingale struck the roofs of its carriages, damaging three of them. The captain of the vessel had apparently failed to comply with the harbour regulations which decreed that the jib booms should be rigged in within 48 hours of the

---

1408 Board of Trade report
1409 *Great Northern Railway Society Newsletter* 81 A1 – A8
1410 *Backtrack* Volume 26, pgs. 270-3
1411 *Belleville Democrat*, Belleville, Kansas, United States, 10th January 1889
1412 *The Times*, London, England, 22nd April 1889
1413 *Goshen Daily News*, Goshen, Indiana, United States, 12th September 1890
1414 *Laredo Times*, Texas, United States, 13th September 1890

ship's docking.[1415]

As noted in the article on the Canonbury collisions in the Tunnel Accidents chapter, triple collisions, that is three separate collisions between different trains occurring one after another, were very rare. Apart from the Canonbury accident only a handful of triple collisions seem to have happened anywhere in the world during the 19th century. One of the others was at East Thompson, Connecticut, United States, on the New York and New England Railroad, early on 4th December 1891.

Early on the morning of the day of the accident a Southbridge way (local or pick-up) freight train was being made up in a siding alongside the westbound main track, which line it was also using, and the trailing points leading into the siding were standing open. The morning was dark and foggy with visibility down to 100 yards, but the conductor of the way freight had failed to send out any flag protection while his train was occupying the main line. What was worst, late-running eastbound freight No. 212 was using the westbound track, having received orders to do so between Putnam and East Douglas, stations on either side of East Thompson. The trailing points into the siding being used by the way freight thus became facing points for train No. 212. Engineer Henry Wildes of train No. 212 saw the switch set for the siding at the last moment, whistled "Down brakes", and then, seeing that a collision could not be avoided, he and his fireman, named Jacob Boyce, jumped, the fireman first.

Four waggons at the head of the freight train; two flat trucks loaded with bricks, one car loaded with apples and one loaded with salt were piled up in a heap in the ensuing collision. The cars of the way freight were derailed and one, loaded with cotton bales, was thrown at right angles across the main lines. Engineer J.A. Page and Fireman Fred Cook also jumped from their engine before the collision and were unhurt. The first collision occurred at 6.40 a.m. Flagmen were sent in each direction but the eastbound Long Island and Eastern States Express (referred to in newspaper reports as 'the Brooklyn train') was very close at hand and the flagman had only covered about 200 yards when it came sweeping around the corner travelling at between 40 and 50 miles an hour. The flag prevented Engineer Harry Taber from seeing the flagman until he was almost on top of him, when he whistled for brakes and reversed his engine. The collision threw the engine into a field and its tender fell onto it, killing Engineer Taber and his fireman, Gerry Fitzgerald. The leading smoking car of the express struck the cotton car of the way freight in turn and was damaged but no one in it was much hurt.

Both the express and the Norwich Steamboat train were running an hour late, and the express had overtaken the boat train at Putnam. The boat train then followed the express and was close behind it at East Thompson. Like Engineer Taber a couple of minutes earlier Engineer W.U. Lowden saw the warning from the flagman at short range, then the Pullman car *Midland* at the rear of the express blocking the line. He reversed his engine and he and Fireman Edward Hurley jumped. They were more fortunate than the crew of the express, escaping with numerous bruises and a severe shaking. It was also fortunate that the passengers in the rear Pullman car had vacated it before the third collision, for it was set alight by coals from the firebox of the boat train engine and was soon well ablaze. The fire also spread to baggage and passenger cars of the boat train and the freight

[1415] *The Evening Journal*, Port Adelaide, South Australia, 16th December 1890 waggons. [1416] [1417]

No passengers were confirmed as having died in the accident, though a number were more or less injured. Mystery surrounded a passenger named Edward Rath who had supposedly been in the *Midland* car, which was reduced to ashes in the fire. No trace of his body could be found in the wreckage despite it having been examined 'with microscopical care', and although his relatives accepted that he had died in the accident there were suggestions that his disappearance was an attempt at an insurance fraud.[1418]

At Putnam at the end of January 1892 Windham county Coroner A.G. Bill found Engineer Wildes and Conductor William Dorman of freight train No. 212 responsible for the accident for 'neglecting or refusing to obey the orders of the proper authority relative to the running of the train.' [1419] Presumably the regulations directed that they should have been running cautiously while operating their train in the wrong direction through a station yard.

If the East Thompson triple collision of 1891 was one of a rare kind, the series of collisions at Brooklyn, New York, United States, on 1st March 1894 was unique. There was a thick fog as engine No. 70 with a train for East New York approached the station at Manhattan Beach, on the Broadway branch of the Union Elevated Railway (which was leased to the Brooklyn Elevated Railway) at about 7.30 a.m. on the day of the accident. Fearful of running into a train stopped at the station and invisible through the fog the engineer slowed his four-car train down to a standstill, whereupon it was run into by another four-car train hauled by engine No. 73, which telescoped the rear car of the first train. The tender and cab of engine No. 73 was also badly damaged, and the rebounding engine telescoped the smoking car of its own train. Engineer Graves had time to jump from engine No. 73 and climbed a tree adjoining the line but his fireman, Eugene Davis, was caught in the cab and suffered a badly cut leg.

A few minutes after the initial collision engine No. 55, with a train of four cars from Broadway ferry, ran into the rear of the second train, 'with more splintering of glass and ripping of woodwork', and its smoker, like that of the previous train, was damaged. Four car trains attached to engines Nos. 53 and 3 collided in turn, then a three-car train drawn by engine No. 7 telescoped the rear of the one drawn by engine No. 3, backed and damaged its own smoking car. Further collisions followed involving trains hauled by engines Nos. 58, 74, 52 and 91, making an incredible nine rear collisions in all over a period of over half an hour. The last four collisions resulted only in slightly damaged platforms, ruined headlights and broken glass.

The wrecked trains now stretched in a long row along the line and the scene was viewed by crowds of people along the pavements and Broadway. Policemen from the 14th precinct tried to telephone the station house but were unable to do so because of broken wires and it was left to Fireman John Bralstead, travelling to work on the train hauled by engine No. 91, to run back 70 yards along the narrow planking bordering the rails and stop yet another train by waving the red-lined lining of his jacket, thus preventing a 10th

---

1416 *Boston Daily Globe*, Boston, Massachusetts, United States, 4th and 5th December 1891
1417 *Boston Post*, Boston, Massachusetts, United States, 5th December 1891
1418 *Boston Daily Globe*, Boston, Massachusetts, United States, 8th December 1891
1419 Ibid, 29th January 1892

collision. A wrecking train then came from Manhattan Beach to sort out the mess and the lines were cleared by 10 a.m. Remarkably no one was killed in all the collisions; in fact, the only person hurt appears to have been hurt was Fireman Davis of engine No. 73.[1420] It would be interesting to know what sort of signalling was employed on the line, if any.

Probably trains were simply operated 'on sight' as on the Chicago elevated railway as late as 1947.[1421]

A freak accident which it would be hard to make up occurred at Bellefontaine, Ohio, United States, on Monday 5th August 1895. The Knights of Pythias (a fraternal organisation and secret society) of Springfield and Dayton had chartered a special train to take them to a camp at Sandusky, Ohio. Already in camp at Johnson's Island, near Sandusky, and breaking camp to go home on the same day was the Third Regiment of the Ohio National Guard, in their own train. In the baggage car of the Pythian's train was a small cannon, with which they were firing off salutes as the train went along! One of these salutes was fired just as the regimental train was passing on the other track, the two trains being of course only a couple of feet apart. The full charge of the train was fired straight into the coach, which had its windows open. About 20 soldiers were more or less seriously injured.[1422] A New York newspaper added that, 'To make matters worse the car was filled with the blinding and suffocating smoke of the powder. With powder burning into some, blood flowing from the wounds of others who had been struck by the heavy wadding, the scene was one of confusion. The hospital corps of the regiment was called into service. Three men were badly cut about the head. Two little boys, "mascots," had their faces shot completely full of powder. Two soldiers were made deaf, probably beyond recovery. Another will lose his eyesight. One man was sitting with his face between his hands, looking out of the window, and received the full charge in the face. Some twenty more had their faces partly filled with powder. The injured were taken care of at Dayton.' [1423]

An earthquake on Saturday 12th June 1897 killed eight people in Calcutta, India, and overturned a train at an unstated location on the Assam – Bengal line. Casualties in the train accident, if any, were not reported.[1424]

As previously noted, Spain and the United States were at war for 10 weeks in the summer of 1898, with hostilities in both the Caribbean and the Pacific. On 6th June 1898 the USS *Dolphin* shelled a train carrying Spanish troops near Santiago de Cuba, Cuba. Many soldiers were reportedly killed.[1425] Another train was shelled by a gunboat two days later.

[1420] *Brooklyn Daily Eagle*, Brooklyn, New York, 1st March 1894
[1421] Webpage – www.Chicago-L.org – Wayside signals
[1422] *Norwalk Daily Reflector*, Norwalk, Ohio, United States, 7th August 1895
[1423] *Brooklyn Daily Eagle*, Brooklyn, New York, United States, 6th August 1895
[1424] *Mid-Surrey Times and General Advertiser*, London, England, 19th June 1897
[1425] *Bloomington Weekly Pantagraph*, Bloomington, Illinois, United States, 10th June 1898

Late on the evening of Saturday 25th June 1898 a large number of passengers who had attended a performance at Clear Lake, Iowa, United States, boarded three four-coach trains of the Clear Lake & Mason City Electric Railway for the trip eastwards to Mason City. The three trains set off at short intervals. Two miles out of Clear Lake the first train came into collision with a westbound train and the two following eastbound trains, unable to pull up, collided with the first train and with each other in another triple collision. Twenty passengers were more or less injured, three seriously.[1426]

An accident which was to be the forerunner of many more in the next century, inasmuch as it involved a train carrying hazardous chemicals, occurred near Portsmouth, New Hampshire, United States, on the Boston and Maine Railroad, on Friday 27th January 1899. A freight train was derailed and several carboys of vitriol (sulphuric acid) were broken and their contents spilt along the track. A wrecking gang set to work to re-rail the waggons, unaware of the danger which to they were exposing themselves, and soon several of them were yelling of burning in their feet and legs, the acid having eaten right through their footwear. Several men were quite badly burned and laid up with their injuries.[1427]

A Reuters correspondent in Valparaiso, Chile, telegraphed to London a report of an accident on the State Southern Railway on 11th March 1899. According to the report this was a triple collision, four trains coming into collision almost simultaneously. The engines attached to the trains were completely wrecked and many carriages caught fire. Perhaps surprisingly in view of the apparent magnitude of the accident only one passenger was reportedly killed and 12 injured. The damage to rolling stock was estimated at $200,000.[1428]

On about 8th June 1899 lightning struck an oil car on a train near Holly, Michigan, United States, starting a fire which burned nine cars and reportedly cremated five tramps.[1429]

A passenger train was hit by a typhoon in Japan on 7th October 1899. British sources gave the location of the accident as Utsu-no-miya, a town 73 miles north of Tokyo. Initially it was reported that six persons had died, and many had been severely injured; a second report stated that the accident had been much more serious than had been initially stated and that 50 passengers had drowned.[1430] A Japanese source stated that the accident had occurred between Yaita and Nozaki stations in Tochigi prefecture on the Tohoku main line of the Nippon Railway. A mixed train of 11 passenger carriages and 11 waggons was blown from a bridge into a river. According to this, probably more reliable source, 20 passengers died and 45 were injured. The bodies of some of the victims were carried away in the flooded river and never recovered.

1426 *Waterloo Daily Courier*, Waterloo, Iowa, United States, 27th June 1898
1427 *Portsmouth Herald*, Portsmouth, New Hampshire, United States, 30th January 1899
1428 *The Times*, London, England, c13th March 1899
1429 *South Haven Sentinel*, South Haven, Michigan, United States, 17th June 1899
1430 *The Argus*, Melbourne, Victoria, 12th October 1899

A civil court case followed the disaster, the relatives of the victims claiming that the railway company should have suspended traffic after receiving weather reports advising of the likelihood of a typhoon. The company lost the first round of the case and agreed to a court-mediated settlement with the relatives in the second round.[1431]

What might justifiably be described as a freak accident occurred to a Steubenville accommodation train of the Panhandle (Pittsburgh Cincinnati Chicago and St Louis) Railroad at 6.40 a.m. on 2nd August 1900 near Carnegie, Pennsylvania, United States. The train was passing under a coal chute of the Boyd Coal Company when the machinery of the chute failed, and 10 tons of coal were precipitated onto the train. The roof and sides of the rear coach of the train were smashed and its many passengers were buried in coal and fragments of the car. A painter named Smith, of Holliday's Cove, was removed to his home terribly crushed and cut, and it was feared that he was fatally injured. Another painter, named Johnson, of Steubenville, was also badly hurt, but it was hoped he would recover. A number of other passengers were less seriously hurt.[1432]

The American state of Texas was devastated by a hurricane in September 1900. A telegram from the Hon. John. H. Pope, a member of the Louisiana state board of education, stated that 85 passengers had lost their lives when a hurricane struck a Gulf and Interstate Railroad train on Saturday 8th September 1900. The train had left Beaumont, Texas, early on Saturday having made a connection with a Southern Pacific Railroad train which had left New Orleans, Louisiana, on Friday night, Mr Pope being one of its passengers. The train reached Bolivar Point (Port Bolivar) at noon and preparations were made to run into onto the ferry for the trip across Galveston Bay to Galveston. The force of the wind was too strong for the ferry to make a landing and after standing on the pier for a few minutes the hurricane began to blow the train back along the track towards Beaumont.

With the wind rapidly increasing in strength as it blew off the open sea the water was soon reaching level with the bottoms of the seats, inducing the passengers to abandon the train and take refuge in a nearby lighthouse. Mr Poe stated that despite all efforts 85 passengers were blown away or drowned, the train being entirely wrecked. The surviving passengers had to spend 50 hours in the lighthouse with almost no food. 'The experience is one that they will remember as one of the most terrible of their whole lives.' [1433]

Another passenger train which left Houston at 7.55 p.m. on that Saturday night was wrecked by a hurricane two miles north of Alvin, Texas, on the Gulf Colorado and Santa Fe Railroad. The whole train was lifted from the track. A Mrs Prather, of Rosenburg, Texas, was thrown across the car and halfway through a window. When she was reached after the accident it was found that her head had ended up under water and she had drowned. Several other passengers were injured.[1434]

[1431] Webpage – Japanese Railway Safety and the Technology of the Day by Masao Saito
[1432] *Evening Express*, Lock Haven, Pennsylvania, United States, 3rd August 1900
[1433] *Bryan Morning Eagle*, Bryan, Texas, United States, 15th September 1900
[1434] *Anaconda Standard*, Anaconda, Montana, United States, 10th September 1900

INDEX

TABLE OF MAJOR RAILWAY ACCIDENTS 1881-1900

Date	Location	Railway Company	Description	Passenger Fatalities
17 Jan 1881	Golden Rock Curve, near Puerto Cortes, British Honduras		Derailment	?
22 Jan 1881	Near Tioga Center, New York, United States	Erie	Derailment, broken axle, fire	5
20 Apr 1881	Near Rock Tunnel, Colorado, United States	Denver and Rio Grande	Derailment	8
21 Apr 1881	Meredosia River Bridge, Illinois, United States	Chicago St Paul and Milwaukee	Bridge collapsed under train	8
24 June 1881	Near Malpas, Mexico		Bridge collapsed under train, fire, explosions	254
4 Jul 1881	Near Covington, Kentucky, United States	Kentucky Central	Head-on collision, single line?	6
8 Aug 1881	Blackburn, England	Lancashire and Yorkshire	Rear collision	7
5 Sep 1881	Charenton, France	Paris Lyon Méditerranée	Rear collision	22
8 Sep 1881	Anchorage, Kentucky, United States	Louisville and Nashville	Collision with cow, derailment, bridge collapsed under train	9

24 Nov 1881	Near Savannah, Georgia, United States		Collision	5
25 Nov 1881	Tayport, Scotland	North British	Head-on collision	5
13 Jan 1882	Spuyten Duyvil, New York, United States	New York Central and Hudson River	Rear collision, fire	8
28 Jan 1882	Between Old Ford and Bow, England	North London	Derailment, defective vehicle, collision	5
21 Mar 1882	Sweet Briar Creek, Dakota Territory, United States	Northern Pacific	Derailment, broken wheel, train fell from bridge, fire	8
24 May 1882	Near Fuerte Pio, Barcelona, Spain		Train diverted onto wrong line, derailment	5
30 May 1882	Heidelberg, Germany		Train diverted onto wrong line, collision	12
15 Jun 1882	Taltal, Chile	Taltal	Runaway train, derailment or buffer stop collision	10
29 Jun 1882	Parkers Creek, New Jersey, United States	New York and Long Branch	Derailment, track, defects, train fell from trestle	5
12 Jul 1882	Between Tcherny and Bastijeur, Russia		Derailment	178
1 Aug 1882	Near South Bend, Indiana, United States	Lake Shore and Michigan Southern	Head-on collision, single line, in fog, fire	6
4 Aug 1882	Between Cranberry Mine and Oil City, Pennsylvania, United States	Buffalo Pittsburgh and Western	Runaway train, derailment, collision	8

3 Sep 1882	Near Hugstetten, Germany	Baden State	Derailment, excessive speed, track defects	64
22 Sep 1882	Essegg, Croatia		Bridge collapsed under train	27
26 Sep 1882	Between Nefiche and Mahsameh, Egypt		Collision	10
6 Oct 1882	Between Picasso and Inca, Chile		Derailment, sabotage	8
27 Nov 1882	Inverythan Bridge, Scotland	Great North of Scotland	Bridge collapsed under train	5
20 Jan 1883	Tehachapi, California, United States	Southern Pacific	Runaway train, derailment, fire	15
14 May 1883	Lockerbie, Scotland	Caledonian	Double collision	5
23 Jun 1883	Elk Creek, Montana, United States	Northern Pacific	Head-on collision, single line?	23
24 Jun 1883	Between Havana and Matanzas, Cuba		Collision	8
2 Jul 1883	Near Rasselas, Pennsylvania, United States	Rochester and Pittsburgh	Runaway vehicles, collision	5
27 Jul 1883	Carlyon, New York, United States	Rome Watertown and Ogdensburg	Collision with waggon	19
21 Aug 1883	Near Rochester, Minnesota, United States	Chicago and North Western	Train struck by tornado	20

3 Sep 1883	Between Carmi and Grayville, Illinois, United States	Louisville and Nashville	Collision with cattle, derailment	9
16 Nov 1883	Otter Creek, Illinois, United States	Chicago Burlington and Quincy	Rear collision, boiler explosion	7
29 Nov 1883	Near Saint-Meen, France		Collision	18
24 Dec 1883	Blue River Bridge, Near Salem, Indiana, United States	Louisville New Albany and Chicago	Bridge collapsed under train, fire	5
2 Jan 1884	Toronto High Park, Ontario, Canada	Grand Trunk	Head-on collision, single line, boiler explosion, fire	28
27 Apr 1884	Alcudia River Bridge, Spain		Bridge collapsed under train, sabotage	59
14 May 1884	Near Connellsville, Pennsylvania, United States	Baltimore and Ohio	Head-on collision, single line, fire	14
3 Jun 1884	Between Downton and Breamore, England	London and South Western	Derailment, excessive speed over poor track	5
16 Jun 1884	Between Hazlehead and Penistone, England	Manchester Sheffield and Lincolnshire	Derailment, broken axle	24
29 Aug 1884	Windsor, Colorado, United States	Colorado and Southern	Fire	10
2 Oct 1884	Arrunghatta, India	Eastern Bengal	Collision	13
14 Nov 1884	Hanau, Germany		Double collision	22

14 Nov 1884	Near Hempstead, Texas, United States	Southern Pacific	Derailment, sabotage	6
25 Jan 1885	Cootamundra, New South Wales		Bridge washed away, train fell into river	6
9 Feb 1885	Between Cromwell and Prescott, Iowa, United States	Chicago Burlington and Quincy	Derailment, broken rail, train fell from bridge	9
11 Jun 1885	Near New River, Tennessee, United States	Cincinnati Southern	Collision with cow, derailment	6
4 Oct 1885	Between Corinth and Kiatka, Greece		Derailment on points, boiler explosion	8
4 Oct 1885	Near French, Minnesota, United States	Northern Pacific	Runaway coaches, collision	5
13 Oct 1885	Omori, Japan		Derailment	7
18 Oct 1885	Hackensack River Bridge, New Jersey, United States	Pennsylvania	Double collision in fog, fire	9
14 Dec 1885	Austell, Georgia, United States	Georgia Pacific	Rear collision	14
12 Jan 1886	Near La Huaca, Peru		Collision with bull, derailment	22
7 Apr 1886	Between Bardwell and West Deerfield, Massachusetts, United States	Fitchburg	Derailment, track subsidence, fire	10

7 Jun 1886	Santee River Trestle, South Carolina, United States	Northeastern	Derailment, track defects, train fell from trestle	6
30 Jun 1886	Between Portadown and Tangeradee, Ireland	Great Northern (Ireland)	Derailment, track distorted by the heat, slight track defects	6
1 Jul 1886	Between Würzburg and Rottendorf, Germany		Head-on collision	13
29 Aug 1886	Mödling, Austria		Collision	7
14 Sep 1886	Near Silver Creek, New York, United States	New York Chicago and St Louis	Head-on collision, single line	15
19 Oct 1886	Between Deweesee and Lawrence, Nebraska, United States	Burlington and Missouri River in Nebraska	Collision with bull, derailment	5
28 Oct 1886	East Rio, Wisconsin, United States	Chicago Milwaukee and St Paul	Derailment on points, fire	17
3 Nov 1886	Between Nikolaiev and Odessa, Russia		Derailment, sabotage	40
12 Nov 1886	Sisteron, France		Train struck by landslide	6
18 Dec 1886	Charkov, Russia		Collision	13
4 Jan 1887	Near Republic, Ohio, United States	Baltimore and Ohio	Head-on collision, single line, fire	15

5 Feb 1887	White River Junction, Vermont, United States	Central Vermont	Derailment, broken rail? broken journal? Part of train fell from bridge, fire	36
14 Mar 1887	Bussey Bridge, Near Roslindale, Massachusetts, United States	Boston and Providence	Bridge collapsed under train, fire	22
27 May 1887	Near Kittaning, Pennsylvania, United States	Pennsylvania	Derailment, broken wheel, collision	6
21 Jun 1887	Peats Ferry, New South Wales		Runaway train, collision	5
15 Jul 1887	St Thomas, Ontario, Canada	London and Port Stanley/ Michigan Central	Side collision on rail/rail level crossing, fire, explosion	13
10 Aug 1887	Near Chatsworth, Illinois, United States	Toledo Peoria and Western	Fire damaged trestle collapsed under train	82
16 Sep 1887	Hexthorpe, England	Manchester Sheffield and Lincolnshire	Rear collision	25
10 Oct 1887	Kouts, Indiana, United States	Erie	Rear collision in fog, fire	9
10 Jan 1888	Between Bradford and Haverhill, Massachusetts, United States	Boston and Maine	Derailment, broken wheel	11
21 Jan 1888	Near San Christobal, Cuba		Head-on collision, single line	?

17 Mar 1888	Allapaha Creek Trestle, Near Blackshear, Georgia, United States	Savannah Florida and Western	Derailment, broken axle, trestle collapsed under train	24
5 Apr 1888	New Hampton, Iowa, United States	Chicago Milwaukee and St Paul	Derailment, obstruction on line	5
13 May 1888	Near Galitsyno, Russia		Runaway coaches, collision	11
19 May 1888	Between San Pol de Mar and Canet de Mar, Spain	Tarragona	Head-on collision, single line?	6
3 Jun 1888	Near Granada, Nicaragua		Derailment, track washed away	5
5 Jun 1888	40 km from Tampico, Mexico	Ferrocarril Central Méxicano	Collision with cow and donkey, bridge collapsed under train, driver intoxicated?	13
12 Jul 1888	Browning Trestle, Between Orange Courthouse and Barbersville, Virginia, United States	Virginia Midland	Trestle collapsed under train	9
14 Jul 1888	Hyde Junction, England	Manchester Sheffield and Lincolnshire	Derailment, broken axle, collision	5
26 Jul 1888	Yutan, Nebraska, United States	Omaha and Republican Valley	Collision between portions of divided train	6
5 Sep 1888	Near Velars, France	Paris Lyon and Méditerranée	Derailment, track defects, collision	8

9 Sep 1888	Waynesville, Ohio, United States	Little Miami	Rear collision	5
10 Sep 1888	Near Wadsworth, Ohio, United States	New York Pennsylvania and Ohio	Rear collision	5
10 Oct 1888	Mud Run, Pennsylvania, United States	Lehigh Valley	Rear collision, fire	64
20 Oct 1888	Near Grassano, Italy		Derailment, train ran into landslide	22
29 Oct 1888	Near Borki, Russia		Derailment, track defects	17
7 Nov 1888	Walker's Crossing, Near Ellaville, Georgia, United States	Savannah and Western	Derailment, broken rail	5
14 Jan 1889	Near Talmadge, Ohio, United States	New York Pennsylvania and Ohio	Head-on collision, single line, in fog, fire	5
3 Feb 1889	Near Groenendael, Belgium		Derailment, overbridge fell onto train	21
27 Feb 1889	St George, Ontario, Canada	Grand Trunk	Derailment, broken tyre, train fell from bridge	10
25 Mar 1889	Chung Liang Cheng, China		Head-on collision, single line, driver intoxicated	20
28 Apr 1889	Near Hamilton, Ontario, Canada	Grand Trunk	Derailment, excessive speed? engine defect? fire	18
12 June 1889	Near Armagh, Ireland	Great Northern (Ireland)	Runaway coaches, collision, runaway coaches	83

26 Jun 1889	Near Latrobe, Pennsylvania, United States	Pennsylvania	Double collision, part of train fell from bridge, fire	12
2 Jul 1889	Near Thaxton, Virginia, United States	Norfolk and Western	Derailment, track washed away, fire, boiler explosion	15
7 Jul 1889	Röhrmoos, Germany		Train diverted onto wrong line, collision	10
22 Aug 1889	Near Flat Creek Gap, Tennessee, United States	Knoxville Cumberland Gap and Louisville	Derailment, track defects? Part of train fell from trestle	6
24 Sep 1889	87th Street, Chicago, Illinois, United States	Chicago Rock Island and Pacific	Collision with coach, driver intoxicated	7
1 Oct 1889	Near Wildpark, Germany		Collision with light engine	8
4 October 1889	Near Longsight, England	London and North Western	Collision with engines and van	6
28 Oct 1889	Hatras, India		Derailment, track defects	18
28 Dec 1889	Jerrys Run, West Virginia, United States	Chesapeake and Ohio	Derailment, broken rail? Broken axle?	9
27 Jan 1890	Carmel, Indiana, United States	Louisville New Albany and Chicago	Derailment, track defects, train fell from trestle, fire	6
1 Mar 1890	Near Penshurst, Victoria		Collision with bullock, derailment	7
6 Mar 1890	Bay View, New York, United States	Lake Shore and Michigan Southern	Collision between portions of divided train	6

30 May 1890	Oakland, California, United States	South Pacific Coast	Train fell from open drawbridge	13
9 June 1890	Near Warrenton, Missouri, United States	Wabash	Head-on collision, single line	7
19 Aug 1890	Quincy, Massachusetts, United States	Old Colony	Derailment, obstruction on line, fire	22
19 Sep 1890	Shoemakersville, Pennsylvania, United States	Philadelphia and Reading	Double collision, fire	21
19 Sep 1890	Near Rinconada, Mexico		Head-on collision, single line	10
27 Sep 1890	Near 18th Street, Chicago, Illinois, United States	Baltimore and Ohio	Rear collision	7
11 Nov 1890	Norton Fitzwarren, England	Great Western	Head-on collision	11
14 Nov 1890	Near Salonica, Macedonia		Derailment, track washed away	37
11 Dec 1890	Near Needles, California, United States	Atlantic and Pacific	Rear collision	5
18 Dec 1890	St Joseph de Levis, Quebec, Canada	Intercolonial	Derailment, broken axle, part of train fell from bridge	7
16 Jan 1891	Near Girard, Iowa, United States	Chicago Milwaukee and St Paul	Head-on collision, single line	12

26 Jan 1891	Near Irving, Kansas, United States	Union Pacific	Head-on collision, single line	8
20 Feb 1891	Park Avenue Tunnel, New York, United States	New York Central and Hudson River	Double collision in tunnel, fire	6
18 Apr 1891	Kipton, Ohio, United States	Lake Shore and Michigan Southern	Head-on collision	6
10 May 1891	Near Austin, Pennsylvania, United States	Sinnemahoning Valley	Fire	5
31 May 1891	Jerrymys Run Trestle, Vaughn, Virginia, United States	Norfolk and Western	Rear collision, trestle collapsed under train	7
14 Jun 1891	Birs River Bridge, Münchenstein, Switzerland		Bridge collapsed under train	71
3 Jul 1891	Ravenna, Ohio, United States	New York Lake Erie and Western	Rear collision, fire	25
4 Jul 1891	Farm, West Virginia, United States	Kanawha and Michigan	Derailment, track defects, part of train fell from fire-damaged trestle	15
8 Jul 1891	Near Warsaw, Russian Poland		Collision	6
11 Jul 1891	Aspen Junction, Colorado, United States	Colorado Midland	Fouling collision	10
22 Jul 1891	Near Presburg, Hungary		Derailment, track washed away	10

26 Jul 1891	St Mandé, France	Chemin de fer de L'Est	Rear collision, fire	51
31 Jul 1891	Near Galion, Ohio, United States		Rear collision	5
6 Aug 1891	Near Montezuma, New York, United States	West Shore	Rear collision in fog, fire	15
17 Aug 1891	Near Zollikofen, Switzerland		Rear collision	14
18 Aug 1891	Near Ostrowo, Russian Poland		Head-on collision, single line? fire	10
27 Aug 1891	Catawba River (Bostian Bridge), North Carolina, United States	Richmond and Danville	Derailment, train fell from trestle, sabotage	18
31 Aug 1891	Between Tell City and Troy, Indiana, United States	Louisville Evansville and St Louis	Derailment, broken wheel	5
18 Sep 1891	Port Neuf, Idaho, United States	Union Pacific	Head-on collision	5
20 Sep 1891	Near Wolbrom, Russian Poland		Head-on collision, single line?	7
23 Sep 1891	Between Burgos and Quintanilleja, Spain		Head-on collision, single line	22
24 Sep 1891	McKims Siding, Near Zelienople, Pennsylvania, United States	Pittsburgh and Western	Collision	8
18 Oct 1891	Kohlfurt, Germany		Sidelong collision, driver intoxicated	6

27 Oct 1891	Thompson, Minnesota, United States	Minneapolis St Paul and Sault Ste Marie	Rear collision	5
5 Nov 1891	Near Nagpur, India		Derailment, broken tyre	12
23 Nov 1891	Near Domnino, Russia		Derailment, broken tyre? Part of train fell from bridge	31
28 Nov 1891	Toledo, Ohio, United States	Lake Shore and Michigan Southern	Rear collision	10
5 Dec 1891	Near Sahiwal, India	North Western	Head-on collision, single line	30
24 Dec 1891	Near Hastings, New York, United States	New York Central and Hudson River	Rear collision	15
6 Feb 1892	Between Bucharest and Braila, Romania		Fire	18
23 Feb 1892	Near Cuyambuca, Brazil	Estrada de Ferro de Recife ao São Francisco	Collision	50
23 Mar 1892	Sonsonate, El Salvador		Derailment	13
27 Apr 1892	Between Sodwalls and Tarana, New South Wales		Derailment, broken rail	8
5 May 1892	Near Revere, Missouri, United States	Atchison Topeka and Santa Fe	Trestle collapsed under train, fire	7

20 May 1892	Near Crooked Bayou, Arkansas, United States	St Louis Southwestern	Collision	7
9 Jun 1892	Esholt Junction, England	Midland	Sidelong collision	5
14 Jun 1892	Bishopsgate, England	Great Eastern	Rear collision	5
25 Jun 1892	Harrisburg, Pennsylvania, United States	Pennsylvania	Rear collision, fire	12
21 Aug 1892	Near Lignan, France		Derailment, track defects	5
10 Sep 1892	West Cambridge, Massachusetts, United States	Fitchburg	Rear collision in fog, fire	8
16 Sep 1892	Between Alcanena and Torres Novas, Portugal		Runaway train, derailment	9
21 Sep 1892	Brown's Cut, Near Wooster, Ohio, United States	Pittsburgh Fort Wayne and Chicago	Head-on collision, single line, fire	9
14 Oct 1892	Near Harrison's Landing, Connecticut, United States	Central Vermont	Head-on collision, single line	5
20 Oct 1892	Near Pensa, Russia		Derailment	20
24 Oct 1892	Flat Rock Tunnel (West Manayunk), Pennsylvania, United States	Philadelphia and Reading	Head-on collision, fire	8

2 Nov 1892	Manor House Siding, England	North Eastern	Rear collision, fire	9
28 Nov 1892	Near Zsolna, Hungary		Collision	5
4 Jan 1893	Kanizsa, Hungary		Sidelong collision	8
c10 Jan 1893	Near Matadi, Congo Free State	Chemin de fer du Congo	Collision with waggon, explosion	50
11 Jan 1893	Between Slatoust and Samara, Russia		Fire	49
12 Jan 1893	South Ripley (Dover) Kentucky, United States	Chesapeake and Ohio	Collision	6
6 May 1893	Tadmore, Ohio, United States	Cincinnati Hamilton and Dayton	Rear collision	5
7 May 1893	Lafayette, Indiana, United States	Cleveland Cincinnati Chicago and St Louis	Runaway train, derailment, fire	6
20 Jun 1893	Parkville Tunnel, New York, United States	New York Brooklyn and Manhattan Beach	Derailment, points moved under train	9
13 Jul 1893	Newburg, New York, United States	West Shore	Derailment on points	7
14 Jul 1893	Between Zumarraga and Bilbao, Spain		Derailment	5
12 Aug 1893	Llantrisant Junction, Wales	Taff Vale	Derailment, engine defect	12
16 Aug 1893	Near Milton, Virginia, United States	Atlantic and Danville	Trestle collapsed under train	7

27 Aug 1893	Bushwick Junction, New York, United States	Long Island	Rear collision	15
31 Aug 1893	Chester, Massachusetts, United States	Boston and Albany	Derailment, inadequate protection of bridgework	9
7 Sep 1893	Near Colehour, Indiana, United States	Pennsylvania	Head-on collision, single line	12
18 Sep 1893	Manteno, Illinois, United States	Illinois Central	Rear collision	12
22 Sep 1893	Near Kingsbury, Indiana, United States	Wabash	Train diverted onto wrong line, collision, boiler explosion	8
13 Oct 1893	Jackson, Michigan, United States	Michigan Central	Rear collision	18
25 Oct 1893	Near Lawrence, New Jersey, United States	Pennsylvania	Derailment, broken axle, collision	5
20 Oct 1893	Battle Creek, Michigan, United States	Chicago and Grand Trunk	Head-on collision, single line, fire	28
3 Nov 1893	Santander, Spain		Train damaged by lineside explosions, fire	?
6 Nov 1893	Near Ramallo, Argentina	Buenos Ayres and Rosario	Collision with 'animal,' derailment	10
8 Nov 1893	71st Street, Chicago, (Near Normal Park), Illinois, United States	Chicago Rock Island and Pacific	Rear collision in fog, fire	12

28 Nov 1893	Limitro. Italy		Collision with shunting goods in fog, fire	43
28 Nov 1893	Near Matadi, Congo Free State	Chemin de fer du Congo	Derailment	10
15 Dec 1893	Near Sheridan, New York, United States	Western New York and Pennsylvania	Trestle collapsed under train	6
15 Jan 1894	Jersey Meadows, New Jersey, United States	Morris and Essex	Rear collision in fog	15
c15 Jan 1894	Near Cumanayagua, Cuba		Collision with cow, derailment	16
c22 Jan 1894	Near Samara, Russia		Collision, fire	15
15 Feb 1894	Near Jelan, Russia		Collision	12
3 May 1894	Between Acajutla and Sonsonate, El Salvador		Derailment, sabotage	200
2 Jul 1894	Between Askwith and Moosehead, Maine, United States	Canadian Pacific	Derailment, trestle collapsed under train, sabotage	5
7 Jul 1894	Near Bilbao, Spain		Runaway train. derailment	13
27 Jul 1894	Near Mihara, Japan		Derailment, track washed away	16
9 Aug 1894	Lincoln, Nebraska, United States	Chicago Rock Island and Pacific	Derailment, train fell from trestle, fire, sabotage	21

1 Sep 1894	Near St Cloud, Minnesota, United States	Eastern Minnesota	Fire	?
1 Sep 1894	Near Pokegama, Wisconsin, United States	St Paul and Duluth	Fire-damaged bridge collapsed under train	?
31 Oct 1894	Redfern, New South Wales		Head-on collision	12
22 Dec 1894	Chelford, England	London and North Western	Collision during shunting operations, collision	14
28 Feb 1895	Between Temamatla and Tenango, Mexico	Ferrocarril Interoceanico del Pacifico	Derailment, excessive speed, driver intoxicated	130
27 Apr 1895	Temamatla, Mexico	Ferrocarril Interoceanico del Pacifico	Derailment	14
9 Jul 1895	Craig's Road, Quebec, Canada	Grand Trunk	Rear collision	11
10 Jul 1895	San Pablo, Corrientes, Argentina		Collision	15
26 Jul 1895	Between Quintin and Plaintel, France	Chemins de fer de l'Ouest	Derailment, excessive speed	8
11 Aug 1895	Bolondron, Cuba		Derailment, sabotage	?
19 Sep 1895	Oederan, Saxony, Germany		Rear collision	10
6 Oct 1895	Near Céroux-Mousty, Belgium		Head-on collision, single line	17
25 Nov 1895	Near Ceinfuego Cuba		Derailment, sabotage	31

30 Dec 1895	Between Dannhausen and Glencoe, Natal		Derailment, excessive speed	37
31 Jan 1896	Near Lawrenceburg Junction, Indiana, United States	Cleveland Cincinnati Chicago and St Louis	Derailment, broken wheel, part of train fell from bridge	5
27 Apr 1896	Between Sikandarabad and Dadri, India		Explosion, fire	6
10 May 1896	Near Adelia, Algeria		Head-on collision, single line	7
11 Jul 1896	Near Logan, Iowa, United States	Chicago and North Western	Head-on collision, single line	31
11 Jul 1896	Near Annotto Bay, Jamaica		Runaway train, collision	17
12 Jul 1896	Near Altonheim, Illinois, United States	Wisconsin Central	Train diverted onto wrong line, collision	5
13 Jul 1896	Between Cidra and Sabanilla, Cuba		Derailment, sabotage	6
25 Jul 1896	Delhi, India	Rajputana-Malwa	Train diverted onto wrong line, collision	12
c28 Jul 1896	Near Spokane, Washington, United States	Great Northern	Derailment, train ran into landslide	12
30 Jul 1896	Meadow Tower, Near Atlantic City, New Jersey, United States	Atlantic City/ West Jersey and Seashore	Side collision on rail/rail level crossing, fire, boiler explosion	59

9 Aug 1896	Near Columbia, Pennsylvania, United States	Columbia and Donegal Electric	Runaway train, derailment	6
30 Sep 1896	Philson, Pennsylvania, United States	Baltimore and Ohio	Runaway train, collision	6
25 Oct 1896	Spring Park (Windsor) (Maramec), Missouri, United States	St Louis San Francisco	Head-on collision, single line	7
27 Dec 1896	Cahaba Bridge, Alabama, United States	Louisville and Nashville	Bridge collapsed under train	17
c10 Mar 1897	Varduari Bridge, Near Salonica, Macedonia		Bridge blown up under train	?
13 May 1897	Near Puka, Russia		Derailment, track washed away	56
18 May 1897	Near Vega Alta, Cuba		Runaway waggons, collision, boiler explosion, fire	5
19 May 1897	Near Gerolstein, Germany		Collision between portions of divided train	9
27 May 1897	American Falls, Idaho, United States	Oregon Short Line	Runaway train, collision	9
9 Jun 1897	Near Campo Florido, Cuba		Derailment, sabotage	7
11 Jun 1897	Welshampton, England	Cambrian	Derailment, excessive speed over poor track	11

26 Jun 1897	Near Missouri City, Missouri, United States	Wabash	Trestle collapsed under train	5
26 Jun 1897	Between Taca Taca and Bacanagua, Cuba		Culvert collapsed under train, sabotage	18
27 Jun 1897	Near Kolomea, Galicia		Bridge collapsed under train	12
11 Jul 1897	Gentofte, Denmark		Rear collision	40
28 Jul 1897	Twenty-Five-Mill Hill, Nevada, United States	Central Pacific	Derailment, track distorted by the heat	6
8 Sep 1897	Between Emporia and Lang, Kansas, United States	Atchison Topeka and Santa Fe	Head-on collision, single line, boiler explosions, fire	6
10 Sep 1897	Near Newcastle, Colorado, United States	Denver and Rio Grande	Head-on collision, single line, fire	30
23 Sep 1897	Near Maddur, India	Mysore State	Bridge washed away, train fell into river	150
24 Oct 1897	Near Garrison, New York, United States	New York and Hudson River	Derailment, track subsidence	19
11 Nov 1897	Between Nuevitas and Puerto Principe, Cuba		Derailment, Sabotage	10
24 Nov 1897	Tournay, France	Chemins de fer du Midi	Runaway train, rear collision	14
30 Nov 1897	Warsaw, Russian Poland		Train diverted onto wrong line, collision	11

c4 Feb 1898	Between Bonita and San Vincent, Cuba		Derailment, sabotage	5
14 Aug 1898	Near Lisieux, France	Chemins de fer de l'Ouest	Derailment, track defects, traction shock	7
17 Aug 1898	Between Mostertshoek and Bavians, Cape Colony	Cape Government	Runaway waggons, collision, fire	30
17 Aug 1898	Trent River Bridge, British Columbia, Canada	Union Colliery Co	Bridge collapsed under train	6
21 Aug 1898	Sharon, Massachusetts, United States	New York New Haven and Hartford	Rear collision	6
2 Sep 1898	Wellingborough, England	Midland	Derailment, obstruction on line, fire	5
17 Oct 1898	Wrawby Junction	Great Central	Collision with derailed waggons	8
15 Nov 1898	Murray Hill, Ontario, Canada	Grand Trunk	Train diverted onto wrong line, collision	9
19 Nov 1898	'33 Miles from Umtali,' Rhodesia	Mashonaland	Derailment, excessive speed	5
9 Jan 1899	Bound Brook, New Jersey, United States	Lehigh Valley	Head-on collision during temporary single line working	18
18 Feb 1899	Forest, Belgium		Rear collision in fog	21

12 May 1899	Exeter, Pennsylvania, United States	Philadelphia and Reading	Rear collision	34
28 May 1899	Sink Creek, Iowa, United States	Burlington Cedar Rapids and Northern	Derailment, track washed away	8
c8 Jun 1899	Near Holly, Michigan, United States		Train struck by lightning, fire	5
5 Aug 1899	Juvisy-sur-Orge	Chemin de fer de Paris à Orléans	Rear collision	18
9 Aug 1899	Ste-Polycarpe, Quebec, Canada	Canadian Atlantic	Derailment, track defects?	6
24 Aug 1899	Mapocho River Bridge, Talagante, Chile		Bridge collapsed under train	20
23 Sep 1899	Between Florence and Reno, Colorado, United States	Denver and Rio Grande	Head-on collision, single line	6
7 Oct 1899	Between Yaita and Nozaki, Japan		Train struck by typhoon and fell from bridge	20
12 Oct 1899	Seven Sisters Road, Cape Colony		Rear collision	9
15 Nov 1899	Near Capelle, The Netherlands	Staatsspoorwegen	Rear collision in fog	8
23 Dec 1899	Wivelsfield, England	London Brighton and South Coast	Rear collision in fog, fire	5
28 Mar 1900	Charing Cross, Glasgow, Scotland	North British	Rear collision in tunnel	7
6 Jun 1900	Staffelbach, Germany		Collision	30

16 Jun 1900	Slough, England	Great Western	Rear collision, fire	5
23 Jun 1900	McDonough, Georgia, United States	Southern	Derailment, track washed away, fire	35
24 Jun 1900	Depere, Wisconsin, United States	Chicago and North Western	Collision with shunting train	9
30 Jul 1900	Near Frederickstad, Transvaal		Derailment, sabotage	15
4 Aug 1900	Between Guanjato and Silac, Mexico		Bridge collapsed under train	6
12 Aug 1900	Near Castel Guibileo, Italy		Rear collision	15
20 Aug 1900	Near Plevna, Bulgaria		Collision	6
23 Aug 1900	Near Baripada, India		Derailment, embankment damaged by floodwater	11
2 Sep 1900	Hatfield, Pennsylvania, United States	Philadelphia and Reading	Rear collision in fog	13
8 Sep 1900	Near Bolivar Point, Texas, United States	Gulf and Interstate of Texas	Train struck by hurricane	85
7 Oct 1900	Karlstor, Germany		Rear collision	6
28 Oct 1900	Dehart, Near Big Timber, Montana, United States	Northern Pacific	Derailment, defective points	7

8 Nov 1900	Between Offenbach and Mullheim, Germany		Rear collision in fog, fire	12
11 Nov 1900	Choisy-le-Roi, France	Chemin de fer de Paris à Orléans	Train diverted onto wrong line, rear collision	6
12 Nov 1900	Near Brüggan, Germany		Derailment	6
15 Nov 1900	St Vincent de Tyrosse, France	Chemins de fer du Midi	Derailment, high speed with light vehicle	9
29 Nov 1900	Between Tamanachi and Symon, Mexico	Ferrocarril Central Méxicano	Head-on collision, single line?	11

ABOUT THE AUTHOR

The author has been interested in railways for many years, ever since he was invited up a signal box at the age of 14 back in 1966. Not long afterwards he 'cabbed' the engine of a goods train which was shunted out of the way of a faster train and its driver showed him a copy of R*ed for Danger*, the late L.T.C. Rolt's book about British railway accidents. This was the first railway book that he ever bought, several months' saved up pocket money going towards a paperback copy of it. Later he came across a copy of *A History of Railroad Accidents, Safety Precautions and Operating Practices* by Robert B. Shaw, dealing with American railroad accidents, and his interest in the subject grew.

When the local signal boxes were all done away with in a modernisation scheme in the mid-1980s he began a more detailed study of world railway accidents. There was no internet back then of course and many hours were spent in the local reference library slogging through bound volumes of *The Times*, writing out longhand in large notebooks details of every accident found. Later it was possible to view the paper on microfilm and print pages, which made things a little easier, and with the internet the amount of information available vastly increased.

Some years ago, the author contributed to *Railway Disasters of the World* by P.W.B. Semmens (which deals only with 20th-century accidents) and more recently the idea came of writing his own book or books on world accidents, starting at the beginning. This book was initially going to cover the whole of the 19th century but there proved to be too much to cover so the period was divided into two, with a previous volume covering the years 1811 to 1880. It is hoped to continue with volumes covering the 20th century and beyond.

Made in the USA
Monee, IL
07 November 2026

68083936R00282